Supplements to the 2nd Edition of

RODD'S CHEMISTRY OF CARBON COMPOUNDS

ELSEVIER SCIENTIFIC PUBLISHING COMPANY
1 Molenwerf
P.O. Box 211, Amsterdam, The Netherlands

ELSEVIER SCIENCE PUBLISHING COMPANY INC.
52 Vanderbilt Avenue
New York, New York, 10017

Library of Congress Card Number: 64-4605

ISBN 0-444-42150-5

Printed in The Netherlands

Supplements to the 2nd Edition of

RODD'S CHEMISTRY OF CARBON COMPOUNDS

VOLUME I

ALIPHATIC COMPOUNDS
★

VOLUME II

ALICYCLIC COMPOUNDS
★

VOLUME III

AROMATIC COMPOUNDS
★

VOLUME IV

HETEROCYCLIC COMPOUNDS
★

VOLUME V

MISCELLANEOUS
GENERAL INDEX
★

Supplements to the 2nd Edition (Editor S. Coffey) of

RODD'S CHEMISTRY OF CARBON COMPOUNDS

A modern comprehensive treatise

Edited by
MARTIN F. ANSELL
Ph.D., D.Sc. (London) F.R.S.C. C. Chem.
Department of Chemistry, Queen Mary College,
University of London (Great Britain)

Supplement to

VOLUME III AROMATIC COMPOUNDS

Part A:
General Introduction; Mononuclear Hydrocarbons and Their Halogeno
Derivatives; and Derivatives with Nuclear Substituents Attached through
Nonmetallic Elements from Group VI of the Periodic Table

ELSEVIER SCIENTIFIC PUBLISHING COMPANY
Amsterdam – Oxford – New York 1983

CONTRIBUTORS TO THIS VOLUME

GRAHAM C. BARRETT, B.Sc., Ph.D., D.Phil., C.Chem., M.R.S.C.

Department of Chemistry, The Polytechnic, Oxford, OX3 OBP

NORMAN B. CHAPMAN, M.A., Ph.D., C.Chem., F.R.S.C.

Department of Chemistry, The University, Hull, HU6 7RX

W. JAMES FEAST, B.Sc., Ph.D.

Department of Chemistry, The University Durham, DH1 3LE

PETER J. GARRATT, M.A., Ph.D., D.Sc.

Department of Chemistry, University College

University of London, London, WC1H OAJ

PHILIP G. HARRISON, B.Sc., Ph.D., D.Sc.

Department of Chemistry, The University, Nottingham, NG7 2RD

HARRY HEANEY, B.A., Ph.D., D.Sc., C.Chem., F.R.S.C.

Department of Chemistry, University of Loughborough

Leicestershire, LE11 3TU

CONTRIBUTORS TO THIS VOLUME

JOHN H. RIDD, Ph.D., D.Sc., F.R.S.C.

Department of Chemistry, University College,

University of London, London, WC1H OAJ

JOHN H.P. TYMAN, B.Sc., Ph.D. C.Chem., F.R.S.C.

Department of Chemistry, Brunel University,

Middlesex, UB8 3PH

GARETH H. WILLIAMS, J.P., Ph.D., D.Sc., C.Chem., F.R.S.C.

Department of Chemistry, Bedford College,

University of London, London, NW1 4NS

NORMAN H. WILSON, B.Sc., Ph.D., M.R.S.C.

Department of Pharmacology, The University,

Edinburgh, EH8 9J2

RAYMOND E. FAIRBAIRN, B.Sc., Ph.D., F.R.S.C.

Formerly of Research Division, Dyestuffs Division,

I.C.I., (INDEX)

PREFACE TO SUPPLEMENT III A

The publication of this volume marks another step in
the supplementation of the second edition of Rodd's
Chemistry of Carbon Compounds, thus ensuring that this
major work of reference does not become out of date. This
supplement covers Chapters 1,2,3,4,5,6 and 7 of Volume III
of the second edition. The preparation of the supplements
covering the remaining chapters of Volume III is well
advanced and they will be published in the near future.
Although the chapters in this book stand on their own, it
is intended that each one should be read in conjunction
with the parent chapter in the second edition.

At a time when there are many specialist reviews,
monographs and reports available, there is still in my
view an important place for a book such as "Rodd", which
gives a broader coverage of organic chemistry. One aspect
of the value of this work is that it allows the expert in
one field to quickly find out what is happening in other
fields of chemistry. On the other hand a chemist looking
for the way into a field of study will find in "Rodd" an
outline of the important aspects of that area of chemistry
together with leading references to other works to provide
more detailed information.

As editor I wish to thank all the contributors for the
very readable critical assessments they have made of the
areas of chemistry covered by their respective chapters.
As an organic chemist I have enjoyed reading the chapters
and profited from learning of advances in fields of
chemistry that are outside my own special interests.

This volume, like the other supplements to Volume III,
has been produced by direct reproduction of manuscripts. I
am most grateful to the contributors for all the care and
effort which they and their secretaries have put into the
production of their manuscripts, including in most cases
the diagrams. I am confident that readers will find this
presentation acceptable. I also wish to thank the staff

X

at Elsevier for all the help they have given me and for
seeing the transformation of authors' manuscripts to
published work.

July 1982   Martin F. Ansell

CONTENTS

VOLUME IIIA

Aromatic Compounds; General Introduction; Mononuclear Hydrocarbons and their halogeno derivatives; and derivatives with nuclear substituents attached through non-metallic elements from Group VI of the Periodic Table

Chapter 1. General Introduction

Chapter 2. Mononuclear hydrocarbons: Benzene and its Homologous
by H. Heaney

Chapter 3. Halogen derivatives of benzene and its homologues
by W.J. Feast

Chapter 4. Nuclear Hydroxy Derivatives of Benzene and
Its homologues
by J.H.P. Tyman

Chapter 5. Mononuclear Hydrocarbons Carrying Substituents Attached
through Sulphur and Selenium: Thiophenols, Sulphides, etc.
by J.H.P. Tyman

*Chapter 6. Mononuclear Hydrocarbons Carrying Nuclear Substituents
Containing Sulphur: Sulphonic Acids, Sulphonic Acids, and Sulphenyl
Compounds of the Benzene Series
by G.C. Barrett*

Chapter 7. Mononuclear hydrocarbons carrying nuclear substituents
containing selenium or tellurium
by P.G. Harrison

OFFICIAL PUBLICATIONS

B.P.	British (United Kingdom) Patent
F.P.	French Patent
G.P.	German Patent
Sw.P.	Swiss Patent
U.S.P.	United States Patent
U.S.S.R.P.	Russian Patent
B.I.O.S.	British Intelligence Objectives Sub-Committee Reports
F.I.A.T.	Field Information Agency, Technical Reports of U.S. Group Control Council for Germany
B.S.	British Standards Specification
A.S.T.M.	American Society for Testing and Materials
A.P.I.	American Petroleum Institute Projects
C.I.	Colour Index Number of Dyestuffs and Pigments

SCIENTIFIC JOURNALS AND PERIODICALS

With few obvious and self-explanatory modifications the abbreviations used in references to journals and periodicals comprising the extensive literature on organic chemistry, are those used in the World List of Scientific Periodicals.

LIST OF COMMON ABBREVIATIONS AND
SYMBOLS USED

A	acid
Å	Ångström units
Ac	acetyl
a	axial; antarafacial
as, asymm.	asymmetrical
at	atmosphere
B	base
Bu	butyl
b.p.	boiling point
C, mC and µC	curie, millicurie and microcurie
c, C	concentration
C.D.	circular dichroism
conc.	concentrated
crit.	critical
D	Debye unit, 1×10^{-18} e.s.u.
D	dissociation energy
D	dextro-rotatory; dextro configuration
DL	optically inactive (externally compensated)
d	density
dec. or decomp.	with decomposition
deriv.	derivative
E	energy; extinction; electromeric effect; Entgegen (opposite) configuration
E1, E2	uni- and bi-molecular elimination mechanisms
ElcB	unimolecular elimination in conjugate base
e.s.r.	electron spin resonance
Et	ethyl
e	nuclear charge; equatorial
f	oscillator strength
f.p.	freezing point
G	free energy
g.l.c.	gas liquid chromatography
g	spectroscopic splitting factor, 2.0023
H	applied magnetic field; heat content
h	Planck's constant
Hz	hertz
I	spin quantum number; intensity; inductive effect
i.r.	infrared
J	coupling constant in n.m.r. spectra; joule
K	dissociation constant
kJ	kilojoule

LIST OF COMMON ABBREVIATIONS

k	Boltzmann constant; velocity constant
kcal	kilocalories
L	laevorotatory; laevo configuration
M	molecular weight; molar; mesomeric effect
Me	methyl
m	mass; mole; molecule; *meta-*
ml	millilitre
m.p.	melting point
Ms	mesyl (methanesulphonyl)
[M]	molecular rotation
N	Avogadro number; normal
nm	nanometre (10^{-9} metre)
n.m.r.	nuclear magnetic resonance
n	normal; refractive index; principal quantum number
o	*ortho-*
o.r.d.	optical rotatory dispersion
P	polarisation, probability; orbital state
Pr	propyl
Ph	phenyl
p	*para-*; orbital
p.m.r.	proton magnetic resonance
R	clockwise configuration
S	counterclockwise config.; entropy; net spin of incompleted electronic shells; orbital state
S_N1, S_N2	uni- and bi-molecular nucleophilic substitution mechanisms
S_Ni	internal nucleophilic substitution mechanisms
s	symmetrical; orbital; suprafacial
sec	secondary
soln.	solution
symm.	symmetrical
T	absolute temperature
Tosyl	*p*-toluenesulphonyl
Trityl	triphenylmethyl
t	time
temp.	temperature (in degrees centigrade)
tert.	tertiary
U	potential energy
u.v.	ultraviolet
v	velocity
Z	zusammen (together) configuration

LIST OF COMMON ABBREVIATIONS

α	optical rotation (in water unless otherwise stated)
$[\alpha]$	specific optical rotation
α_A	atomic susceptibility
α_E	electronic susceptibility
ε	dielectric constant; extinction coefficient
μ	microns (10^{-4} cm); dipole moment; magnetic moment
μB	Bohr magneton
μg	microgram (10^{-6} g)
λ	wavelength
ν	frequency; wave number
χ, χ_d, χ_μ	magnetic, diamagnetic and paramagnetic susceptibilities
$\sim$	about
(+)	dextrorotatory
(−)	laevorotatory
(±)	racemic
$\ominus$	negative charge
$\oplus$	positive charge

Chapter 1

1. AROMATIC CHARACTER

P.J. GARRATT

(a) General

The general criteria applied to judge whether or not a molecule is aromatic are unaltered (G.S. Chandler and D.P. Craig, Rodd IIIA, p 5), and the previous structure on attempting any precise general definition of aromatic character remains (*ibid*, p 30). The last decade has seen a major interest in calculations aimed at reproducing the heats of formation of conjugated hydrocarbons from atoms. Such methods, when coupled with calculations on a suitable model system, can provide resonance energies. Acyclic polyenes, in which the energy is an additive function of the individual bond energies, have now been generally used to provide the reference energies for the model nonconjugate cyclic polyene (R. Breslow and E. Mohacsi, J. Amer. chem. Soc., 1963, 85, 431; A.L.H. Chung and M. J.S. Dewar, J. chem. Phys., 1965, 42, 756; Dewar and G.J. Gleicher, J. Amer. chem. Soc., 1965, 87, 685,692; Dewar and C. de Llano, *ibid*, 1969, 91, 789). Initially Dewar *et al.* used a self-consistent Field (SCF) treatment, but it was later shown that similar results would be obtained by the Hückel method if the same reference model were used (L.J. Schaad and B.A. Hess, *ibid*, 1972, 94, 3068; J. chem. Educ., 1974, 51, 640). A comparison of the heats of formation obtained by the two methods, together with the experimentally observed values are shown below. The resonance energies in this way have been termed Dewar Resonance Energies (DRE) (N.C. Baird, *ibid*, 1971, 48, 509).

Graph theory has been used to determine resonance energies, using the same reference system, and these agree with the DRE's (J.-I. Aihara, J. Amer. chem. Soc., 1976, 98, 2750; 1977, 99, 20; I. Gutman, M. Milun, and N. Trinajstic, *ibid*, 1977, 99, 1692). This method can also be applied to radicals and ions and is free from empirical parameters. Renewed interest has been shown in weighed structure counting (C.F. Wilcox, *ibid*, 1969, 91, 2732; Croat. Chem. Acta, 1975, 47, 87; W.C. Herndon,

Comparison of Calculated and Observed Heats of Formation of Conjugated Hydrocarbons from Atoms.

Compound	ΔH(eV)		
	Calc[1] PPP	Calc[2] Hückel	Observed
Benzene	57.16	57.14	57.16
Naphthalene	90.61	90.59	90.61
Anthracene	123.89	123.95	123.93
Phenanthrene	124.22	124.14	124.20
Azulene	89.47	90.13	89.19
Pyrene	138.62	138.60	138.88
Triphenylene	157.94	157.79	157.76
Biphenyl	109.75	–	109.76
Biphenylene	104.87	104.67	102.00

1. Values from Dewar and de Llano, *loc. cit.*
2. Values from Schaad and Hess, *loc. cit.*

J. Amer. chem. Soc., 1973, 95, 2404; J. chem. Educ., 1974, 51, 10) and resonance energies have been calculated using only Kekule structures (Herndon, *loc. cit.*). Various indices have been proposed for estimating aromaticity (H.P. Figeys, Tetrahedron, 1970, 26, 5225; J. Kruszewski and T.M. Krygowski, Canad. J. Chem., 1975, 53, 945) and the "Facts and Theories of Aromaticity" discussed (D. Lewis and D. Peters, Macmillan, London, 1975). For other general reviews see "Aromatic and Heteroaromatic Compounds" (ed. C.W. Bird and G.W.H. Cheeseman, Chem. Soc. Spec. Publ. 1973-8, Vol 1-5), "Aromaticity" (P.J. Garratt, McGraw Hill, Maidenhead, 1971), "Theoretical Aromatic Chemistry" (I. Agranat in "MTP International Review of Science" ed H. Zollinger, Butterworths, London, 1973, Vol 3, p 139) and "Aromaticity" (Garratt in "Comprehensive Organic Chemistry", ed. D.H.R. Barton and W.D. Ollis, Pergamon, Oxford, 1979, Vol 1, p 215).

(b) Two-electron Systems

Although the parent cyclobutadienyl dication (1, X = Y = H) has not been prepared, a number of substituted derivatives have been obtained in strongly acid media (G.A. Olah and J.S. Starval, J. Amer. chem Soc., 1976, 98, 6290). The decoupled ^{13}C NMR spectra of these ions suggest that they are delocalised, doubly charged species.

(1) X = Y = H
X = Ph, Y = Ph, H, F
X = Y = Me

(2)

The sulphur analogue (2) of the squaric acid dianion has been prepared and an X-ray crystallographic study indicates that it has a symmetrical, delocalised structure (R. Allmann, *et al.*, Chem. Ber., 1976, 110, 2208). Various related dianions with mixtures of sulphur, oxygen and selenium substituents have also been prepared (D. Eggerding and R. West, J. org. Chem., 1976, 41, 3904; A.H. Schmidt, W. Ried, and P. Pustoslensek, Chem-Ztg, 1977, 101, 154; *idem*, Angew. Chem. internat. Ed., 1976, 15, 704; G.H.R. Seitz and G. Arndt, Synthesis, 1976, 445).

(c) Four-electron Systems

Cyclobutadiene (3) has been prepared in matrices at low temperature (*ca.* 8-20 K) by a number of workers (C.Y. Lin and A. Krantz, J. chem. Soc. Chem. Comm., 1972, 1111; Kranz, Lin, and M.D. Newton, J. Amer. chem. Soc., 1973, 95, 2744; S. Masamune *et al.*, J. chem. Soc. Chem. Comm., 1972, 1268; O.L. Chapman, C.L. McIntosh, and J. Pacansky, J. Amer. chem. Soc., 1973, 95, 614; G. Maier and B. Hoppe, Tetrahedron Letters, 1973, 861). All methods employ photoirradiation and a variety of precursors have been used (Maier, Angew. Chem. internat. Ed., 1974, 13, 425; Masamune, Pure and appl. Chem., 1975, 44, 861). The original IR spectra obtained were incompatible with either a square planar or rectangular structure, but many of these difficulties were resolved by Maier, H.-G. Hartan, and T. Sayrac (Angew. Chem. internat. Ed., 1976, 15, 26) who showed that the 653 cm^{-1} band was not due to free cyclobutadiene. Only two intense bands (1240, 570 cm^{-1}) could now definately be attributed to cyclobutadiene (Maier, Hartan, and Sayrac, *loc. cit.*; R.G.S. Pong *et al.*, J. Amer. chem. Soc., 1977, 99, 4153) until Masamune *et al.* (*ibid*, 1978, 100, 4889) observed two further weak bands at 1523 and 723 cm^{-1}. Tetradeuteriocyclobutadiene was also prepared (Masamune *et al.*, *loc. cit.*) and the shifts of the IR bands used to assign vibrational modes to these bands. It was concluded that the spectra is consistent with the molecule having a rectangular geometry with D$_2$h

symmetry. Maier, Hartman and Sayrac (*loc. cit.*) had favoured
a square singlet ground state, as predicted by W.T. Borden (J.
Amer. chem. Soc., 1975, 97, 5968) in contrast to all other
predictions (see W.J. Hehre and J.A. Pople, *ibid*, 1975, 97,
6941). The electronic spectrum (Masamune *loc. cit.*; Maier and
Hoppe, *loc. cit.*)shows only end absorption beginning at 290 nm.
On thawing the matrices containing (3) the dimer, syn-tricyclo-
[4.2.0.0^{2,5}]octa-3,7-diene, is obtained as the sole product.

(3) 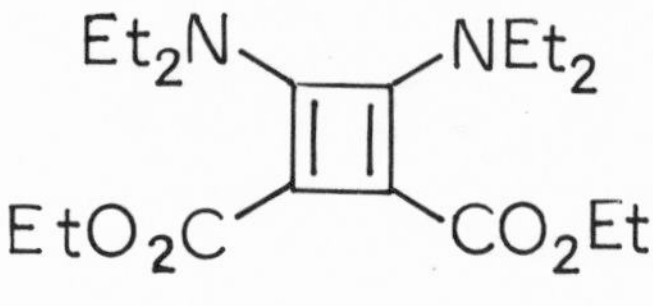   (4) 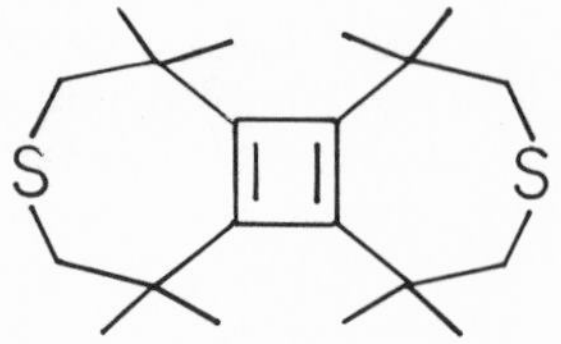   (5)

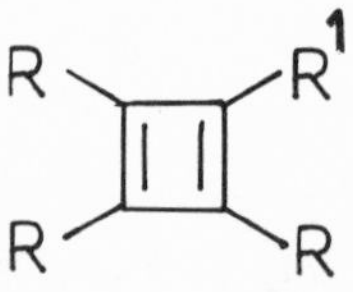

(6) a R=Bu^t, R^1=H   (7)

b R= Bu^t, R^1=CO$_2$Me

c R = R^1 = Bu^t

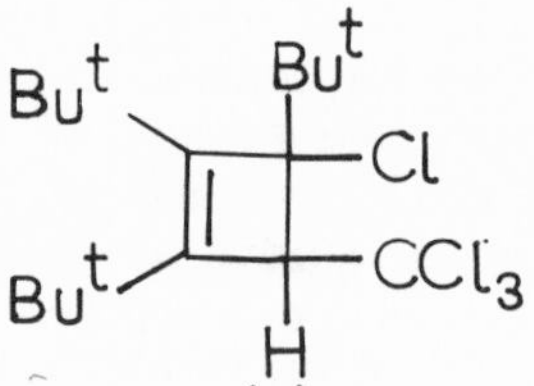

A variety of cyclobutadiene derivatives have now been prepared
which are more stable than the parent compound. The first type
to be isolated is stabilised by push-pull conjugation, the
four-membered ring bearing appropriately arranged electron-
withdrawing and donating substituents, as for example in (4)
(R. Gompper and G. Seybold, "Topics in Nonbenzenoid Aromatic
Chemistry", ed. T. Nozoe *et al.*, Hirokaou, Tokyo, 1977, Vol 2,
p 29; Gompper *et al.*, J. Amer. chem. Soc., 1973, 95, 8480).
Subsequently cyclobutadienes stabilised only by steric hindrance have been obtained, of which the tricyclic system (5) was
the first to be prepared (H. Kimling and A. Krebs, Angew. Chem.,

internat. Edn, 1972, 11, 932). Compound (5) has been shown to have a rectangular structure by an X-ray analysis (H.Irngart-inger and H. Rodewald, *ibid*, 1974, 13, 740) and the sulphur atoms do not interact with the π-system (G. Lauer *et al.*, *ibid*, 1974, 13, 544). Photoirradiation of 2,3,4-tri-tert-butylcyclo-pentadienone at -196°C in a perdeuteriated methylcyclohexane glass gave tri-tert-butylcyclobutadiene (6a) which survives thawing to room temperature (Maier, *loc. cit.*). The ^{1}H NMR spectrum [δ 5.35 (1H), 1.14 (9H), 1.05 (18H)] shows the proton of the cyclobutadiene ring at somewhat higher field than a simple alkene proton, presumably due to the paratropicity of the system. Compound (6a) was also prepared by Masamune and co-workers (J. Amer. chem. Soc., 1973, 95, 8481) via the ester (6b), which they obtained by ring expansion of the appropriately substituted cyclopropylmethylazide. An X-ray crystallographic determination of the structure of the ester (6b) (L.T.J. Del-baere *et al.*, *ibid*, 1975, 97, 1973) showed it to be rectangular with two long (151, 155 pm) and two short (141, 138 pm) bonds. Tri-tert-butylcyclobutadiene reacts with carbon tetrachloride to give the adduct (7) in a triplet like reaction (Maier and W. Sauer, Angew. Chem., internat. Edn, 1977 16, 51). It also behaves both as a diene and a dienophile (Maier, *loc. cit.*), as a nucleophile (Maier and Sauer, *loc. cit.*) and as an electr-ophile (Maier and F. Köhler, *ibid*, 1979, 18, 308). This multi-plicity of behaviour can be ascribed to a pseudo-Jahn-Teller effect providing both a low lying LUMO and a high energy HOMO, the system readily accepting or donating electrons. Tetra-tert-butylcyclobutadiene (6c) has been prepared by heating tetra-tert-butylprismane (8) at 130°C in cyclosilane (Maier *et al.*, *ibid*, 1978, 17, 520), (8) having been prepared by photoirradiation of tetra-tert-butylcyclopentadienone at -100°C. Compound (8), the first prismane to be prepared, is a crystalline solid which melts at 135°C! Tetra-(trifluoromethyl)-cyclobutadiene (9) was prepared by photoirradiation of the ozonide of hexa(trifluoromethyl)benzvalene (Masamune, T. Ma-chiguchi, and M. Arantani, J. Amer. chem. Soc., 1977, 99, 3524) and was shown to have a singlet, rectangular structure.

The currently available evidence on cyclobutadiene and its derivatives suggests that these compounds all have singlet, rectangular ground states.

Another 4 π-electron system, the cyclopentadienyl cation (10), has been prepared by treatment of 5-bromo- or 5-chloro-cyclo-pentadiene with antimony pentafluoride at 78 K (M. Saunders *et al.*, *ibid*, 1973, 95, 3017). The e.s.r. spectrum shows that (10) has a triplet ground state and the spectral parameters

(8)   (9)   (10)

are in accord with it being a planar pentagon. The previously
prepared pentachloro derivative also had a triplet ground state
whereas the pentaphenyl derivative had the triplet state 5.4
kJ mol^{-1} above the ground state (Breslow in "Topics in Nonbenzenoid Aromatic Chemistry" ed. T. Nozoe *et al.*, Hirokawa,
Tokyo, 1973, Vol 1, p 81).

(d) Six-electron Systems

Some evidence for the occurrence of the cyclobutadienyl
dianion (11) was obtained by treating 3,4-dichlorocyclobutene
with sodium naphthalide and quenching the product with MeOD,
when a low yield of 3,4-dideuteriocyclobutene was obtained
(J.S. McKennis *et al.*, J. chem. Soc. Chem. Comm., 1972, 365).
The disubstituted dianion (12) has been well characterised,
but its spectral and physical properties indicate that it is
neither diatropic nor stabilised by comparison to the corresp-
ondingly substituted cyclobutanyl dianion (Garratt and R.
Zahler, J. Amer. chem. Soc., 1978, 100, 7753).

(11)   (12)   (13)

A derivative, (13), of the cyclooctatetraenyl dication has
been obtained by treatment of 1,3,5,7-tetramethylcycloocta-
tetraene with SbF$_5$-SO$_2$ClF at -78°C (Olah, Starval, and L.A.
Paquette, *ibid*, 1976, 98, 1267). The ^{1}H NMR spectra showed
signals at δ 4.27 and 10.80 (3.1), clearly indicating that (13)

is diatropic. At -20°C it rearranges to the corresponding bicyclo[3.3.0]octadienyl dication.

(e) Eight-electron Systems

Paquette and co-workers (J. Amer. chem. Soc., 1975, 97,
3538; 1976, 98, 4936) have dramatically demonstrated the difference between cyclo-octatetraene and benzene by isolating a
tetramethylcyclooctatetraene as two constitutional isomers, 1,
2,3,4-tetramethylcyclooctatetraene (14) and 1,2,3,8-tetra-
methylcyclooctatetraene (15). The methyl groups clearly inhibit
the bond shift and inversion processes which occur in cyclo-
octatetraene, the planar form being further destabilised.

(14)   **(15)**

(f) Ten-electron Systems

A second isomer of the cyclononatetraenyl anion has been
prepared (G. Boche, H. Böhme, and D. Martens, Angew. Chem.,
internat. Edn., 1969, 8, 594; Boche, Martens, and W. Davies,
ibid, 1969, 8, 894) by treatment of anti-9-methoxybicyclo-
[6.1.0]nonatriene with potassium, the presumed initial cyclopropyl anion opening in a conrotatory manner to give the mono-
trans anion (16). In the presence of potassium (16) rearranges
to the all-*cis* anion at room temperature, but in the pure state
this rearrangement is slow, the rate of topomerisation, equil-
ibrating the inner with the outer protons, being much faster.
The anion (16) is diatropic, the outer protons resonating (δ
7.3-6.4) at a similar position to the all-*cis* isomer (δ 7.0),
while the inner proton is at high field (δ -3.52) (Boche, A.
Bieberbach, and H. Weber, *ibid*, 1975, 14, 562; Boche and Bieberbach, Tetrahedron Letters, 1976, 1021).

Earlier studies in the photochemical ring opening of 9,10-
dihydronaphthalene (T.L. Burkoth and E.E. van Tamelen, in
"Nonbenzenoid Aromatics", ed. J.P. Snyder, Academic Press, New
York, 1969, Vol 1, p 63; van Tamelen, Burkoth and R.H. Greely,
J. Amer. chem. Soc., 1971, 93, 6120) have been clarified and
extended by Masamune *et al.* (*ibid*, 1971, 93, 4966; A.V. Kemp-

8

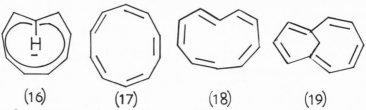

(16)   (17)   (18)   (19)

Jones and Masamune, in "Topics in Nonbenzeniod Aromatic Chemistry", ed. T. Nozoe *et al.*, Hirokawa, Tokyo, 1973, Vol 1,
p 121) who isolated both the all- *cis* isomer (17) and the
mono-*trans* isomer (18) at low temperature. The spectral properties of both isomers indicate that these are non-planar,
atropic systems. The inner hydrogen of (18) equilibrates with
the outer hydrogens at -40°C. The all-*cis* isomer (17) rearranges to *cis*-9,10-dihydronaphthalene at -14°C and the mono-*trans*
(18) rearranges to *trans*-9,10-dihydronaphthalene at -40°C, in
both cases the product predicted by orbital symmetry considerations being obtained.

A second bridged[10]annulene, 1,5-methano[10]annulene (19)
has been synthesized (Masamune *et al.*, J. Amer. chem. Soc.,
1976, 98, 8277; Masamune and D.W. Brooks, Tetrahedron Letters,
1977, 3239; L.T. Scott and W.R. Brunsvold, J. Amer. chem. Soc.,
1978, 100, 4320) and the ^{1}H NMR spectrum shows it to be diatropic, the bridge protons appearing as a double doublet (δ
-0.50, -0.95). The electronic spectrum has absorption extend-
ing into the visible region and differs from that of 1,6-meth-
ano[10]annulene in a manner reminiscent of the difference
between naphthalene and azulene.

(g)   Systems with more than ten electrons

General reviews on the annulenes (F. Sondheimer in "Proceedings of the Robert A. Welch Foundation Conferences on Chemical Research", XII, Organic Synthesis, ed. W.O. Milligan,
Houston, Texas, 1969, p 125; Accounts chem. Res., 1972, 5, 81;
Garratt and K. Grohmann, in "Methoden der Organische Chemie"
(Houben-Weyl) ed. E. Müller, Verlag Chemie, Stuttgart, 1972,
Vol 5/1d, p 527; G. Schröder, Pure and appl. Chem., 1975, 44,
925), on the bridged annulenes (E. Vogel, in "Proceedings of
the Robert A. Welch Foundation Conferences on Chemical Research", XII, Organic Synthesis, ed. W.O. Milligan, Houston,
Texas, 1969, p 215; Pure and appl. Chem., 1971, 28, 355;
V. Boekelheide in "Topics in Nonbenzenoid Aromatic Chemistry",
ed. T. Nozoe *et al*, Hirokawa, Tokyo, 1973, Vol 1, p 47; Pure

and appl. Chem., 1975, 44, 751) and on the ^{1}H NMR spectra of the annulenes (R.C. Haddon, V.R. Haddon, and L.M. Jackman, Current Topics chem. Res., 1971, 16, 105) have appeared. All of the annulenes from cyclobutadiene to [24]annulene have now been prepared and, except for [8] and [10]annulene, they all show the expected alternation of properties, the (4n + 2) members being diatropic and the 4n members paratropic. A general synthesis of dehydroannulenes has been introduced by Nakagawa and his co-workers (M. Nakagawa in "Topics in Nonbenzenoid Aromatic Chemistry" ed. T. Nozoe *et al*., Hirokawa, Tokyo, 1973, Vol 1, p 191; Pure and appl. Chem., 1975, 44, 885; Angew Chem., internat. Edn., 1979, 18, 202). Analogues of naphthalene, anthracene and phenanthrene have been synthesized in which some or all of the rings are macrocyclic (U.E. Meissner, A. Gensler, and H.A. Staab, *ibid*, 1976, 15, 365; *idem*, Tetrahedron Letters, 1977, 3; R.H. Mitchell and R.J. Carruthers, *ibid*, 1975, 4331; H. Günther *et al*., Angew. Chem., internat. Edn., 1973, 12, 243; Nakagawa, *loc. cit.*; T.M. Cresp and Sondheimer, J. Amer. chem. Soc., 1975, 97, 4412; *ibid*, 1977, 99, 194; T. Kashitani *et al*., *ibid*, 1975, 97, 4424; S. Akiyama, M. Iyoda, and Nakagawa, *ibid*, 1976, 98, 6410). [16]Annulene can be reduced to the corresponding dianion and oxidized to the corresponding dication, both diatropic systems but having different configurational structures (J.F.M. Oth *et al*., Angew. Chem., internat. Edn., 1973, 12, 327; Tetrahedron Letters, 1968, 6265). Numerous other macrocyclic ions have been prepared (see Garratt in "Comprehensive Organic Chemistry", ed. Barton and Ollis, Pergamon, 1979, Vol 1, p 361).

(h) Homoaromaticity and Bicyclohomoaromaticity

Recent developments in homoaromaticity and antihomoaromaticity have been extensively reviewed (P.M. Warner, "Topics in Nonbenzoid Aromatic Chemistry", ed. T. Nozoe *et al*., Hirokawa, Tokyo, 1977, Vol 2, p 283; Paquette, Angew. Chem., internat. Edn., 1978, 17, 106). Calculations have been made in an attempt to assess the magnitude of homoaromatic interaction (W.J. Hehre, J. Amer chem. Soc., 1974, 86, 5207; Haddon, Tetrahedron Letters, 1976, 863; W.L. Jorgensen, J. Amer. chem. Soc., 1976, 98, 6784). Bishomocations such as (20) (Warner and S. Winstein, *ibid*, 1971, 93, 1284; Paquette *et al*., *ibid*, 1973, 95, 3386), (21) (P. Ahlberg, D.L. Harris and Winstein, *ibid*, 1970, 92, 2146, 4454) and (22) (Schröder *et al*., Tetrahedron Letters, 1970, 351; Ahlberg, *loc. cit.*) and trishomocations such as (23) (Masamune *et al*., Canad. J. Chem., 1974, 52, 855), (24) (R.M. Coates and E.R. Fretz, J. Amer chem. Soc., 1975, 97, 2538) and

(25) (Masamune *et al.*, Angew. Chem., internat. Edn., 1973, 12, 769) have been well characterized. Bishomoaromatic anions have been observed (2nd Edn., Vol IIIA, p 44) but trishomoaromatic anions have not yet been detected. Some evidence for neutral bishomoaromatic systems has been adduced.

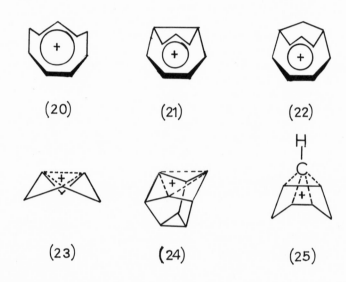

(20)   (21)   (22)

(23)   (24)   (25)

Some evidence for bicycloaromaticity in bicyclo[3.2.2]nonatrienyl anions has been advanced (S.W. Staley and D.W. Reichard, J. Amer chem. Soc., 1969, 91, 3998) but the effect appears to be quite small when compared to the homoaromatic stabilization of the corresponding bicyclo[3.2.2]nonadienyl anion (M.V. Moncur and J.B. Grutzner, *ibid*, 1973, 95, 6449; M.J. Goldstein and S. Natowsky, *ibid*, 1973, 95, 6451).

2. ELECTROPHILIC AROMATIC SUBSTITUTION

J.H.Ridd

Since the review of this subject in the second edition a
general survey of the kinetics of electrophilic aromatic
substitution has appeared (R.Taylor, "Comprehensive Chemical
Kinetics", ed. C.H.Bamford and C.F.H.Tipper, Elsevier, 1972,
Vol. 13, Chap.1) and accounts of nitration (J.G.Hoggett,
R.B.Moodie, J.R.Penton and K.Schofield, "Nitration and
Aromatic Reactivity", Cambridge University Press, 1971),
halogenation (P.B.D.de la Mare, "Electrophilic Halogenation",
Cambridge University Press, 1976) and heteroaromatic
substitution (J.H.Ridd, "Physical Methods of Heterocyclic
Chemistry", ed. A.R.Katritzky, Academic Press, 1971, Vol.4,
Chap.3). The chapters on electrophilic substitution in the
Chemical Society series of Specialist Periodical Reports on
"Aromatic and Heteroaromatic Chemistry" provide a detailed
account of work from ca. 1970 and the A.C.S. Symposium on
Industrial and Laboratory Nitrations (A.C.S. Sym.Series
1975, 22) provides summaries of much recent work.

Almost all of the review in the second edition remains true
today but several new aspects of electrophilic aromatic
substitution have become important in recent years
particularly in consequence of studies on *ipso*-attack,
diffusion control and electron transfer reactions. A large
part of this recent work concerns nitration but the conclusions are relevant to other electrophilic aromatic
substitutions.

(a) Ipso-attack

The term '*ipso*-attack' was introduced by C.L.Perrin and
G.A.Skinner (J.Amer.chem.Soc., 1971, 93, 3389) to denote
attack by a reagent on an aromatic carbon atom carrying a
substituent. Such reactions have long been recognised in
connection with the displacement of the substituent by the
attacking electrophile but the current interest derives
from other potential consequences of *ipso*-attack. This
subject has been reviewed by R.B.Moodie and K.Schofield
(Acc.chem.Res., 1976, 9, 287) and by S.R.Hartshorn (Chem.
Soc.Rev., 1974, 3, 167).

One of the implied assumptions in the earlier discussions of substituent effects was that the isomeric proportions obtained in electrophilic substitution could be taken as a measure of the intrinsic reactivity of the different ring positions to electrophilic attack, provided that the products were stable. The discovery that *ipso*-attack can be followed by specific intramolecular rearrangements has shown that this assumption is not necessarily true.

The nitration of the alkyl benzenes in acetic anhydride provides direct evidence of the extent of *ipso*-attack for, under these conditions, the initial ions (e.g. (1)) are captured effectively quantitatively by acetate ions derived from the medium : thus, the nitration of *ortho*-xylene gives 59% of the adduct (2) (A.Fischer and G.J.Wright, Aust.J.Chem., 1974, 27, 217). These 1,4-adducts lose nitrous acid readily by a 1,4-elimination to yield the corresponding aryl acetates and only the latter were detected in the earlier experiments. The relative amounts of the adducts formed in the nitration of di- and poly-methyl benzenes accord reasonably well with the additivity principle when a partial rate factor (i_f) is assigned for *ipso*-attack. The value of the *ipso* partial rate factor is about three times that for *meta* substitution. The products from the nitration of toluene in acetic anhydride are in approximate agreement with this ratio for the yield of *para*-tolyl acetate derived from *ipso*-attack (3.1 mole %) exceeds that for nitration at one *meta*-position (1.4 mole %) by a factor of 2.2 (Fischer and Wright, *loc.cit.*). The reason why a methyl group should activate the ring more to *ipso*-attack than to *meta*-attack is not clear for this does not accord with recent experimental and theoretical results on the gas phase protonation of toluene (J.L.Devlin, J.F. Wolf and R.W.Taft, J.Amer.chem.Soc., 1976, 98, 1990).

(1)   (2)   (3)

Nitration of the alkyl benzenes in 70-100% sulphuric acid does not lead to adduct formation presumably because of the lower nucleophilicity of the medium : instead, *ipso*-attack then results in rearrangement. Part of the evidence for this comes from the reaction of the 1.4-adduct (2) in 70-80% sulphuric acid to form 3-nitro-*ortho*-xylene with < 0.5% of the 4-nitro-isomer (P.C.Myhre, J.Amer.chem.Soc., 1972, 94, 7291). This reaction is considered to occur through alkyloxygen cleavage to give the *ipso*-intermediate (1). Since the nitration of *ortho*-xylene gives substitution at both the 3- and 4-positions, the absence of the 4-nitro-isomer in the rearrangement is very significant: it shows that the rearrangement occurs without separation of the nitronium ion from the aromatic substrate and without even a return to the encounter pair (ArH.NO$_2^+$). This 1,2-rearrangement of the *ipso*-intermediate (1) is presumably analogous to the degenerate 1,2 shifts of the nitro group in the ion (3); n.m.r. studies show that random transfers of the nitro-group between carbon atoms do not occur in this system (V.A.Koptyug *et al.,* Zhur.org.Khim., 1974, 10, 2487).

The above work (Myhre, *loc.cit.*) has provided an elegant explanation of why the product composition in the nitration of *ortho*-xylene depends on the concentration of sulphuric acid. Recent studies show that the yield of the 3-nitro isomer rises from 11% to 57% over the range 55-75% sulphuric acid. At low acidities, the ion formed by *ipso*-attack (1) appears to add water to form phenolic products but at high acidities rearrangement to the 3-nitro isomer supervenes (H.W.Gibbs, R.B.Moodie and K.Schofield, J.chem.Soc.Perkin II, 1978, 1145). The predominant formation of the 3-nitro isomer at high acidities is not therefore because this is the most reactive position for electrophilic attack but because two independent reaction paths lead to the 3-nitro isomer and only one reaction path to the 4-nitro isomer.

A number of nitration reactions are now known in which the product composition indicates a significant contribution from *ipso*-attack and rearrangement. The reactions involve di- or polysubstituted benzenes in which *ipso*-attack at one substituent is favoured by the directing effect of another substituent. The directing substituent may be Me,Cl,Br, or OMe, and the *ipso*-substituent Me,Cl or Br (R.B.Moodie, K.Schofield and J.B.Weston, J.chem.Soc.Perkin II, 1976, 1089; R.B.Moodie, K.Schofield and G.D.Tobin, J.chem.Soc.chem.Comm.,

1978, 180). There is spectroscopic evidence for the formation of the Wheland intermediate from *ipso*-attack at methyl substituents when the directing group is OH (R.G.Coombes, J.G.Golding and P.Hadjigeorgiou, J.chem.Soc. Perkin II, 1979, 1451) and NMe_2 (K.Fujiwara, J.C.Giffney and J.H.Ridd, J.chem.Soc.chem.Comm., 1977, 301) but the latter reaction cannot involve attack by the nitronium ion since nitrous acid catalysis is required. Where the directing groups are OMe, OH or NMe_2, the formation of the *ipso* Wheland intermediate is followed by a 1,3-migration of the nitro-group and there is evidence that this may occur by two mechanisms depending on the substrate and conditions: homolytic dissociation of the $C-NO_2$ bond leading to a radical pair (C.E.Barnes and P.C.Myhre, J.Amer.chem.Soc., 1978, 100, 973) and heterolytic dissociation leading to a return to the encounter pair (Coombes, Golding and Hadjigeorgiou, *loc.cit.*). A 1,3-migration of a nitro-group in an *ipso*-intermediate can also occur by two successive 1,2-migrations: this is particularly important when the substrate contains an alkyl substituent adjacent to the point of *ipso*-attack, for *ipso-ipso* migration is then faster than migration to an un-substituted position (C.E.Barnes and P.C.Myhre, J.Amer.chem. Soc., 1978, 100, 975).

The above discussion has omitted some important side re-actions deriving from *ipso*-attack followed by proton loss from a side-chain: e.g. those involving the ion (3) (H.Suzuki, T.Mishira and T.Hanafusa, Bull.chem.Soc.Japan, 1979, 52, 191). Such reactions are included in the review by Hartshorn *loc.cit.* and the synthetic aspects have been reviewed by Suzuki (Synthesis, 1977, 217). A great deal of work has also been done on the chemistry of the adducts (e.g. 2) formed by nitration in acetic anhydride (cf. A.Fischer and K.C.Teo, Canad. J.Chem., 1978, 56, 258) : thus the treatment of the adduct (2) with dry hydrochloric acid in non polar media gives a useful route to 4-chloro-*ortho*-xylene (R.C.Hahn, H.Shosenji and D.L.Strack, A.C.S.Sym.Ser., 1975, 22, 95). The synthetic value of such reactions has been discussed (K.S.Feldman and P.C.Myhre, J.Amer.chem.Soc., 1979, 101, 4768).

There is considerable recent evidence for *ipso*-attack in other electrophilic substitutions including halogenation (E.Baciocchi and G.Illuminati, Tetrahedron Letters, 1975, 2265; P.B.D.de la Mare, Acc.chem.Res., 1974, 7, 361), alkylation (A.H.Jackson, P.V.R.Shannon and P.W.Taylor, J.chem.Soc. chem.Comm., 1978, 734; B.E.Leach, J.org.Chem., 1978, 43, 1794), diazocoupling (A.H.Jackson, P.V.R.Shannon and A.C.Tinker, J.chem.Soc.chem.Comm., 1976, 796), and the coupling of aromatic amines with 9-methylacridinium iodide (O.N.Chupakhin, V.N.Charushin and E.O.Sidorov, Dokl.Akad. Nauk S.S.S.R., 1978, 239, 614) but in general these re- actions have been studied in less detail than nitration.

At present, the following conclusions appear valid. The extent of *ipso*-attack at alkyl or halogen substituents is important only when the electrophile is directed to the *ipso*-position by another substituent or by the relative reactivity of positions in polycyclic or hetero-aromatic systems. When *ipso*-attack is followed by dissociation to reform the reactants, there is no observable consequence unless the intermediate ion (e.g. 1) can be detected spectroscopically. When *ipso*-attack is followed by dis- sociation to the encounter pair and subsequent substitution elsewhere in the system, the reactivity of the system should be enhanced by *ipso*-attack but the orientation of sub- stitution should be unaffected. When *ipso*-attack is followed by specific 1,2-rearrangements, nucleophilic addition or side reactions, both the overall reactivity and the product composition are affected. The conventional additivity principle clearly needs modification to allow for the con- sequences of *ipso*-attack. Some attempts have been made to quantify the extent of *ipso*-attack by the calculation of *ipso* partial rate factors (Perrin and Skinner, *loc.cit.*) but much more work needs to be carried out in this area.

(b) Diffusion control of reaction rates

Three aspects of the diffusion control of aromatic sub- stitution have become important in recent years. One concerns what can be called macroscopic diffusion control: the distortion of reaction rates and product compositions produced when the rate of reaction of chemical species is comparable to the rate of mixing of the reactant solutions. Another, termed microscopic diffusion control, concerns the limitation of reaction rates when reaction occurs on every encounter between the reactants. The third, termed pre- association, concerns the description of chemical reactions when the life-time of the postulated intermediates is too short to permit reaction to occur by diffusion.

(i) Macroscopic diffusion control

Part of the interest in this area comes from the long controversy over the significance of the experiments of Olah and his co-workers concerning the addition of nitronium salts in organic solvents to mixtures of aromatic compounds in organic solvents. The disagreement concerned whether the similar reactivity of toluene and benzene in such experiments derived from the high reactivity of the electrophile or the fact that the rate of reaction was comparable to the rate of mixing. The two arguments have been summarised (G.A.Olah, Acc.chem.Res., 1971, 4, 240, J.H.Ridd, *ibid.*, 1971, 4, 248) and the second interpretation appears now to be generally accepted (L.M.Stock, Prog.phys.org.Chem., 1976, 12, 21). The resolution of this matter was important because the first interpretation required that such highly reactive electrophiles deviated from Brown's selectivity relationship in giving low substrate selectivity but high positional selectivity. Similar deviations from Brown's selectivity relationship had been reported for other reactions including benzylations (G.A.Olah, *loc.cit.*) but recent work suggests that some of the benzylation results involve errors (F.P.DeHaan et al., J.Amer.chem.Soc., 1978, 100, 5944) and a new examination of the selectivity relationship has given no evidence for a systematic deviation depending on the reactivity of the electrophile (C.Santiago, K.N.Houk and C.L.Perrin, J.Amer.chem.Soc., 1979, 101, 1337).

The mathematical treatment of macroscopic diffusion control is complex but for certain types of reaction a relationship has now been obtained between the product composition and the efficiency of mixing as measured by the size of the resulting eddies. (P.Rys, Angew. Chem., 1977, 16, 807; Acc.chem.Res., 1976, 9, 345). The theory is applicable to reactions where an aromatic substrate can undergo both mono- and disubstitution with a given electrophile and those where two aromatic substrates compete in reaction with a given electrophile. A quantitative test of the theory is difficult because of the lack of independent evidence on the size of the eddies but a series of papers have been published in which the theory is developed and applied to consecutive nitrations and azo-coupling reactions (J.R.Bourne, E.Crivelli and P.Rys, Helv.chim.Acta, 1977, 60, 2944; and references therein). This approach illustrates the dangers in the competitive method of determining reactivities when fast reactions are involved.

(ii) Microscopic diffusion control

The first clear evidence that the rate of encounter can become rate-determining in aromatic substitution came from the limiting rates of nitration in sulphuric and perchloric acids observed by R.G.Coombes, R.B.Moodie and K.Schofield (J.chem.Soc.(B), 1968, 800). For nitration in 68% sulphuric acid, this limiting rate is ca. 38 times the rate of nitration of benzene and is reached with compounds of re-activity equal or greater than that of the xylenes. At this limiting rate, the molecular second-order rate coefficient calculated from the concentrations of the aromatic substrate and the nitronium ion is ca. 2×10^8 $mol^{-1}s^{-1}dm^3$. This is a little lower than would be expected from the rate of encounter of the nitronium ion and the aromatic substrate as calculated from the viscosity of the medium but the difference is within the errors involved in estimating the concentration of nitronium ions.

These authors have now studied the limiting rate of nitr-ation in a large number of media (R.B.Moodie, K.Schofield et $al.$, J.chem.Soc.Perkin II, 1979, 747; and earlier parts of this series). In general, the lower the viscosity of the media, the greater the reactivity required to reach the limiting rate and the greater the ratio of this rate to the rate of nitration of benzene. The viscosity is not the only factor involved but the general pattern accords with reaction on encounter. Part of the importance of this work is that it shows the necessity for including the encounter pair $ArH.NO_2^+$ as a step in aromatic nitration so that the reaction path has now to be written as below:

$$NO_2^+ + ArH \rightleftharpoons ArH.NO_2^+ \longrightarrow Ar{\overset{\displaystyle +}{\underset{\displaystyle H}{<}}}^{NO_2} \longrightarrow ArNO_2 + H^+$$

The above interpretation has not been universally accepted, for the limiting rate of nitration has also been ascribed to the reactivity required for electron transfer to form the radical pair $ArH^{+\cdot}$ $NO_2^{\cdot}$ (C.L.Perrin, J.Amer.chem.Soc., 1977, 99, 5516). However, on this interpretation, it is difficult to see why the degree of aromatic reactivity required to reach the limiting rate should depend on the viscosity of the solvent.

No clearly defined limiting reaction rate has been reported for other electrophilic aromatic substitutions but the effects of substituents on the molecular bromination of aromatic amines are unexpectedly small (ρ = -2.19) which may indicate an approach to a limiting reaction rate (R.Uzan and J.E.Dubois, Bull.Soc.chim.France,1971, 598).

(iii) Pre-association

One of the older problems in electrophilic aromatic substitution concerns the disagreement between the kinetic evidence for the involvement of such species as Br^+ or H_2OBr^+ in positive halogenation and the thermodynamic evidence that the equilibrium concentration of these species is too low to explain the observed reaction rate even if reaction with the aromatic substrate occurs on every encounter. Recent kinetic evidence (H.M.Gilow and J.H.Ridd, J.chem. Soc.Perkin II, 1973, 1321) has provided support for the earlier suggestion (E.A.Shilov, F.M.Vainshtein and A.A.Yasnikov, Kinetika i Kataliz, 1961, 2, 214) that the effective electrophile is formed after the preliminary association of hypobromous acid with the aromatic substrate. The general importance of this type of reaction path involving the pre-association of the substrate with the precursor of the electrophile compared with the more conventional reaction path involving the diffusion together of the substrate and the pre-formed electrophile has been discussed in a recent review (J.H.Ridd, Adv.phys.org.chem., 1978, 16, 1). The pre-association path becomes more important as the lifetime of the electrophile decreases.

(c) Mechanisms involving electron transfer

The conventional representation of electrophilic substitution as shown above for nitration (section 2 b ii), involves no single electron transfers. However, following the earlier arguments of S.Nagakura (Tetrahedron, 1963, 19(Suppl.2), 361), a number of authors have recently been considering the possible importance of electron transfer leading to aromatic radical cations in substitution reactions. The present position has been reviewed by Z.V.Todres (Usp.Khim., 1978, 47, 260).

There is now considerable evidence for substitution reactions in which an aromatic compound is first oxidised to a
radical cation (ArH$^{+\cdot}$) by a powerful oxidising agent and then
reacts with a nucleophile (X$^-$) as shown below:

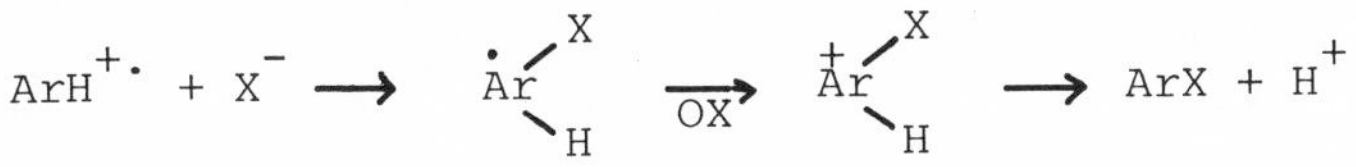

$$ArH^{+\cdot} + X^- \longrightarrow \overset{\cdot}{Ar}\!\!\diagup^{X}_{\diagdown H} \xrightarrow{OX} \overset{+}{Ar}\!\!\diagup^{X}_{\diagdown H} \longrightarrow ArX + H^+$$

Recent examples of these reactions include hydroxylation
by aqueous peroxydisulphate ions (M.K.Eberhardt, J.org.Chem.,
1977, 42, 832), acetoxylation by co-ordinated silver (II)
peroxydisulphate in acetic acid (K.Nyberg and L.-G.Wistrand,
Acta chem.Scand.(B), 1975, 29, 629) and trifluoroacetoxylation
by cobalt (III) trifluoroacetate in trifluoroacetic acid
(J.K.Kochi, R.T.Tang and T.Bernath, J.Amer.chem.Soc., 1973,
95, 7114). These reactions appear to pass through the same
σ-complex as those involving electrophilic attack by the
species X$^+$. The peroxydisulphate-initiated aromatic
halogenation by chloride ions in the presence of copper (II)
chloride was previously believed to involve intermediate
radical cations but recent work points to chlorination by
molecular chlorine (R.Filler and R.C.Rickert, J.chem.Soc.
chem.Comm., 1976, 133).

Another frequently proposed reaction involves the oxidation
of an aromatic compound to a radical cation by an electrophile followed by the combination of the resulting radicals :

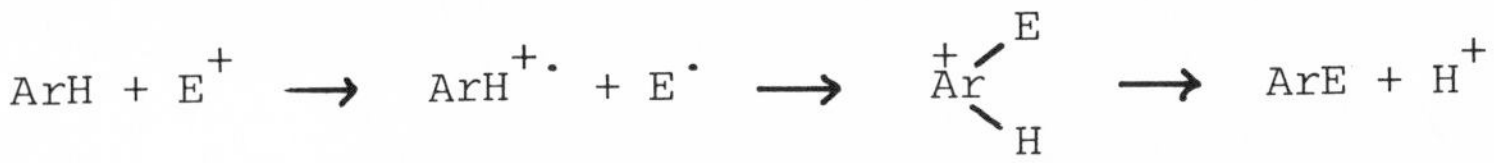

$$ArH + E^+ \longrightarrow ArH^{+\cdot} + E^{\cdot} \longrightarrow \overset{+}{Ar}\!\!\diagup^{E}_{\diagdown H} \longrightarrow ArE + H^+$$

This mechanism was suggested for the azo-coupling of phenols
on the grounds that the ^{15}N n.m.r. signals in the product
were enhanced by CIDNP (N.N.Bubnov *et al.*, J.chem.Soc.chem.
Comm., 1972, 1058). However, later work has shown that this
enhancement arises through the nuclear polarisation produced
in a homolytic pre-equilibrium involving, the diazonium ions
and is carried over into the products formed by a normal
heterolytic substitution (E.Lippmaa, T.Pehk, T.Saluvere and
M.Magi, Org.mag.Res., 1973, 5, 441). A related reaction path
in which the aromatic radical cation loses a proton from the
side-chain has been suggested to explain the facile side-
chain chlorination of certain alkyl benzenes (J.K.Kochi,
Tetrahedron Letters, 1975, 41) but a consideration of the
product composition makes a heterolytic reaction involving

ipso-attack more probable (E.Baciocchi and G.Illuminati, Tetrahedron Letters, 1975, 2265).

Many authors have suggested that the initial step of nitration involves electron transfer from the aromatic compound to the nitronium ion. There is no evidence for CIDNP effects in the nitration of benzene (I.P.Beletskaya, S.P.Rykov and A.L.Buchachenko, Org.mag.Res., 1973, 5, 595) and so the evidence for reaction through an aromatic radical cation is said to come from the rates and orientation of nitration (E.B.Pedersen *et al.*, Tetrahedron, 1973, 29, 579), the detection of a radical cation by e.s.r. (in the nitration of 2,6-di-tert-butylphenol) (V.D.Pokhodenko *et al.*, Zhur.org. Khim., 1975, 11, 1873), evidence for the reaction of cation radicals with nitrogen dioxide (A.S.Morkivnik, O.Yu.Okhlobystin and E.Yu.Belinskii, Zhur.org.Khim., 1979, 15, 1565) and arguments based on the energetics of the electron transfer and the retention of intramolecular selectivity in reaction on encounter (C.L.Perrin, J.Amer. chem.Soc., 1977, 99, 5516).

For several reasons, it is difficult to obtain clear-cut mechanistic conclusions in this type of work. The possible co-existence of heterolytic and homolytic pathways greatly complicates the mechanistic analysis with the result that some of the evidence previously advanced for the radical cation path now seems invalid (L.Eberson, L.Jonsson and F.Radner, Acta chem.Scand.(B), 1978, 32, 749). There is also some confusion over the definition of the radical cation path for we have to distinguish between (a) reaction of the aromatic compound and the nitronium ion through a transition state involving significant biradical character, (b) reaction through an intermediate radical pair $ArH^{+\cdot}NO_2^{-}$ all of which recombines to form the nitro product and (c) reaction involving independent cation radicals $ArH^{+\cdot}$ At the moment it appears that (a) is probably true for most nitrations (T.Takabe *et al.*, Chem.Phys.Letters, 1976, 44, 65), (b) may be true for the nitration of compounds more reactive than toluene (Perrin, *loc.cit.*) and (c) is true only under some special conditions, e.g. in some side-reactions accompanying nitrous acid catalysed nitration (J.C.Giffney and J.H.Ridd, J.chem.Soc.Perkin II, 1979, 618).

3. NUCLEOPHILIC AROMATIC SUBSTITUTION

N.B. CHAPMAN

(a) General

Five valuable general reviews are now available: (a) by S.D.
Ross in Comprehensive Chemical Kinetics, ed. C.H. Bamford and .
C.F.H. Tipper, Elsevier, Amsterdam, 1972, Vol.13, p.407; (b)
by J.H. Ridd in Physical Methods in Heterocyclic Chemistry,
ed. A.R. Katritzky, Vol.4, p.55, as regards the reactivity of
heterocyclics towards nucleophiles, (c) by P. Rys, P. Skrabal,
and H. Zollinger (Angew. Chem. internat. Edn., 1972, 11, 874),
which, although mainly concerned with electrophilic substitution, has a useful section entitled "Instrumental Analysis of
Accumulating Adducts in Electrophilic and Nucleophilic Substitution Reactions",(d) by C.F. Bernasconi in the M.T.P. International Review of Science, Organic Chemistry, Series 1, 1973,
3, 33, entitled Mechanisms and Reactivity in Aromatic Nucleophilic Substitution, which deals only with the ArS_N2 mechanism,
(e) by J.R. Beck, Tetrahedron, 1978, 34, 2057, which is concerned solely with the nitro-group as a leaving group (its
nucleofugacity may equal or surpass even that of the fluoro-
group), and which lays emphasis on the use of this type of
reaction in the synthesis of novel benzene derivatives and of
heterocyclic compounds.

Of considerable general interest are reports that aromatic
nitro-compounds may be alkylated in the *ortho* and *para* positions by reaction with alkyl-lithium or alkyl Grignard reagents
(F. Kienzle, Helv. chim. Acta, 1978, 61, 449), and of 'vicar-
ious substitution' in benzene derivatives in which hydride ion
is formally displaced in reactions involving a nucleophile
which itself contains a leaving group as exemplified below

(J. Golinski and M. Makosza, Tetrahedron Letters, 1978, 223,
227; A.I. Meyers, R. Gabel, and E.D. Mihelich, J. org. Chem.,
1978, 43, 1372). Displacement of hydrogen in nucleophilic
aromatic substitution has been recently reviewed (O.N.
Chupakhin and I. Ya·Postovskii, Russ. chem. Rev., 1976, 45,
454).

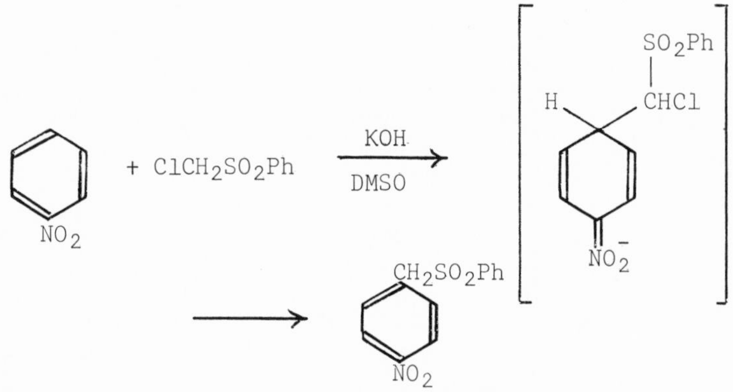

Reviews have also appeared annually in "Organic Reaction
Mechanisms", John Wiley and Sons, Chichester, by M.R. Crampton
and others.

(b) Mechanism of substitution

(i) Unimolecular mechanism

The homogeneous uncatalysed thermal decomposition of arene diazonium salts in solution in the dark has continued to be the
main focus of interest since 1969. (This class of reactions
can be regarded as belonging to a wider class known as denitro-
genations or dediazoniations). A review by H. Zollinger (Acc.
chem. Res.,1973, 6, 335) deals briefly with this subject. In
1975, however, a definitive paper by C. Gardner Swain, J.E.
Sheats, and K.G. Harbison (J. Amer. chem. Soc., 1975, 97, 783)
appeared which provides (a) a comprehensive survey of possible
mechanisms, (b) an extensive set of references, (c) clear evidence for the exclusion of the benzyne mechanism, (d) crucial
evidence for "the (singlet) phenyl cation as an intermediate
in reactions of benzenediazonium salts in solution", to quote
the title of the paper. The well-known low selectivity of the
reaction between nucleophiles and arenediazonium cations

requires a highly reactive intermediate. In the hydrolysis of PhN$_2^+$ the constancy of the first-order rate coefficient, k_1 (within 4%) when H$_2$O is replaced by D$_2$O as the solvent demonstrates clearly that water is not involved as a nucleophile in the rate-determining step, a conclusion powerfully supported by the large positive entropy of activation (+10.5 cal. mole^{-1} deg^{-1}). Measurement of k_1 for a very diverse group of solvents ranging from 0.001M-H$_2$SO$_4$ to oleum (21M-H$_2$SO$_4$) and including acetic acid, dichloromethane, and dioxan shows barely a four-fold range of values, and over a more limited range of solvents (14-21M-H$_2$SO$_4$, acetic acid, and CH$_2$Cl$_2$) k_1 remains constant (within 10%). This implies a common rate-determining step over this last range of solvents. Moreover k_1 changes by only 2% on going from 14M (80%) H$_2$SO$_4$ to 21M (105%) H$_2$SO$_4$ despite the facts that (a) a_{H_2O} decreases by more than 1000-fold from 80 to 98% H$_2$SO$_4$, (b) HSO$_4^-$ is a much poorer nucleophile than H$_2$O, (c) even $\left[\text{HSO}_4^-\right]$ drops rapidly above 100% H$_2$SO$_4$. These facts exclude all mechanisms except the following.

$$C_6H_5\overset{+}{N}\equiv N \rightarrow C_6H_5^+ + N_2 \ (r.d.s)$$

$$C_6H_5^+ + Y \rightarrow C_6H_5Y^+$$

This paper and accompanying papers (*idem.*, *ibid.*, 1975, 97, 791, 796) cover substituent effects and nitrogen and hydrogen isotope effects, and in the author's judgment supersede other papers mentioned in the article to which this is a supplement. We now have "a really convincing demonstration of a reaction pursuing the simple limiting S$_N$1 mechanism" *i.e.* ArS$_N$1 in solution, despite evidence of the great difficult of formation of the phenyl cation in the gas phase (Ann. Rep. B, 1976, 73, 59). Moreover, M.C.R. Symons *et al.* (J. Amer. chem. Soc., 1978, 100, 4779) claim, on the basis of esr studies of photolysis at 77K of arenediazonium salts having 4-dialkylamino substituents, as powders, or solutions in polymer films or in aqueous LiCl glass, to have provided "irrefutable evidence for the existence of these carbonium ions" *viz.* triplet aryl cations.

Further evidence relevant to Swain's conclusions has been supplied by R.G. Bergstrom, R.G.M. Landells, G.H. Wahl, jun., and H. Zollinger (J. Amer. chem. Soc., 1976, 98, 3301) through ^{15}N studies and by showing that it is likely that C$_6$H$_5^+$ can react with external N$_2$ from nitrogen under 300 atm pressure, *i.e.* the heterolysis of $Ph\overset{+}{N}_2$ is reversible. Recently, Zollinger and his co-workers (J. Amer. chem. Soc., 1978, 100, 2811, 2816) have studied solvent effects on isotopic rearrangements in the dediazoniation of $Ph\overset{+}{N}_2$ and $2,4,6\text{-Me}_3C_6H_2\overset{+}{N}_2$ and claim that their results require for rational interpretation the intervention in the reaction under the conditions they employed of at least two intermediates: (a) a tight molecule-ion pair, *e.g.* Ph^+/N_2, (b) a (solvated) aryl cation, free of or solvent-separated from N_2.

A useful review has recently appeared (H.B. Ambroz and T.J. Kemp, Chem. Soc. Reviews, 1979, 8, 353).

(ii)   The elimination-addition mechanism - arynes and hetarynes

Aryne chemistry has continued to attract much attention. The present account is necessarily confined to brief supplementary information concerning the preparation of arynes and heterarynes and to their intermediacy in nucleophilic aromatic substitution (*cf.* 2nd Edn. p.96), now only a small but significant part of aryne chemistry as a whole.

1-(2-Carboxyphenyl)-3,3-dimethyltriazine,

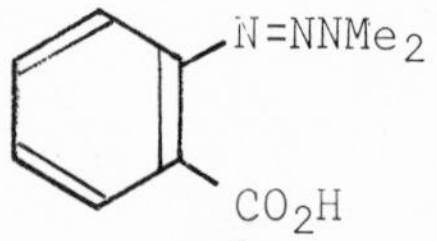

a relatively stable non-explosive compound, gives a high yield of benzyne when heated under reflux in chlorobenzene (J. Nakayama, O. Simamura, and M. Yoshida, Chem. Comm., 1970, 1222). C.W. Rees and his co-workers have provided a low-temperature route to benzyne by decomposition in polar aprotic solvents of the highly unstable lithium salt shown below (J. Chem. Soc. Perkin I, 1972, 1315).

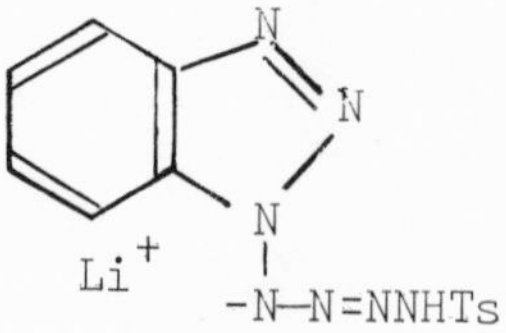

Benzyne has also been generated by deprotonation of $[Ph\overset{+}{N}_2]SbCl_5^-$

with various phosphoryl compounds (Von H. Teichmann,
M. Jatkowski, and G. Hilgetag, J. prakt. Chem., 1972, 314, 129).
Aryne formation by vapour-phase thermolysis of heterocycles
has come to the fore recently (Ann. Rep. B, 1973, 70, 177),
and even more interesting is the generation of immobilised
benzyne on a polymer support, which gives it extended lifetime
and suppresses dimerisation (P. Jayalekshmy and S. Mazur, J.
Amer. chem. Soc., 1976, 98, 6710). Moreover aryldiazonium
carboxylates, formed by heating N-nitroso-N-acylanilines in
CCl$_4$, may also serve as aryne precursors (cf. also J.I.G.
Cadogan and his co-workers, J. chem. Soc. Perkin I, 1972,
2563). However, the formation of the precursor is often a
complex reaction (Ann. Rep. B, 1970, 67, 341).

The chemistry of hetarynes has been reviewed (T. Kauffmann
and R. Wirthwein, Angew. Chem. internat. Edn., 1971, 10, 20).
Of special relevance to nucleophilic aromatic substitution is
a discussion of the competition between the elimination-
addition and the addition-elimination mechanism in the reactions of hetaryl halides. The existence of five-membered
hetarynes containing one heteroatom had not been demonstrated
by 1971. Reactions of aryl halides with nucleophilic reagents
involving the intermediacy of arynes have been reviewed (J.F.
Bunnett, Acc. chem. Res., 1972, 5, 139), and the use of
sodamide-containing complex bases in aryne chemistry has also
been reviewed (P. Caubere, Acc. chem. Res., 1974, 7, 301),
cf. also idem., Topics in current Chem., 1978, 73, 49.
Bunnett's review also covers potassium anilide catalysed rearrangements of trihalogenobenzenes, the so-called 'base-
catalysed halogen dance'.

What was thought to be a stable nickel complex of benzyne
(2nd Edn. p.98, E.W. Gowling, S.F.A. Kettle and G. Sharples,
Chem. Comm., 1968, 21), produced by reaction of Ni(CO)$_4$ and
o-C$_6$H$_4$I$_2$, has since been shown not to be a benzyne complex
(N.A. Bailey, S.E. Hull, R.W. Jotham, and S.F.A. Kettle, Chem.
Comm., 1971, 282).

(iii) The S_N2 mechanism, "activated" nucleophilic substitution
and Meisenheimer complex formation

In a review entitled Anionic σ-Complexes, M.J. Strauss (Chem.
Rev.,1970, 667) deals specifically with the nature of intermediates in the S$_N$Ar2 mechanism. Further reviews include those
of P. Buck (Angew. Chem. internat. Edn., 1969, 8, 120),and C.F.

Bernasconi in MTP International Review of Science, Organic
Chemistry, Series 1, Vol. 3, 1973, pp. 56-59.

Much detailed work on the general chemistry of and the
kinetics and thermodynamics of the formation and decomposition
of Meisenheimer complexes continues to appear, without any-
thing very novel in principle emerging. Recent noteworthy
examples of such investigations include those of V. Gold and
J. Toullec, (J. Chem. Soc. Perkin II, 1979, 596) and of M.R.
Crampton, (*ibid*, 1977, 1442). A recent paper by C.F. Bernasconi, M.C. Muller, and P. Schmid (J. org. Chem., 1979, 44,
3189) deals with solvent effects on the rate-limiting proton
transfer in the formation of Meisenheimer complexes between
1,3,5-trinitrobenzene and amines; leaving group effects are
also discussed. This paper includes a comprehensive set of
references. Illuminati and his co-workers have studied the
influence of heterocyclic aza-groups on Meisenheimer complex
formation (P. Bemporad, G. Illuminati and F. Stegel, J. Amer.
chem. Soc., 1969, 91, 6742), a thiophen-derived complex has
been reported (G. Doddi, G. Illuminati, and F. Stegel, J.
org. Chem., 1971, 36, 1918), and an analogous selenophen
derivative (F. Terrier, A.-P. Chatrousse, R. Schael, C. Paul-
mier, and P. Pastour, Tetrahedron Letters, 1972, 1961).
Meisenheimer complexes with alkyl groups attached to the benzene ring have been isolated (R.P. Taylor, J. org. Chem.,
1970, 35, 3578). There has also been extensive application of
fast-reaction techniques (stopped-flow and temperature jump)
by Fendler and his co-workers among others (J. org. Chem.,
1971, 36, 1749; J. chem. Soc. Perkin II, 1972, 1403) to
Meisenheimer complex formation. Bernasconi (J. org. Chem.,
1971, 36, 1671) has provided evidence of the formation of the
complex formulated below in the reaction of TNT with MeONa/
MeOH.

Meisenheimer complexes are readily formed from 1,3,5-trinitro-
benzene and anions in benzene solution in the presence of crown
ethers: some of the anions *e.g.* F^- do not react thus in
aqueous solution - (A.R. Butler, J. chem. Soc. Perkin I, 1975,
1557). A stable *gem*-difluoro-Meisenheimer complex has been
detected by n.m.r. spectroscopy in the interaction in acetonitrile of phenyl fluoride and F^-/crown ether (G. Ah-Kow, M-J.
Poult, and M-P. Simoninin, Tetrahedron Letters, 1976, 227, *cf.*
S. Ohsawa, Nippon Kaguku Kaishi, 1976, 456, for a similar *gem*-
diamino-complex).

(b)   New mechanisms

(i)   The $S_{RN}1$ mechanism

   This is a radical-chain mechanism discovered by J.F. Bunnett
and his co-workers, a fascinating and comprehensive account of
which with all necessary references has been given by Bunnett
himself (Acc. chem. Res., 1978, 11, 413).   A typical initiation
stage is as follows (eq 1):

$$e^-_{NH_3} + ArX \rightarrow [ArX]^{-\cdot} \qquad (1)$$

The propagation stages, in which most of the reaction occurs
if the chains are long (as indeed they may be) are formalated
below.

$$[ArX]^{-\cdot} \rightarrow Ar^{\cdot} + \overline{X}$$

$$Ar^{\cdot} + Y^- \rightarrow [ArY]^{-\cdot}$$

$$[ArY]^{-\cdot} + ArX \rightarrow ArY + [ArX]^{-\cdot}$$

Three types of reaction intermediate are involved, the radical
$Ar^{\cdot}$ and the radical ions $ArX^{-\cdot}$ and $ArY^{-\cdot}$:   generation of any
one of these will start the propagation cycle.

   The mechanism was discovered during an investigation of the
action of potassamide in liquid ammonia on various halogeno-
pseudocumenes (5- and 6-halogeno-1,2,3-trimethylbenzenes),
which give proportions of rearranged and unrearranged products
not readily understood in terms of the aryne mechanism, and it
was confirmed by the suppression of the non-rearranging component of the reaction by a radical scavenger ($Ph_2N.NPh_2$), and
its promotion to 100% of the reaction by use of metallic

potassium in liquid ammonia (to give $K^+_{NH_3}$ + $e^-_{NH_3}$) as the reagent. A survey of solvents suitable for this mechan-
ism is available (Bunnett, R.G. Scamehorn, and R.P. Traber,
J. org. Chem., 1976, 41, 3677): DMSO, DMF, and acetonitrile
may each replace liquid ammonia.

The $S_{RN}1$ mechanism has also been shown to be involved in the
recently discovered electrochemical induction in liquid
ammonia of nucleophilic aromatic substitution, and competition
with electron transfer has also been studied (J. Pinson and
J.M. Saveant, J. chem. Soc. chem. Comm., 1974, 933, *idem*,
J. Amer. chem. Soc., 1978, 100, 1506; Pinson *et al.*, *ibid.*,
1979, 101, 6012).

In 1978 the formation and probable intermediacy in substitution of an anion radical in the displacement of one nitro
group from *p*-dinitrobenzene by OH⁻ in aqueous DMSO was demonstrated (visible and esr spectroscopy, kinetic methods) by
T. Abe and Y. Ikegami (Bull. chem. Soc. Japan, 1978, 51, 196).

(ii)   The S_N (ANRORC) mechanism

This mechanism, which so far has been shown to apply only
to heterocyclic aromatic substrates, *e.g.* halogenopyrimidines,
involves Addition of the Nucleophile, Ring Opening, and Ring
Closure, hence the notation used in the sub-title. A full
account, with extensive references, has been given by Henk
C. van der Plas (Acc. chem. Res., 1978, 11, 462), *cf.* also
H.C. van der Plas, "Ring Transformations of Heterocycles",
Academic Press, London, 1973. The year 1978 was clearly
something of an *annus mirabilis* for the summary exposition of
novel mechanisms of nucleophilic aromatic substitution! Again
the mechanism emerged in a study of the generation of aryne
intermediates, *viz*. pyrimidynes, and a very brief account of
it now follows. [The first example of trapping of a 4,5-di-
dehydropyrimidine has recently been reported by D. Christophe,
R. Promel, and M. Maeck (Tetrahedron Letters, 1978, 4435).]
First the following reaction system was discovered.

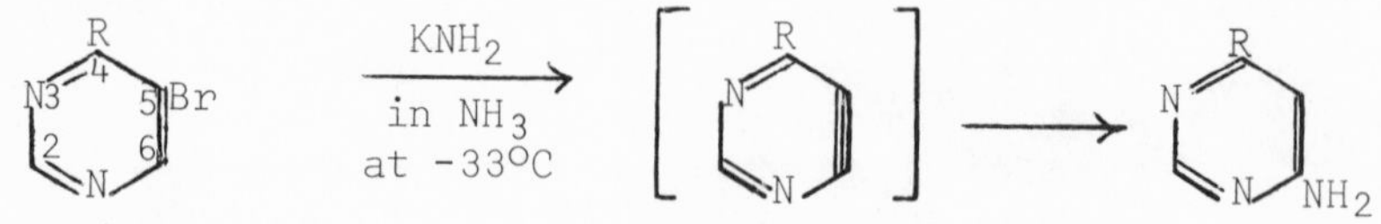

(R = Bu^t, Ph, Me, OMe, OH or NH$_2$)   and no 5-amino-compound

Moreover the isomeric 6-bromo-compounds (R = Bu^t or Ph) gave
only the corresponding 6-amino-compounds in rapid reactions
at -75 °C. However 6-bromo-4-R-5-deuteriopyrimidine (R = Bu^t)
gave a 6-amino-product containing no deuterium, despite the
fact that under the reaction conditions it gave little D/H
exchange. This surprising result, implying a preference for
the aryne mechanism, led to an investigation of the reactions
of another strongly basic reagent, *viz.* Li piperidide in
piperidine/ether: the 6-bromopyrimidine derivatives (R = Bu^t
or Ph) gave no trace of 6-piperidinopyrimidines but instead
2-aza-4-cyano-1-piperidinobuta-1-3-dienes, by a proposed
reaction sequence, also formulated for the reaction of NH$_2^-$,
as follows. The reaction of Li piperidide yields (3a) in
analogous fashion.

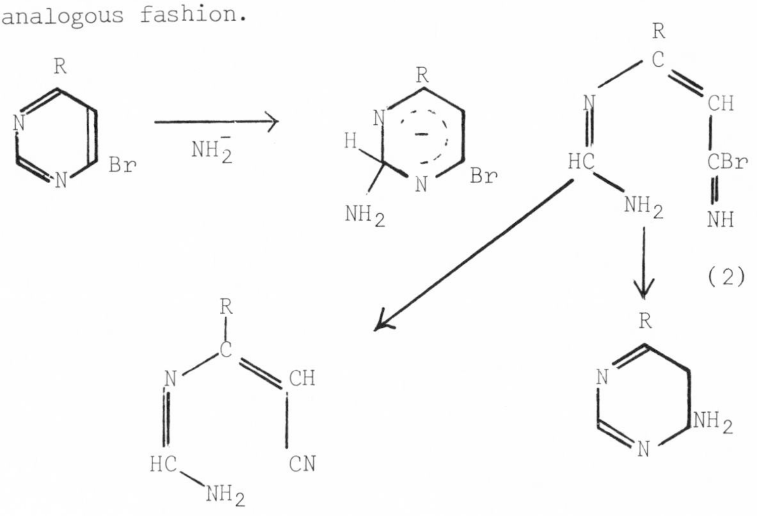

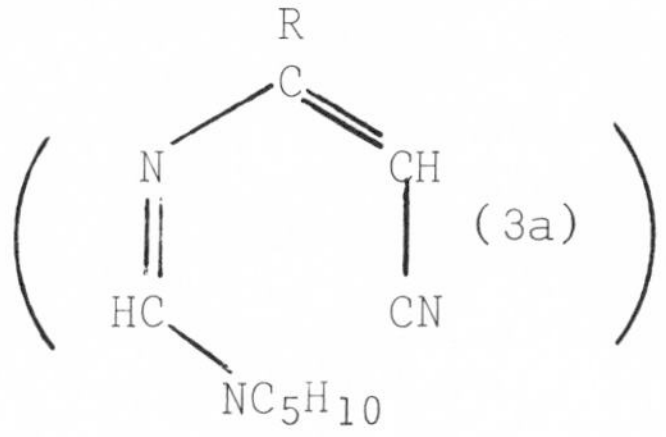

The key intermediate is the highly reactive imidoyl bromide (2),
and the loss of D mentioned above is ascribed to rapid D/H
exchange in (3) through the tautomeric equilibrium shown below.

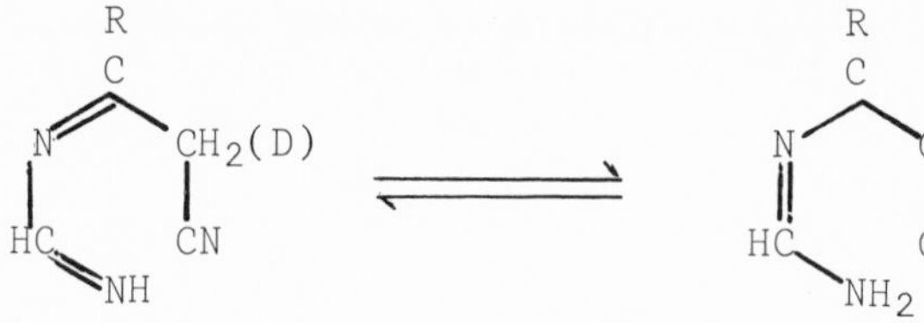

Proof of the correctness of this hypothesis came from a study of monolabeled 6-bromo-4-phenyl [1(3)-^{15}N] pyrimidine (4) containing 6.0% of ^{15}N excess, scrambled over both nitrogen atoms. If the mechanism via the imidoyl bromide is correct the 6-amino product should have the same percentage excess of ^{15}N as in (4). It was found that (5) had 6.0% of ^{15}N excess and (7) had 2.5% of ^{15}N excess.

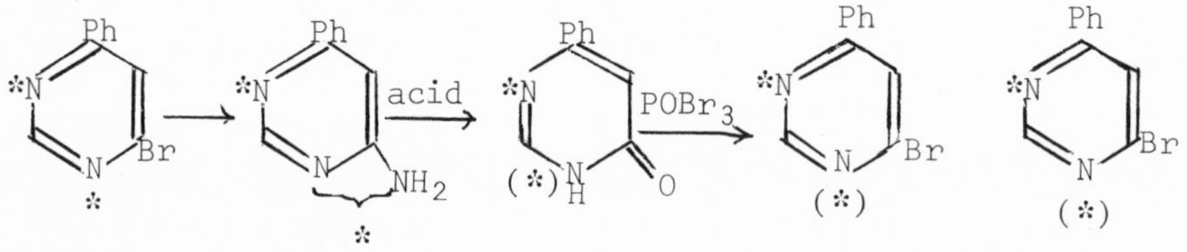

(4)   (5)   (6)   (7)

(c)   Photochemical reaction

A review of photochemical reactions of aromatic compounds by J. Cornelisse and E. Havinga (Chem. Rev., 1975, 75, 353) contains many examples of nucleophilic substitution.

In an important series of publications the focus of which is the S$_{RN}$1 mechanism [Section (b)(i)], J.F. Bunnett and his co-workers (J. org. Chem., 1973, 38, 1407, 3020; 1974, 39, 382, 3173; 1976, 41, 1702; 1977, 42, 1457; 1978, 43, 1867, 1873, 1877; J. Amer. chem. Soc., 1977, 99, 4690) have discussed the scope, the preparative value (which is very considerable), and the mechanism of photostimulated reactions of aryl iodides and related compounds with nucleophiles, particularly enolate anions and other carbanions, arenethiolate ions, and dialkyl-phosphate anions. The reactions of the last reagent especially have been subjected to much quantitative study.

Unfortunately lack of space precludes extensive review of

these reactions here: they afford many novel, mechanistically interesting, and preparatively useful nucleophilic aromatic substitutions. A related thermally induced reaction has also been discovered (J. org. Chem., 1977, 42, 1449).

Photosubstitution of aryl fluorides by primary and secondary amines gives both conventional and cine-substitution, probably by an addition-elimination mechanism (D. Bryce-Smith, A. Gilbert, and S. Krestonosuh, J. chem. Soc. Comm., 1976, 405). This reaction of nucleophiles in general, without substitution has been known for some years, *cf.* J. Cornelisse, G.P. de Gunst, and E. Havinga, "Advances in Physical Organic Chemistry", ed. V. Gold, Academic Press, London, 1975, Vol.11, p.225.

(d) Gas phase reactions

In the gas phase MeO^- and EtO^- react slowly with PhF to give PhOR (R = Me or Et) but OH^- reacts as shown below.

$$OH^- + PhF \rightarrow H_2O + C_6FH_4^-$$

The former reactions, therefore, seem unlikely to proceed via benzyne. With C_6F_6 and RO^- (R = Me, Et, or Pr^i) the reaction is as follows.

$$RO^- + C_6F_6 \rightarrow C_6F_5O^- + RF$$

It is concluded from the above and other evidence that (a) gas-phase nucleophilic substitution by alkoxide ions occurs only when the acidity of the aromatic compound is less than or comparable to that of an alcohol, (b) electron-attracting substituents do not speed up substitution of halogen, but increase overall reactivity through formation of a substituted phenoxide. The proposed mechanism is

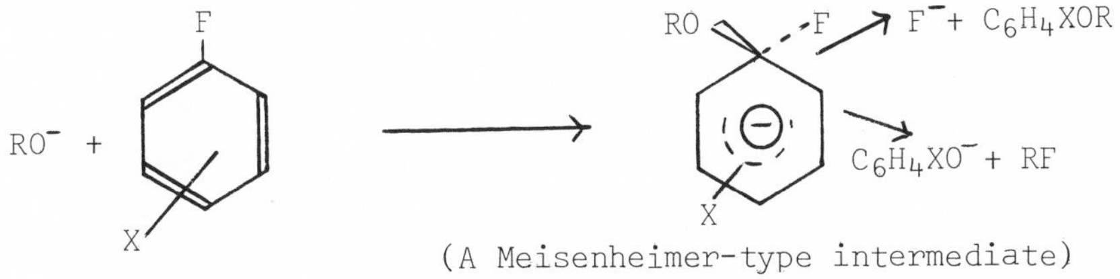

(A Meisenheimer-type intermediate)

(S.M.J. Briscese and J.M. Riveros, J. Amer. chem. Soc., 1975, 97, 230). (See also J.H. Bowie, Acc. chem. Res., 1980, 13, 79.)

(e) Miscellany

(a) The Smiles and related rearrangements are discussed by
W.E. Truce, E.M. Kreider, and W.W. Brand in Organic Reactions,
1970, 18, 99-215. This article provides a detailed amplific-
ation of the brief summary included in the article to which
this is a supplement, and in particular pyridyl systems are
covered, also formation of phenazine, phenothiazine, and
phenoxazine derivatives by Smiles rearrangements. Other novel
features of the Smiles rearrangement are reported in Synthesis,
1977, pp 31-35.

(b) An example of facile nucleophilic substitution in a ring
system lacking an electron-withdrawing group has been discovered (M.J. Perkins, Chem. Comm., 1971, 231). This is the
reaction of EtO^- with 5-chloroacenaphthylene, in which the
addition intermediate formulated below may be stabilised by
formation of a cyclopentadienide-type ion.

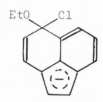

(c) In this supplementary article no additional material
relating to two topics discussed in the previous article,
viz. Substituent Effects in the ArS_N2 mechanism and the influence of the nucleophilic reagent, has been included. The
quantification of nucleophilicity, in a general way, has
recently been reviewed by Claude Duboc in Correlation Analysis
in Chemistry, eds. N.B. Chapman and J. Shorter, Plenum. New
York, 1978. As regards the former topic, there has been some
accumulation of details without novel principles being developed.

HOMOLYTIC AROMATIC SUBSTITUTION

G.H. WILLIAMS

Several important reviews of the subject, and of various aspects of it have appeared. The whole subject has been reviewed by M.J. Perkins ("Free Radicals" Vol. I, J. Kochi ed. John Wiley and Sons, New York, 1973, 231). Homolytic substitution in heteroaromatic compounds has been comprehensively reviewed by K.C. Bass and P. Nababsing ("Advances in Free Radical Chemistry", Logos Press, London 1972, 4, 1), and polar effects in synthetically useful homolytic aromatic and heteroaromatic substitution reactions by F. Minisci and A. Citterio, "Advances in Free Radicals Chemistry", Heyden and Son, London, 1980, 6, 65). Synthetic aspects of these reactions have been discussed by D.I. Davies and M.J. Parrott ("Free Radicals in Organic Synthesis", Springer Verlag, Berlin, Heidelberg and New York, 1978, 24). Other reviews pertinent to particular reactions are cited as appropriate in the following sections.

(a) Alkylation

Simple cyclic and bridgehead alkyl radicals are now generally recognised to be appreciably nucleophilic in character and Minisci and his co-workers have capitalised on this property in developing the homolytic alkylation of heteroaromatic bases in acidic media as a useful and versatile synthetic method. The protonated bases are alkylated much more rapidly and with much greater regioselectivity than the free bases (see Minisci, Synthesis, 1973, 5, 1, Minisci and Citterio *loc. cit.* and references cited therein). Thus, for example, whereas methylation of 4-methylpyridine with methyl radicals from the thermolysis of t-butyl peroxide gives the 2- and the 3-methylated product in the ratio 65:35, the analogous methyl-

ation of the 4-methylpyridinium cation gives the 2-isomer almost exclusively, reflecting the enhanced reactivity to nucleophiles of the 2-position in the cation. The alkylation of
such protonated bases in aqueous acidic solution is conveniently accomplished by silver-catalysed decarboxylation of the
appropriate carboxylic acids by peroxydisulphate ion (Minisci
$et.al.$,Tetrahedron 1972, 28, 2403).

$$S_2O_8^{2-} \; + \; Ag^+ \longrightarrow SO_4^{\overline{\cdot}} \; + \; SO_4^{2-} \; + \; Ag^{2+}$$

$$SO_4^{\overline{\cdot}} \; + \; Ag^+ \longrightarrow SO_4^{2-} \; + \; Ag^{2+}$$

$$RCO.OH \; + \; Ag^{2+} \longrightarrow RCO.O\cdot \; + \; Ag^+ \; + \; H^+$$

$$RCO.O\cdot \longrightarrow R\cdot \; + \; CO_2$$

Substituted alkyl radicals such as hydroxy-, alkoxy- or amino-
alkyl radicals (conveniently obtained by oxidation of alcohols,
ethers and amines. e.g. by $S_2O_8^{2-}$) are also nucleophilic because the substituent groups are electron-releasing, and can
be used to bring about analogous substitutions in protonated
heteroaromatic compounds. The 2-acetoxyethyl radical has
likewise been shown to be nucleophilic ($\rho = 0.24$) in its attack on monosubstituted benzenes (M. El Zein, C. Gardrat and
B. Maillard, Bull. Soc. chim. Belg., 1978, 87, 67).

On the other hand, substituted alkyl radicals containing
electron-withdrawing groups (e.g. $\cdot CH_2CO_2H$, $CH_3CO.CH_2\cdot$,
$\cdot CH_2NO_2$) can be electrophilic, as indicated by the partial
rate factors for their attack on benzene derivatives (Minisci
and Citterio, $loc.cit.$).

(b) Amination

Useful reviews by G. Sosnovsky and D.J. Rawlinson ("Advances in Free Radical Chemistry", Logos Press, London, 1972,
4, 213) and Minisci and Citterio ($loc.cit.$) have been published.
Numerous references are cited. The most useful reagents are
the powerfully nucleophilic amino-radical-cations, obtained by
reaction of chloroamines with transition metal ions [Fe(II),
Ti(III), Cr(II), Cu(I)] in acid solution(Minisci $et.al.$,
J. chem. Soc., Perkin II, 1974, 416). The aromatic substitutions are thought to proceed according to the following scheme and display a high degree of substrate and positional selectivity arising from the nucleophilicity of the reagent, as
indicated by partial rate factors.

$$\overset{+}{R N H C l} \ + \ Fe(II) \ \longrightarrow \ R_2\overset{\bullet+}{NH} \ \overset{C_6H_6}{\longrightarrow} \ C_6H_5\overset{\bullet}{<}\overset{H}{\underset{\overset{+}{N}HR_2}{}}$$

$$\overset{+}{R_2NHCl} \ \underset{\text{chain step}}{\longrightarrow} \ \text{[ring structure: } Cl, \ \overset{+}{N}HR_2, \ H, \ H\text{]} \ \overset{-HCl}{\longrightarrow} \ C_6H_5\overset{+}{N}HR_2$$

(c) Hydroxylation and related reactions

Oxygen-centred radicals (e.g. HO$\cdot$, SO$_4^{\bullet}$, ArCO.O$\cdot$ and ROCO.O$\cdot$) have been used extensively in aromatic substitutions and are electrophilic as indicated by ρ-values, although the reactions may be complicated by electron-transfer and reversible processes (Minisci and Citterio, *loc.cit.*, and references cited therein).

Whereas Fenton's reagent [H$_2$O$_2$ + Fe(II)] gives phenols by homolytic aromatic hydroxylation only in low yields, phenols are major products if S$_2$O$_8^{2-}$/Fe(II) in the presence of Cu(II) is the reagent (C. Walling and D.M. Camaioni, J. Amer. chem. Soc., 1975, 97, 1603). The rather complex mechanism suggested involves initial attack by the SO$_4^{\bullet}$ radical-anion.

Extensive studies of the benzoyloxylation which accompanies phenylation by benzoyl peroxide have shown that the yield of esters is increased by the presence of oxygen (T. Nakata, K. Tokumaru and O. Simamura, Tetrahedron Letters, 1967, 3303) or copper salts (M.E. Kurz and P. Kovacic, J. org. Chem., 1968, 33, 1950). In both cases the increased yield arises from more efficient dehydrogenation of the sigma-complex intermediate. The latter workers, as well as M.E. Kurz and M. Pellegrini (J. org. Chem., 1970, 35, 990) have demonstrated the electrophilicity of the benzoyloxy-radical. J.Saltiel and H.C. Curtis (J. Amer. chem. Soc., 1971, 93, 2056) have shown by a study of deuterium isotope effects that the addition of benzoyloxy-radicals to benzene is reversible, and the CIDNP technique has been used to establish the reversibility of the addition of the pentafluorobenzoyloxy-radical to the aromatic nucleus (J. Bargon, J. Amer.chem. Soc., 1971, 93, 4630; R. Kaptein, R. Freeman and H.D.W. Hill, Chem. Comm., 1975, 953).

(d) Thienylation

M. Tiecco and A. Tundo (Int. J. sulphur Chem., 1973, 3,295)

have reviewed homolytic substitution reactions in the thiophen nucleus and aromatic substitution by thienyl radicals.

(e) Arylation

This, the most extensively studied homolytic substitution reaction, has been reviewed by G.H. Williams ("Essays in Free-Radical Chemistry", Chemical Society Special Publication No.24, 1970, 25) and by M.J. Perkins ("Free Radicals" Ed. J. Kochi, John Wiley and Sons, New York, 1973, Vol.2, p.231).

Formation of Aryl Radicals

Some of the reactions which give rise to homolytic aryl-ation have been subjected to extensive mechanistic investig-ation.

The Gomberg-Bachmann and acylarylnitrosamine reactions. The second of these reactions has excited much interest, and some of the conclusions concerning its mechanism can also be applied to other arylations involving diazo-compounds such as the Gomberg-Bachmann reaction and the reaction of primary aro-matic amines with pentyl nitrite. Several reviews of this apparently very complex reaction have been published but the most recent and comprehensive (J.I.G. Cadogan, "Advances in Free-Radical Chemistry", Heyden and Son, London, 1980, 6, 185) gives all the relevant references.

Cadogan, R.M. Paton and C. Thomson (Chem. Comm., 1969, 614; J. chem. Soc., (B), 1971, 583) showed by e.s.r. spectro-scopy that both radicals previously thought to be key interm-ediates in these reactions (PhN=NO· and ·O(Ph)N-N(Ph)COCH$_3$ from nitrosoacetanilide itself: cf. 2nd Edn. Vol. IIIA, pp.128, 129) are actually present but that the phenyldiazoxyl radical is a σ- rather than a π-radical as previously thought. It is now considered, however, that the previously suggested mechan-istic schemes for arylation based on these radicals as the re-agents responsible for conversion of σ-complexes (ArAr'H·) in-to biaryls (ArAr') by hydrogen-abstraction are not the main contributors to the mechanism. They do, however, take part in chain initiation processes, whereby aryl radicals can be form-ed by processes involving them.

$$\underset{\substack{| \\ ArNAc}}{ArNOH} \xrightarrow{ArN_2^+OAc^-} \underset{\substack{| \\ ArNAc}}{ArNON_2Ar} \longrightarrow Ar\cdot + N_2 + \underset{\substack{| \\ ArNAc}}{ArNO\cdot}$$

$$(1)$$

$$2ArN_2^+OAc^- \longrightarrow ArN_2ON_2Ar + Ac_2O \longrightarrow Ar\cdot + N_2 + ArN_2O\cdot \quad (2)$$

Sigma-complexes are then formed by addition to the substrate nucleus.

$$Ar\cdot + C_6H_5X \longrightarrow$$

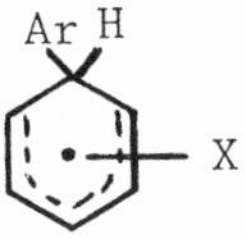

The radicals (1) and (2) may dehydrogenate a few σ-complexes but the main reaction proceeds by the following chain mechanism.

$$ArAr'H\cdot + ArN_2^+ \longrightarrow ArAr'H^+ + N_2 + Ar\cdot$$

$$ArAr'H^+ \xrightarrow{AcO^-} ArAr' + HOAc \qquad Ar\cdot \xrightarrow{Ar'H} ArAr'H\cdot$$

Thus at each stage a new aryl radical is formed directly from a diazonium *ion*, the ion-pair being in equilibrium with the diazo-acetate, as was elegantly demonstrated by H. Suschitzky and his co-workers (Angew. Chem. internat. edn., 1967, 6, 596).

$$ArN{:}NOAc \rightleftharpoons ArN_2^+OAc^-$$

The postulate of homolysis of the diazo-ester is thus no longer necessary, and that of the participation of acyloxy-radicals is avoided. It is probable that analogous mechanisms operate in similar reactions involving diazo-compounds, such as the Gomberg-Bachmann reaction.

The complexity of the acylarylnitrosamine reaction does not end here, however, since another mode of decomposition to arynes, which can be trapped with tetracyclone can also occur.

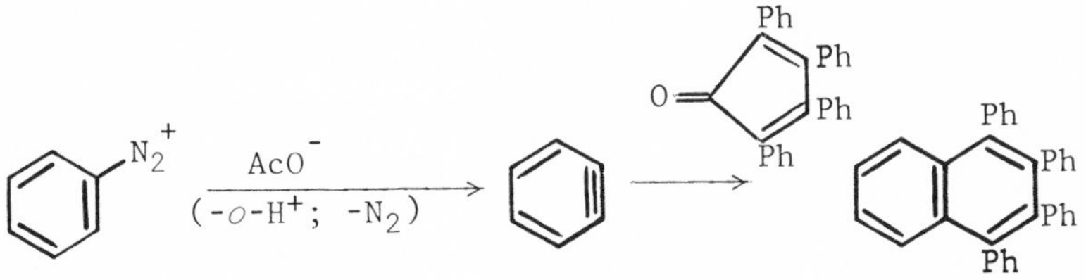

The presence of water inhibits benzyne-formation by solvating the acetate ions, and hence rendering them insufficiently basic to remove the *ortho*-proton. Hence, arynes cannot be obtained by an analogous process in which diazonium ions are formed by diazotization of aniline or a derivative thereof with pentyl nitrite in the presence of acetate ions because of the formation of water in this reaction.

$$ArNH_2 \ + \ RONO \ + \ Ar'H \longrightarrow ArAr' \ + \ ROH \ + \ N_2 \ + \ H_2O$$

Arynes are, however, formed in this system if acetic anhydride is present to remove this water.

An improvement of the Gomberg-Bachmann procedure by the use of phase transfer catalysts has been described (S.H. Korzeniowsky, L. Blum and G.W. Gokel, Tetrahedron Letters, 1977, 1871). Apparently in the presence of potassium acetate and 18-crown-6 in non-polar solvents diazonium ions give diazoanhydrides which then decompose, and the aryl radicals react with arenes to give biaryls.

E.s.r. spectroscopy and the "spin-trapping" technique have been used to study arylation by the thermolysis of diazoethers. These experiments, isomer ratio measurements, and CIDNP, all indicate that aryl radicals are involved (R.M.Paton and R.U. Weber, J. chem. Soc. chem. Comm., 1977, 769). The spin-trapping technique has also been used to demonstrate the formation of phenyl radicals in the radiolysis of benzene (F.P. Sargent and E.M. Gardy, J. chem. Phys., 1977, 67, 1973).

Aroyl peroxides – The mechanism of the arylation reaction with aroyl peroxides as radical-sources has been discussed in detail in the reviews cited earlier. Yields are greatly improved by the presence of various additives (cf. 2nd edn. Vol IIIA, p.12) of which copper(II) and iron(III) benzoate are particularly convenient. These additives provide favourable routes for the dehydrogenation of σ-complexes, preventing their loss by dimerisation.

e.g. $ArAr'H \ + \ Fe^{3+} \longrightarrow ArAr' \ + \ H^+ \ + \ Fe^{2+}$

$ArCO \cdot O \cdot \ + \ Fe^{2+} \longrightarrow ArCO \cdot O^- \ + \ Fe^{3+}$

$ArCO \cdot O^- \ + \ H^+ \longrightarrow ArCO_2H$

The following scheme represents the mechanism of phenylation by benzoyl peroxide (reactions leading to benzoyloxylation by addition of benzoyloxy-radicals to the arene, which accompany those leading to phenylation, and which normally occur to the extent of a few per cent, are omitted for simplicity).

(a) $(PhCO.O)_2$ $\longrightarrow$ $2PhCO.O\cdot$

(b) $PhCO\cdot O\cdot$ $\longrightarrow$ $Ph\cdot + CO_2$

(c) $Ph\cdot + ArH$ $\longrightarrow$ $PhArH\cdot$

(d) $PhArH\cdot + (PhCO\cdot O)_2$ $\longrightarrow$ $PhAr + PhCO_2H + PhCO\cdot O\cdot$

(e) $PhArH\cdot + PhCO\cdot O\cdot$ $\longrightarrow$ $PhAr + PhCO_2H$

(f) $2PhArH\cdot$ $\longrightarrow$ $PhAr + PhArH_2$

(g) $2PhArH\cdot$ $\longrightarrow$ $(PhArH)_2$

In arenes such as benzene, fluorobenzene and chlorobenzene in which the induced decomposition has a kinetic order of 1.5, biaryls are formed only in reactions (d) and (f). Reaction (e) does not occur in these arenes so yields of biaryls are low and much residue $(PhArH)_2$ is formed in reaction (g). In solvents such as bromobenzene and nitrobenzene, in which first-order induced decomposition becomes important, and in solvents of the former group in the presence of additives such as iron(III) benzoate, chains are terminated by reaction (e) or its equivalents, reactions (f) and (g) occur to a much reduced extent, and high yields of biaryls are obtained.

Polyhalogenophenyl radicals

Pentafluorophenyl radicals which arylate arenes to give derivatives of 2,3,4,5,6-pentafluorobiphenyl $(C_6F_5C_6H_4X)$ are formed by reaction of pentafluoroaniline with pentyl nitrite, by oxidation of pentafluorophenylhydrazine with silver oxide or bleaching powder, by the thermolysis of pentafluorophenyl-azotriphenylmethane , by the photolysis of pentafluorohalo-genobenzenes and of *bis*(pentafluorophenyl) mercury, by the thermolysis of pentafluorobenzoyl peroxide and from some other sources. (R. Bolton and G.H. Williams, "Advances in Free Radical Chemistry", Logos Press, London, 1975, 5, 1) and references cited therein). Pentafluorobenzoyl peroxide arylates benzene in good yield, but with some substituted benzenes, phenyl pentafluorobenzoate is formed by aroyloxyl-ation at the 1-position. Several of the other sources give the expected arylation products in good yields.

Pentachlorophenyl radicals formed from pentachloroaniline and pentyl nitrite, by oxidation of pentachlorophenylhydrazine and by photolysis of pentachloroiodobenzene (R. Bolton, E.P. Mitchell and G.H. Williams, J. chem. Research, 1977, (S), 223 (M) 2618) and pentabromophenyl radicals formed by analogous processes (M.I. Bhanger, R. Bolton and G.H. Williams, unpublished observations) arylate arenes in moderate yield, but with the concomitant formation of penta- and hexa-chloro- or bromo-benzenes.

Properties of aryl radicals ; the arylation process

There has been considerable discussion of the possibility that addition of aryl radicals to arenes is reversible (cf. M.J. Perkins, "Free Radicals" Vol. I, J. Kochi ed., John Wiley and Sons, New York, 1973, 231). The process has been considered to be irreversible by J. Saltiel and H. Curtis, (J. Amer. chem. Soc., 1971, 93, 2056) because of the lack of an isotope effect in the phenylation of benzene-d and on thermodynamic grounds by R.A. Jackson (Chem. Comm., 1974, 573). D.J. Atkinson, M.J. Perkins and P. Ward, (J. chem. Soc., (C), 1971, 3149) showed that phenyl radicals were formed by decomposition of the σ-complex PhPhH· (which was prepared by a method not involving phenyl radicals) only at elevated temperatures. Consequently the formation of the σ-complex is not appreciably reversible at the more moderate temperatures at which arylation reactions are usually conducted. The addition of phenyl radicals to chlorobenzene is, however, reversible in the gas phase at 500 °C (R. Louw, J.W. Rothuizen and R.C.C. Wegman, J. chem. Soc., Perkin II, 1973, 1635).

On the other hand D.C. Nonhebel $et.al.$,(Chem. Comm., 1974, 987; Tetrahedron Letters, 1975, 3855, 3857) have interpreted observed temperature effects on relative yields of products as being due to the reversibility of the addition. An isotope effect observed in the reaction in dimethylsulphoxide (M. Kobayashi $et.al.$,Bull. chem. Soc., Japan, 1969, 42, 2738; 1972, 45, 2042) was interpreted as being due to reversible addition arising in this case from a solvent effect of DMSO on the behaviour of the radicals. The matter of this possible reversibility cannot therefore be regarded as completely settled.

Polar effects arising from the electrophilicity or nucleo-philicity of appropriately substituted aryl radicals have been discussed by Minisci and Citterio ("Advances in Free Radical Chemistry", Heyden & Son, London, 1980, 6, 65). They can be interpreted in terms of a modified Hammett relationship (O. Simamura *et.al.*, Tetrahedron, 1965, 21, 955. Williams, ("Essays in Free-Radical Chemistry", Chemical Society Special Publication No.24, 1970, 25) and the following ρ-values have been given.

Reaction	Reaction constant (ρ)
Pentafluorophenylation	-0.81
p-Nitrophenylation	-0.81
p-Chlorophenylation	-0.27
Phenylation	+0.05
p-Methylphenylation	+0.03
p-Methoxyphenylation	+0.09

These polar differences lead to predictable changes in partial rate factors for arylation. The slight nucleophilic-ity of the phenyl radical and the pronounced electrophilicity of the pentafluorophenyl radical are noteworthy. Pentafluoro-phenylation of nitrobenzene unusually gives the 3-isomer as the main product. Hammett ρ-values of +0.09 for *meta-*, and +0.31 for *meta-* and *para-* phenylation have been given by W.A. Pryor *et.al.*, (J. Amer. chem. Soc., 1973, 95, 6993).

E.G. Janzen and C.A. Evans (J. Amer. chem. Soc., 1975, 97, 205) have used spin-trapping in e.s.r. spectroscopy to give a measure of the absolute rate of addition of phenyl radicals to benzene at 30° as 7.8×10^4 $1 \text{ mol}^{-1}\text{s}^{-1}$.

A number of measurements of relative rates and partial rate factors have been reported, often under high-yield cond-itions using catalysed aroyl peroxide reactions. These results demonstrate that the broad picture of the reaction presented by the partial rate factors given in the second edition (Vol.IIIA, Table 2, p.123) is correct. However the most reliable data, given below, are those obtained from such high-yield reactions, (Bolton, K. Hirakubo, K.H. Lee and Williams, unpublished observations).

Arene	Relative Rate (K, C_6H_6 = 1)	Partial Rate Factors		
		F_o	F_m	F_p
PhF	1.3	2.0	1.1	1.2
PhCl	1.5	2.5	1.1	1.5
PhBr	1.5	2.5	1.3	1.4
$PhCH_3$	1.4	2.7	1.0	1.3
$PhBu^t$	0.8	0.5	1.2	1.2
$PhOCH_3$	1.7	3.4	1.0	1.2
PhCN	1.8	3.0	0.9	3.0
$PhCO_2Me$	2.0	3.4	1.0	3.5
PhCOPh	2.0	3.4	1.0	3.5

The beneficial effects of additives (catalysts) in giving more reliable quantitative results in the phenylation of anisole have been noted by R.A. McClelland, R.O.C. Norman and C.B. Thomas, (J. chem. Soc. Perkin I, 1972, 570) of pyridinium chloride by J.M. Bonnier and J. Court (Bull. Soc. chim. France, 1972, 1834) and of γ-picoline by H.J.H. Dou, G. Vernin and J. Metzger (Tetrahedron Letters, 1968, 953) and S. Vidal, J. Court and J.H. Bonnier, J. chem. Soc. Perkin II, 1973, 2071; 1976, 497; Tetrahedron Letters, 1976, 2023).

A further complication which may lead to unreliability of measured relative rates of arylation has been pointed out by Bolton and and Williams ("Advances in Free Radical Chemistry", Logos Press, London, 1975, 5, 1) and by Bolton, J.P.B. Sandall and Williams (J. fluorine Chem., 1974, 4, 355; J. chem. Research, 1977, (S), 24, (M), 373). This is the possible formation of charge-transfer complexes between the substrate and the radical or its precursor, which can become serious if there is a large polar difference between these two components. In such cases, intermolecular selection, which determines relative rates, is influenced by the relative stabilities of the complexes formed by two competing solvents, and measured yields are not true indices of intrinsic relative reactivities. Intramolecular selection, which determined isomer ratios, is not so affected. This complication produces notable anomalies in apparent relative rates of pentafluoro-phenylation of simple arenes, of phenylation of polyfluoro-arenes ($q.v.$) and possibly to a lesser extent of di- and tri-substituted benzene derivatives (D.I. Davies, D.H. Hey and B. Summers, J. chem. Soc., (C), 1971, 2681; Bolton, Hirakubo, Lee and Williams, unpublished observations). Its influence in the arylation of monosubstituted arenes, however, is not significant.

(*f*) *Silylation*

C. Eaborn, R.A. Jackson *et. al.,*(Chem. Comm., 1967, 920;
J. organometall. Chem., 1968, $\underline{16}$, 17; 1971, $\underline{28}$, 59; J. chem.
Soc. Perkin II, 1972, 55) have described the silylation of
arenes with photochemically produced silyl radicals,

$$(Me_3Si)_2Hg \quad \xrightarrow{h\nu} \quad Me_3Si\cdot \quad \xrightarrow{C_6H_6} \quad C_6H_5SiMe_3 .$$

H. Sakurai *et.al.,*(Tetrahedron Letters, 1969, 1755; J. Amer.
chem. Soc., 1971, $\underline{93}$, 1709) obtained trialkylsilyl radicals
by the reaction of t-butoxy-radicals with trialkylsilanes.

$$tBuO\cdot + R_3SiH \quad \longrightarrow \quad tBuOH + R_3Si\cdot \quad \xrightarrow{ArH} \quad ArSiR_3 .$$

E.P. Mikheev (Doklady Akad. Nauk. U.S.S.R., 1964, $\underline{155}$, 1361)
accomplished the trichlorosilylation of benzene by photolysis
of a mixture of chlorine, trichlorosilane and benzene. Silyl
radicals display a pronounced nucleophilic character in arylation reactions.

(*g*) *Acylation*

Protonated heterocyclic bases are easily acylated by nucleo-
philic acyl radicals (G.P. Gardini and F. Minisci, J. chem.
Soc., (C), 1970, 929; Minisci and Citterio ("Advances in Free
Radical Chemistry", Heyden and Son, London, 1980, $\underline{6}$, 109).

(*h*) *Ipso-substitution*

Replacement of atoms other than hydrogen is called *ipso*-sub-
stitution, and a number of homolytic reactions of this type
have been described recently.
    Alkylation. 1-Adamantyl and other alkyl radicals replace
the acyl groups in 2-acylbenzothiazoles. The 1-adamantylation
is facilitated by electron withdrawing substituents, thus
showing the 1-adamantyl radical to be nucleophilic. (M.Tiecco
et al., J. chem. Soc., Perkin II, 1977, 1679). The 1-adamantyl
radical also replaces the nitro-group in compounds of the type
$pXC_6H_4NO_2$ (L. Testaferri, M. Tiecco and M. Tingoli, J. chem.
Soc., Chem. Comm., 1978, 93). Some pentafluorophenylcyclo-
hexane has been shown by V.L. Salenko, L.S. Kobrina and
G.G. Yakobson (Zhur. org. Khim., 1978, $\underline{14}$, 1646) to be formed
by the reaction of cyclohexyl radicals with hexafluorobenzene,
in which fluorine is the atom displaced.

Arylation. The replacement of fluorine in polyfluoro-
arenes by aryl radicals has been extensively studied. The
subject has been reviewed (Bolton and Williams "Advances in
Free Radical Chemistry", Logos Press, London 1975, 5, 1). Aryl
radicals, from the corresponding aroyl peroxides, give derivatives of 2,3,4,5,6-pentafluorobiphenyl in reasonable yield by
their reaction with hexafluorobenzene.

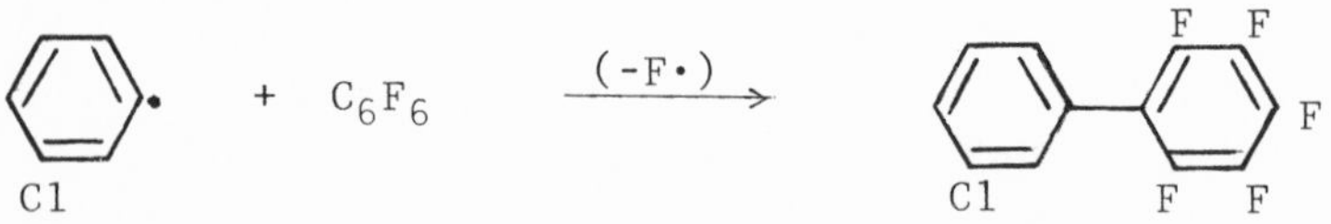

The mechanism of defluorination of the σ-complex has been
problematic since it cannot be analogous to the dehydrogenation of the σ-complex for straightforward aromatic substitution. The mechanism in fact involves benzoic acid as the
defluorinating agent.

$$PhC_6F_6\cdot \ (\sigma_F\cdot) + PhCO_2H \longrightarrow PhC_6F_5 + PhCO.O\cdot + HF$$

Benzoic acid is formed to a relatively small extent as a
result of a rearrangement of $\sigma_F\cdot$ which leads ultimately to the
formation of a little 2,2',3,4,5,6-hexafluorobiphenyl.

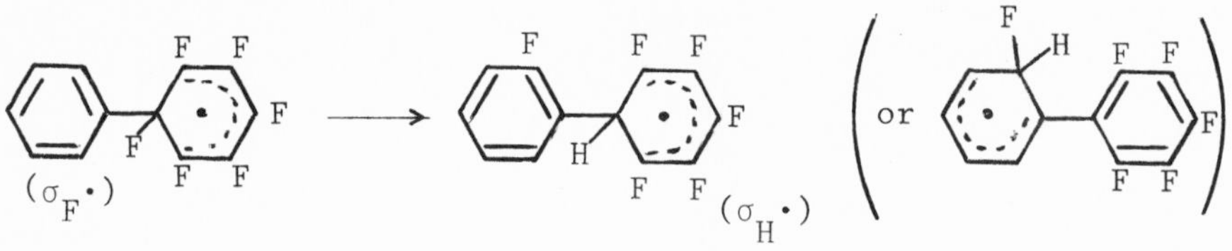

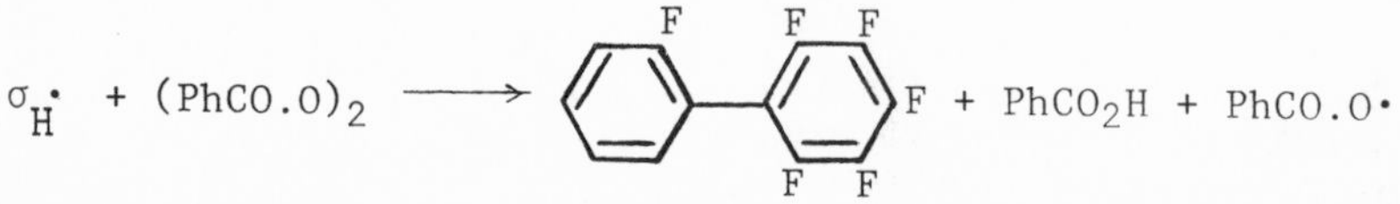

This small amount of benzoic acid is recycled by transesterification reactions involving the hydrogen fluoride
formed in the above defluorination step and the tertiary
benzoate esters which are formed by dimerisation or other
reactions of the σ-complexes formed by addition of benzoyloxy-
radicals to hexafluorobenzene. For example, the reaction
with σ-complex dimers is as follows.

$$PhCO.O\cdot + C_6F_6 \longrightarrow PhCO.OC_6F_6\cdot \xrightarrow[\text{etc}]{\text{dimerisation}} (PhCO.OC_6F_6)_2$$

$$(PhCO.OC_6F_6)_2 + 2HF \longrightarrow (C_6F_7)_2 + 2PhCO_2H$$

This mechanism was confirmed by R. Bolton, W.K.A. Moss, J.P.B. Sandall and G.H. Williams (J. fluorine Chem., 1976, 7, 597).

With chloro- and bromo-pentafluorobenzene, the 2-, 3- and 4-fluorine atoms are replaced and partial rate factors are similar to those for phenylation of chloro- and bromo-benzenes, although the relative rates of these reactions are probably influenced by complexation of the fluorinated substrate and the phenyl radical or its precursors. Fluorine is replaced only one-tenth as easily as hydrogen in 1,3,5-trifluorobenzene. The rate difference is smaller in more highly fluorinated poly-fluorobenzenes, although it is possible that these results are distorted by complex-formation (Bolton, Sandall and Williams, J. fluorine Chem., 1978, 11, 591).

Pentafluorophenylation of hexafluorobenzene has been observed by V.L. Salenko, L.S. Kobrina and G.G. Yakobson (Izvest. Sib. Otdel Akad. Nauk. S.S.S.R. Ser. Khim. Nauk., 1978, 97, 103), although it is not a favoured reaction.

An intramolecular *ipso*-arylation in which chlorine or bromine is replaced has also been reported (L. Benati, P.C. Montevecchi and A. Tundo, J. chem. Soc. Chem Comm., 1978, 530).

(R = Cl or Br)

The radicals were obtained from the corresponding amines.

Chlorination. Photolysis of chlorine in the presence of p-bromonitrobenzene gives both p-chloronitrobenzene and p-chlorobromobenzene, indicating that both chlorodenitration and chlorodebromination occur (C.R. Everly and V.G. Traynham, J. Amer. chem. Soc., 1978, 100, 4316). The formation of a small amount of 1-chloro-2-bromo-4-nitrobenzene indicates a rearrangement of a σ-complex similar to those which occur in the phenyldefluorination of hexafluorobenzene.

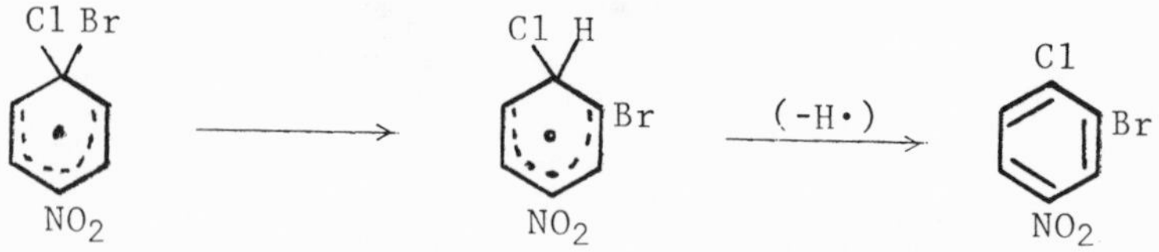

Aroyloxylation. The formation of phenyl pentafluorobenzoate by reaction of benzoyloxy-radicals with chloro-, bromo-
or nitro-benzene at the *ipso*-position has been reported
(P.H. Oldham, G.H. Williams and B.A. Wilson, J. chem. Soc.,
(B), 1970, 1346).

5. FORMATION AND FISSION OF THE BENZENE NUCLEUS

N.H. Wilson

This area of organic chemistry has advanced considerably over the past ten years. The Chemical Society Specialist Periodical Reports on Aromatic and Heteroaromatic Chemistry cover the 1970 decade in considerable detail.

(a) Formation of the benzene nucleus

(ι) Pyrolysis, photolysis, condensation and polymerisation reactions

The transformation of aliphatic compounds into aromatics is of practical and theoretical interest. High energy methods have given way to catalytic techniques both in industry and in the laboratory. n-Hexane can be aromatised into benzene over chromium catalysts (2nd edn., Vol IIIA p 131) and platinum catalysts can also be employed, (M. Ichikawa, Chem. Comm., 1976, 11). Ichikawa has also investigated the trimerisation of acetylene to benzene over nickel clusters at room temperature (*ibid*, 26). Acetylenes are particularly useful as precursors of aromatic rings and there are several reviews on the subject. (E.L. Muetterties, Bull. Soc. Chim. Belges, 1975, 84, 959 and K.P.C. Vollhardt, Acc. chem. Res., 1977, 10, 1).

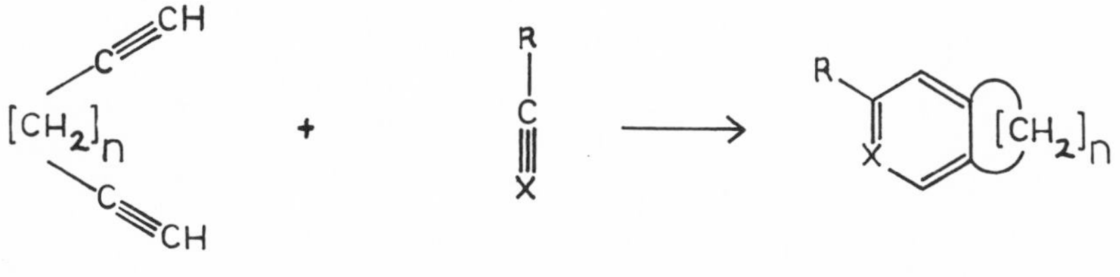

Benzene (x = CH) and pyridine (x = N) rings can be obtained. Papers on the mechanism of alkyne to aromatic transformations are by D.R. McAlister, J.E. Bercaw and R.G. Bergman, J. Amer. chem. Soc., 1977, 99, 1666 and P. Caddy, Angew. Chem., internat. Edn., 1977, 16, 648. The use of metal cluster catalysis for acetylene trimerisation in general has

been explored. (E.L. Muetterties , J. Amer. chem.
Soc., 1977, 99, 743). Dienediynes have been shown
to be useful in preparing benzocyclobutenes,
(K.P.C. Vollhardt and L.S. Yee, J. Amer. chem.
Soc., 1977, 99, 2010. See also J.M. Riemann and
W.S. Trahanovksy, Tetrahedron Letters, 1977, 1863;
R.J. Spangler, B.G. Beckmann and J.H. Kim, J. org.
Chem., 1977, 42, 2989; P.D. Brewster et al., ,
Tetrahedron Letters, 1977, 4573; R.L. Hilliard
tert. and K.P.C. Vollhardt, Angew. Chem., internat.
Edn., 1977, 16, 399; R. Victor, Transition metal
Chem. (Weinheim), 1977, 2, 2). Perhaps of greatest
general utility is the rhodium complex route to
aromatics due to E. Muller et al. , (Ann., 1971,
754, 64; 1972, 758, 16; 1975, 761, 1435; Tetra-
hedron Letters, 1972, 1035).

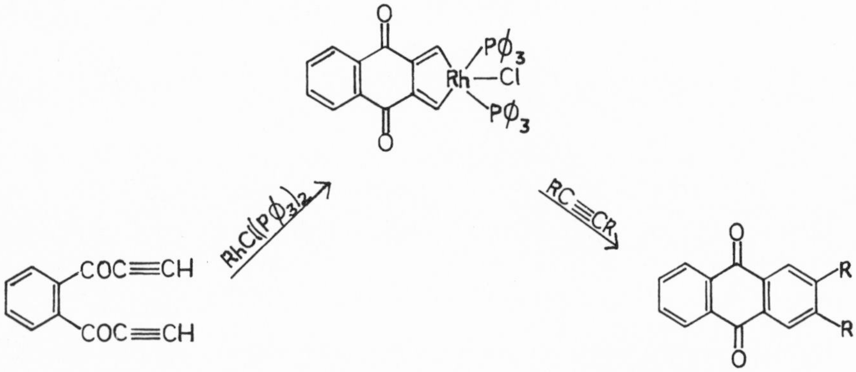

There are many variations of this method. Similar
reactions mediated by heat alone or by other reagents, (usually base), are discussed in the reviews given above.

Octa-3,5-dien-1,7-diyne spontaneously rearranges
to benzocyclobutene. However the fragmentation of
-diazidobenzene, with loss of nitrogen, gives the
dinitrene which collapses to cis-cis-muconitrile.
Other routes to the dinitrene intermediate are
depicted below. The similar reaction of 4-phenyl-
benzene-1,2,3-triazine provides 2-phenylbenzazete
which is isolable at -80°C. (C.W. Rees et al .,

J. chem. Soc., Perkin I, 1975, 33; Chem. Comm.,
1976, 411; Tetrahedron Letters, 1976, 3931, 4647;
cf. 2nd edn., Vol IIIA, p 148).

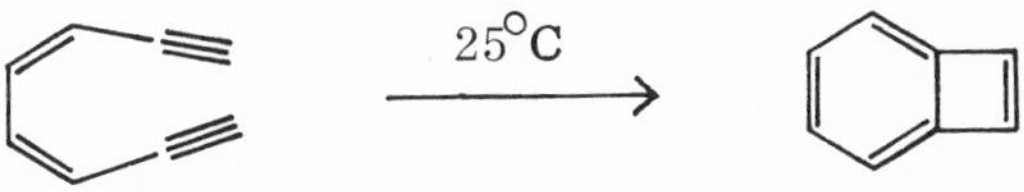

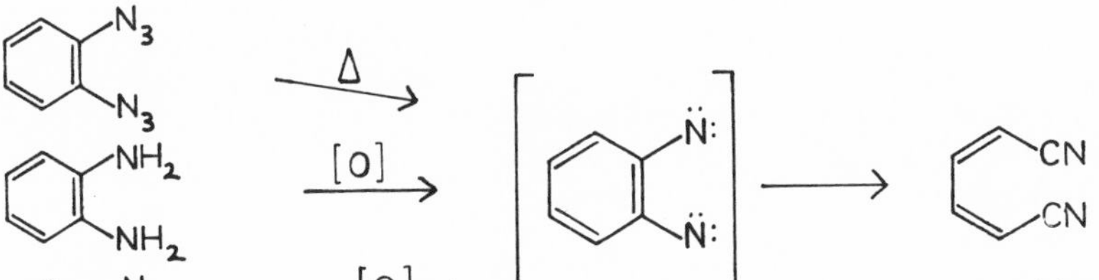

muconitrile

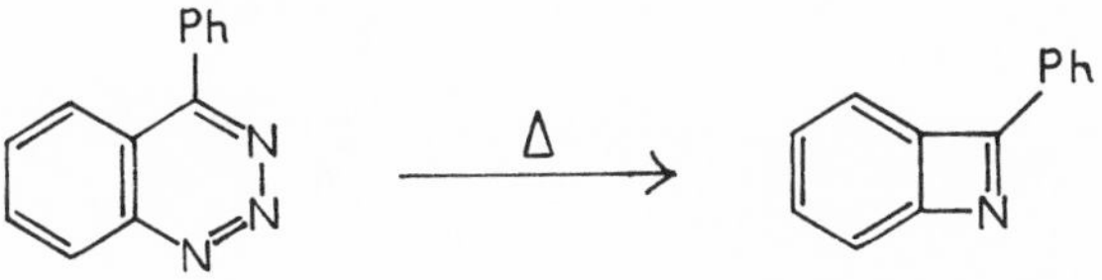

2-phenylbenzazete

The degenerate Cope rearrangement of hexa-3-en-
1,5-diyne can occur on heating and appears to
involve 1,4-dehydrobenzene (*p*-benzyne) as intermediate (R.R. Jones and R.G. Bergmann, J. Amer.
chem. Soc., 1972, 94, 660). The intermediate is
able to abstract hydrogen from the solvent to form
benzene, or halogen atoms from a halogenated solvent
to yield *p*-dihalobenzene. The labelling study
depicted is further evidence for this mechanism.

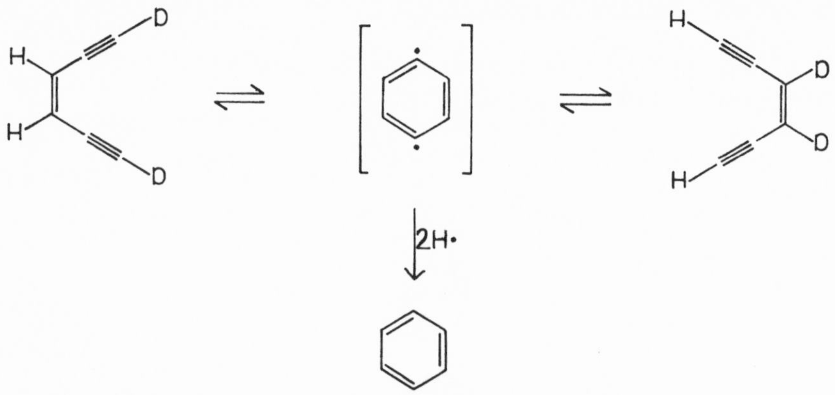

This rearrangement was originally reported by
S. Masamune *et al.* , (Chem. Comm., 1971, 1516) who
prepared " *meso*-anthracyne" analogously. This 9,
10-dehydroanthracene has also been prepared by decarbonylation of the cis-keten. (O.L. Chapman,
C.C. Chang and J.L. Kolc, J. Amer. chem. Soc.,
1976, 98, 5703). All these 1,4-dehydroaromatic
systems behave as free radicals. (For a review see
R.G. Bergmann, Acc. chem. Res., 1973, 6, 25).
There appear to be two minima on the energy surface
of *p*-benzyne, (M.J.S. Dewar and W.K. Li, J. Amer.
chem. Soc., 1974, 96, 5569), and the generation of
another form of *p*-benzyne, the butalene form, has
been claimed by R. Breslow, J. Napierski and I.C.
Clark (J. Amer. chem. Soc., 1975, 97, 6275). A
labelling study supports the formation of this
intermediate (R. Breslow and P.L. Kharma, Tetrahedron Letters, 1977, 3429).

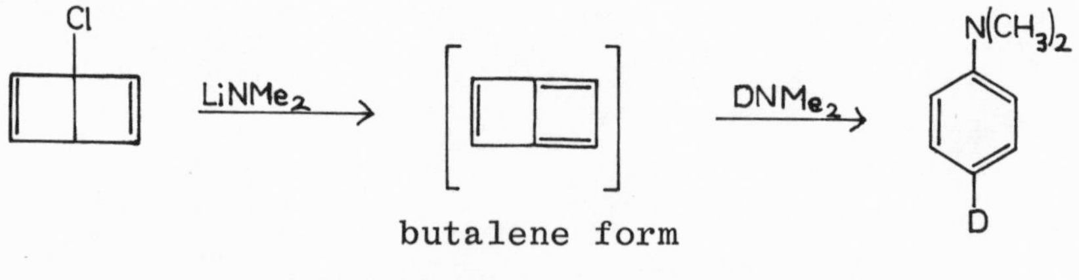

butalene form

This last aromatic ring formation occurs from a
Dewar benzene and this leads to a consideration of
the formation of aromatic systems by generation of

valence isomers.

Prior to 1957 benzene was thought to be photo-
stable. However, it has now been shown that
benzene can be converted into several valence
isomers. (2nd edn., Vol IIIA, p 133 *et seq.* ;
E.E. van Tamelen, Acc. chem. Res., 1972, 5, 186;
D. Bryce-Smith and A. Gilbert, Tetrahedron, 1976,
32, 1309; L.T. Scott and M. Jones, Chem. Rev.,
1972, 72, 181).

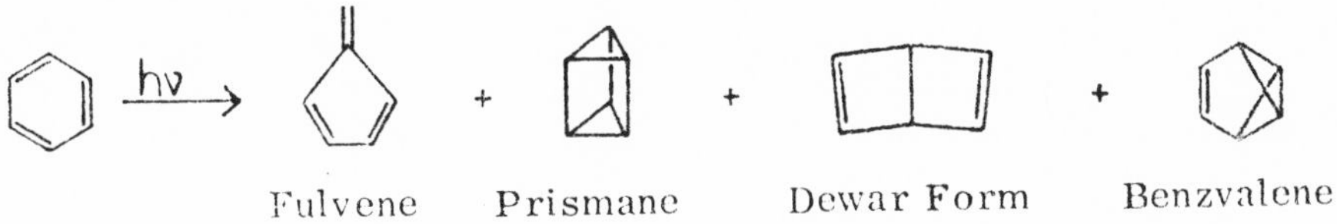

Fulvene   Prismane   Dewar Form   Benzvalene

Thus fully aromatic systems can be synthesised by
the generation of the valence isomers, as in the
above example via Dewar benzene, (see also
R. Breslow, J. Napierski and A.H. Schmidt, J. Amer.
chem. Soc., 1972, 94, 5906).

An unusual synthesis from a cyclobutadiene metal
complex which occurs via a Dewar benzene has been
reported, (R.H. Grubbs, T.A. Pancoast and R.A. Grey,
Tetrahedron Letters, 1974, 2425).

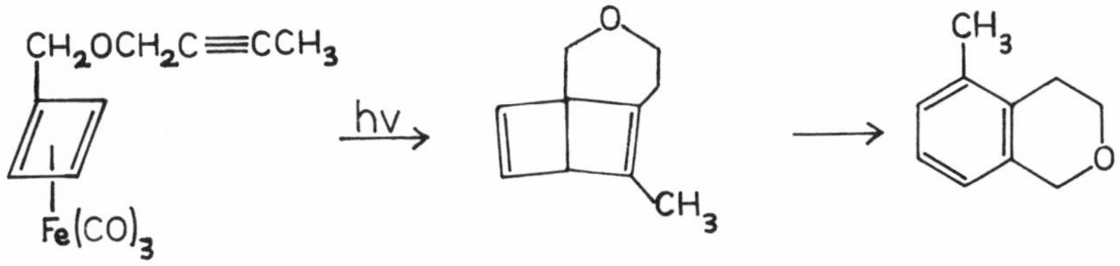

The generation of benzenes from biscyclopropenes
via Dewar forms has been the subject of much
research.

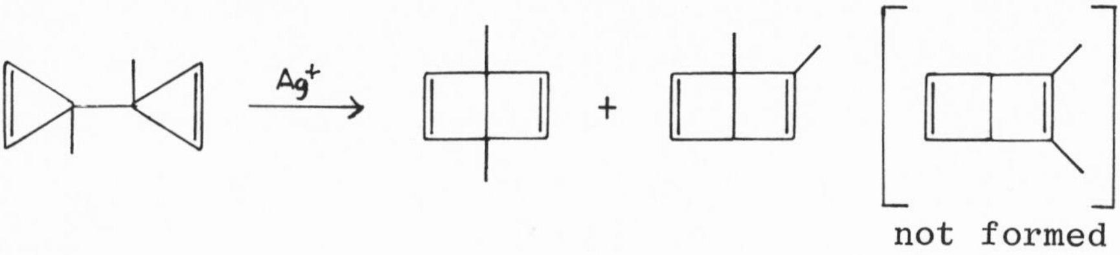

not formed

A prismane form is excluded in the above reaction as one possible isomer is not formed, (W.H. de Wolf, J.W. von Staten and F. Bickelhaupt, Tetrahedron Letters, 1972, 3209). This reaction, mediated by radiation, is thought to occur via a diradical cyclopropene dimer or p-benzyne type intermediate, (R. Weiss and H. Kolbl, J. Amer. chem. Soc., 1975, 97, 3222; W.H. de Wolf, I.J. Landheer and F. Bickelhaupt, Tetrahedron Letters, 1975, 179, 349; R.G. Bergmann, J.H. Davis and K.J. Shea, J. Amer. chem. Soc., 1977, 99, 1499 and ref. therein. Other Dewar benzenes and their rearrangements are described by L.A. Paquette, Synthesis, 1975, 347; I.J. Landheer et al ., Tetrahedron Letters, 1975, 4187, 4499; H. Wynberg et al ., J. Amer. chem. Soc., 1975, 97, 216; Tetrahedron Letters, 1975, 4297; Chem. Comm., 1975, 972; M.G. Barlow, R.N. Haszeldine and M.J. Kershaw, Tetrahedron, 1975, 31, 1649; R.N. Haszeldine et al ., Chem. Comm., 1975, 323.

Benzene has been produced from prismane itself which was obtained, for the first time, by a rational synthesis from 7,8-diazatetracyclo-[3.3.0.0^{2,4}.0^{3,6}] oct-7-ene, (T.J. Katz and N. Acton, J. Amer. chem. Soc., 1973, 95, 2738; 1976, 98, 4320; N.J. Turro and V. Ramamurthy, Rec. trav. Chim., 1979, 98, 173).

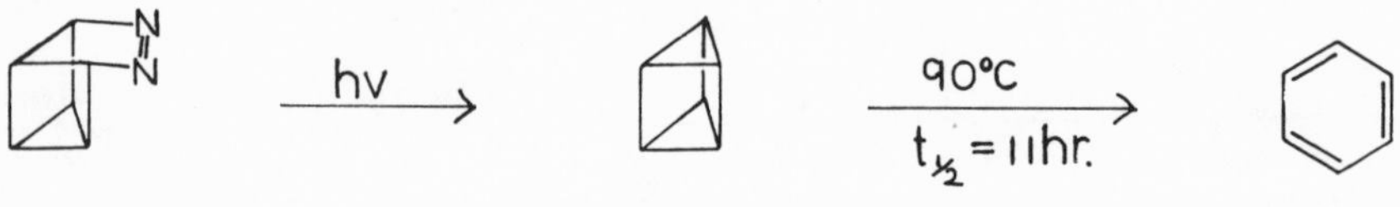

A reaction, yielding an aromatic system, from a
similar starting material, diazabasketene, has been
reported by J.P. Snyder, L. Lee and D.G. Farnum,
(J. Amer. chem. Soc., 1971, 93, 3816).

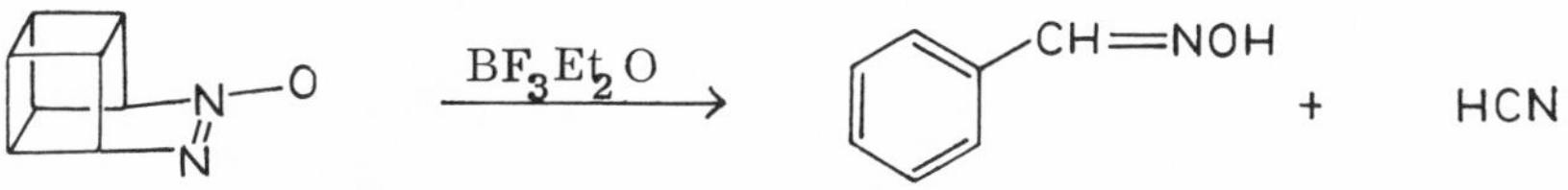

Syntheses of aromatic rings via the remarkable
benzvalene system are rare, but at least one has
been claimed, (E.B. Hoyt *et al*., Tetrahedron
Letters, 1972, 1579; see also R. Criegee *et al*.,
Ber., 1973, 106, 857).

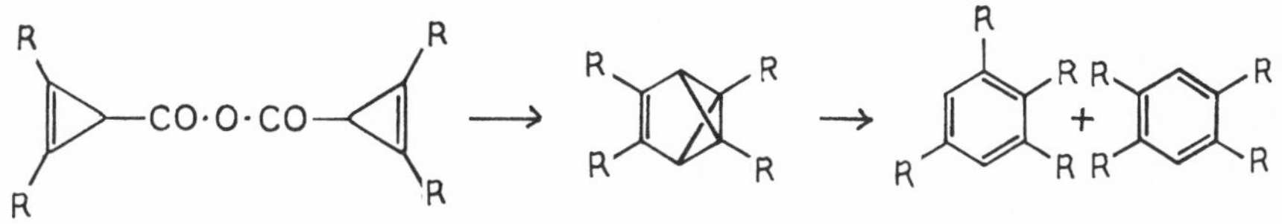

Benzvalenes can also be synthesised by carbene
additions to cyclopentadiene carbanions, (T.J. Katz,
E.J. Wang and N. Acton, J. Amer. chem. Soc., 1971,
93, 3783). For a discussion of the bonding and the
energy levels in benzvalenes see M.J.S. Dewar and
S. Kirschner, (J. Amer. chem. Soc., 1975, 97, 2932)
and T.J. Katz *et al*., (*ibid*., 2568).

Much effort has gone into the generation of
aromatic carbenes or nitrenes, and many of the
methods employed constitute a synthesis of benzenoid
rings.

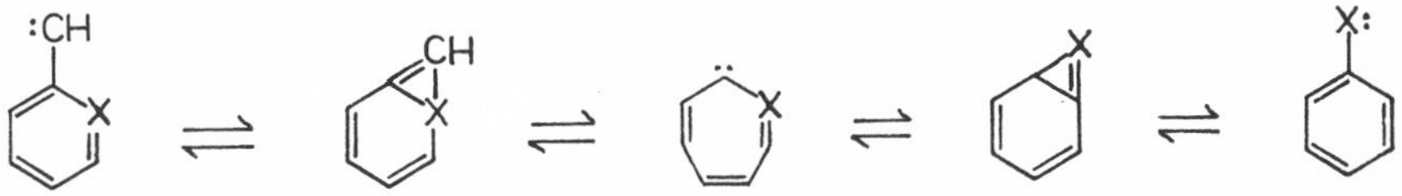

Such reactions have been reviewed by C. Wentrup,
(Topics current Chem., 1976, 62, 173). This review discusses energetics and label (position)
scrambling by this process (e.g. X = ^{13}CH in above
diagram). These carbene intermediates can be
obtained by pyrolysis of phthalides, (see below),
or benzocyclopropenes, (U.E. Wiersum and
T. Nieuwenhuis, Tetrahedron Letters, 1973, 2581;
C. Wentrup and P. Miller, *ibid* ., 2915; W.D. Crow
and M.N. Paddon-Row, Austral. J. Chem., 1973, 26,
1705; W.E. Billups and L.E. Reed, Tetrahedron
Letters, 1977, 2239).

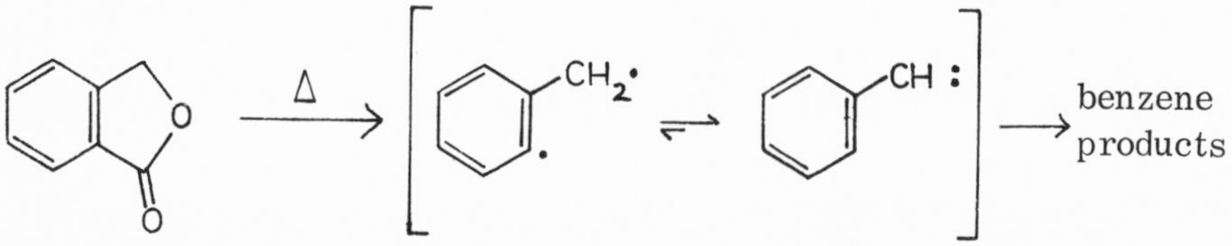

Excellent reviews of benzocyclopropene chemistry
are available, (B. Halton, Chem. Reviews, 1973,
73, 113; W.E. Billups, Acc. chem. Res., 1978, 11,
245). A reaction where the carbene is generated
by the Bamford-Stevens decomposition of a tosylhy-
drazone salt is shown below, (K.E. Krajca,
T. Mitsuhashi and W.M. Jones, J. Amer. chem. Soc.,
1972, 94, 3661).

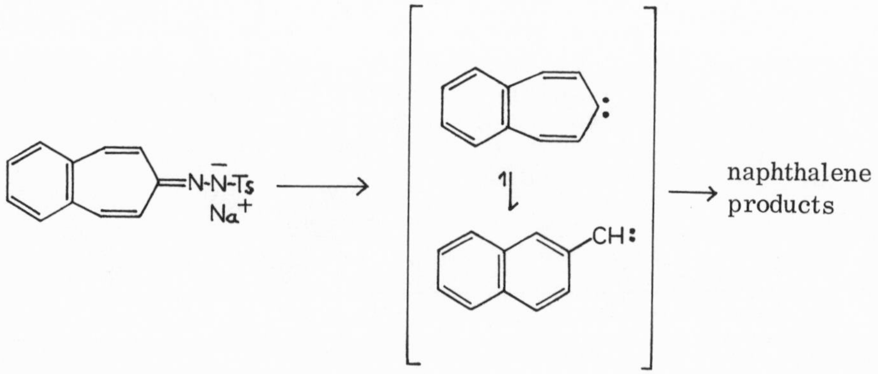

Acid or base catalysed ring closures of carbonyl compounds, or similarly activated species, to arenes give rise to a very large number of syntheses. An example from the polyketide field is given below, (P.J. Wittek and T.M. Harris, J. Amer. chem. Soc., 1973, 95, 6865; 1975, 95, 3270).

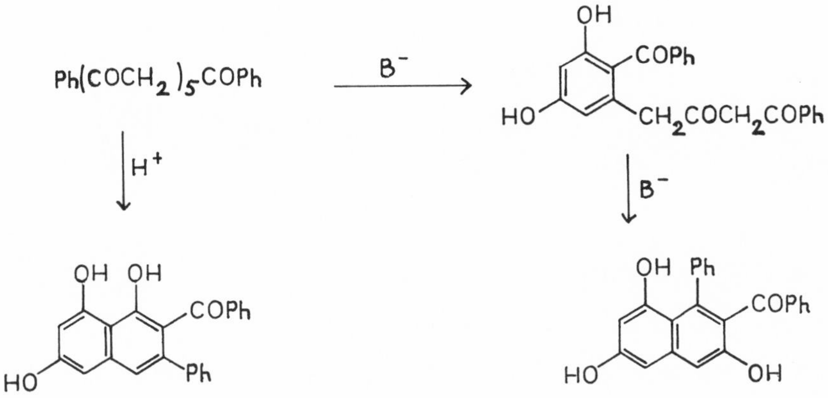

Shikimic acid routes to aromatic amino acids in higher plants have been studied by P.O. Larsen, D.K. Onderka and H.G. Floss, (Chem. Comm., 1972, 842) and shikimate routes to menaquinones (naphthaquinones) have been investigated by R.M. Baldwin, C.D. Snyder and H. Rapoport, (J. Amer. chem. Soc., 1973, 95, 276). Further examples and other facets of shikimic acid biosynthetic routes are discussed in a review by B. Ganem, (Tetrahedron, 1978, 34, 3353). Synthetic applications of acid or base catalysed cyclisations to aromatic rings are too numerous for specific mention here. Many examples and related reactions occur throughout this section and in sections dealing with specific ring formations.

(ii) Aromatisation of 6-membered ring compounds.

M.S. Newman and W.M. Hung, (J. org. Chem., 1973, 38, 4073), have revived the Semmler-Wolff reaction, in the dehydrative transformation of 1-tetralone oxime to 1-acetamidonaphthalene. (cf. 2nd edn., IIIG p 184).

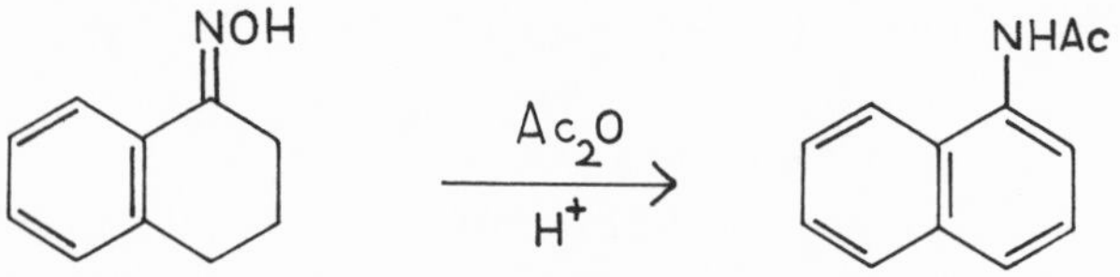

2-Tetralone oxime has also been made to undergo a similar reaction, (W.E. Rosen and M.J. Green, J. org. Chem., 1963, 28, 2797). (See also next section on dienone-phenol rearrangement).

High potential quinones continue to be frequently used to effect dehydrogenation, thus o-chloranil aromatises alicyclic nitro compounds. Yields in these reactions are generally low, but the reaction is useful in orientation studies since the starting materials can be obtained unambiguously by peracid oxidation of oximes, (H.F. Andrew, N. Campbell and N.H. Wilson, J. chem. Soc., Perkin I, 1972, 755).

A relatively novel and useful development is aromatisation by use of strong base to deprotonate the substrate, followed by hydride removal by an acceptor ketone. The reaction is used when reactive arenes are involved which can undergo Diels-Alder reactions with quinone reagents. An example where other dehydrogenation methods are also of little value is afforded by the synthesis of 9,10-dialkylanthracenes, tetracenes or similar systems.

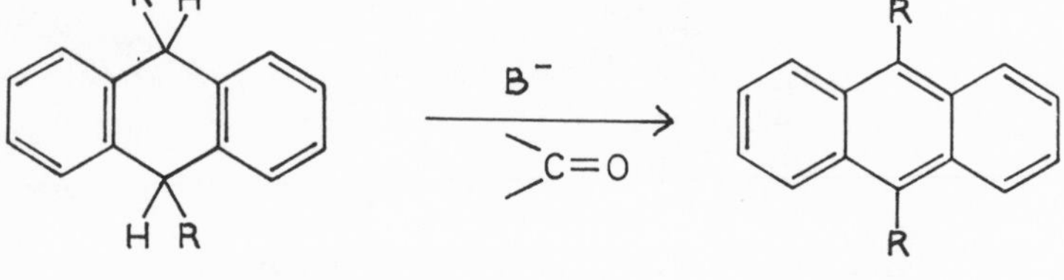

(M.T. Reetz and F. Eibach, Angew. Chem. internat.
Edn., 1978, 17, 278). In some cases it is advanta-
geous to use a very powerful base, (Bu Li/tetra-
methylenediamine), to generate the dianion before
final oxidation to the arene, (R.G. Harvey,
L. Nazareno and H. Cho, J. Amer. chem. Soc., 1973,
95, 2376).

A method, employing neutral/acid conditions,
involves hydride removal first then deprotonation.
Trityl esters are the reagents of choice. This
method, due to W. Bonthrone and D.H. Reid, (J. chem.
Soc., 1959, 2773), has been made very convenient
using trityl trifluoroacetate, ($Ph_3COH + CF_3COOH$),
by P.P. Fu and R.G. Harvey, (Tetrahedron Letters,
1974, 3217). A selection of typical aromatisations
have been performed by this technique.

Sulphur, selenium and noble metals are still
popular dehydrogenation reagents, (see 2nd edn.,
Vol IIIA, p 137 for examples).

A comprehensive review of aromatisation reactions
is available, (P.P. Fu and R.G. Harvey, Chem. Reviews, 1978, 78, 317), which discusses all methods,
including the use of soluble noble metal complexes
to effect aromatisation in an extension of the
heterogeneous methods mentioned above.

(iii) Isomerisation of alicyclic compounds.

The dienone-phenol rearrangement, (2nd edn.,
Vol IIIA, p 143), has been reviewed (B. Miller,
Acc. chem. Res., 1975, 8, 245).

The dienol-benzene rearrangement is really a dehydration, but is included here due to its similar-
ity to the dienone-phenol rearrangement and to the
arene oxide aromatisation, (see below). The term is
used with reference to cyclohexadienol rings with
geminal disubstitution and hence aromatisation
requires a substituent migration.

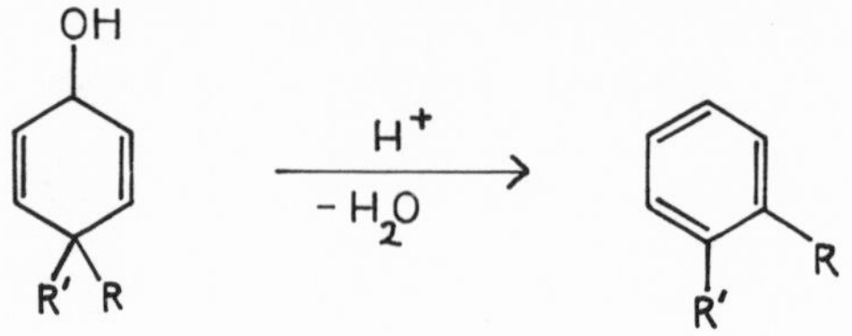

This reaction including a mechanistic discussion
is reviewed by V.P. Vitullo and M.J. Cashen,
(Tetrahedron Letters, 1973, 4823). Related to the
above reactions is the aromatisation of arene
oxides. These compounds are of interest since they
appear to be formed in *in vivo* oxidations by the
liver of aromatic hydrocarbons. It is suggested
that arene oxides are toxic species leading to
harmful effects in living organisms, presumably due
to their ability to alkylate DNA, RNA and proteins.
Examples of bio-effects include liver damage,
depression of bone marrow function leading to
leukaemia and carcinogenesis. This last bio-
activity is confirmed by the observation that the
5,6-oxide of benz[α]anthracene is highly carcino-
genic in rodents. A review of the metabolic
activation of polycyclic aromatic hydrocarbons, and
the structural features correlating to carcinogen-
icity is available, (Wing-Sum Tsang and G.W. Griffin,
" Metabolic Activation of Polynuclear Aromatic
Hydrocarbons" , Pergamon Press, Oxford, 1979). The
aromatisation of these oxides is depicted below,
and includes the "NIH shift" in which there is
often retention of a heavy isotope of hydrogen
label [H*], due to a kinetic isotope effect,
(J.W. Daly, D.M. Jerina and B. Witkop, Experientia,
1972, 28, 1129).

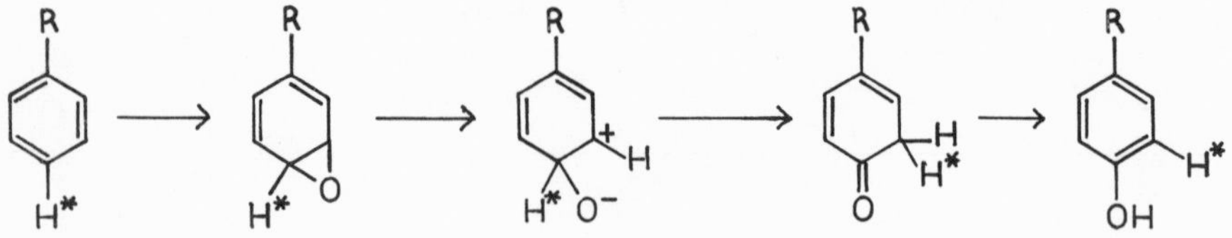

In a tautomeric system similar to that of a dienone-phenol, nitrosophenols can be formed *via* their quinone monoxime tautomers from acyclic precursors, (E. Yu Belyaev *et al*., Chem. Abs., 1978, **88**, 104824). A rarer rearrangement is the cyclopropylcarbinyl-allylcarbinyl rearrangement, (J.E. Baldwin and R.E. Peavy, J. org. Chem., 1971, **36**, 1441).

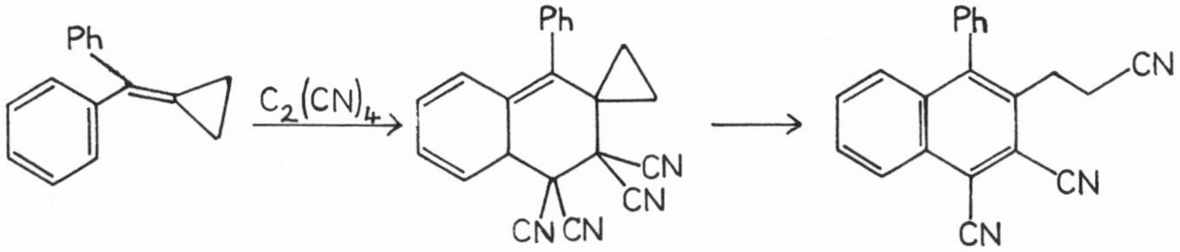

Aromatic ring formation by rearrangements in polycyclic aromatic hydrocarbons has been briefly reviewed, (N. Campbell and N.H. Wilson, Comm. Roy. Soc. Edin., 1979, 204).

(iv) Aromatisation of 3,4,5,7, and 8-membered rings.

(a) Ring expansion.

Cyclopropyl rings can be used as synthons to form benzenoid systems, (M.P. Doyle and T.R. Bade, Tetrahedron Letters, 1975, 3031). A similar photo-reaction is recorded by A.P. Marchand and C.T. Change, (J. chem. Soc., Perkin I 1973, 1948).

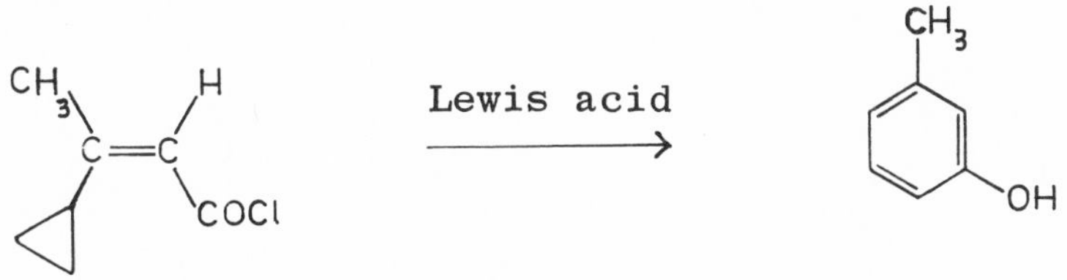

Cyclobutanes can also undergo ring expansion, (W.H. Starnes, D.A. Plank and J.C. Floyd, J. org.

Chem., 1975, <u>40</u>, 1124).

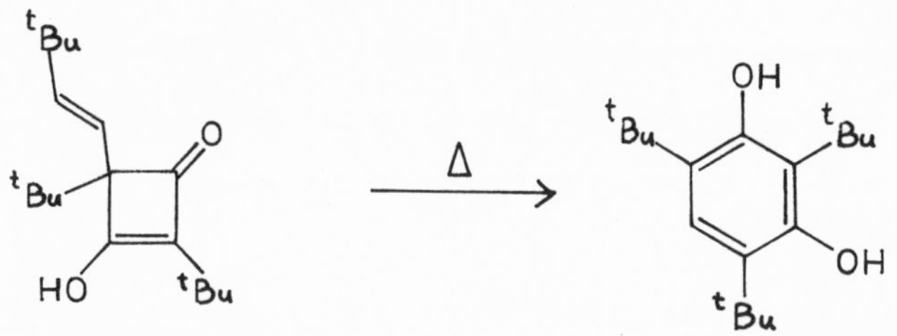

Silver ions catalyse the aromatisation-rearrangement of spironorcaradiene systems, (H. Durr and H. Kober, Ber., 1973, <u>106</u>, 1565; see also J.I.G. Cadogan *et al*., Chem. Comm., 1975, 618).

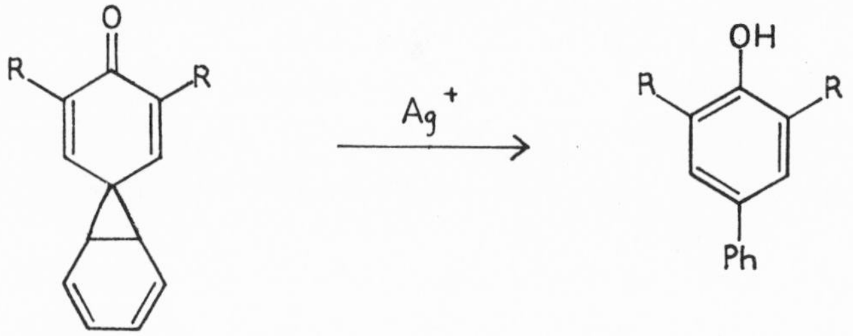

A reaction involving the reaction of cyclopropenones with phosphorus ylides yields 1-naphthols, (R. Breslow, M. Oda and J. Pecoraro, Tetrahedron Letters, 1972, 4415).

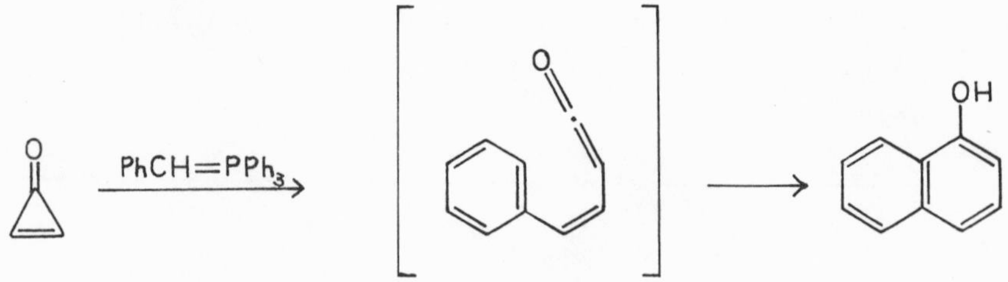

Reactions involving cyclobutene rings or ketene dimers can also lead to naphthalenes, (Z. Zubovics and H. Wittman, Ann., 1972, 760, 171; 765, 15; cf. W. Ried, *et al* ., Tetrahedron Letters, 1972, 3885).

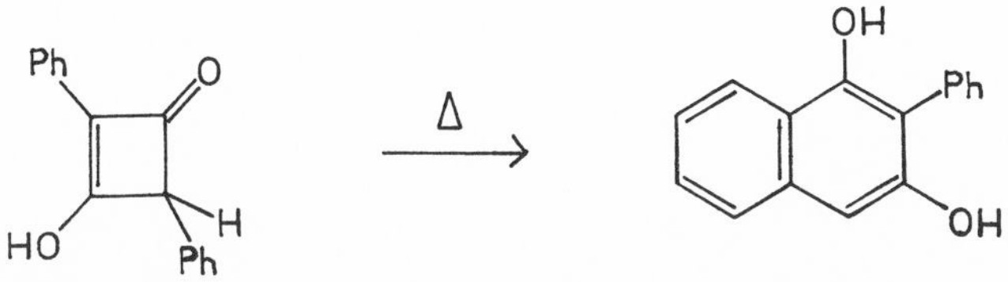

There are also examples of nitrogen elimination from heterocycles leading to ring expansion into benzenoid rings,(W. Ried *et al* ., Ber., 1975, 108, 1413; H. Durr and H. Schmitz, Angew. Chem. internat. Edn., 1975, 14, 647).

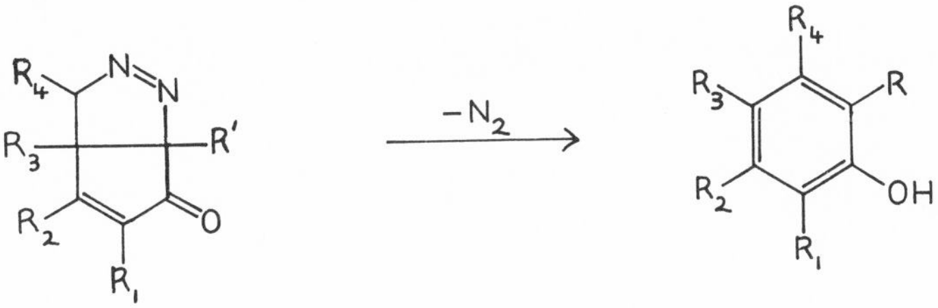

(b) Ring contraction.

Ring contractions leading to benzenoid rings are fairly common and attention has been drawn to this in regard to tropones and tropolones (2nd. edn., Vol IIIA, p 144). Examples are given below.

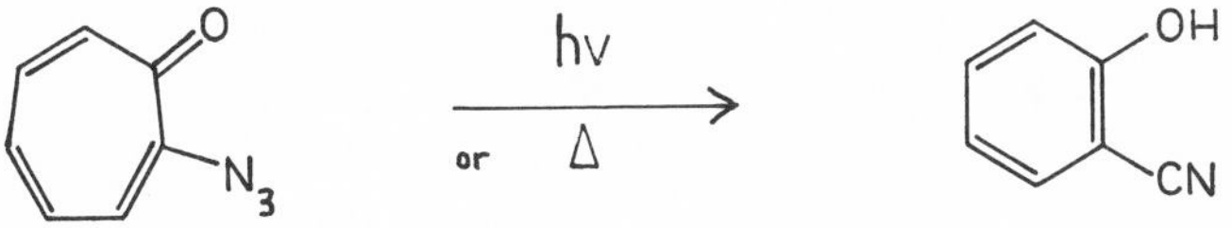

62

(J.D. Hobson and J.R. Malpass, Chem. Comm., 1966, 141).

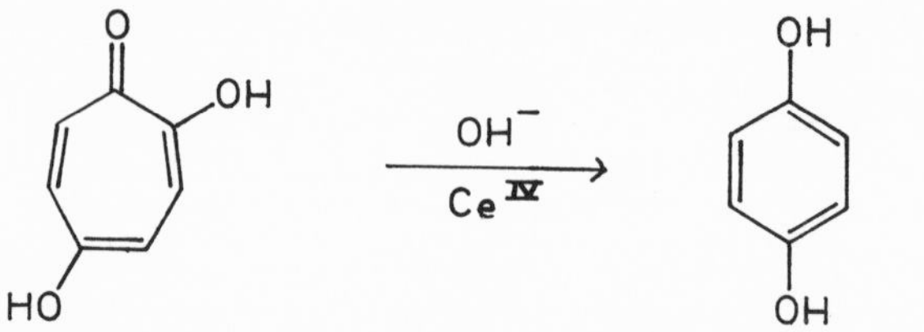

(W.T. Dixon and D. Murphy, J. chem. Soc., Perkin II, 1964, 1430).

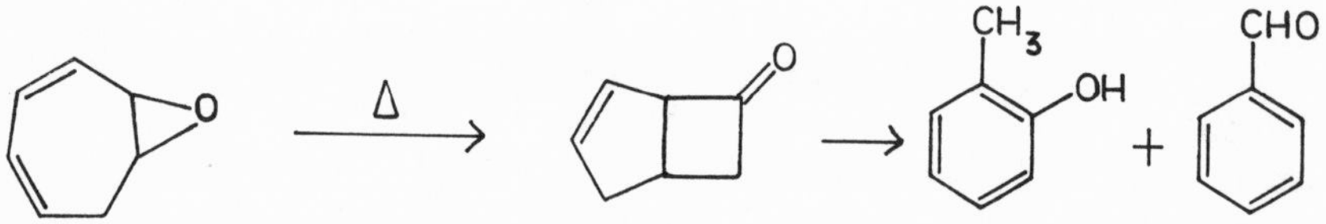

(P. Schiess and M. Wisson, Helv., 1974, 57, 1692; *cf.* R. Miyamoto, T. Teyuka and T. Mukai, Tetrahedron Letters, 1975, 891).

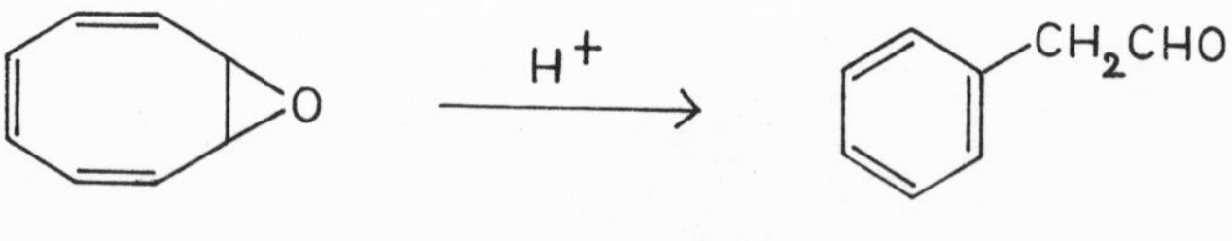

(M. Ogawa *et al.*, Tetrahedron, 1975, 31, 299).

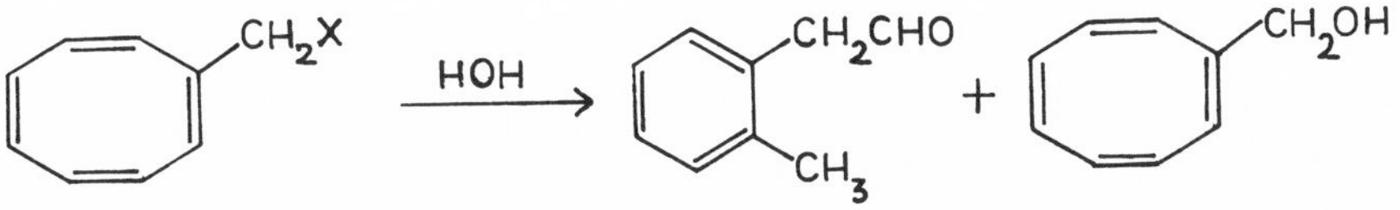

(L.A. Paquette and K.A. Henzel, J. Amer. chem. Soc., 1973, 95, 2725; G.M. Iskander, J. chem. Soc., Perkin I, 1973, 2202).

(v) Benzenoid rings from hetrocycles, and related reactions.

Extrusion of fragments causing aromatisation is a very large field, and only a few examples can be included here. In particular, reactions such as the Diels-Alder or "ene" reaction can lead to many aromatic systems. Often the final step requires the loss of a stable fragment. These methods especially with acetylenes such as dimethyl acetylenedicarboxylate, (DMAD), can yield many arene systems. (For short reviews see H.G. Viehe, "Chemistry of Acetylenes", Marcel Dekker Inc., New York 1969, p 477 *et seq.*, and T.F. Rutledge, "Acetylenes and Allenes", Reinhold, New York, 1969, p 262 *et seq.*).

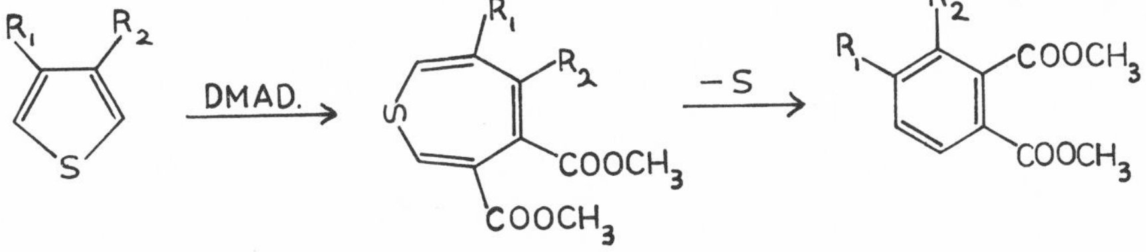

The above reaction involves a sulphur extrusion, (D.N. Reinhoudt and C.G. Kouwenhoven, Chem. Comm., 1972, 1232), and many examples of this are known, (see also I. Murata, T. Tatsuoka and Y. Sugihera, Angew. Chem. internat. Edn., 1974, 13, 142; D.L. Coffen, Y.C. Poon and M.L. Lee, J. Amer. chem. Soc., 1971, 93, 4627). Another example is the photochemical extrusion of sulphur from a spiro-ring yielding a fully aromatic system, (K. Praefcke and Ch. Weichsel, Tetrahedron Letters, 1976, 1787).

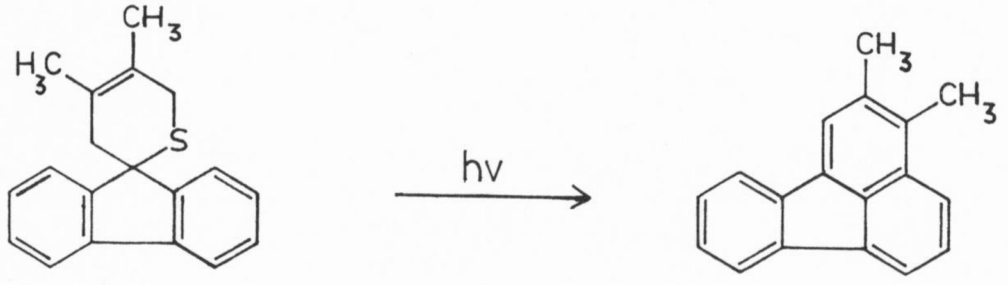

Many extrusions are in fact retro-Diels-Alder reactions. The formation of benzocyclopropenes is a very interesting example, (E. Vogel, W. Grimme and S. Korte, Tetrahedron Letters, 1965, 3625; E. Vogel et al ., Angew. Chem. internat. Edn., 1968, 7, 289).

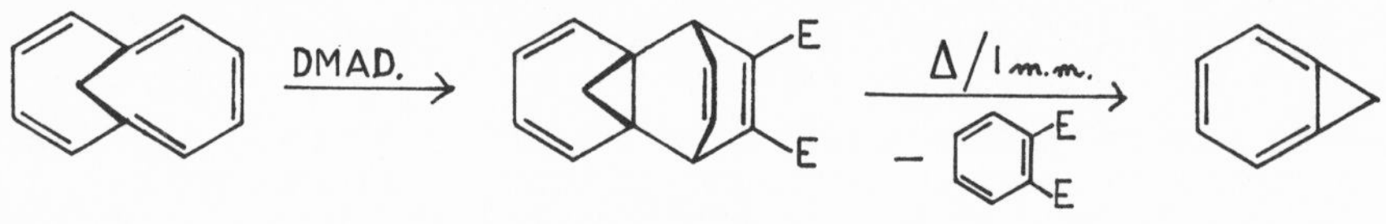

A more usual example of this type of reaction is shown below and it is this type of reaction which gives rise to the vast literature on "DMAD" Diels-Alder reactions, (W.J. Feast, W.K.R. Musgrave and R.G. Weston, J. chem. Soc., (C) 1971, 1547).

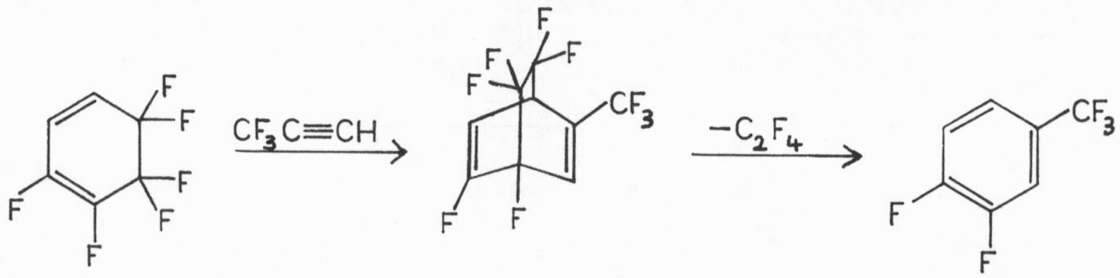

Examples of carbon monoxide extrusion are reported by A. Krebs and D. Byrd (Annalen, 1967, 707, 66), A.S. Langley and M.A. Ogliaruso, (J. org. Chem., 1971, 36, 3339), and are of the form depicted below.

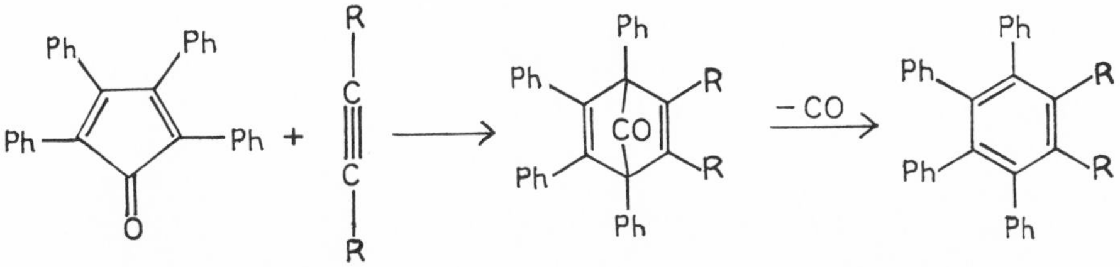

Elimination of carbon dioxide from adducts of coumalic acid or l-pyrone are useful routes to aromatic systems, (2nd edn., Vol IIIA, p 146; T. Jaworski and S. Kwiatkowski, Roczniki Chem., 1975, 49, 63). The literature also abounds with similar extrusions of nitrogen molecules or sulphur dioxide, (2nd edn., Vol IVA, p 226). There are also rarer examples of extrusion of unstable fragments where the driving force of the reaction is even more dependent on the high stability of the forming arene. These reactions are of the general type shown for sulphur extrusion previously, where the sulphur atom is replaced by the leaving fragment, and are illustrated below:

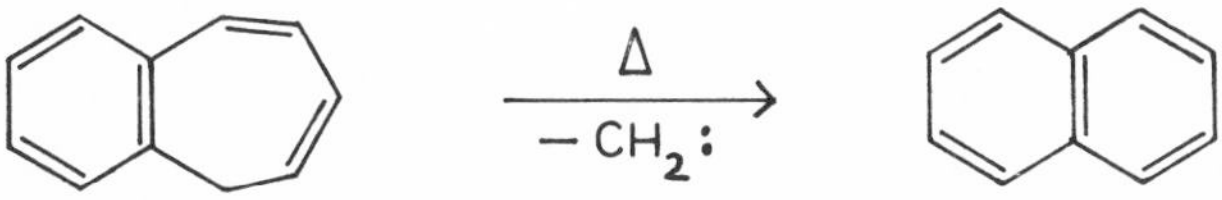

(M. Pomerantz and A.S. Ross, J. Amer. chem. Soc., 1972, 94, 1403).

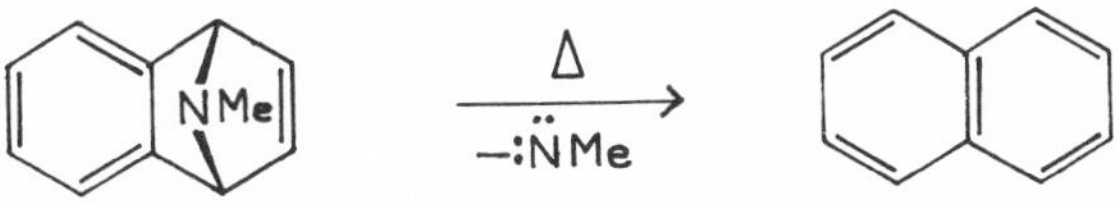

(L.J. Kricka and J.M. Vernon, Chem. Comm., 1971, 942; G.W. Gribble and R.W. Allen, Tetrahedron Letters, 1976, 3673).

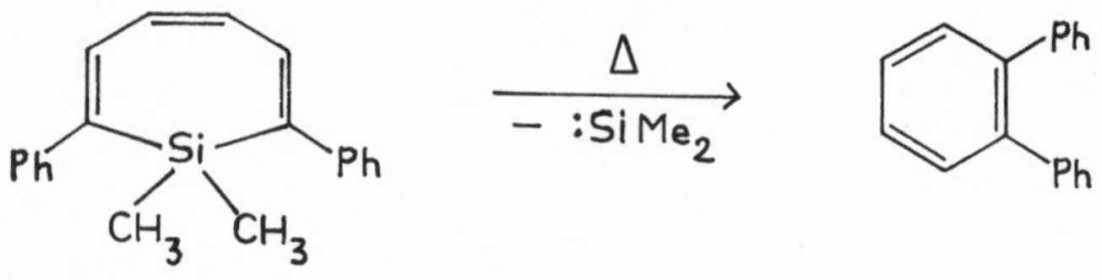

(T.J. Barton, R.C. Kipperham and A.J. Nelson, J. Amer. chem. Soc., 1974, 96, 2272).

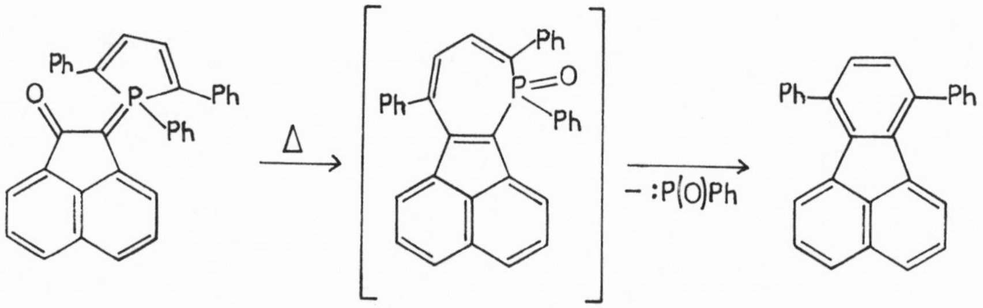

The loss of phosphorus fragment is thought to occur *via* the seven-membered ring shown, (J.I.G. Cadogan, A.G. Rowley and N.H. Wilson, Annalen, 1978, 74; J.I.G. Cadogan *et al*., J. chem. Soc., Perkin I 1977, 1044).

The conversion of pyrylium salts into carbocyclic aromatic rings has been mentioned, (2nd edn., Vol IIIA, p 146). This reaction of these positively charged heterocycles with carbanions has seen further use and extension to thiapyrylium salts has been made, (G.A. Reynolds and J.A. van Allan, J. heterocyclic Chem., 1971, 8, 301). Similar ring conversions can be applied to pyridinium rings, (T. Eicher and E. von Angerer, Ann., 1971, 746, 120), and even to tropylium systems, (B. Föhlisch and E. Haug, Ber., 1971, <u>104</u>, 2324).

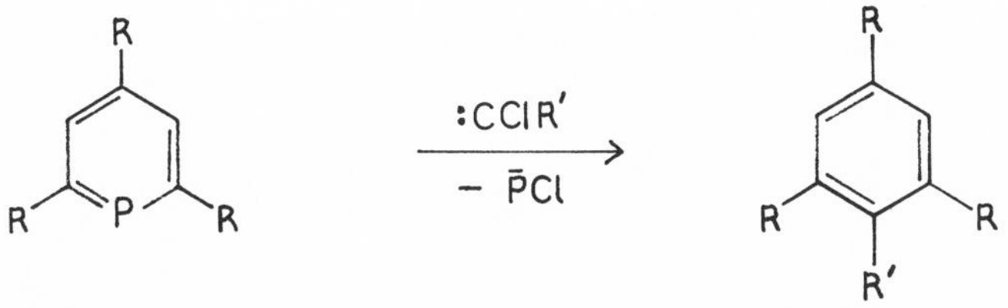

The displacement of the phosphorus atom in phospha-
benzenes has similarly been accomplished by reaction with carbenes, (G. Markl and A. Merz, Tetrahedron Letters, 1971, 1269).

(b) Fission of the benzene ring.

Excluded from this review are disruptions of the
aromatic nucleus involving additions to the ring
thereby destroying aromatic character, unless a
cleavage of the ring follows immediately. Not
covered therefore are Birch reduction, nitrene and
carbene attack, Meisenheimer-type complexes and
photoadditions of, e.g. olefins to the arene (for
a review of this last topic see D. Bryce-Smith and
A. Gilbert, Tetrahedron, 1977, 33, 2459, *cf*. page
4)

Nearly all fission reactions involve oxidation
and several new photoreactions are of particular
interest, (L. Kaplan, L.A. Wendling and K.E.
Wilzbuch, J. Amer. chem. Soc., 1971, <u>93</u>, 3819;
M. Luria and G. Stein, J. phys. Chem., 1972, <u>76</u>,
165; J. Irma and K.C. Kurien, Chem. Comm., 1973,
738).

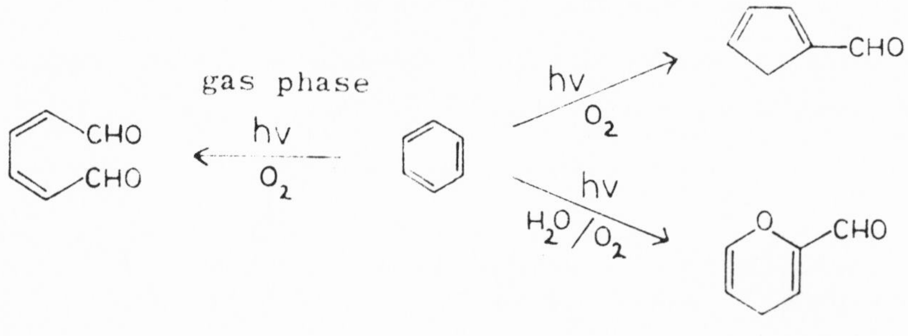

Related reactions with species other than oxygen have led to ring opened products, (D. Bryce-Smith, A. Gilbert and C. Manning, Angew. Chem. internat. Edn., 1974, 13, 341).

$$C_6H_6 + t\text{-BuNH}_2 \longrightarrow CH_2:CHCH:CHCH:CH\text{-NH-Bu-}t$$

Benzene has also been shown to react with mono-oxygen radicals produced in various ways. The intermediates of such reactions appear to be un-saturated aldehydes which polymerise rapidly, (J.N. Pitts and B.J. Finlayson, Angew. Chem. internat. Edn., 1975, 14, 1; J.S. Gaffney, R. Atkinson and J.N. Pitts, J. Amer. chem. Soc., 1976, 98, 1828).

Singlet oxygen ($^1\Delta O_2$) produced photochemically or otherwise, can cause aromatic cleavage *via* the 1,4-epidioxide, (J. Rigaudy, R. Dupont and N.K. Cuong, Compt. Rend., 1971, 272C, 1678; 1973, 276C, 1215).

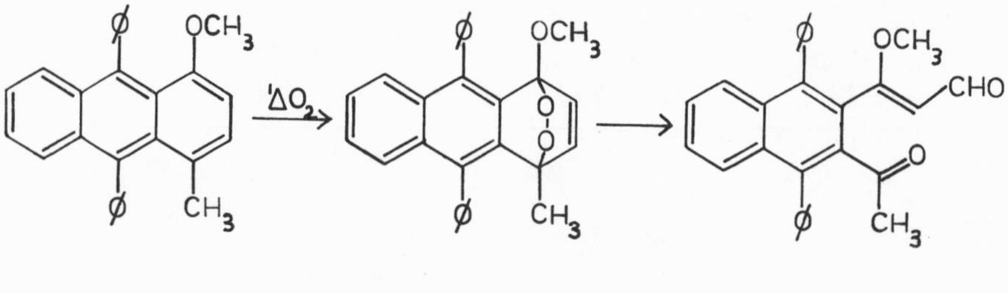

A related reaction producing a similar fission of
1,4-dimethoxynaphthalene by photolysis in presence
of nitrobenzene has been described by I. Saito,
M. Takami and T. Matsuura, (Bull. chem. Soc. Japan,
1975, 48, 2865). A dioxetane intermediate has been
observed in the singlet oxygen cleavage of 9,10-
dimethoxyphenanthrene to diphenic ester, (G. Rio
and J. Bethelot, Bull. Soc. Chim. France, 1972,
822).

The fission of benzofurazan by deoxygenation or
by photolysis has been performed (see also page 49)
(T. Mukai, M. Nitta and T. Oine, Chem. Abs., 1972,
76, 85109; M. Geogarakis, H.J. Rosenkranz and
H. Schmidt, Helv., 1971, 54, 819).

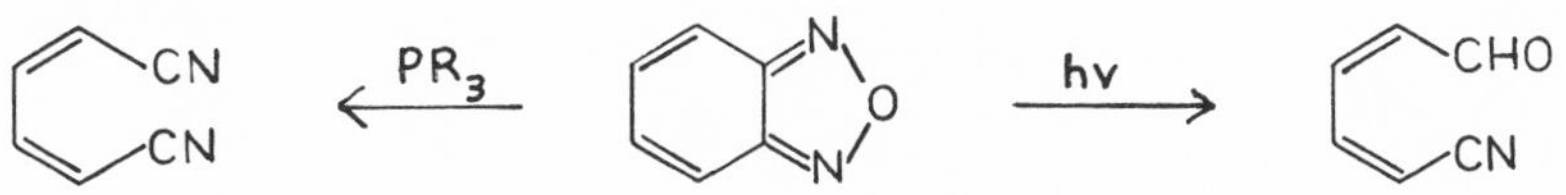

Similar reactions to those cited previously producing *cis, cis*—muconic acid derivatives have been
studied in other aromatic systems and 2,3-diazido-
naphthalene on thermolysis yields *trans*—1,2-di-
cyanobenzocyclobutene, (M.E. Peek, C.W. Rees and
R.C. Storr, J. chem. Soc., Perkin I 1974, 1260;
D.S. Pearce, M.S. Lee and H.W. Moore, J. org. Chem.,
1974, 39, 1362). A general method for fission of
o-diaminoaromatic rings to dinitriles has been
developed, (T. Kajimoto, H. Takahashi and J. Tsuji,
J. org. Chem., 1976, 41, 1389).

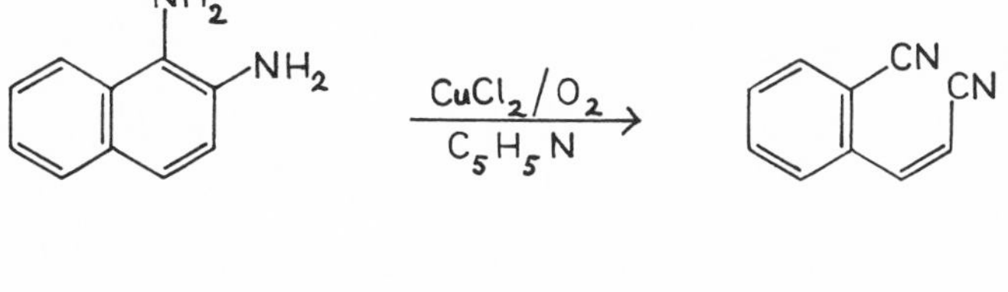

This reagent also cleaves phenols presumably via
the catechols, oxidative fissions of which are
fairly common, (2nd edn., Vol IIIA p 148;
J. Tsuji and H. Takahashi, Tetrahedron Letters,
1976, 1365; J. Amer. chem. Soc., 1974, 96, 7349).
Catechols can be cleaved similarly by nickel
peroxide, (Fr.Pat. 2054699, Chem Abs., 1972, 76,
24699), peracetic acid, (J.C. Farrand and D.C.
Johnson, J. org. Chem., 1971, 36, 3606;
A. Nishinaga, T. Itahara and T. Matsuura, Bull.
chem. Soc. Japan, 1974, 47, 1881), and perchloric
acid, (L. Kalvoda, Coll. Czech. Comm., 1972, 37,
4046). Muconic acids or lactones derived therefrom are the products of these reactions.

Gas phase, high temperature fission of phenols
gives cyclopentadienylidenketen by an elimination
followed by Wolff rearrangement. Loss of carbon
monoxide gives cyclopentadiene, (O.A. Mamer, Canad.
J. Chem., 1971, 49, 3602; H.F. Grutzmacher and
J. Hubner, Ann., 1971, 748, 154; Tetrahedron
Letters, 1971, 1455; M. Saito *et al.*, Chem. Abs.,
1972, 77, 25657).

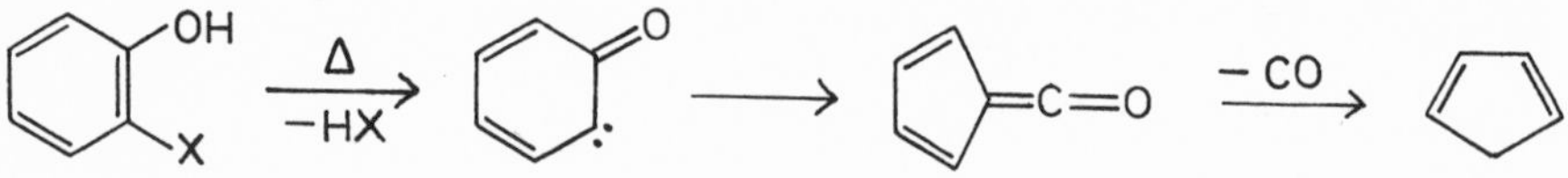

The reaction of the related 1-iminocarbene is known
to give cyanocyclopentadiene, (C. Thetaz and
C. Wentrup, J. Amer. chem. Soc., 1976, 98, 1258).
Benzene itself can be cracked by chlorine and
sulphur at high temperature to yield carbon tetrachloride, (H. Krekeler *et al.*, Chem. Abstr., 1973,
78, 110540).

Powerful chemical oxidants such as ozone, osmium
and ruthenium tetroxides have been discussed previously (2nd edn., Vol IIIA, p 148, 150), and work
with the latter has continued with cleavage of
naphthols and naphthalenes to phthalic acid derivatives, (D.C. Ayres and A.M.H. Hossain, Chem. Comm.,

1972, 428). The mechanisms of these reactions and
further procedures are recorded by U.A. Spitzer
and D.G. Lee, (J. org. Chem., 1974, 39, 2468;
Canad. J. Chem., 1975, 53, 2865).

High temperature hydrogenation of 2-methyl-naph-
thalene yields alkylbenzenes, (E. Oltay, J.M.L.
Penninger and P.G.J. Koopman, Chimia (Switz), 1973,
27, 318).

The remarkable cleavage of 6 methylsalicyclic
acid by *Pen. Urticae* leads to the formation of
patulin, a highly toxic metabolite of several
fungi, (P.I. Forrester and G.M. Gaucher, Biochem.,
1972, 11, 1102).

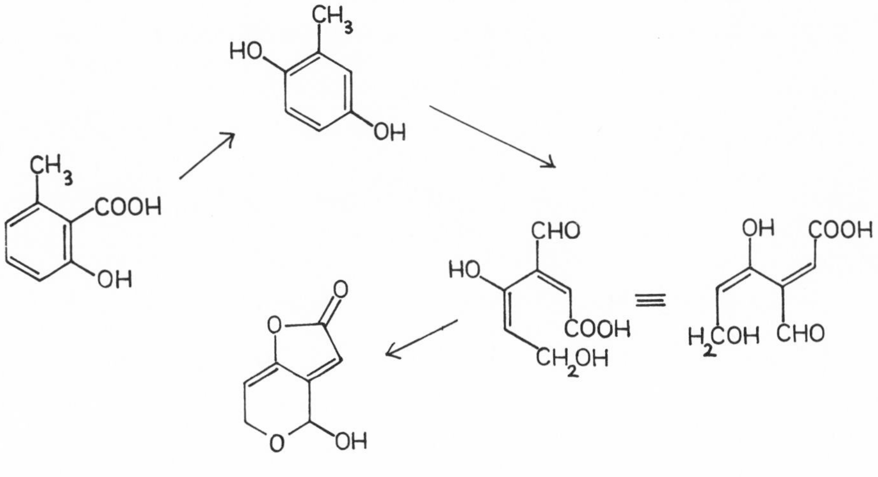

patulin

Chapter 2

MONONUCLEAR HYDROCARBONS: BENZENE AND ITS HOMOLOGUES

H. HEANEY

1. *Nuclear Magnetic Resonance Spectroscopy*

The use of nuclear magnetic resonance (n.m.r.) spectroscopy as a probe for aromaticity is well known. The variations in chemical shift from the positions for benzene
(δ_H = 7.27; δ_C = 128.5 p.p.m. from tetramethylsilane) have
been studied in considerable detail (see L. M. Jackman and
S. Sternhell, 'Nuclear Magnetic Resonance Spectroscopy in
Organic Chemistry', Pergamon, Oxford, 1969 and G. E. Maciel
in 'Topics in Carbon-13 N.M.R. Spectroscopy', Ed. G. C. Levy,
Vol. 1, p. 53, Wiley-Interscience, New York, 1974), and
tabulations of substituent shift effects are available.
Reproductions of representative spectra (see G. J. Pouchert
and J. R. Campbell, 'The Aldrich Library of N.M.R. Spectra',
Aldrich, Milwaukee, Vol. 4, 1974 and L. F. Johnson and
W. C. Jankowski, 'Carbon-13 N.M.R. Spectra', Wiley-Interscience, New York, 1972) and extensive tabulations of [13]C
N.M.R. data are also available (see E. Breitmaier, G. Haas,
and W. Voelter, Heyden, London, 1979, and W. Bremser,
L. Ernst, B. Franke, R. Gerhards, and A. Hardt, Verlag
Chemie, Weinheim, 1979). Lanthanide induced shifts of
aromatic protons have been observed in the presence of silver
fod (T. J. Wenzel, *et al.*, J. Amer. chem. Soc., 1980, <u>102</u>
5903).

2. *The Formation of Benzene derivatives by Cyclisation*
Reactions

A number of interesting developments involving the
cyclotrimerisations of acetylenes have been published. A
wide range of catalysts have been employed, the majority of
which involve transition metals. Ziegler catalysts have
also been used and a report of the use of ethylaluminium
dichloride has appeared (J. H. Lukas, F. Baardman, and
A. P. Kouwenhoven, Angew. Chem. internat. Edn., 1976, <u>15</u>, 369).

The report of the conversion of diphenylacetylene into hexaphenylbenzene using $Fe_3(CO)_{12}$ (W. Hübel and C. Hoogzand, Ber., 1960, 93, 103) has been followed by the preparation of hexakis-(cyclopropyl)benzene in 20% yield from dicyclopropyl-acetylene in the presence of the same catalyst. The other important product in this latter reaction is tetracyclopropyl cyclopentadienone (ca. 30% yield) (V. Usieli, R. Victor, and S. Sarel, Tetrahedron Letters, 1976, 2705). Ynamines undergo cyclotrimerisation in moderate yields using Ni^O or Ni^{II} catalysts (J. Ficini, J. d'Angelo, and S. Falou, ibid., 1977, 1645).

α,ω-Diacetylenes have been found to undergo trimerisations in the presence of Ziegler catalysts. Unfortunately the major pathway in these reactions leads to polymers in which trisubstituted benzenes are the likely intermediates. However, benzene derivatives are formed very efficiently from acetylenes using π-cyclopentadienyl cobalt dicarbonyl $(CpCo(CO)_2)$. Hexa-1,5-diyne gives 1,2-di-(4-benzocyclo-butenyl)-ethane (K. P. C. Vollhardt and R. G. Bergman, J. Amer. chem. Soc., 1974, 96, 4996). Of greater general interest is the observation that α,ω-diacetylenes and mono-acetylenes can be co-oligomerised. Bis-trimethylsilyl-acetylene has proved to be particularly useful as the mono-acetylene component in reactions involving $CpCo(CO)_2$ (K. P. C. Vollhardt, Acc. chem. Res., 1977, 10, 1; R. L. Funk and K. P. C. Vollhardt, Chem. Comm., 1976, 833). Not only do steric problems preclude self-trimerisation, but also the replacement of the trimethylsilyl residues using a wide variety of electrophiles gives control over the sub-stitution pattern in the final product.

n=0 ca. 60%
n=1   82%
n=2   85%

The bulky nature of the two *ortho*-trimethylsilyl residues results in steric acceleration in the step involving the removal of the first of these groups. Thus two different electrophiles can be used sequentially. A mechanistic interpretation of the trimerisation reactions has now been published (D. R. McAlister, J. E. Bercaw, and R. G. Bergman, J. Amer. chem. Soc., 1977, 99, 1666).

Aryl ethynylketones trimerise efficiently when heated in dimethylformamide (DMF) (K. K. Balasubramanian, S. Selvaraj, and P. S. Veukataramani, Synthesis, 1980, 29).

Diels-Alder reactions using dimethyl acetylene-dicarboxylate as the dienophile allow the formation of benzene derivatives after an oxidation step and removal of the methoxycarbonyl groups. This approach is exemplified below for the preparation of 1,2,4-triphenylbenzene (M. Ballester, J. Castñer, and J. Morell, Anales de Quim, 1977, 73, 439). Reactions of buta-1,3-dienyl-1-amines and their enol ether analogues with acetylenic dienophiles afford benzene derivatives after the loss of an amine or alcohol (S. Tanimoto, *et al.*, Tetrahedron Letters, 1977, 2899).

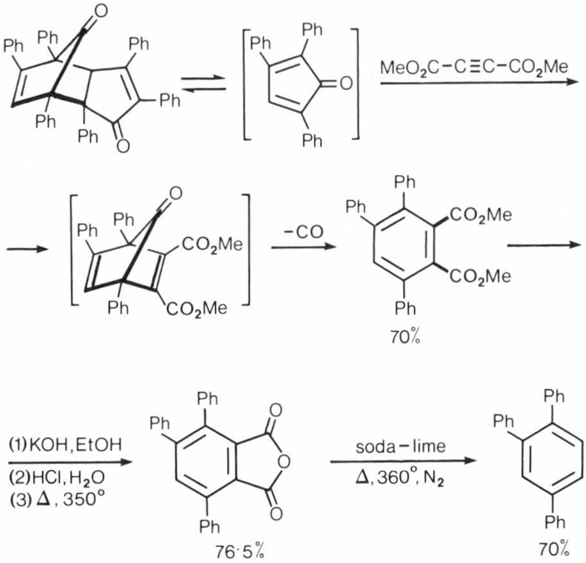

Although the annelation of ketones is well established
(M. E. Jung, Tetrahedron, 1976, 32, 3) the benzoannelation
of ketones having an adjacent methylene group has been
reported only recently. The conversion of heptan-4-one
into 3-n-propylhexa-1,3-diene by reaction with vinylmagnesium
bromide followed by dehydration, allows the formation of a
1,4-dihydrobenzene derivative by Diels-Alder cycloaddition
with dimethyl acetylenedicarboxylate. Dehydrogenation using
DDQ and decarboxylation of the derived phthalic acid
derivative affords 2-n-propylethylbenzene. This type of
reaction sequence may be used to prepare benzopinane and
2-t-butyl-1,2,3,4-tetrahydronaphthalene (L. A. Paquette,
W. P. Melega, and J. D. Kramer, Tetrahedron Letters, 1976,
4033).

The use of 2-$\underline{H}$-pyran derivatives as the diene component,
for example 2,2,4,6-tetramethyl-2-$\underline{H}$-pyran, allows the
formation of substituted benzenes after the cyclo-elimination
of a carbonyl compound from the Diels-Alder adduct obtained
when using an acetylenic dienophile (R. G. Salomon, J. R.
Burns, and W. J. Dominic, J. org. Chem., 1976, 41, 2918).

The use of potassium fluoride as the catalyst in the
dimerisation of pentan-2,4-dione in dimethylformamide, leads
to the formation of 4,6-dimethyl-2-hydroxyacetophenone in
good yield (J. H. Clark and J. M. Miller, J. chem. Soc.
Perkin I, 1977, 2063). The Lewis acid catalysed titanium
(IV) chloride (TiCl$_4$) dimerisation of ethyl 3-ethoxy-but-2-
enoate in dichloromethane at 0° affords ethyl 6-ethoxy-4-
methyl-salicylate in 35% yield (G. Declercq, G. Montardiev,
and P. Mastagli, Compt. Rend., 1975, 28(C), 279).

The ring expansion of five-membered rings into benzene
derivatives provides some interesting chemistry, involving
carbene or carbenoid reagents. The cyclopentadienyl anion
reacts with methylene chloride and methyl lithium at -45°
to afford benzvalene (24%) and benzene (6%) (T. J. Katz,
J. J. Cheung, and N. Acton, J. Amer. chem. Soc., 1970, 92,
6643; T. J. Katz, E. J. Wang, and N. Acton, ibid., 1971,
93, 3782). The same reagents, at -20°, react with 6,6-
dimethylfulvene to afford α-methylstyrene (24%) (A. Amaro
and K. Grohmann, J. Amer. chem. Soc., 1975, 97, 3830).

3. *Benzocyclopropenes and benzocyclobutenes*

The dehydrochlorination of appropriate bicyclo[4.1.0] heptane derivatives yields benzocyclopropenes in reasonable yields, exemplified in equations [1] (W. E. Billup, A. J Blakeney, and W. Y. Chow, Chem. Comm., 1971, 1461), [2] (A. Kumar, S. R. Tayal, and D. Devaprabhakara, Tetrahedron Letters, 1976, 863), and [3] (B. Halton and P. J. Milsom, Chem. Comm., 1971, 814).

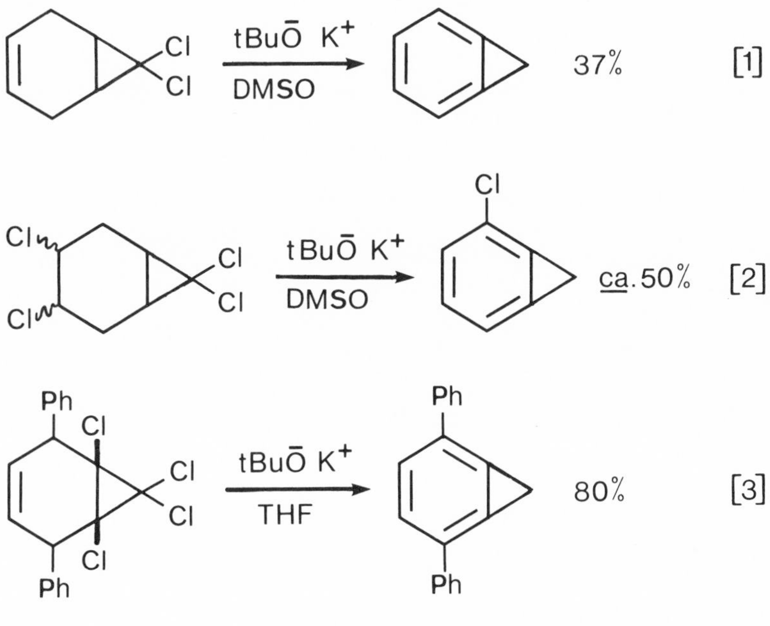

7,7-Dichloro-2,5-diphenylbenzocyclopropene undergoes exchange with silver fluoride at room temperature to afford the difluoro analogue in 85% yield (P. Müller, Chem. Comm., 1973, 895). The reductive removal of chlorine or fluorine

from 1,1-dihalobenzocyclopropenes using freshly prepared
aluminium hydride yields the parent compounds (P. Müller,
Helv., 1974, 57, 704). A particularly simple method, which
gives benzocyclopropene in 30% yield involves the reaction
between n-butyl lithium and o-bromobenzyl methyl ether,
initially at -40° (P. Radlick and H. T. Crawford, Chem.
Comm., 1974, 127). The flash vacuum pyrolysis of tolyl-
propargyl ethers gives aromatic ring methylated benzocyclopropenes in low yields (J. M. Riemann and W. S. Trahanovsky,
Tetrahedron Letters, 1977, 1863).

Benzocyclopropene dimerises thermally at 80° to afford
9,10-dihydrophenanthrene. On the other hand the silver (I)
catalysed reaction at 0° gives 9,10-dihydroanthracene. In
the presence of buta-1,3-diene the silver ion catalysed
reaction proceeds as shown (W. P. Billups, W. Y. Chow, and
C. V. Smith, J. Amer. chem. Soc., 1974, 96, 1979).

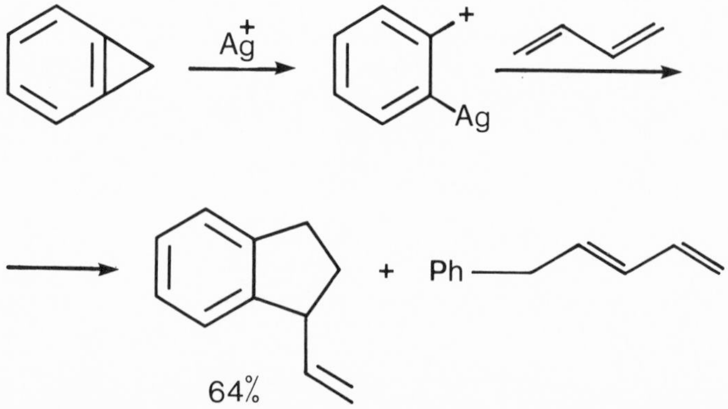

Benzocyclopropene can function as a dienophile in Diels-
Alder reactions. The reaction with, for example, 4,5-dibromo-
o-benzoquinone gave the adduct in 65% yield. This reaction
was the first step in the synthesis of a bridged-10-π-troponoid
(E. Vogel, J. Ippen, and V. Buch, Angew. Chem. internat. Edn.,
1975, 14, 566).

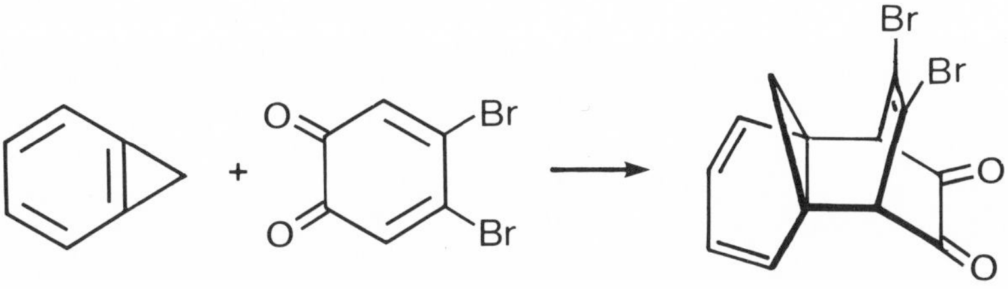

In addition to the preparation of benzocyclobutene already mentioned, the dehalogenation of 1,2-dibromobenzocyclobutene using tri-n-butyl chlorostannane and lithium aluminium hydride is noteworthy. Vicinal dibromides are normally converted into the corresponding alkene using this method (A. Sanders and W. P. Giering, J. org. Chem., 1973, 38, 3055). The well known extrusion of sulphur dioxide from 1,3-dihydroisobenzothiophen-2,2-dioxides which gives benzocyclobutenes has been utilised in the synthesis of dicyclobutenobenzenes (E. Giovannini and H. Vuilleumier, Helv. 1977, 60, 1452). Dicyclobuteno[a,c]benzene has also been prepared by the oxidative decarboxylation reaction shown in equation [4] (R. P. Thummel, J. Amer. chem. Soc., 1976, 98, 628).

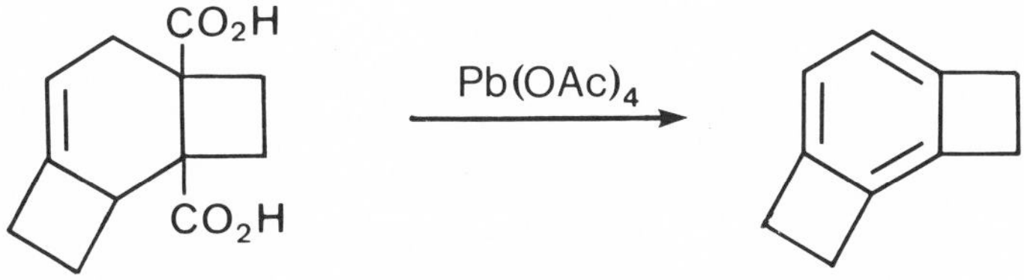

The reaction of α,α,α',α'-tetrabromo-*o*-xylene with iron pentacarbonyl results in the formation of mainly *trans*-1,2-dibromobenzocyclobutene (R. Victor, Transition Metal Chem. (Weinheim),

1977, <u>2</u>, 2). Benzocyclobutenones are obtained in good yields by flash vacuum pyrolysis of 2-methylbenzoyl chlorides at 600° (O.3 s) (P. Schiess and M. Heitzmann, Angew. Chem. internat. Edn., 1977, <u>16</u>, 469).

4. *Electrophilic Addition-with-Elimination Reactions*

A number of important general features of the reactions of electrophiles with arenes will be exemplified in this section.

The section is sub-divided according to the group of the periodic table from which the electrophile is derived. An excellent account of some new synthetic reagents has appeared (G. A. Olah, Acc. chem. Res., 1980, <u>13</u>, 330).

(a) *Protonation Reactions*

These reactions may be divided into two types. Those which are hydrogen exchange reactions and those in which a substituent other than hydrogen is displaced.

A considerable amount of kinetic work has been reported in connection with acid-catalysed hydrogen exchange reactions of arenes. This work has been concerned both with the determination of the mechanism of the reaction and also with an evaluation of substituent effects (R. Taylor, in 'Comprehensive Chemical Kinetics', ed. C. H. Bamford and C. F. H. Tipper, Elsevier, Amsterdam, 1972, Vol. 13, Chapter 1; R. Taylor, in 'M.T.P. International Review of Science, Organic Chemistry Series One', Butterworths, London, 1973, Vol. 3 (ed. H. Zollinger), Chapter 1). This type of reaction can be used to deuteriate and tritiate aromatic rings selectively. The general mechanism involving the intermediacy of σ-complexes (Wheland intermediates) has been confirmed. In favourable cases σ-complexes can be isolated. Thus at low temperatures mesitylene reacts with hydrogen fluoride in the presence of boron trifluoride to afford a yellow σ-complex that reverts to starting materials at higher temperatures (-15°). Using the less nucleophilic hexafluoroantimonate as the counter-ion, the mesitylene salt does not decompose until 51°. The n.m.r. spectrum of this latter salt showed resonances at δ_H 2.7 (*o*-CH$_3$), 2.8 (*p*-CH$_3$), 4.5 (CH$_2$), and 7.6 (CH) (G. A. Olah, *et al.*, J. Amer. chem. Soc., 1972, <u>94</u>, 2034). The parent benzenium ion has been prepared by the protonation of benzene with SbF_5-HSO_3F in SO_2ClF-SO_2F_2 at $-120°$. Alternatively solutions of the benzenium ion may be made by the ionisation of 3-bromo-1,4-cyclohexadiene or by the dealkylation of t-butylbenzene. At $-80°$ n.m.r. spectra of the benzenium ion (100 MHz, ^1H) show $\delta_H = 8.09$ and $\delta_C = 144.8$ which result from time averaging the degenerate set of ions [1(a) → 1(f)]. The coalescence temperature is, of course, raised by determining the spectra at higher field (L. M. Jackman and F. A. Cotton, ed., 'Dynamic Nuclear Magnetic Resonance Spectroscopy', Academic Press, New York, 1975). Spectra taken at 270 MHz (^1H) and 67.9 MHz (^{13}C) at $-140°$ and $-135°$ respectively showed that even under those conditions some equilibration is observed. The $\delta_H = 5.6$ (CH_2), 8.6 (m-H's), $ca.$ 9.3 (p-H), and 9.7 (o-H's) and $\delta_C = 55.2$ (t), 136.9 (d), 178.1 (d), and 186.6 (d) p.p.m. values are obtained. A complete line shape analysis of the 100 MHz ^1H n.m.r. spectra near coalescence gives an Arrhenius activation energy of $ca.$ 10 K cal mol^{-1} for the equilibration process (G. A. Olah, et $al.$, J. Amer. chem. Soc., 1978, 100, 6299).

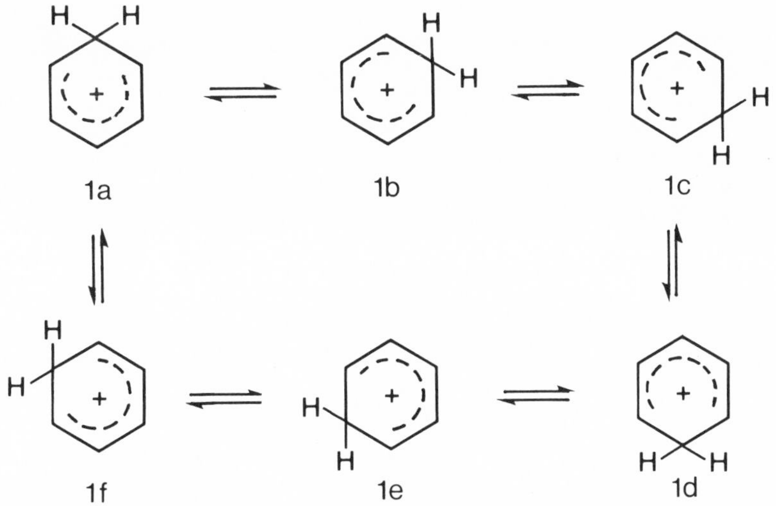

More basic arenes are protonated much more easily with acids that are weaker than HF-SbF_5. Protonation does not normally occur at a site bearing a methyl group (c.f. mesi-

tylene above). On the other hand hexamethylbenzene is
ca. 50% protonated in sulphuric acid (90%) at room temperature.

The reversibility of sulphonation and the Friedel-Crafts
alkylation reaction is well known. In this section proto-
demetallation reactions are considered. Proto-desilylation
has been studied in considerable detail (C. Eaborn,
J. organometal. Chem., 1975, 100, 43) as also have reactions
involving germanium, tin, and lead analogues. Primary kinetic
isotope effects in the range $k1_H/k2_H = 1.55$-3.05 are observed
using deuterium chloride in deuterium oxide. These results
confirm the operation of the addition-with-elimination
mechanism involving a σ-complex. The facility with which
protonation occurs at the *ipso*-position (a position bearing
a substituent) is evidently due to the polarity of the
carbon-metal bond which increases the electron density at
carbon as compared with the metal. The rate of cleavage of
ArMEt$_3$ compounds in aqueous methanolic perchloric acid
increases in the series (M=) Si<Ge<<Sn<<Pb (R. W. Bott,
C. Eaborn, and P. M. Greasley, J. chem. Soc., 1964, 4804).
The preparation of specifically deuteriated compounds can be
achieved by, for example, deuterio-destannylation of aryl
trimethylstannanes. This method appears to be preferable
to the hydrolysis of Grignard- or organolithium- reagents if
high isotopic purity (>98%) is required. It also has an
obvious advantage when functional groups are present that
are incompatible with the formation of Grignard reagents.
Although not strictly relevant to this section it is worth
noting that if microscopic incorporation is required (e.g. ^{3}H),
it is achieved, with the above reservation concerning other
functional groups, by allowing an aryl halide to react with
n-butyl-lithium at -70° in ether that has been wetted with
^{3}H$_2$O (R. Taylor, Tetrahedron Letters, 1975, 435).

(b) *Metallation Reactions*

Electrophilic metallations involving arenes are observed
using, for example, mercury, thallium, and lead salts. These
reactions should not be confused with reactions using, for
example, n-butyl-lithium. In these latter reactions the
most acidic hydrogen is removed.

The direct thalliation of many aromatic compounds is
rapid at room temperature using thallium (III) trifluoroacetate (E. C. Taylor and A. McKillop, Acc. chem. Res., 1970,

3, 338). Using thallium (III) salts in trifluoroacetic acid - acetic acid mixtures, toluene has been found to be seven times as reactive as benzene and undergoes predominantly *para*-substitution (87%). The thalliation reactions are freely reversible. The reaction of 2-phenylethanol with thallium (III) trifluoroacetate proceeds at room temperature under kinetic control to yield the *ortho*-substitution product, possibly by intramolecular delivery of the electrophile. On the other hand, reaction at 73° gives predominantly the *meta*-isomer under thermodynamic control.

Very useful synthetic transformations can be achieved using aryl-thallium intermediates. A number of typical examples are shown in equations [5]-[7], [8] (S. Uemura, Y. Ikeda, and K. Ichikawa, Tetrahedron, 1972, 28, 3025), [9] (E. C. Taylor, R. H. Danforth, and A. McKillop, J. org. Chem., 1973, 38, 2088), [10] (B. Davies and C. B. Thomas, J. chem. Soc. Perkin I, 1975, 65), and [11] (R. B. Herbert, Tetrahedron Letters, 1973, 1375). The reactions of potassium fluoride with arylthallium bis-(trifluoroacetates) afford arylthallium difluorides which yield fluoroarenes on treatment with boron trifluoride (E. C. Taylor, *et al.*, J. org. Chem., 1977, 42, 362).

$$ArTl(OCOCF_3)_2 \xrightarrow{\text{KI}} ArI \qquad [5]$$

$$ArTl(OCOCF_3)_2 \xrightarrow{\text{Pb(OAc)}_4} ArOH \qquad [6]$$

$$ArTl(OCOCF_3)_2 \xrightarrow{\text{LiAlD}_4} ArD \qquad [7]$$

$$ArTl(OAc)(OCOCF_3) \xrightarrow{\text{CuCN,Py}} ArCN \qquad [8]$$

$$ArTl(OCOCF_3)_2 \xrightarrow{\text{2HCl}} ArTlCl_2 \xrightarrow{\text{NOCl}} ArNO \quad [9]$$

$$ArTl(OCOCF_3)_2 \xrightarrow{\text{NO}_2} ArNO_2 \qquad [10]$$

$$ArTl(OCOCF_3)_2 \xrightarrow{\text{NaBH}_4, \text{EtOD}} ArD \qquad [11]$$

It is not always necessary to isolate the arylthallium reagent. This is illustrated by the formation of 4-bromobiphenyl in 93% yield using bromine in the presence

of thallium (III) acetate. Polyiodination of benzene and
mesitylene occurs using thallium (III) trifluoroacetate and
iodine (N. Ishkikawa and A. Sekiya, Bull. chem. Soc. Japan,
1974, 47, 1680). The electrophile involved has low
reactivity but a high steric requirement (A. McKillop,
D. Bromley, and E. C. Taylor, J. org. Chem., 1972, 37, 88).

Rather similar reactions occur using lead (IV) trifluoroacetate and with a range of benzene derivatives (but
not nitrobenzene) it is possible to obtain good yields of
arenetrifluoroacetate. In the case of the reaction using
fluorobenzene, the intermediate, p-fluoro-phenyl-lead
tris-trifluoroacetate, has been isolated (J. R. Campbell,
et al., Tetrahedron Letters, 1972, 1763). The available
methods that allow the preparation of arenelead tris-
carboxylates have been summarised (H. C. Bell, *et al.*,
Austral. J. Chem., 1972, 32, 1521). In reactions using
p-substituted aryl trimethylsilanes the expected *ipso*-attack
gave almost quantitative yields [equation 12] (J. R. Kalman,
J. T. Pinhey, and S. Sternhell, *ibid.*, 1972, 5369). Reactions
with arenes give biaryls (H. C. Bell, *et al.*, Austral. J.
Chem., 1979, 32, 1531).

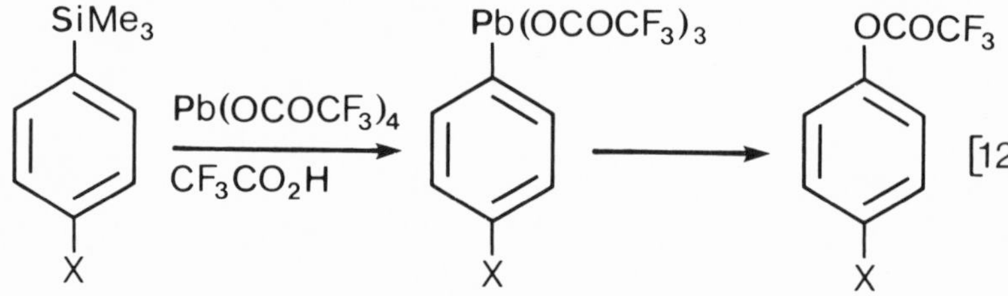

(c)   *Group IV Electrophiles*

This sub-section is concerned with advances in the
Friedel-Crafts and related reactions (G. A. Olah, 'Friedel-
Crafts Chemistry', Wiley-Interscience, New York, 1973).

(i)   *Alkylation Reactions*

Complex mixtures of products are frequently obtained in

reactions that are carried out using classical procedures involving Lewis acid catalysts. These complexities arise due to polyalkylation and isomerisation-disproportionation processes. The Du Pont ion-membrane resins (e.g. Nafion-H) or long chain perfluorinated alkanesulphonic acids can be used to achieve clean reactions under heterogeneous conditions in the gaseous phase. The alkylation of arenes using these catalysts, or if necessary in the presence of Lewis acids such as antimony pentafluoride (SbF_5), can be achieved using alkenes, alcohols, alkyl halides, and esters (G. A. Olah, *et al.*, J. org. Chem., 1977, 42, 3046; 1978, 43, 3147; 1978, 43, 3142; Nouveau J. Chim., 1979, 3, 269.

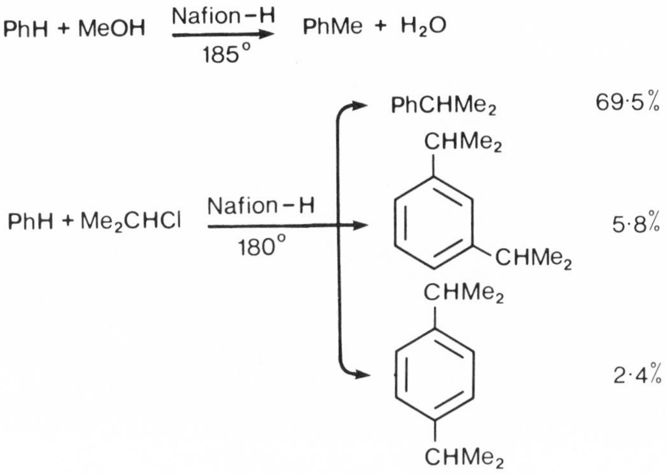

$$PhH + MeOH \xrightarrow[185°]{Nafion-H} PhMe + H_2O$$

PhH + Me$_2$CHCl $\xrightarrow[180°]{Nafion-H}$

PhCHMe$_2$ 69·5%

CHMe$_2$ ... CHMe$_2$ 5·8%

CHMe$_2$ ... CHMe$_2$ 2·4%

Transalkylation reactions of di- and poly-alkylbenzenes can also be carried out using the same catalyst systems (G. A. Olah and J. Kaspi, Nouveau J. Chim., 1978, 2, 581). The alkylation of arenes has also been achieved using alkyl trifluoromethanesulphonates and related compounds in the presence of a catalytic quantity of the corresponding acid (B. L. Booth, R. N. Haszeldine, and K. Laali, J. chem. Soc Perkin I, 1980, 2887). These latter reactions as well as a number of other Friedel-Crafts alkylations proceed efficiently in the presence of SbF_5 (G. A. Olah and J. Nishimura, J. Amer.

chem. Soc., 1974, <u>96</u>, 2214).

In the absence of hetero-atom substituents, the
C-alkylation of arenes can be achieved using dialkylhalonium-
and trialkyloxonium-salts (G. A. Olah, *et al.*, *ibid.*, 1974,
<u>96</u>, 884). The former reactions are carried out in sulphuryl
chloride fluoride and the latter in magic acid (FSO_3H — SbF_5).
The $o:p$ ratios obtained in methylation reactions of toluene
using dimethylhalonium salts are very similar to those
observed in conventional Friedel-Crafts reactions using
methyl halides and Lewis acids. However, a significant
reduction in the $o:p$ ratios was found when diethylhalonium
salts were used. The decrease in the $o:p$ ratio is related
to an increase in the steric requirements for alkylation
using diethylhalonium salts. These results suggest that
the active ethylating agents in reactions using diethylhalonium salts and ethyl halide — Lewis acid combinations
are different and that the reactions proceed through an
S_N2 like transition state. The results obtained with
trialkyloxonium salts are similar to those obtained using
dialkylhalonium salts.

The generation of the methoxycarbenium ion ($CH_3OCH_2^+$)
either by dechlorination of chlorodimethyl ether or by
dechlorodecarbonylation of methoxyacetyl chloride using
hexafluoroantimonic acid provides a new alkylation system for
arenes (G. A. Olah and J. J. Svoboda, Synthesis, 1973, 52).

Although the cation tricarbonyl [cyclohexadienyl]iron
(1+) does not react with benzene and a number of simple
derivatives the alkylation of aniline has been reported
(Y. Becker, A. Eisenstadt and Y. Shvo, Tetrahedron Letters,
1972, 3183). Evidently high electron density on ring
carbon(s) is required. This principle has been used in
reactions of a range of aryl-trimethylsilanes and aryl-
trimethylstannanes which give the corresponding diene substituted aromatic compounds after *ipso*-attack and re-aromatisation by loss of the trimethylsilyl- or trimethylstannyl-
residue (G. R. John, L. A. P. Kane-Maguire, and C. Eaborn,
Chem. Comm., 1975, 481).

Reactions of α-chloronitrones with nucleophilic arenes,
such as p-methoxytoluene, in the presence of silver (I) salts,
lead to substitution and hence affords β-aryl aldehydes
(S. Shatzmiller, *et al.*, 1973, <u>56</u>, 2961).

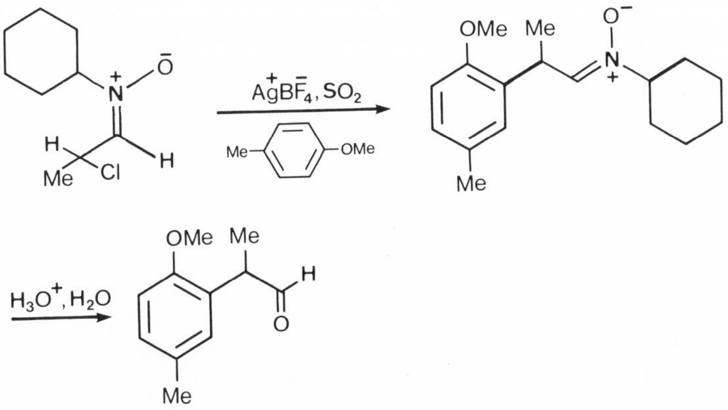

The cyclodehydration of arylalkanols proceeds efficiently in the presence of a sulphonic acid ion-exchange resin (Amberlist-15) as shown in equation [13] (W. M. Harms and E. J. Eisenbraun, Org. prep. Proc. Internat., 1971, 3, 239).

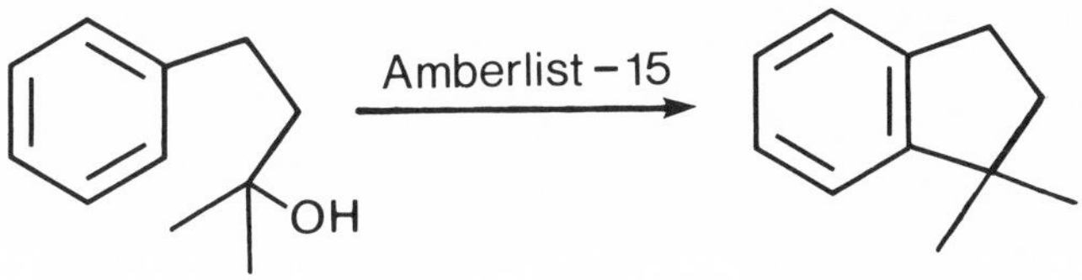

[13]

The generation of 'hot' carbenium ions by the de-diazoniation of alkanediazonium ions or by the dechloro-decarboxylation of alkanechloroformates has led to a number of interesting observations. The generation of 'hot' carbenium ions from chloroformates with the assistance of silver (I) ions leads to clean reactions (P. Beak, Acc. chem. Res., 1976, 9, 230). Thus 1-apocamphyl chloroformate reacts with silver hexafluoroantimonate in chlorobenzene to give a mixture of 1-chlorophenyl-apocamphanes (*o:m:p* = 39:35:26) in 81% yield. The cation is so 'hot' that reaction occurs even with nitrobenzene and affords 1-(*m*-nitrophenyl)apocamphane in 24% yield. A point of general interest emerges from a study of the *C*-alkylation of anisole with silver (I) ion reactions of chloroformates. The reaction of [$^2H_3$]methyl chloroformate with anisole in the presence of silver hexafluoroantimonate gave results that showed the initial product to be dimethylphenyloxonium hexafluoroantimonate which rearranges intermolecularly to afford the final products (methoxytoluenes).

Transalkylation of products formed under kinetic control can lead eventually to the most thermodynamically stable product: the *meta*-isomer. This feature may be applied in a useful way. Thus the reaction of t-butylbenzene with t-butyl chloride in the presence of aluminium chloride (initially at -40°) affords 1,3,5-tri-t-butylbenzene in about 80% yield (P. C. Myhre, T. Rieger, and J. T. Stone, J. org. Chem., 1966, 31, 3425). The same product is obtained by the t-butylation of 1,4-di-t-butylbenzene. [^3H] Labelling studies indicate predominant, if not exclusive, intermolecular transfer of t-butyl groups. The complete removal of t-butyl groups by proto-dealkylation can be achieved in suitably activated systems by the use of aqueous trifluoroacetic acid under reflux, as exemplified in equation [14] (J. F. W. McOmie and S. A. Saleh, Tetrahedron, 1973, 29, 4003).

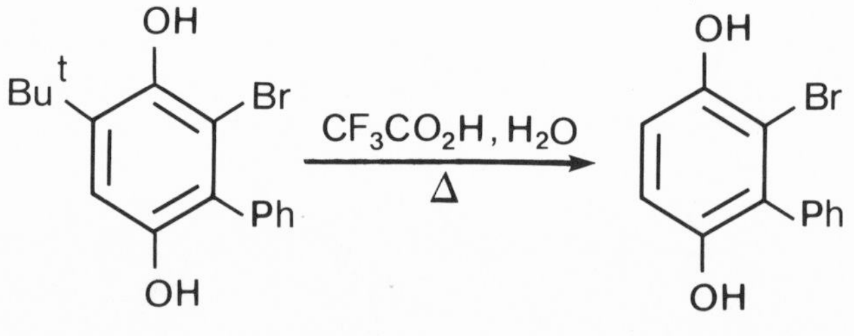

t-Butylation of some arenes can be achieved using t-butyl trifluoroacetate in trifluoroacetic acid at room temperature. Toluene, for example, gives 4-t-butyltoluene in ca. 60% yield (U. Svanholm and V. D. Parker, J. chem. Soc. Perkin I, 1973, 562).

(ii) *Chloromethylation Reactions*

1-Chloro-4-chloromethoxybutane is a particularly valuable reagent in decreasing the secondary side reactions (e.g. the formation of diarylmethanes) that sometimes prove troublesome in chloromethylation reactions. Reaction occurs *via* the *O*-chloromethyltetrahydrofuranium ion and the leaving group, tetrahydrofuran, coordinates with the Lewis acid catalyst, (G. A. Olah *et al.*, Synthesis, 1974, 560; J. org. Chem., 1976, <u>41</u>, 1627).

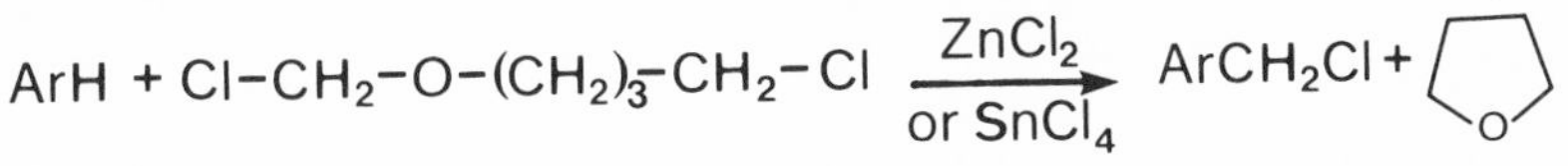

$$ArH + Cl-CH_2-O-(CH_2)_3^--CH_2-Cl \xrightarrow[\text{or SnCl}_4]{ZnCl_2} ArCH_2Cl + \underset{O}{\bigcirc}$$

(iii) *Acylation Reactions*

Mixed carboxylic-sulphonic anhydrides (F. Effenberger and G. Epple, Angew. Chem. internat. Edn., 1972, <u>11</u>, 299; 300), especially those derived from trifluoromethanesulphonic acid, are particularly effective acylating agents: reaction with benzene proceeds smoothly in the absence of a catalyst. The mixed anhydrides can be formed *in situ* by the interaction of acyl halides with trifluoromethanesulphonic acid. Thus benzoyl trifluoromethanesulphonic anhydride reacts with benzene to give benzophenone in 90% yield. Pivaloyl chloride and trifluoromethanesulphonic acid give 4-methoxy-pivalophenone (54%) on reaction with anisole.

Acylations can be effected using aroyl chlorides and an arene in the presence of Nafion-H, but deprotonation of the acetylium ion gives ketene preferentially (G. A. Olah *et al.*, Synthesis, 1978, 672). Arenes react with bromo- and chloro-acetylium hexafluoroantimonates to give high yields or aryl halomethyl ketones (G. A. Olah, H. C. Lin, and A. Germain, *ibid.*, 1974, 895). Work on Friedel-Crafts acylation reactions (see the reviews by R. Taylor, in

'Comprehensive Chemical Kinetics', ed. C. H. Bamford and C. F. H. Tipper, Elsevier, Amsterdam, 1972, Vol. 13, Chapter 1; and in 'M.T.P. International Review of Science, Organic Chemistry Series One', Butterworths, London, 1973, Vol. 3 (ed. H. Zollinger), Chapter 1) allows a rationalisation of some apparent anomalies and reflects the duality of mechanism. At one extreme free acylium ions are involved while at the other end of the spectrum the arene attacks a 1:1 complex formed between the Lewis acid and, for example, the acyl halide. Thus the acetylation of toluene in nitrobenzene affords more 4-methylacetophenone than when the reaction is conducted in carbon disulphide. These results evidently reflect a lower steric demand in the reaction carried out in carbon disulphide. Similarly naphthalene gives 1-acetylnaphthalene in carbon disulphide but mainly 2-acetylnaphthalene in nitrobenzene. The use of nitromethane as a solvent instead of nitrobenzene in Friedel-Crafts acylations results in significant improvements. In the reaction of methylsuccinic anhydride with p-xylene the isomer shown was obtained in 82% yield (E. J. Eisenbraun, et $al.$, J. org. Chem., 1971 $\underline{36}$, 2480).

Friedel-Crafts acylation reactions are normally irreversible. However, the isomerisation of 1,2-dihydro-phenanthren-4-(3H)one to 3,4-dihydroanthracen-1-(2H)one in 25% yield has been shown to proceed in poly-phosphoric acid at 140° (I. Agranat and Y. -S. Shih, Synthesis, 1974, 865).

(iv) *Formylation Reactions*

In the classical Gattermann-Koch formylation procedure carbon monoxide and hydrogen chloride, in the presence of copper (I) chloride, function as if formyl chloride is formed.

This is then assumed to ionise to give the formylium ion.
Formyl chloride is unstable unless kept at low temperatures.
It can be generated at -65° by the reaction of hydrogen
chloride with N-formylimidazole in methylene chloride. The
use of formyl fluoride (G. A. Olah and S. J. Kuhn, J. Amer.
chem. Soc., 1960, 82, 2380; G. A. Olah, M. Nojima, and
I. Kerekes, Synthesis, 1973, 487) or formic anhydride
(G. A. Olah, *et al.*, Angew. Chem. internat. Edn., 1979, 18,
614) allows a wide range of formylations of arenes to be
carried out using boron trifluoride catalysis. Remarkably,
although naphthalene is reported not to react under the
classical Gattermann-Koch conditions, the use of formyl
fluoride affords 1-naphthaldehyde in 73% yield.

The **Vilsmeier** reaction, in which formylation is achieved
using a disubstituted formamide and phosphoryl chloride or
phosgene, is particularly useful when using nucleophilic
aromatic substrates. Although there is abundant evidence
which shows that phosphoryl chloride reacts with N,N-dimethyl-
formamide to give a phosphorus-free electrophile (G. J. Martin
and S. Poignant, J. chem. Soc. Perkin II, 1972, 1964) reactions
using N-N-dimethylthioformamide show markedly different
reactivities - as judged by the yields of the products
obtained (J. G. Dingwall, D. H. Reid, and K. Wade, J. Chem.
Soc. (C), 1969, 913). In the writer's opinion the mechanism
proposed by Zollinger (H. H. Bosshard and H. Zollinger,
Helv., 1959, 42, 1659), in which the effective electrophile
is *not* a phosphorus-free reagent, has *not* been convincingly
disproved.

Formylation of arenes can also be achieved using dichloro-
methyl methyl ether in CH_2Cl_2 in the presence of a Lewis
acid such as $TiCl_4$. Thus mesitaldehyde is obtained in
ca. 90% yield from mesitylene (A. Rieche, H. Gross, and
E. Höft, Org. Synth., 1967, 47, 1).

(d) *Group V Electrophiles*

    (i) *Nitration Reactions*

Studies of these important reactions (K. Schofield,
'Aromatic Nitration', Cambridge University Press, 1980) have
made major contributions to our understanding of the
mechanistic detail of electrophilic addition-with-elimination
reactions of arenes. Indeed some controversy has existed

recently concerning the relative importance of σ-complexes
(Wheland intermediates) in these reactions (G. A. Olah,
Acc. chem. Res., 1971, 4, 240; P. Rys, P. Skrabal, and
H. Zollinger, Angew. Chem. internat. Edn., 1972, 11, 874).
A number of other important reviews have been published
(J. H. Ridd, Acc. chem. Res., 1971, 4, 248; G. A. Olah in
'Industrial and Laboratory Nitrations', A.C.S. Symposium
Series, No. 22, ed. L. F. Albright and C. Hanson, American
Chemical Society, Washington, 1976; L. M. Stock, Progr.
phys. Org. Chem., 1976, 12, 21).

The involvement of the nitronium ion (NO_2^+) has been
established for some time. The nitration of aromatic
compounds that have reactivities similar to that of benzene
still proceed in the presence of water at levels such that
the nitronium ion cannot be detected. That these reactions
also involve the nitronium ion is indicated by the identity
of the rate of nitration and the rate of ^{18}O exchange
between the medium and the nitric acid.

Nitrations have been used to establish the relative
reactivities of a large number of arenes, and partial rate
factors (the statistically corrected positional relative
rates) are available from relative rate data. These data
($k_{C_6H_5R}/k_{C_6H_6}$) are obtained either from kinetic studies or
from competition reactions, using isomer distributions found
under the same experimental conditions. In a practical
sense the use of preformed nitronium salts leads to a number
of advantages. In conventional nitrations using mixed acids
the water formed during the reaction dilutes the acid and
hence reduces its efficacy. In addition the strongly
oxidising nature of the sulphuric acid/nitric acid mixture
limits its applicability.

The nitration of benzene using nitronium fluoroborate
in fluorosulphuric acid leads to the formation of s-trinitro-
benzene (G. A. Olah and H. C. Lin, Synthesis, 1974, 444).
This reaction indicates that even very weakly nucleophilic
aromatic substrates can be nitrated. Regioselectivity is
maintained in the nitration of toluene, anisole, and o-xylene
regardless of the reactivity of the nitrating system as long
as the solvent and more particularly the temperature is kept
constant (G. A. Olah, et al., Proc. nat. Acad. Sci. U.S.A.,
1978, 75, 545). A change in positional selectivity can be
achieved by a correct choice of reaction conditions. Thus

the nitration of toluene in sulpholan at 25° using nitronium fluoroborate gives an isomer distribution $o:m:p$ = 65.4:2.8: 31.8 (G. A. Olah, S. J. Kuhn, and S. H. Flood, J. Amer chem. Soc., 1961, 83, 4571). Using nitronium trifluoromethane-sulphonate in CH_2Cl_2 at -60°, toluene is mono-nitrated efficiently and gives an isomer distribution $o:m:p$ = 62.0: 0.5:37.5 (C. L. Coon, W. G. Blucher, and M. E. Hill, J. org. Chem., 1973, 38, 4243). This latter result clearly reflects the expected kinetic effect.

A number of new nitration systems have been described. The mono-nitration of polymethylbenzenes proceeds efficiently without side-reactions using alkyl nitrates. Methyl nitrate, using BF_3 catalysis, give a 97% yield with durene and a 99% yield using pentamethylbenzene (G. A. Olah and H. C. Lin, Synthesis, 1973, 488). Nafion-H catalysis with butyl nitrate gave a 90% yield using mesitylene and acetone cyanohydrin nitrate (ACN) with benzene gives an 85% yield. Nitration of p-xylene (60%) is achieved using a strongly acidic resin and fuming nitric acid with azeotropic removal of the water formed, (G. A. Olah, R. Malhotra and S. C. Narang, J. org. Chem., 1978, 43, 4628). In this latter reaction it is clear that dinitrogen tetroxide (N_2O_4) formation must be suppressed: toluene and N_2O_4 in the presence of Nafion-H affords some phenylnitromethane. ACN in the presence of boron trifluoride etherate is an efficient nitrating agent, p-xylene gives a 90% yield of the mono-nitrocompound (S. C. Narang and M. J. Thompson, Austral. J. Chem., 1978, 31, 1839). The efficacy of ACN may well be related to the presence of an excellent leaving group.

A similar feature is also observed in two other types of system that have been described recently. These involve, for example, dimethylnitrosulphonium (G. A. Olah, *et al*., Proc. nat. Acad. Sci. U.S.A., 1978, 75, 1045; G. A. Olah,

B. G. B. Gupta, and S. C. Narang, J. Amer. chem. Soc., 1969, 101, 5317) and N-nitropyridinium- and N-nitroquinolinium- ions (G. A. Olah, et al., ibid., 1980, 102, 3507).

The results of studies of the reaction of benzene-toluene mixtures with preformed nitronium salts, and similar experiments using m-xylene and mesitylene, have raised a number of mechanistic queries. The meta partial rate factor for toluene was apparently lower than for an individual position in benzene! This is a consequence of the apparent relative rate values being closer to unity than is the case using for example nitric acid in acetic acid. It has been pointed out that reactivities obtained from competition experiments can be invalid if significant reaction occurs before the reagents are adequately mixed. This question has been examined by studying the nitration of bibenzyl. The nitration of bibenzyl using nitric acid in acetic anhydride gives ca. 12% of dinitrobibenzyls. This clearly indicates that the presence of the first nitro-group deactivates the other ring towards the introduction of the second nitro-group. When the nitration of bibenzyl in sulpholan is studied using nitronium fluoroborate, using a number of different concentrations and mixing conditions, the amount of disubstituted product always considerably exceeds 25% - the amount predicted if there is no transmission of the substituent effect between the two rings. Mixing rates evidently do have a major influence.

Kinetic studies of the nitration of arenes that are more reactive than benzene have also produced interesting results. Although mesitylene might be expected to be more reactive than benzene by a factor of ca. 1.6×10^4, in fact the majority of studies show that a limiting factor of about 40 is observed for polymethylbenzenes. An exception to this generalisation is observed in competition reactions using N-nitro-2,4,6-collidinium fluoroborate at 25°. Values for k_{ArH}/k_{PhH} include 4.4×10^2 for m-xylene and 1.2×10^3 for anisole (G. A. Olah, et al., J. Amer. chem. Soc., 1980, 102, 3507). In the cases which show a limiting rate factor it has been suggested that the rate determining step involves the formation of an encounter complex.

Additional apparent anomalies obtained in studies using polymethylbenzenes have further complicated our understanding of the nitration reactions. In particular, nitrations carried

out with nitric acid in acetic anhydride were known to afford
acetoxylation products and only serve to confuse the issue
by suggesting that electrophilic acetoxylation occurred.
This is now known not to be the case. Nitric acid reacts
with acetic anhydride to afford acetyl nitrate which can
fragment to afford the nitronium and the acetate ions
(N. C. Marziano, *et al.*, J. chem. Soc. Perkin II, 1977, 1361).
The nitration of *p*-xylene gives a mixture of 2,5-dimethyl-
nitrobenzene and 2,5-dimethylphenyl acetate. The isolation
of adducts in which acetyl nitrate has reacted at positions
already substituted (*ipso* attack) and the fact that these
adducts decompose apparently by an intramolecular shift,
supports the mechanism shown in equation [15]. Much of the
early work on *ipso*-nitration and the isolation of *ipso*-adducts
has been reviewed (R. B. Moodie and K. Schofield, Acc. chem.
Res., 1976, 9, 287; S. R. Hartshorn, Chem. Soc. Rev., 1974,
3, 167).

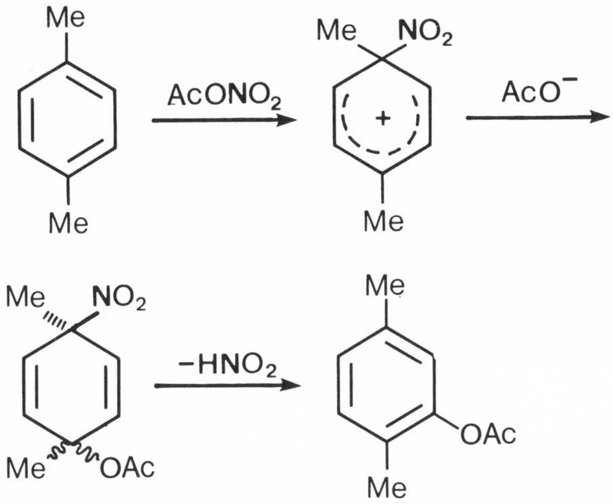

[15]

The solvolytic behaviour of, for example, 4-nitro-3,4,5-
trimethylcyclohexa-2,5-dienyl acetate has been reported
(T. Banwell, *et al.*, J. Amer. chem. Soc., 1977, 99, 3042).
The nitration of *o*-xylene in sulphuric acid gives mixtures
of 2,3-dimethyl- and 3,4-dimethyl- nitrobenzenes in yields
which depend upon the acidity of the system. Thus nitration
ortho- to an alkyl group may well involve a significant
amount of initial *ipso* attack. The solvolysis of 1,2-dimethyl-

4-hydroxy-1-nitrocyclohexa-2,5-diene is known to give
1,2-dimethyl-3-nitrobenzene. 3,5-[^{2}H$_2$]1,2-dimethyl-4-hydroxy-
1-nitrocyclohexa-2,5-diene in 85 wt % H$_2$SO$_4$ generates the
ipso ion, 3,5-[^{2}H$_2$]1,2-dimethyl-1-nitrocyclohexadienylium,
which rearranges to afford, after loss of a proton, a 1:1
mixture of 4,6-[^{2}H$_2$]1,2-dimethyl-3-nitrobenzene and
5-[^{2}H]1,2-dimethyl-3-nitrobenzene. With the assumption that
the isotopomers result from steady state intermediates, and
that secondary isotope effects are negligible, the results
indicate that the rearrangement to the equivalent *ipso*-
position occurs at *ca.* 50 times the rate of migration to an
adjacent site bearing hydrogen (C. E. Barnes and P. C. Myhre,
J. Amer. chem. Soc., 1978, 100, 974). The solvolysis of this
type of nitrodienol in aqueous ethanol containing urea gives
the corresponding phenol. The corresponding acetates
solvolyse more slowly and importantly, whereas intramolecular
hydrogen migration (the N.I.H. shift) is observed using the
nitrodienols, none is observed using the dienyl acetates
(K. S. Feldman, A. McDermott, and P. C. Myhre, *ibid.*, 1979,
101, 505). The acidolysis shown in equation [16] affords
the mixture of constitutional isomers indicated (R. C. Hahn
and M. B. Groen, *ibid.*, 1973, 95, 6128).

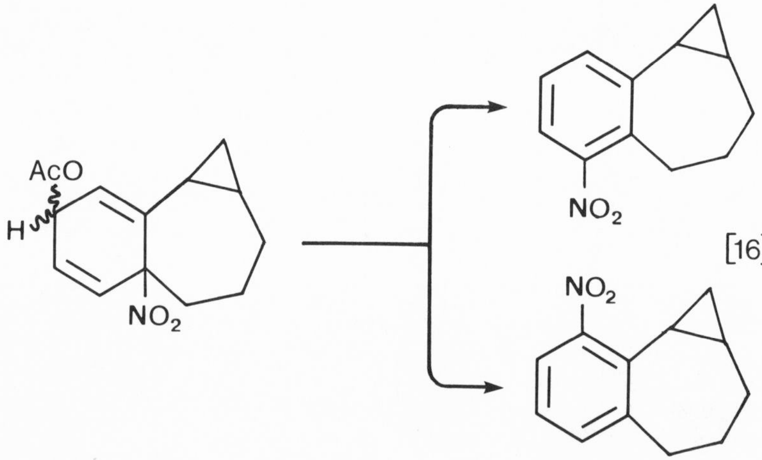

[16]

Three modes of migration of a nitro-group in an *ipso*
arenium ion have been considered. In intramolecular
migrations the nitro-group never becomes sufficiently free

from the carbon skeleton to achieve more than a 1,2-shift. No well authenticated examples of successive intramolecular 1,2-shifts across an unsubstituted position are known although a series of 1,2-shifts can be envisaged as exemplified by equation [16].

In some rearrangements the nitro-group becomes free enough to be able to distinguish and select from among the available positions on the arene. This type of rearrangement can either be intermolecular — in which case the nitronium ion diffuses into the solvent and is thus available for interaction with a competing substrate — or it can be extramolecular. An extramolecular rearrangement is defined as one where the nitronium ion becomes sufficiently free from the arene for σ-basicity effects to become important but where it does not leave the encounter complex from which it was originally formed.

The rearrangements which occur during the aromatisation reactions of 4-nitrocyclohexa-2,5-dienones have been investigated by means of double labelling experiments using ^{15}N and ^{2}H. The results indicate a significant inter-molecular component in a radical dissociation-recombination reaction. The dissociation step is rate determining (C. E. Barnes and P. C. Myhre, J. Amer. chem. Soc., 1978, 100, 973).

The direct observation of an *ipso*-Wheland intermediate under normal nitration conditions has been reported using ^{1}H and ^{13}C n.m.r. spectroscopy with *N,N*-dimethyl-*p*-toluidine (K. Fukiwara, J. C. Giffney and J. H. Ridd, Chem. Comm., 1977, 301).

Nitration *ipso* to an ethyl group occurs in *p*-diethyl-benzene, *p*-ethyltoluene, and almost exclusively in ethyl-mesitylene (A. H. Clemens, *et al.*, Austral. J. Chem., 1977, 30, 103). *Ipso* attack also occurs in a number of reactions carried out using nitric acid in aqueous trifluoroacetic acid (R. B. Moodie, K. Schofield, and G. D. Tobin, J. chem. Soc. Perkin II, 1977, 1688). It is noteworthy that the rate of nitration of benzene, chlorobenzene, toluene, the xylenes, and the trimethylbenzenes, using this system, is not dependent on the aromatic substrate.

Ipso-nitration adducts have been used as sources of

carbon electrophiles that allow the regiospecific synthesis of a wide range of products including the introduction of the following functions: methoxy-, amino-, nitro-, and cyano (K. S. Feldman and P. C. Myhre, J. Amer. chem. Soc., 1979, 101, 4768).

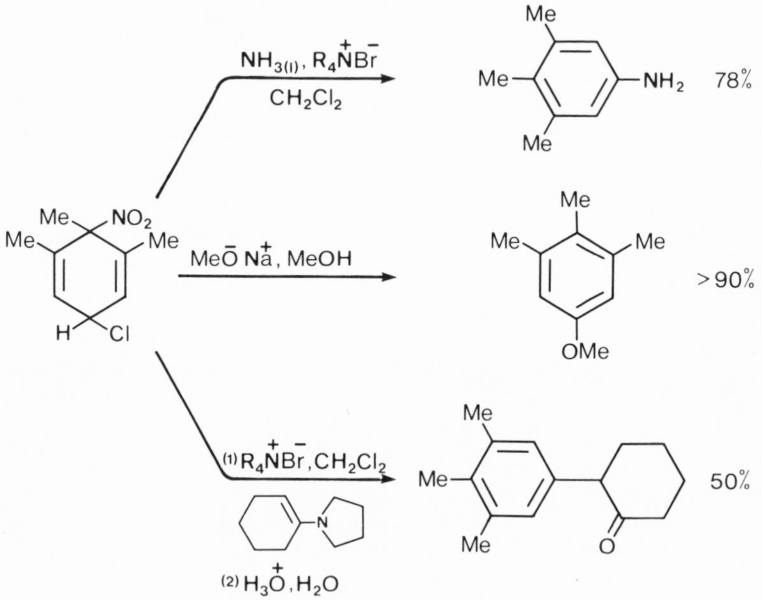

Many reactions are known in which a substituent is replaced by a nitro-group, but little is known about the mechanisms of many of them. The reactions include dealkylation (particularly with secondary or tertiary alkyl groups), deacylation, desulphonation, decarboxylation, dediazoniation, dehalogenation, and desilylation. The conversion of 1,4-bis-trimethylsilylbenzene into 4-nitro-trimethylsilylbenzene, using nitric acid in acetic anhydride, is synthetically useful (C. Eaborn, J. organometal. Chem., 1978, 144, 271).

Nitro-dechlorination has rarely been observed. On the other hand nitro-debromination is more common, although even

in these reactions some may involve initial nitroso-
debromination. In the 'absence' of nitrous acid, i.e. using
nitric acid in acetic anhydride containing urea, p-bromo-
anisole gave p-nitroanisole in 31% yield. The nitration of
2,6-dimethyl-4-iodoanisole has been shown to proceed by
nitroso-deiodination followed by oxidation; no displacement
of iodine occurs in the absence of nitrous acid (A. R. Butler
and A. P. Sanderson, J. chem. Soc. Perkin II, 1974, 1784).
The Reverdin rearrangement, in which 4-iodoanisole gives
2-iodo-4-nitroanisole, is a more complex example of the same
type (A. R. Butler and A. P. Sanderson, J. chem. Soc.
Perkin II, 1972, 99). Once again, no displacement of iodine
occurs in the absence of nitrous acid.

The side-chain nitration of polymethylarenes may well
involve the intermediacy of *ipso* adducts. The nitration of
1,4-dimethylnaphthalene, which can give 1-methyl-4-nitromethyl-
naphthalene has been interpreted in this way (R. Robinson,
J. chem. Soc. B, 1970, 1289). One of the products formed
when 1,4-dimethyl-4-nitro-2,5-cyclohexadienyl acetate is
solvolysed in acetic acid — acetic anhydride is p-tolylnitro-
methane (J. N. Ramsay, Canad. J. chem., 1974, 52, 3960).
The side reactions which can accompany aromatic nitrations
have been reviewed (H. Suzuki, Synthesis, 1977, 217).

(ii) *Nitrosation Reactions*

Nitrosation normally requires particularly nucleophilic
aromatic substrates because the nitrosonium ion is much more
stable, and hence much less reactive, than the nitronium ion.
Kinetic studies have shown that the nitrosonium ion is at
least 10^{14} times less reactive than the nitronium ion
(B. C. Challis and A. J. Lawson, J. chem. Soc. Perkin II,
1972, 1831; B. C. Challis and R. J. Higgins, *ibid.*, p 2365).
The nitroso group can be introduced into an arene that is
normally too unreactive by nitroso-destannylation (E. H.
Bartlett, C. Eaborn, and D. R. M. Walton, J. chem. Soc.(C.),
1970, 1717). The increased electron density on the aromatic
carbon to which the metal is attached promotes *ipso* attack
by the electrophile. The trimethylstannyl group is an
excellent leaving group.

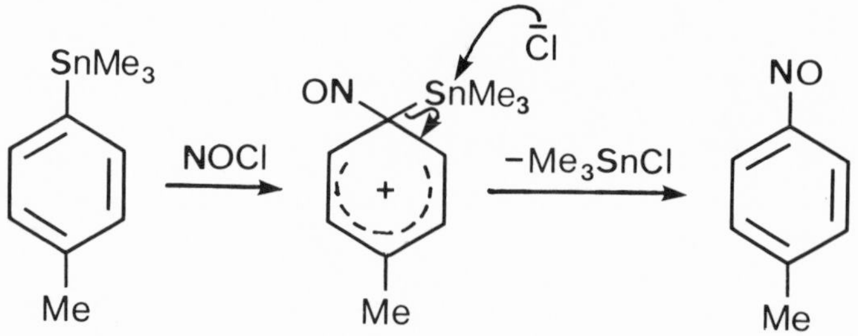

(iii) *Azo-Coupling Reactions*

Arenediazonium ions are too weakly electrophilic to couple with arenes unless the arene has a number of electron releasing alkyl groups (e.g. mesitylene) and the arenediazonium ion has a number of electron withdrawing groups (e.g. 2,4,6-trinitrobenzenediazonium) (H. Zollinger, 'Azo and Diazo Chemistry', Interscience, New York, 1961). The method does have some unexplored potential as a mild method of introducing an amino group into nucleophilic (and hence easily oxidised) aromatic compounds.

(e) *Group VI Electrophiles*

(i) *Hydroxylation Reactions*

The hydroxylation of certain substituted arenes can be achieved using trifluoroperacetic acid in the presence of boron trifluoride (H. Hart, Acc. chem. Res., 1971, $\underline{4}$, 337). The reaction with mesitylene is exothermic at -40° and gives mesitol in almost quantitative yield. Interestingly, 1,2,3,4-tetramethylbenzene and durene both give, in addition to the expected phenols, products derived by 1,2-methyl shifts. Advantage has been taken of these reactions, where *ipso* attack is followed by an alkyl shift, to prepare cyclohexadienones that would be otherwise difficult to prepare (equation [17]).

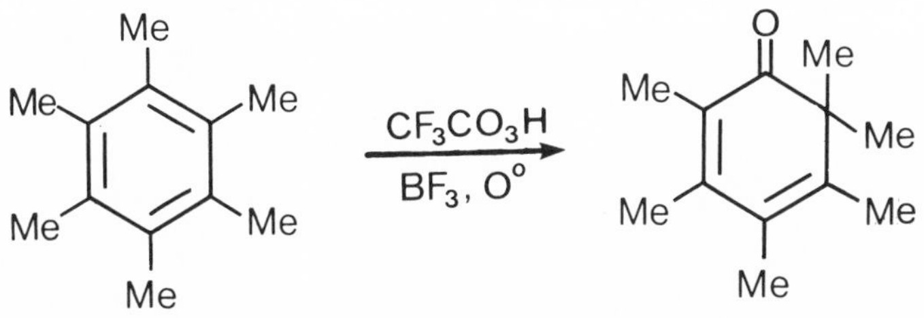

[17]

Reactions of arenes with t-butyl hydroperoxide in the presence of aluminium chloride have been studied. The phenols that are finally isolated are presumed to arise by the cleavage of t-butyl aryl ethers. The electrophile was suspected of being the t-butoxy cation (S. Hashimoto and W. Koike, Bull. chem. Soc. Japan, 1970, 43, 293). This explanation is, however, rather unlikely since the t-butoxy cation would be expected to rearrange rapidly to the dimethyl methoxycarbenium ion (equation [18]).

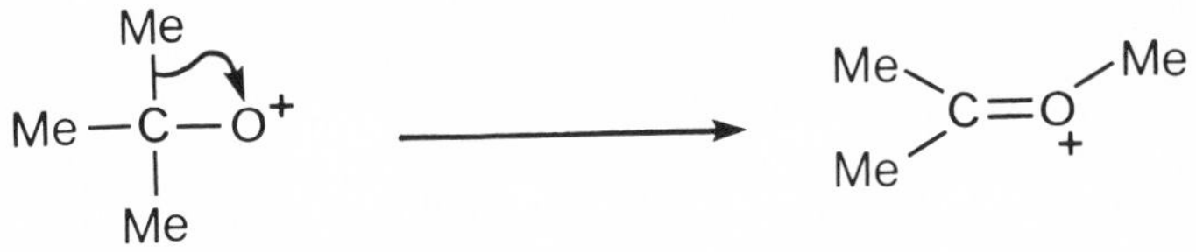

[18]

t-Butyl hydroperoxide has been shown to rearrange rapidly, in the presence of magic acid, to afford the dimethyl methoxycarbenium ion (G. A. Olah, *et al.*, J. Amer. chem. Soc., 1976, 98, 2245.) and no evidence for the presence of the t-butoxy cation was observed in the n.m.r. spectra. In addition it has been shown that ozone reacts rapidly with the trimethylcarbenium ion with the formation of the dimethyl methoxycarbenium ion (G. A. Olah, N. Yoneda, and D. G. Parker, J. Amer. chem. Soc., 1976, 98, 2251).

   The hydroxyl cation has been suggested as the electrophilic species involved in reactions of benzene and simple alkylbenzenes with hydrogen peroxide that are catalysed by hydrogen fluoride. In reactions with mesitylene (equation [19]), mesitol and 2,4,6-trimethylresorcinol are obtained in 74 and 25 mol % respectively (J. A. Veseley and L. Schmerling, J. org. Chem., 1970, 35, 4028). The product ratios obtained in the reactions with toluene are very similar to those obtained using Fenton's reagent (A. J. Davidson and R. O. C. Norman, J. chem. Soc., 1964, 5404) and may reflect radical involvement. Aluminium chloride has also been used as a catalyst in the hydroxylation of arenes with hydrogen peroxide (M. E. Kurz and G. J. Johnson, J. org. Chem., 1971, 36, 3184).

$$\text{Me} \overset{\text{Me}}{\underset{\text{Me}}{\bigcirc}} \quad \xrightarrow{\text{H}_2\text{O}_2,\text{HF}} \quad \text{Me} \overset{\text{OH}}{\underset{\text{Me}}{\bigcirc}} \text{Me} \quad + \quad \text{Me} \overset{\text{OH}}{\underset{\text{Me}}{\bigcirc}} \text{Me} \quad [19]$$

   Arenes such as benzene and alkylbenzenes are also readily hydroxylated with hydrogen peroxide in superacid media (G. A. Olah and R. Ohnishi, J. org. Chem., 1978, 43, 865). The formation of some 2,6-dimethylphenol in the reaction using o-xylene is good evidence for cationic hydroxylation.

(ii) *Sulphur Electrophiles*

   The reactions of aromatic substrates with sulphur electrophiles have been much studied over a long period of time. Many reviews exist and only the more recent work will be summarised here (H. Cerfontain, 'Mechanistic Aspects in Aromatic Sulphonation and Desulphonation', Interscience,

New York, 1968; H. Cerfontain and C. W. F. Kort, Internat. J. sulfur. Chem. (C), 1971, $\underline{6}$, 123).

In fuming sulphuric acid at concentrations up to about 104% sulphuric acid, the reagent is thought to be $H_3S_2O_7^+$, while at higher concentrations (i.e. more SO_3) it is $H_2S_4O_{13}$ (A. Koeburg-Telder, H. Cerfontain, J. chem. Soc. Perkin II, 1973, 633, and references cited therein). Unlike most of the other reactions discussed, sulphonation is reversible. Primary kinetic isotope effects have been observed in the sulphonation of $[1,3,5-^2H_3]$benzene and the sulphonation of anthracene at the 9-position, with the dioxan-sulphur trioxide complex at 40° in dioxan, proceeds with a maximal substrate kinetic isotope effect $(k_{1_H}/k_{2_H} = 6.8)$ (A. Koeberg-Telder and H. Cerfontain, Rec. trav. Chim., 1972, $\underline{91}$, 22). The kinetics of sulphonation of benzene, toluene, and p-xylene over a range of sulphuric acid concentrations have been reported (M. Sohrabi, T. Kaghazchi, and C. Hanson. J. appl. Chem. Biotechnol., 1977, $\underline{27}$, 453). Isomer distributions and product stabilities in the sulphonation of polyethyl- and polyisopropyl-benzenes have been reported for reactions carried out in aqueous sulphuric acid (A. Koeberg-Telder and H. Cerfontain, J. chem. Soc. Perkin II, 1977, 717; H. Cerfontain, A. Koeberg-Telder, and C. Ris, *ibid.*, p. 720).

New methodology for sulphonylation has also been reported. The mixed sulphonic anhydrides obtained by reaction of an arenesulphonyl bromide with silver trifluoromethanesulphonate, sulphonylate arenes in high yield (F. Effenberger and K. Huthmacker, Angew. Chem. internat. Edn., 1974, $\underline{13}$, 409). Alkanesulphonyl bromides also afford mixed anhydrides and hence aryl alkylsulphones (K. Huthmacker, G. Konig, and F. Effenberger, Ber., 1975, $\underline{108}$, 2947). The sulphonylation of arenes using trifluoromethanesulphonic anhydride, in the presence of aluminium chloride, affords aryl trifluoromethylsulphones in good yield (J. B. Hendrickson and K. W. Bair, J. org. Chem., 1977, $\underline{42}$, 3875).

(f) *Group VII Electrophiles*

Three general procedures are available for the introduction of halogen into an aromatic compound by means of an electrophilic reagent. In order of increasing reactivity these are (i) molecular halogen, (ii) molecular halogen in the presence of a Lewis acid catalyst, and (iii) positive halogen, usually associated with a carrier, e.g. the hypochlorous acidium ion. Which of these is the method of choice depends on the nucleophilicity of the aromatic substrate. Thus, for example, although bromine reacts with benzene in solvents that are polar or acidic, the reaction is very slow. On the other hand the direct reaction of bromine with aniline leads to s-tribromoaniline. The reaction can be moderated to give a high yield of 4-bromoaniline by heating aniline hydrobromide in dimethylsulphoxide (DMSO) (P. A. Zoretic J. org. Chem., 1975, 40, 1867). A mechanism involving the adduct of bromine with DMSO was suggested. However, the following mechanism may be more plausible, particularly in view of the observed regioselectivity ($o:p$ = 1:12).

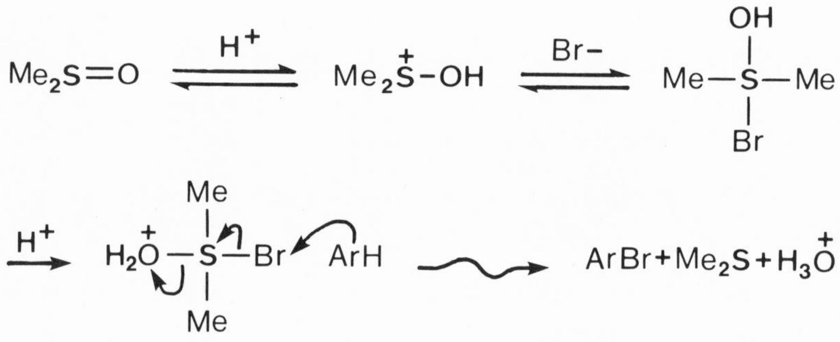

Bromodimethylsulphonium bromide has been prepared (N. Furukawa, T. Inoue, T. Aida, and S. Oae, Chem. Comm., 1973, 212) and has been shown (e.g. G. A. Olah, Y. D. Vankar, and M. Arvanaghi, Tetrahedron Letters, 1979, 3653) to be a convenient source of electophilic bromine. [13]C N.m.r. studies in CH_2Cl_2 have shown that the adduct of tetrahydrothiapyran with bromine is a simple molecular complex (J. B. Lambert, et al., J. Amer. chem. Soc., 1976, 98, 3778).

2,4,4,6-Tetrabromocyclohexa-2,5-dienone is a mild and selective brominating agent for aryl-amines. Aniline, in

CH_2Cl_2 at -20°, is converted into *o*- and *p*-bromoanilines in 87% yield (*o:p* 24:1) (V. Caló, *et al.*, J. chem. Soc. (C), 1971, 3652).

For convenience the following discussion will be separated into three divisions: fluorination., chlorination and bromination , and iodination .

(i) *Fluorination Reactions*

The controlled fluorination of benzene and a number of its derivatives has been achieved using molecular fluorine in acetonitrile at temperatures in the range -15 to -75° (V. Grakauskas, J. org. Chem., 1970, 35, 723). The orientation of the substitutions is that expected for electrophilic fluorination. A radical-cation mechanism has been proposed for the fluorination of benzene and fluoro-benzene using xenon difluoride and a trace of hydrogen chloride (M. J. Shaw, H. H. Hyman, and R. Filler, J. Amer. chem. Soc., 1969, 91, 1563). The fluorination of *o*-xylene using xenon difluoride and hydrogen fluoride gives a mixture of 3- and 4-fluoro-*o*-xylenes (B. Sket and M. Zupan, J. org. Chem., 1978, 43, 835).

The use of fluoroxytrifluoromethane at low temperatures in halogenated solvents allows the direct introduction of fluorine into suitably activated aromatic rings (D. H. R. Barton, *et al.*, J. chem. Soc. Perkin I, 1974, 732). Benzene only gives a 17% yield of fluorobenzene, which is increased to 65% when the reaction mixture is photolysed (J. Kollonitsch, L. Barash, and G. A. Doldouras, J. Amer. chem. Soc., 1970, 92, 7494).

(ii) *Chlorination and Bromination Reactions*

Considerable research effort has been devoted to the development of systems that allow the controlled chlorination and bromination of arenes. Positively charged species have been particularly important targets. The relevant evidence has been reviewed (E. Berliner, J. chem. Ed., 1966, 43, 124). Whereas molecular bromine reacts very slowly with benzene, hypobromous acid can, at sufficiently high acidity, brominate nitrobenzene. The mechanisms available for bromination reactions using molecular bromine have been reviewed (N. H. Briggs, P. B. D. de la Mare, and D. Harl, J. chem. Soc.

Perkin II, 1977, 106).

Bromination using bromine and an appropriate silver (I) salt in an acid involves a positively charged species since aromatic compounds that are normally unreactive are attacked. It is not clear whether the electrophile is Br^+ or a co-ordinated species, such as $AgBr_2^+$ or $\overset{+}{B}rSO_3$. Reactions involving 'positive chlorine' are similar to those involving 'positive bromine'. However, 'positive chlorine' is thermo-dynamically less favourable. The steric requirements of 'positive chlorine' are relatively small (P. B. D. de la Mare and L. Main, J. chem. Soc. (B), 1971, 90).

N-Halo-imides have found considerable use as halogenating agents in the presence of acids. Dibromo-isocyanuric acid is a very powerful reagent: nitrobenzene is converted rapidly into pentabromonitrobenzene in oleum at room temperature (W. G. Gottardi, Monatsh., 1968, 99, 815; 1969, 100, 42).

The importance of *ipso* attack has been exemplified in a number of the earlier sections. This type of reaction is also met in chlorinations and brominations. Thus the reaction of bromine and silver (I) salts with 1,3,5-tri-t-butylbenzene in acetic acid results in products, including 3,5-di-t-butyl-bromobenzene and 3,5-di-t-butylphenyl acetate, that clearly implicate the intermediacy of *ipso* adducts (P. C. Myhre, G. S. Owen, and L. L. James, J. Amer. chem. Soc., 1968, 90, 2115). On the other hand the reaction of a solution of bromine in trimethyl phosphate with 1,3,5-tri-t-butylbenzene gives 2,4,6-tri-t-butylbromobenzene in an isolated yield of 59% (D. E. Pearson, *et al.*, Synthesis, 1976, 621). The solution of bromine in trimethyl phosphate is stable if it is protected from light. The reaction is rapid, even at 25°, and although no hydrogen bromide is evolved, methyl bromide is apparently formed. The mechanism of this reaction is far from clear but may involve the sequence shown below:

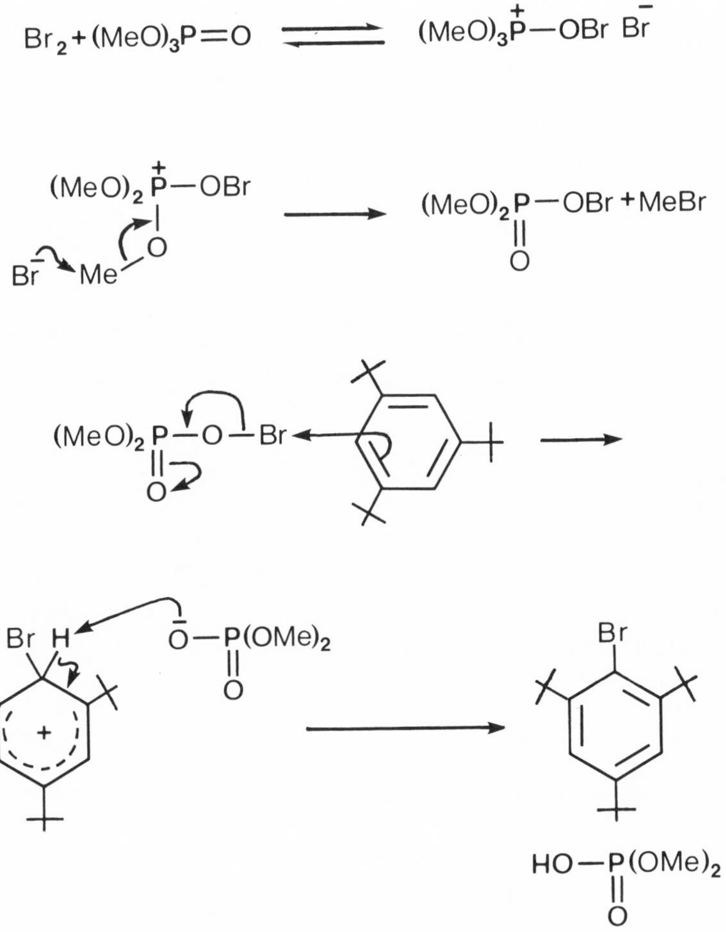

$$Br_2 + (MeO)_3P = O \rightleftharpoons (MeO)_3\overset{+}{P} - OBr \ \overset{-}{Br}$$

(iii) *Iodination Reactions*

It is only with particularly nucleophilic aromatic compounds that iodination can be achieved using elemental iodine. Normally an oxidising agent is used to generate a positively charged iodinating agent. Thus benzene and iodine give a good yield of iodobenzene in the presence of nitric acid. Silver trifluoromethanesulphonate and iodine in benzene

give iodobenzene quantitatively at room temperature
(Y. Kobayashi, I. Kumadaki, and T. Yoshida, J. chem. Res. (S),
1977, 215).

There is no evidence that protodeiodination is a rapid
reaction other than in those cases, such as the iodophenols,
where the aromatic compound is very nucleophilic. Even nitro-
dehalogenation in p-haloanisoles, where the attacking electrophile is stronger than a proton, proceeds more slowly than
nitrodeprotonation. The relative rates using nitric acid in
acetic anhydride were found to be 0.18 (I), 0.079 (Br), 0.061
(Cl), and 1.0 (H) (C. L. Perrin and G. A. Skinner, J. Amer.
chem. Soc., 1971, 93, 3389). The frequently expressed opinion,
that an oxidising agent is added to the iodine in order to
oxidise the hydrogen iodide and hence preclude the reverse
reaction, is evidently incorrect (A. R. Butler, J. chem. Ed.,
1971, 48, 508).

5.   *Additions to Arenes*

   (i)   *The Reduction of Arenes and Functional Groups*

   The reduction of arenes proceeds in the presence of a
wide range of 'homogeneous' catalysts (B. R. James,
'Homogeneous Hydrogenation', Wiley, New York, 1973). High
stereoselectivity has been reported in the reduction of both
o- and m-xylene to the corresponding cis-dimethylcyclohexanes
(E. L. Muetterties and F. J. Hirsekorn, J. Amer. chem. Soc.,
1974, 96, 4063; F. J. Hirsekorn, M. C. Rakowski, and
E. L. Muetterties, *ibid.*, 1975, 97, 237; E. L. Muetterties
and J. R. Blecke, Acc. chem. Res., 1979, 12, 324).

   Reviews of the Birch reduction continue to appear and
indicate the continuing importance of this method (A. J. Birch
and G. Subba Rao, Adv. org. Chem., 1972, 8, 1; R. G. Harvey,
Synthesis, 1970, 161; E. M. Kaiser, *ibid.*, 1972, 391). The
reduction of 1-cyanobenzocyclobutene, first using lithium in
ammonia in the presence of propan-2-ol, and then with the same
reagents together with THF gives 3,6-dihydrobenzocyclobutene
in 80% yield (P. Radlick and L. R. Brown, J. org. Chem., 1973,
88, 3412). The reduction of t-butylbenzene with lithium
dissolved in a mixture of ethylenediamine and morpholine
affords ca. 84% of 1-t-butylcyclohexene (R. A. Benkeser,
R. K. Agnihotri, and T. J. Hoogeboom, Synthesis, 1971, 147).

An interesting variation that gives cyclohexa-1,4-diene in ca. 55% yield from benzene involves the use of lithium in THF in the presence of trimethylchlorosilane. The intermediate 1,4-bis-(trimethylsilyl)cyclohexa-2,5-diene is then allowed to react with potassium hydroxide at 0 — 5° (J. Dunoguès, R. Calas, and N. Ardoin, J. organometal. Chem., 1972, 43, 127).

The reduction of functional groups or their reductive removal has also continued to attract attention in recent years. p-Alkylbenzaldehydes are reduced in liquid ammonia and THF by means of lithium. In one procedure t-butanol is added. The rapid reactions are quenched by the addition of an excess of ammonium chloride. p-t-Butylbenzaldehyde gives p-t-butyltoluene in high yield (S. S. Hall, A. P. Bartels, and A. M. Engman, J. org. Chem., 1972, 37, 670; S. S. Hall, ibid., 1973, 38, 1738). Phenyl cyclopropylketone is reduced to benzylcyclopropane (S. S. Hall and C. -K. Sha, Chem. and Ind., 1976, 216). Benzaldehyde is reduced to toluene in 84% yield using polymethylhydrosilane in ethanol in the presence of Pd-C (J. Lipowitz and S. A. Bowman, J. org. Chem., 1973, 38, 162).

The selective reduction of polycarboxylic acids can be achieved using trichlorosilane together with a tertiary amine. Aromatic carboxylic acid groups are reduced to methyl groups whereas esters are unaffected (R. A. Benkeser and D. F. Ehler, J. org. Chem., 1973, 38, 3660).

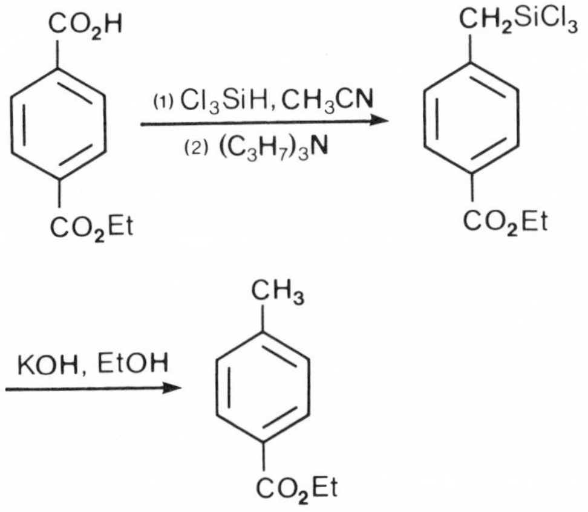

The reductive methylation of aromatic carbonyl compounds can be achieved by treatment with methyl-lithium in ether followed sequentially by the addition of lithium wire and ammonium chloride (S. S. Hall and S. D. Lipsky, Chem. Comm., 1971, 1242). Catalytic transfer reduction of carbonyl compounds using cyclohexene or limonene in the presence of 10% Pd-C and a Lewis acid gives excellent yields of the hydrocarbon. Alcohols are intermediates (G. Brieger and T. -H. Fu, Chem. Comm., 1976, 757).

The reductive removal of a phenolic hydroxy-group has continued to be a desirable objective. The hydrogenolysis of 1-phenyl-5-(4-acetylaminophenoxy)-tetrazole is reported to give acetanilide in good yield (K. Kratzl and F. W. Vierhapper, Monatsh., 1971, 102, 224). The hydrogenolysis of arylcarbamates (J. D. Weaver, E. J. Eisenbraun, and L. E. Harris, Chem. and Ind., 1973, 187), O-aryl-isoureas (E. Vowinkel and C. Wolff, Ber., 1974, 107, 907; 1739; E. Vowinkel and H. J. Baese, *ibid.*, 1974, 107, 1213), aryl methanesulphonates (K. Clauss and H. Jensen, Angew. Chem. internat. Edn., 1973, 12, 918), and 2,4,6-tri(aryloxy)-s-triazines (A. W. van Muijlwijk, A. P. G. Kieboom, and H. van Bekkum, Rec. trav. Chim., 1974, 93, 204) all give the corresponding arenes in good yield.

The denitration of nitro polyhaloarenes can be achieved using sodium borohydride in DMSO. For example, penta-chloronitrobenzene gave pentachlorobenzene in 93% yield (D. W. Lamson, P. Ulrich, and R. O. Hutchins, J. org. Chem., 1973, 38, 2928).

The removal of halogen, particularly bromine and chlorine, has been much studied. The hydrogenolysis of aryl halides has been reported using a 0.2 M solution of potassium hydroxide in ethanol at room temperature and hydrogen (1 atmosphere) in the presence of Raney nickel (A. J. de Koning, Org. prep. Proc. Internat., 1975, 7, 31). Sodium borohydride, in the presence of palladium chloride, has been used to remove bromine and chlorine. Bromine is removed more easily than chlorine and the use of sodium borodeuteride gives a site specific method for deuteriation (T. R. Bosin, M. G. Raymond, and A. R. Buckpitt, Tetrahedron Letters, 1973, 4699; and references cited therein). The addition of potassium tri-s-butylborohydride to copper (I) iodide produces a hydride species, presumed to be $(KCuH_2)_n$ that

reduces aryl halides to the arene (T. Yoshida and
E. -I. Negishi, Chem. Comm., 1974, 762).

Isopropylmagnesium halides remove bromine in the
presence of manganese (II) chloride; p-bromoanisole gave
anisole in 84% yield (G. Cahiez, A. Masuda, and J. F. Normant,
J. Organometal. Chem., 1976, 113, 99; 107). Isopropylmagnesium bromide reduces aryl bromides in the presence of
η^5-dicyclopentadienyltitanium dichloride. Chlorides are not
affected. Thus, p-bromo-chlorobenzene gives an almost
quantitative yield of chlorobenzene. The authors consider
that Cp_2Ti-H is involved (E. Colomer and R. Corriu,
J. Organometal. Chem., 1974, 82, 367). The complete removal
of four chlorine atoms from a range of tetrachlorobenzobarrelene derivatives can be achieved in high yield using
sodium wire in THF containing t-butanol (N. J. Hales,
H. Heaney, J. H. Hollinshead, and P. Singh, Org. Synth.,
1980, 59, 71).

(ii) *Thermal Cycloaddition Reactions*

The Diels-Alder reaction (H. Wollweber, 'Diels-Alder
Reaktion', Thieme, Stuttgart, 1972) only proceeds with monocyclic arenes using very electrophilic dienophiles. Even
then, it is exceptional for good yields to be obtained with
benzene. Electron releasing substituents on the arene
increase the electron density in the ring and lead to an
increase in yield under standardised conditions. Hexafluoro-
but-2-yne gives a cycloadduct with benzene (at 180°) in 8%
yield (R. S. H. Liu, J. Amer. chem. Soc., 1968, 90, 215) but
at 200° with toluene and p-xylene (equation [20]) the yields
are 21% and 57%, respectively.

A reaction with durene affords a 41% yield of a single isomer, the thermodynamically more stable possibility (C. G. Krespan, B. C. McKusick, and T. L. Cairns, J. Amer. chem. Soc., 1960, 82, 1515). Similar results are obtained in reactions using dicyanoacetylene as the dienophile, but those reactions are strongly accelerated by Lewis acid catalysts (R. C. Cookson and J. Dance, Tetrahedron Letters, 1962, 879; C. D. Weis, J. org. Chem., 1963, 28, 74; E. Ciganek, Tetrahedron Letters, 1967, 3321). The reaction of dicyanoacetylene with p-xylene (equation [21]) gives the two products [2] and [3] in the ratio 2:8 using aluminium chloride and 7:3 (in low yield) in the absence of a catalyst.

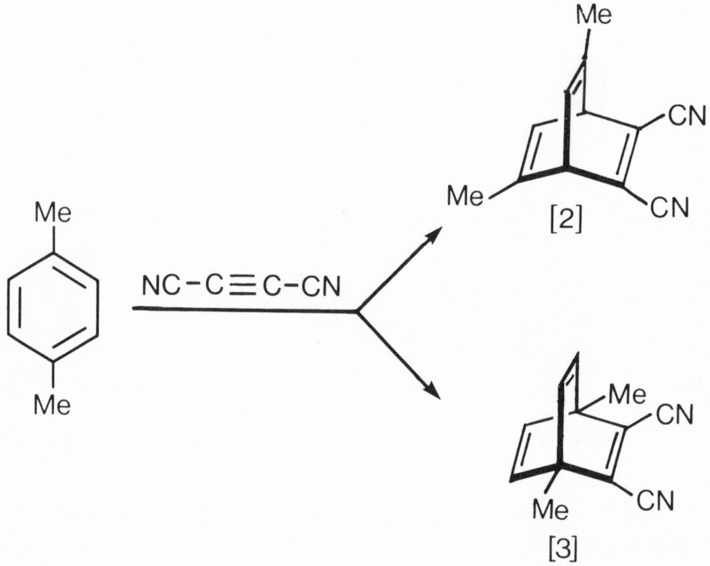

These results suggest that the transition states involved in the non-catalysed reactions are product-like but that the catalysed reactions may proceed in a step-wise sequence.

The reaction of benzyne with benzene is also relatively inefficient and the highest yield (14%) of the 1,4-cycloadduct (benzobarrelene) was obtained by the decomposition of benzene-diazonium-2-carboxylate in benzene at very high dilution (H. E. Zimmerman, et al., J. Amer. chem. Soc., 1976, 98, 7680). However, the electrophilicity of arynes can be

increased in the same manner as has been exemplified above
with acetylene (H. Heaney, Fortschr. chem. Forsch., 1970,
16, 35). Thus, when pentachlorophenyl-lithium is allowed to
decompose in the presence of an excess of benzene a 60-65%
yield of tetrachlorobenzobarrelene is obtained. As is the
case in the reactions of hexafluorobut-2-yne with p-xylene
and durene, the reactions of the tetrahalogenobenzynes
proceed efficiently but no adduct with methyl groups at
bridgehead positions is detected. Arguments can be adduced
in favour of orbital-symmetry-controlled concerted reactions,
in which the transition state is product-like. Proof is, of
course, impossible. In some cases it is certain that the
transition state cannot be entirely symmetrical. The reaction
of tetrafluorobenzyne with t-butylbenzene affords a mixture
of the two possible 1,4-cycloadducts in an almost statistical
ratio. On the other hand the adduct with a t-butyl group at
a bridgehead position shows, in its ^{1}H n.m.r. spectra, good
evidence for steric hindrance to rotation of the t-butyl
group.

The thermal cycloaddition (6% yield) of the 2-methoxyallyl
cation to benzene has been reported (H. M. R. Hoffmann and
A. E. Hill, Angew. Chem. internat. Edn., 1974, 13, 136).

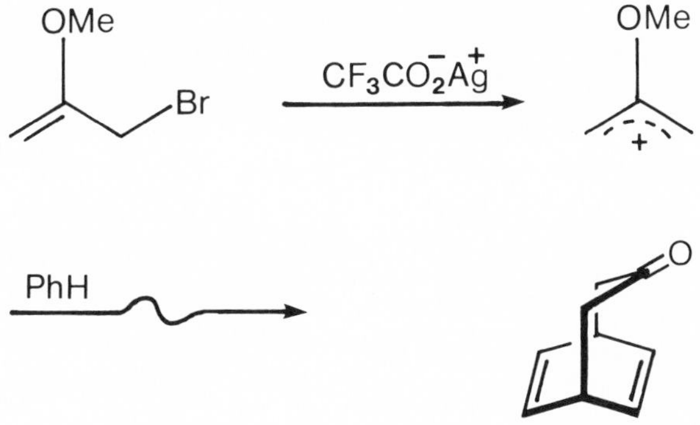

(iii) *Photocycloaddition Reactions*

The photoaddition and photoisomerisation reactions of arenes with a variety of unsaturated substrates have been studied in detail and a number of comprehensive reviews have been published (D. Bryce-Smith, Pure appl. Chem., 1973, 34, 193; D. Bryce-Smith and A. Gilbert, Tetrahedron, 1976, 32, 1309; 1977, 33, 2459.). The photoaddition of maleic anhydride to benzene has now been completely explained.

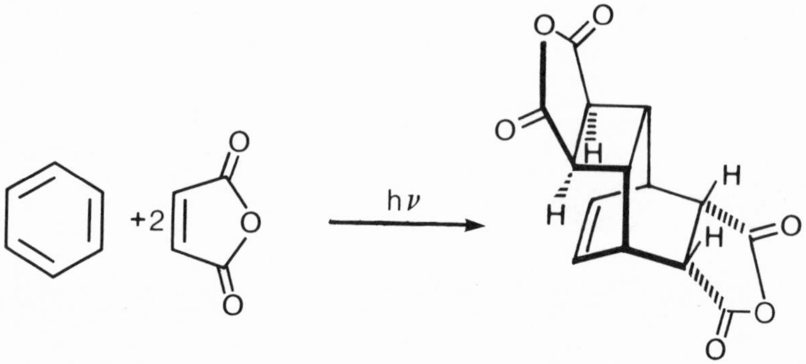

The structure of the product may suggest that the reaction proceeds by two concerted reactions; a photochemical $_\pi2_s + _\pi2_s$ reaction followed by a thermal $_\pi2_s + _\pi4_s$ reaction. However, when the reaction is carried out in the presence of proton donors the reaction is diverted and phenylsuccinic anhydride can be isolated. This result suggests that the first step proceeds from an S_1 complex to a zwitter-ion intermediate.

The photoaddition of n-butylmaleimide to benzene also affords a similar 2:1 adduct, and the addition of tetracyano-ethylene (TCNE) to the initial reaction mixture resulted in the formation of a 1:1:1 adduct (D. Bryce-Smith, Pure appl. Chem., 1968, 16, 47). These reactions are shown on the next page.

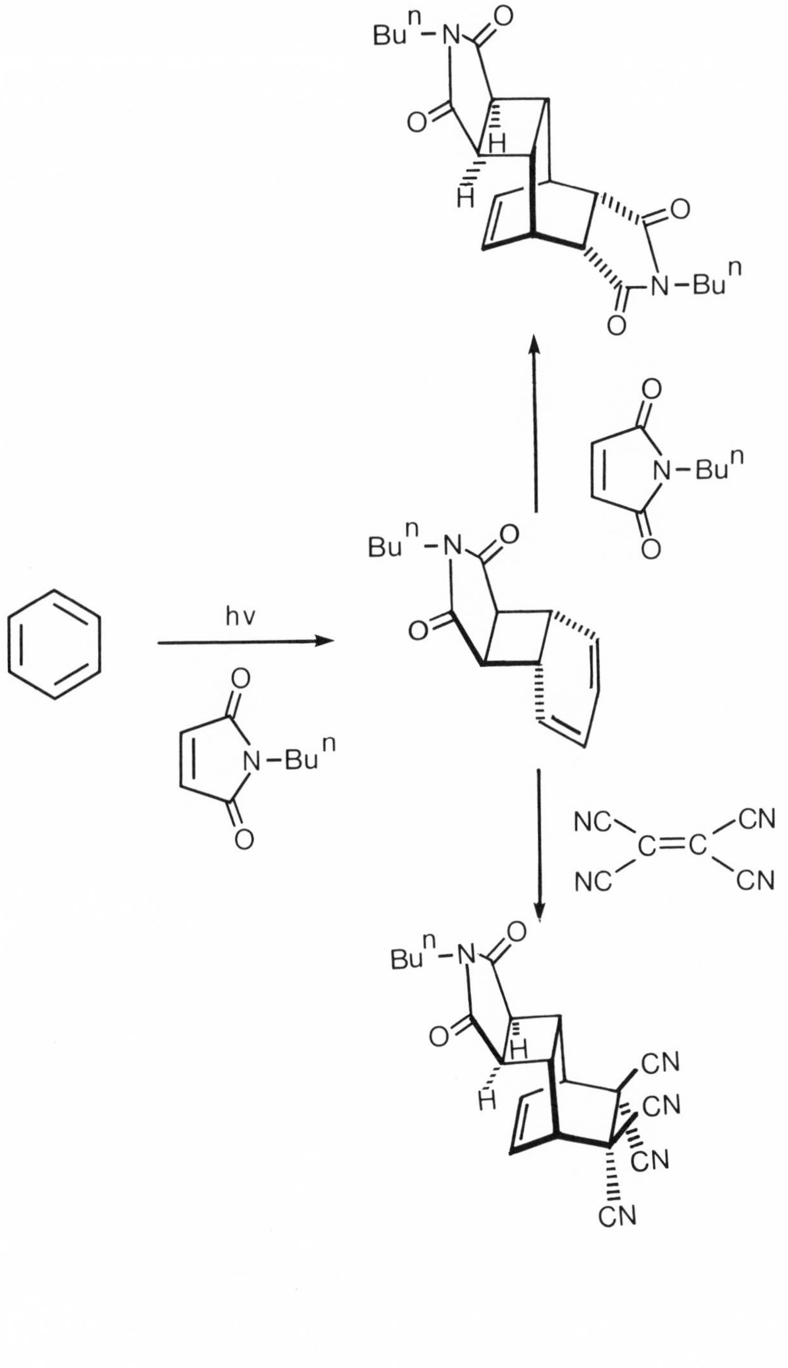
Bu^n–N
O
O
H
H
O
N–Bu^n
O
O
N–Bu^n
O
Bu^n–N
O
O
hv
O
N–Bu^n
O
NC
CN
C=C
NC
CN
Bu^n–N
O
O
H
H
CN
CN
CN
CN

On the other hand, the irradiation of a solution of maleic anhydride in benzene containing TCNE was originally reported not to give a 1:1:1 adduct. This result only served to confuse the interpretation and suggested that a second molecule of maleic anhydride (functioning as a nucleophile) formed a 1,4-cycloadduct with the zwitter-ionic intermediate before the collapse to the final product took place. The postulated bicyclic intermediate has been prepared by an alternative route and has been shown to be remarkably unreactive towards TCNE. Presumably TCNE, being a tetra-substituted dienophile, is more subject to steric problems than is maleic anhydride. In addition, the original experiments using maleic anhydride and TCNE in benzene were carried out at concentrations such that the TCNE exists wholly as its charge transfer complex with benzene. This considerably reduces the dienophilic reactivity of TCNE.

When this information was available and the properties of the 1:1:1 adduct were known it was possible to design alternative experiments to test the possible involvement of the zwitter-ion in the second cycloaddition step.

Using more suitable concentrations of maleic anhydride and TCNE in benzene it has been possible to show that the zwitter-ion intermediate does collapse to the predicted bicyclic intermediate which then undergoes the thermal Diels-Alder reaction (W. Hartmann, H. -G. Heine, and L. Schrader, Tetrahedron Letters, 1974, 883; 3101). Conclusive proof was obtained by studying the photoaddition of a solution of maleic anhydride in benzene containing N-phenyl maleimide. N-Phenyl maleimide had been shown previously to be a very inefficient photo-addend towards benzene (D. Bryce-Smith and M. A. Hems, Tetrahedron Letters, 1966, 1895) but it does capture the bicyclic intermediate efficiently (D. Bryce-Smith, R. Deshpande, and A. Gilbert, Tetrahedron Letters, 1975, 1672).

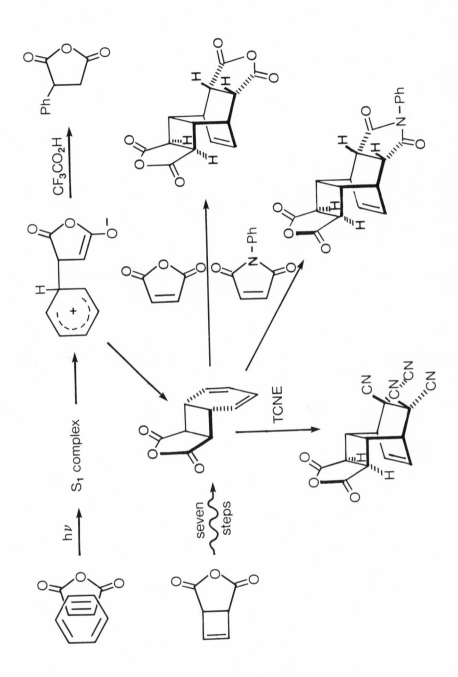

In spite of the above discussion, 1,2-photoadducts have been isolated from reactions of arenes with olefins. Electrophilic olefins, such as acrylonitrile, normally afford the *exo*-adduct whereas nucleophilic olefins such as *cis*-but-2-ene (K. E. Wilzbach and L. Kaplan, J. Amer. chem. Soc., 1971, 93, 2073) and dihydropyran (A. Gilbert and G. N. Taylor, Tetrahedron Letters, 1977, 469) (equation [22]) normally give the *endo*-adduct. Other examples have also been reported (D. Bryce-Smith, A. Gilbert, B. Orger, and H. Tyrrell, Chem. Comm., 1974, 334).

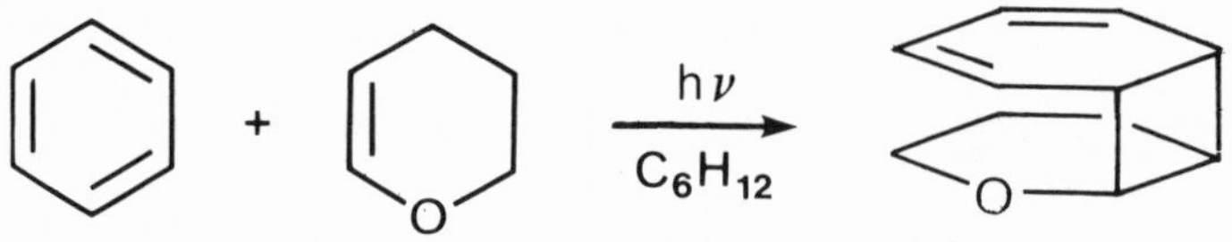

Mixtures of the *endo*- and *exo*-adducts are obtained sometimes, for example, in the photo-cycloadditions of methyl acrylate and methyl methacrylate to benzene (R. J. Atkins, *et al.*, Tetrahedron Letters, 1977, 3597) in which the *exo:endo* ratios were ca. 2:1.

The photoreaction of benzene with dimethyl acetylene-dicarboxylate also proceeds *via* a zwitter-ion intermediate. In this case the 1,2-cycloadduct that is formed by collapse of the zwitter-ion, is captured easily by powerful dienophiles. In the absence of such reagents, valence isomerisation to the cyclo-octatetraene derivative occurs.

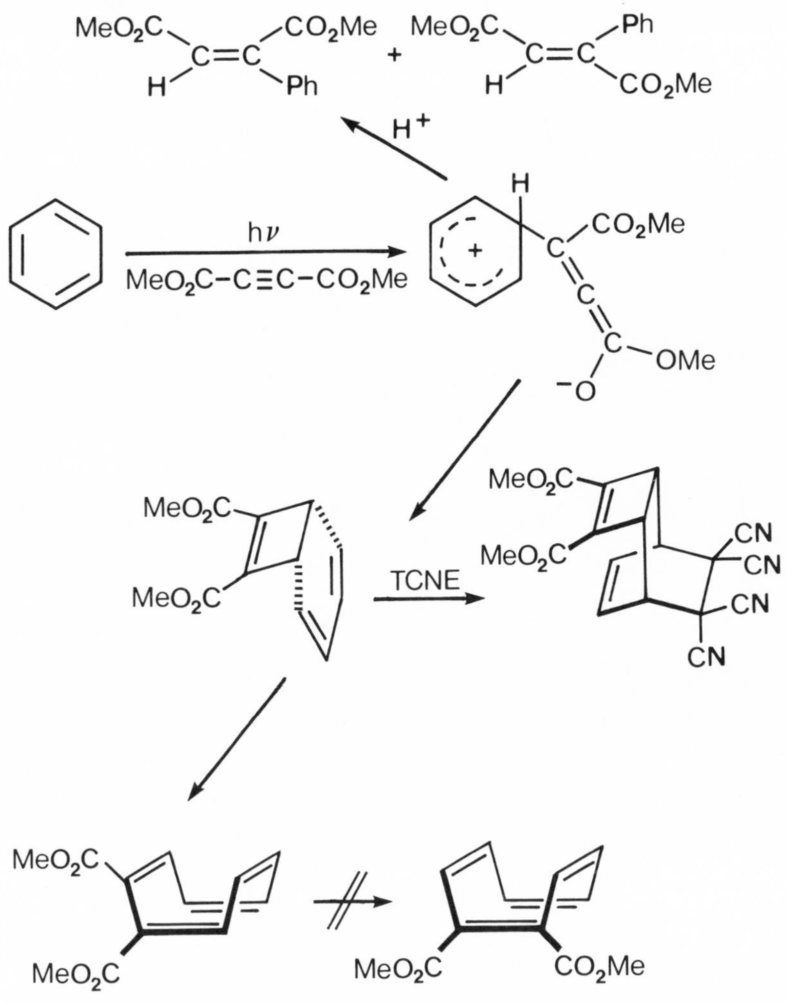

Methods have been published for preparative scale formation of cyclo-octatetraenes (G. L. Grunewald and J. M. Grindel, Org. photochem. Synth., 1976, 2, 20; L. A. Paquette and R. S. Beckley, ibid., p. 45).

The isolation of the initial 1,2-photocycloadduct, a bicyclo [4.2.0] octatriene derivative has been reported in reactions of hexafluorobenzene and a number of derivatives of phenylacetylene (B. Sket and M. Zupan, J. Amer. chem. Soc., 1977, 99, 3504). Using 1-phenyl-3,3-dimethylbut-1-yne the 1,2-adduct has been isolated in 86% yield. The photo-addition of methyl phenylpropiolate to benzene gives a product of another structural type when the irradiation is conducted using a Pyrex filter.

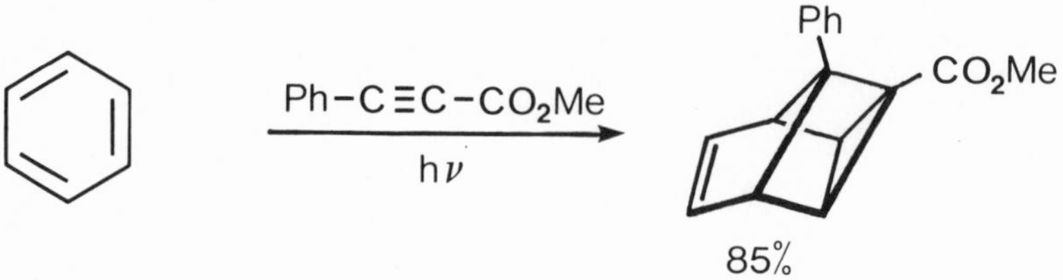

The anticipated cyclo-octatetraene derivative is also isolated and both are interconverted, presumably *via* a bicyclo[4.2.0] intermediate (A. H. A. Tinnemans and D. C. Neckars, J. Amer. chem. Soc., 1977, 99, 6459).

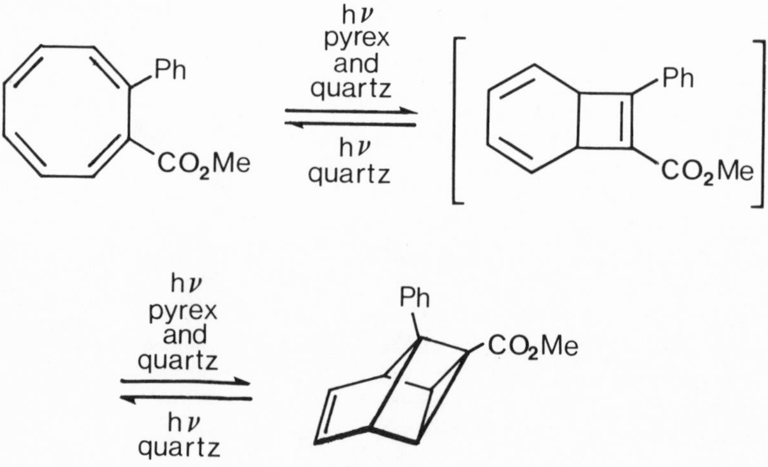

The predominant photochemical process in the reactions of simple olefins with benzene and alkylbenzenes involves the formation of 1,3-cycloadducts (see J. D. Downer in 'Rodd's Chemistry of Carbon Compounds', 2nd Edition, ed. S. Coffey, Elsevier, Amsterdam, 1971, Vol. III A, p. 199). The stereochemistry of the olefin is preserved in the products, for example using *cis-* and *trans-*cyclooctene and *cis-* and *trans-*but-2-ene. On the other hand, the amounts of the *exo-* and *endo-*isomers produced depends on the structure of the olefin. In many cases the *endo* isomer predominates. The reaction looks 'as if' the singlet pre-fulvene (4) is involved but there appears to be no firm evidence on this point. Thus, although reactions involving toluene would be expected to afford sustitution derived from (4a) or (4b) the majority do not.

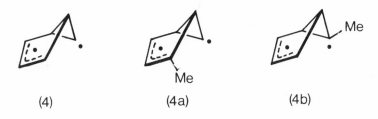

(4)   (4a)   (4b)

In an alternative interpretation an *endo* exciplex has been regarded as product determining (J. Cornelisse, V. Y. Merritt, and R. Srinivasan, J. Amer. chem. Soc., 1973, 95, 6197). The olefin and the singlet excited arene are assumed to form a sandwich structure in which the π-electrons of the olefin are below two atoms in the arene that are in a 1,3-relationship. With small ring olefins such as cyclobutene, maximum overlap is favoured in an *endo* orientation but with increasing ring size this becomes less and less favoured and increasing amounts of the *exo-*adduct are formed. With a mono-substituted arene such as toluene, the formation of an *exo-* adduct has not been detected. Bonding between the olefin and the arene is thought to be concerted and would be followed by a puckering of the six-membered ring and ring closure to form the cyclopropane ring between either of the pairs of atoms indicated by the dotted lines as shown in the equation on the next page. Both possible products that would result from this last step are formed in reactions using *m*-xylene.

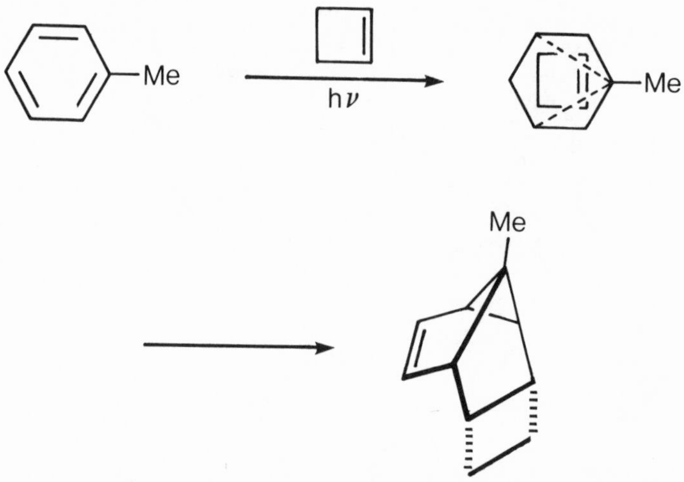

The 1,3-cycloaddition of functionalised olefins such as vinyl acetate (A. Gilbert and M. W. bin Samsudin, Angew. Chem. internat Edn., 1975, 14, 552), ethyl vinyl ether (with anisole) (A. Gilbert and G. Taylor, Chem. Comm., 1977, 242), and vinylene carbonate (H. G. Heine and W. Hartmann, Angew. Chem. internat. Edn., 1975, 14, 698) have also been reported. These examples are shown in the *scheme* on the next page.

A knowledge of the regio- and stero- selectivity of the 1,3-photocycloadditions can be put to good effect in synthesis. Thus the reaction of cyclopentene with 1,2,3,4-tetrahydro-naphthalene gives the expected adduct (equation [23]) which, in the presence of a catalytic amount of hydrogen chloride, gives the [4.3.3] propelladiene shown (C. S. Angadiyavar, *et al.*, Tetrahedron Letters, 1973, 4407).

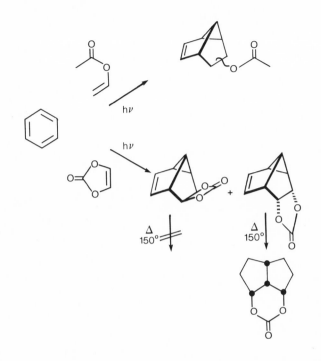

[23]

Originally, it was thought that 1,3-cycloaddition was
the almost exclusive process involved in the photoadditions
of simple olefins with benzene. This is now known not to be
the case. 1,2-Cycloadditions as well as 1,4-cycloaddition
and the photo-ene reaction have been observed. Examples of
the first two alternatives are given in equations [24]
(H. -D. Scharf and J. Mattay, Tetrahedron Letters, 1977, 401)
and [25] (R. Srinivasan, J. Amer. chem. Soc., 1972, 94, 8117).

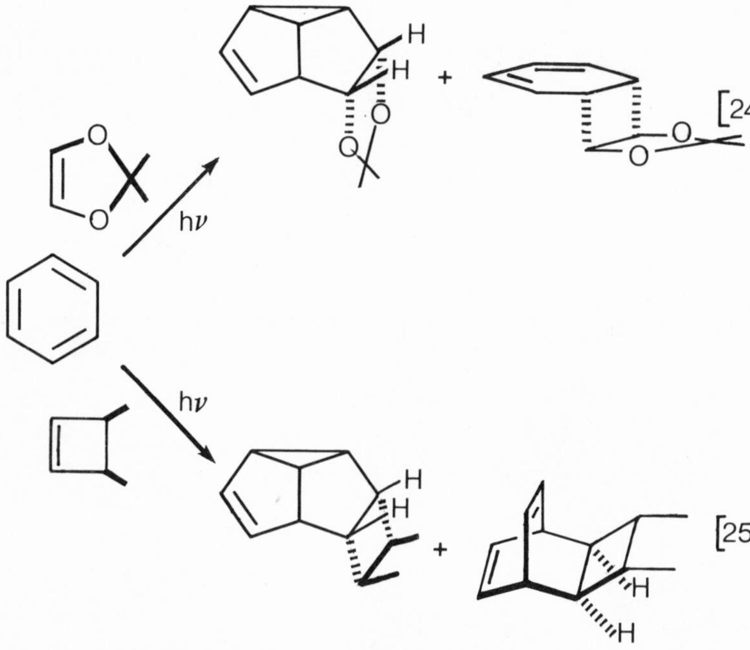

Only in reactions of allenes (equation [26]) is 1,4-photo-
cycloaddition dominant (D. Bryce-Smith, B. E. Foulger and
A. Gilbert, Chem. Comm., 1972, 664).

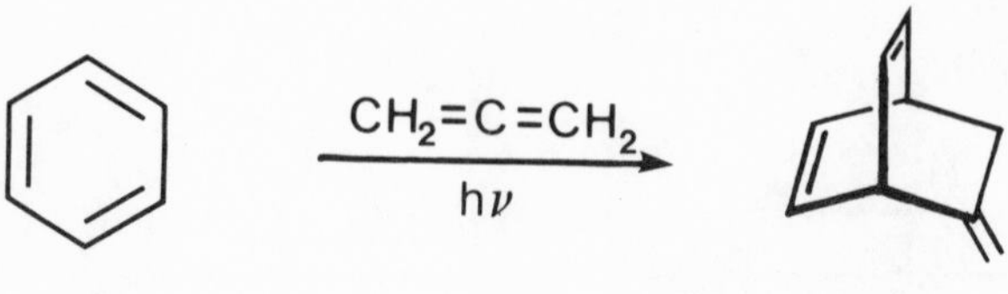

The photoaddition reactions of 2,3-dimethylbut-2-ene with benzene exemplifies the photo-ene reaction (D. Bryce-Smith, *et al.*, Chem. Comm., 1971, 794). Three major 1:1 adducts were obtained. The minor product was shown to be the 1,2-cyclo-adduct (5), since it formed the Diels-Alder adduct (6) with maleic anhydride. The product of intermediate abundance was the 1,3-cycloadduct (7). The major product, which was eight times as abundant as the minor adduct, was shown to be the ene-product (8), a derivative of 1,4-dihydrobenzene. This latter product is formed non-stereospecifically and its formation is promoted by proton donors, presumably by way of a polarised intermediate, of which the zwitter-ion shown is an extreme structure. The 1,4-cycloadduct (9) was also formed.

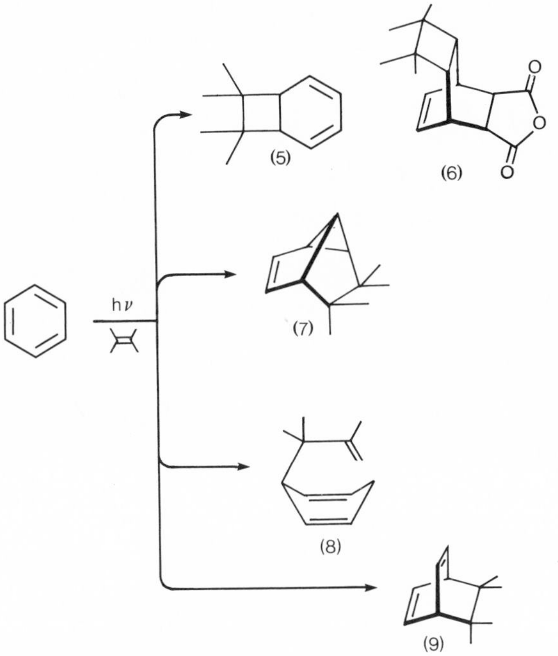

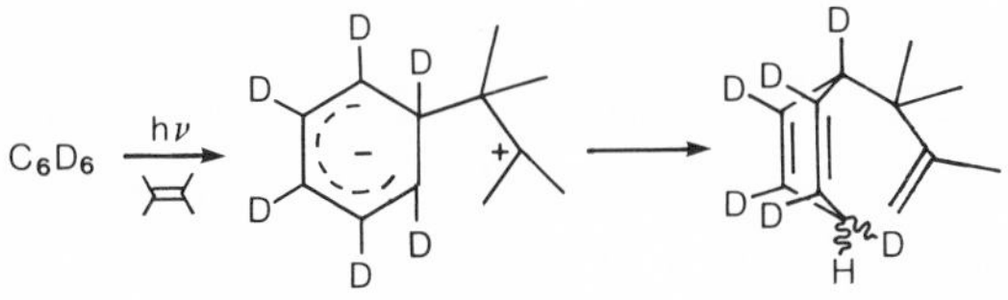

6. *Valence Isomers of Arenes*

A number of reviews on the valence isomers of benzene
and its derivatives have been published (E. E. van Tamelen,
Acc. Chem. Res., 1972, 5, 186; L. J. Scott and M. Jones Jr.,
Chem. Rev., 1972, 72, 171; A. H. Schmidt, Chem. unserer Zeit.,
1977, 11, 118).

A very efficient thermal synthesis of benzvalene involves
the interaction of cyclopentadienyl-lithium with dichloromethane and an alkyl-lithium reagent (T. J. Katz, E. J. Wang,
and N. Acton, J. Amer. chem. Soc., 1971, 93, 3782). The
availability of benzvalene has enabled its intermediacy in
the photohydration of benzene to be confirmed. The use of
deuterium oxide gives the stereoisomer shown.

Benzvalene has also been shown to undergo a degenerate
rearrangement that is catalysed by silver (I) ion. This
reaction is in competition with the known catalysed aromatisation (U. Burger and F. Mazenod, Tetrahedron Letters, 1977,
1757).

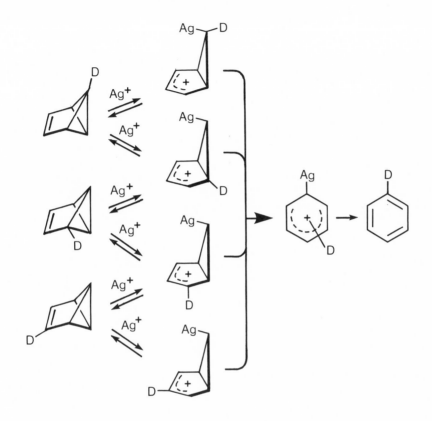

The irradiation of liquid benzene at 254 nm gives a mixture of fulvene and benzvalene *via* the first exited singlet state, but no Dewar benzene. In contrast, the use of an oxygen lamp (165-200 nm) does afford some Dewar benzene. The prismane to benzene isomerisation, although exothermic by *ca.* 90 k cal. mol^{-1}, is thermally forbidden by the principle of conservation of orbital symmetry. The half life of hexamethylprismane is 2 h at $100°$. It has been suggested that aromatisation proceeds *via* the Dewar and benzvalene isomers.

128

The photoisomerisation of substituted benzenes also
includes a second category of reactions. This involves the
formation of constitutional isomers in which a substituent
has *apparently* migrated to a new position. In fact the
substituent does not migrate. The 1,2-alkyl shifts which
result from the transposition of the ring carbon atoms (for
example with the xylenes and the di-t-butylbenzenes) have
been suggested as proceeding by the re-aromatisation of
benzvalene derivatives.

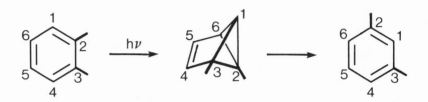

It has been pointed out that the intermediacy of benzvalenes, Dewar benzenes, and prismanes has been proposed more
frequently than their participation has been established.
An analysis of ring-carbon transpositions in terms of twelve
possible ring-permutation patterns has been suggested
(J. A. Barltrop, *et al.*, Chem. Comm., 1975, 177; 729).

A number of rational syntheses of Dewar benzene derivatives
have been published. The parent compound has been prepared
in 20% yield by the oxidative decarboxylation of bicyclo-
[2.2.0]-hex-5-ene-2,3-dicarboxylic acid, using lead (IV)
acetate (E. E. van Tamelen, S. P. Pappas and K. L. Kirk,
J. Amer. chem. Soc., 1971, 93, 6092). Anodic oxidation has
been used in the preparation of 1-chloro- and 1,4-dichloro-
Dewar benzene (R. Breslow, J. Napierski, and A. H. Schmidt,
J. Amer. chem. Soc., 1972, 94, 5906).

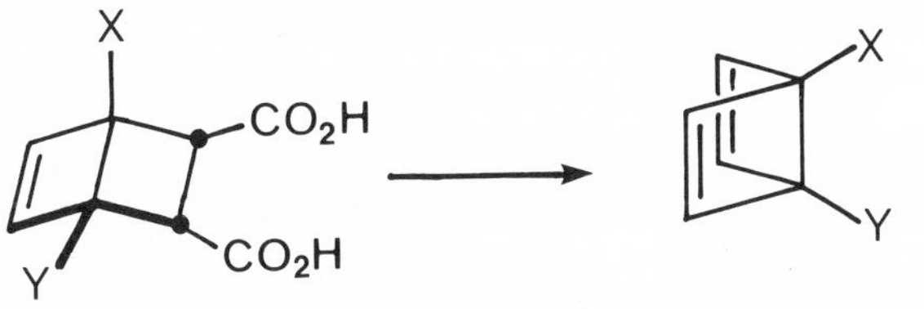

X = Cl; Y = H
X = Y = Cl

The monochloro compound is thermally very unstable, despite the anti-aromatic transition state involved in the disrotatory ring opening. A 'push-pull' effect that results in a polar transition state has been suggested to explain this result. The dichloro-compound is very much more stable than Dewar benzene.

The formation of hexamethyl Dewar benzene by the aluminium chloride catalysed trimerisation of but-2-yne (W. Schäfer and H. Hellmann, Angew. Chem. internat. Edn., 1967, 6, 518) is known to proceed via the tetramethylcyclobutadiene-aluminium chloride adduct, which has been characterised by n.m.r. spectroscopy (J. B. Koster, G. J. Timmermans, and H. van Bekkum, Synthesis, 1971, 139). The latter authors prepared dimethyl tetramethyl Dewar benzene-2,3-dicarboxylate by the interaction of dimethyl acetylenedicarboxylate with the complex. The reaction is thought to involve a Diels-Alder reaction of tetramethylcyclobutadiene. The same approach was used in the first synthesis of a chiral Dewar benzene derivative, methyl 2-phenyltetramethyl Dewar benzene-3-carboxylate. Hydrolysis to the carboxylic acid allowed the resolution of the enantiomeric acids to be achieved (J. H. Dopper, B. Greijdanus, and H. Wynberg, J. Amer. chem. Soc., 1975, 97, 216). The Dewar benzene derivative is converted into the expected biphenyl derivative at 250°. Photo-isomerisation to the prismane derivative is achieved in very high yield (>90%) by irradiation at 350 nm. This result is undoubtedly due to the very large difference in the $\lambda_{(max)}$ values of the starting material and product, a condition which does not occur with other Dewar benzenes that have been studied.

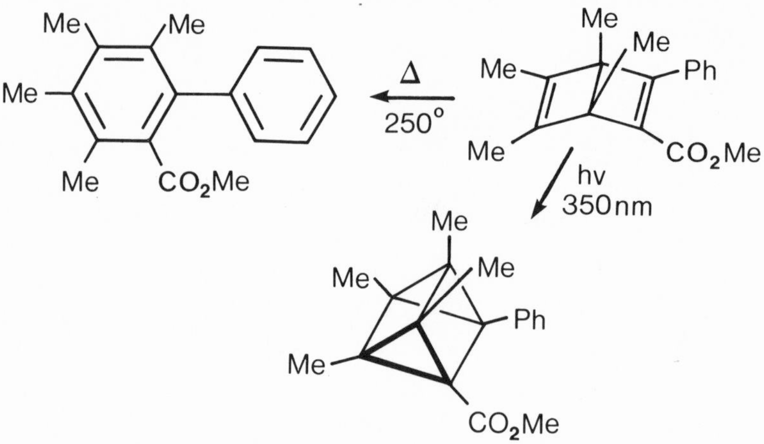

In a pair of benzene-Dewar benzene isomers it is possible to arrange the substituents such that the thermodynamic relationships of the two isomers help to stabilise the non-aromatic structure. Thus two bulky *ortho* substituents favours the non-planar product that is less subject to steric over-crowding (equation [27]) (E. E. van Tamelen, and S. S. Pappas, J. Amer. chem. Soc., 1962, 84, 3789; I. E. Den Besten, L. Kaplan, and K. E. Wilzbach, *ibid.*, 1968, 90, 5868).

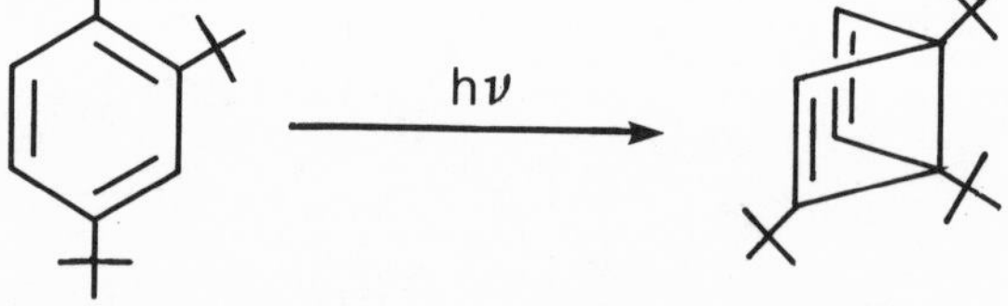

A similar rationale has been used in the synthesis of a number of 1,4-methylene-bridged derivatives of Dewar benzene (propella[n.2.2]dienes). A good review of silver (I) catalysed rearrangements in cyclic systems has been published (L. A. Paquette, Synthesis, 1975, 347). Propelladienes have been prepared in which the largest bridging group has 3-, 4-,

5-, or 6-methylene groups. Thus 1,4-trimethylene- and
1,2-trimethylene-Dewar benzenes are produced as shown in
equation [28] (I. J. Landheer, W. H. de Wolff, and
F. Bickelhaupt, Tetrahedron Letters, 1975, 349). The 1,4-
bridged isomer is exceptionally stable.

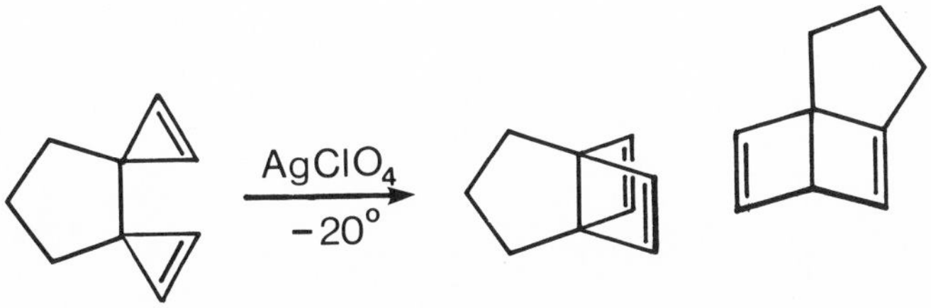

$$\text{AgClO}_4 \quad -20° \qquad \qquad \qquad [28]$$

The tetramethylene-bridged Dewar benzenes have been made
by a similar route (I. J. Landheer, W. H. de Wolff, and
F. Bickelhaupt, Tetrahedron Letters, 1974, 2813) and in very
low yield by an alternative approach (K. Weinges and
K. Klessing, Ber., 1974, 107, 1915). Whereas 1,2-tetramethyl-
ene-Dewar benzene rearranges to 1,2,3,4-tetrahydronaphthalene
easily ($t_{\frac{1}{2}}$ 58 min at 20°), the 1,4-bridged isomer is much more
stable ($t_{\frac{1}{2}}$ 26 min at 140°) and shows no tendency to ring open
to [4]paracyclophane. Flash vacuum pyrolysis results in the
ejection of ethylene and the formation of poly-p-xylylene.
    1,4-Pentamethylene-Dewar benzene (J. W. van Straten,
I. J. Landheer, W. H. de Wolff, and F. Bickelhaupt, Tetrahedron
Letters, 1975, 4499) undergoes a rather remarkable rearrangement
at 120° to give benzocycloheptene!

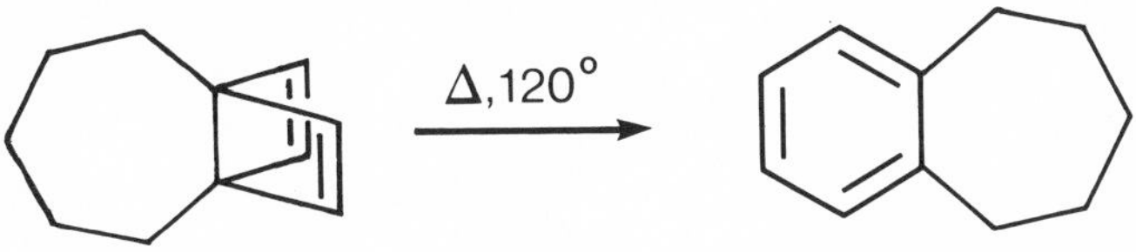

$$\Delta, 120°$$

The point at which the methylene-bridged Dewar benzene derivatives are interconvertible with the [n] paracyclophanes occurs when [n] = 6 (S. L. Kammula, *et al.*, J. Amer. chem. Soc., 1977, <u>99</u>, 5815). The aromatic isomer is the more stable structure.

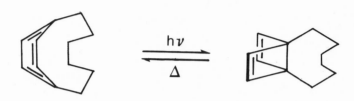

7. *Aromatisation and Functional Group Modification Reactions*

A number of aromatisation reactions have been reported that utilise alkali metal derivatives of ethylenediamine as exemplified below (R. J. Crawford, J. org. Chem., 1972, <u>37</u>, 3543).

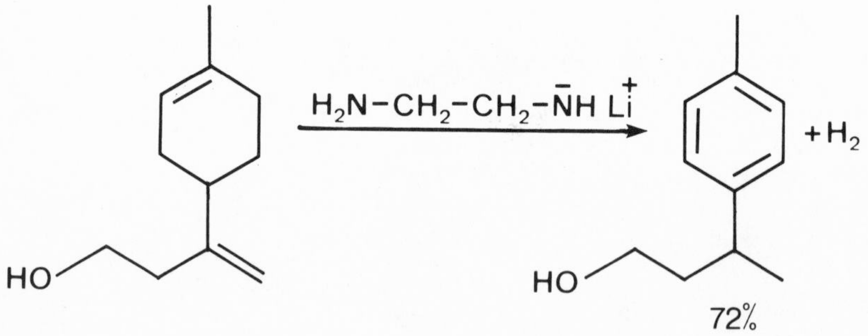

$$H_2N-CH_2-CH_2-\bar{N}H\ \overset{+}{Li}$$

72%

$+H_2$

The potassium salt of ethylenediamine is much more efficient than the lithium salt (C. A. Brown, J. Amer. chem. Soc., 1973, <u>95</u>, 982). Whereas limonene is slowly converted into *p*-cymene using the lithium salt at 90° (2 h) the potassium reagent gives a 95% yield in 5 min at 25°.

Copper (II) bromide in dimethoxyethane has been used to effect the monobromination of the enolates of cyclic β-diketones. Dehydrobromination to the resorcinol derivative is then achieved in DMF (R. S. Marmor, J. org. Chem., 1972, 37, 2901). In the generalised examples shown the overall yields were 75-80%.

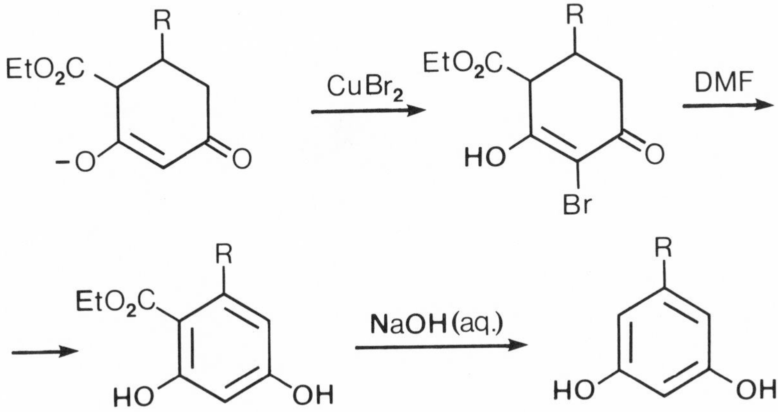

Methyl 2-aryl-6-oxocyclohex-1-enylacetates are aromatised to 3-arylphenols in yields of *ca.* 70% with concomitant loss of the acetate side chain using sodium hydride in DMF at 100° (D. Nasipuri, S. R. R. Choudhury, and A. Bhattacharya, J. chem. Soc. Perkin I, 1973, 1451).

1-Alkyl-cyclohexa-2,5-diene-1-carboxylic acids form bromo-lactones on reaction with bromine: treatment with diaza-bicycloundecane results in dehydrobromination and decarboxy-lation to the alkyl-benzene (W. E. Barnett and L. L. Needham, J. org. Chem., 1975, 40, 2843; G. W. Holbert, L. B. Weiss, and B. Ganem, Tetrahedron Letters, 1976, 4435). The oxidative decarboxylation of the cyclohexadiene-carboxylic acid shown proceeds efficiently (A. J. Birch and J. Slobbe, Tetrahedron Letters, 1976, 2079).

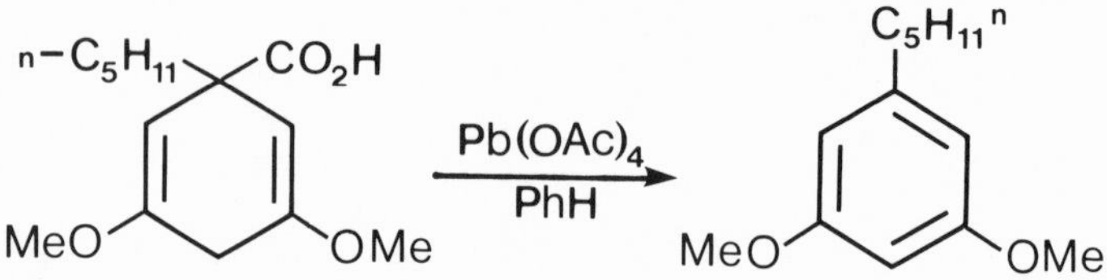

Thermolysis of the bis-(spirocyclopropyl)cyclohexadiene affords 2-ethylstyrene in 87% yield. Thermolysis at 100° (residence time 10 s) over a gold surface gives 2-ethylstyrene (10%) and 1,2,3,4-tetrahydronaphthalene (90%) (L. -U. Meyer and A. de Meijere, Tetrahedron Letters, 1976, 497).

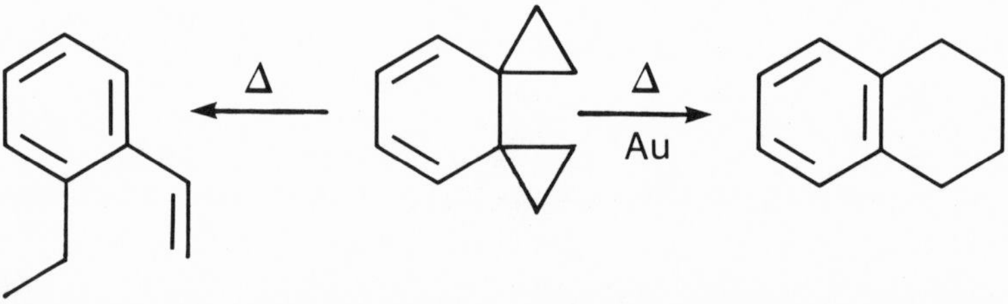

The formation of carbon-carbon bonds using π-allylnickel compounds has been reviewed (M. F. Semmelhack, Org. Reactions, 1972, 19, 115). Methallylbenzene is formed in about 70% yield (M. F. Semmelhack and P. M. Helquist, Org. Synth., 1972, 52, 115).

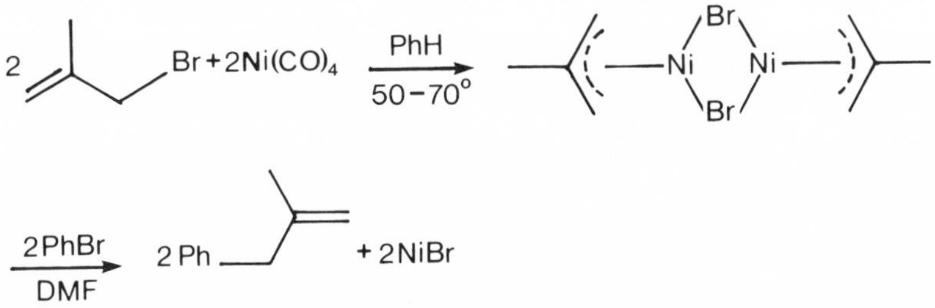

The conversion of aryl iodides and bromides to the methylarenes can be achieved using methyltris(triphenylphosphine) rhodium (I). The rhodium containing by-product can be reconverted into the reagent (M. F. Semmelhack and L. Ryono, Tetrahedron Letters, 1973, 2967). Homophthalic acids are obtained from 2-bromobenzoic acids by reactions with β-keto-esters in the presence of copper catalysts (A. Bruggink and A. McKillop, Tetrahedron, 1975, 31, 2609). The complete retention of stereochemistry is achieved in the reaction of *trans*-β-bromostyrene with tetrakis (triphenylphosphine)palladium followed by treatment with methyl-lithium (M. Yamamura, I. Moritani, and S. -I. Murahashi, J. organometal. Chem., 1975, 91, C39). A 98% yield of *trans*-β-methylstyrene was obtained. Prenylation can be effected using isoprene epoxide. The latter with phenyl-lithium gives, as the major product, Z-2-methyl-4-phenyl-but-2-en-1-ol after ring opening and allylic rearrangement (G. C. M. Aithie and J. A. Miller, Tetrahedron Letters, 1975, 4419).

A simplified Wittig synthesis of styrenes from benzylic halides has been published (R. Broos and M. Auteunis, Synth. Comm., 1976, 6, 53). In the reactions of secondary alcohols with triphenylphosphine and carbon tetrachloride overall dehydration to alkenes can occur if the reaction is carried out at high temperatures (R. Appel and H. -D. Wihler, Ber., 1976, 109, 3446).

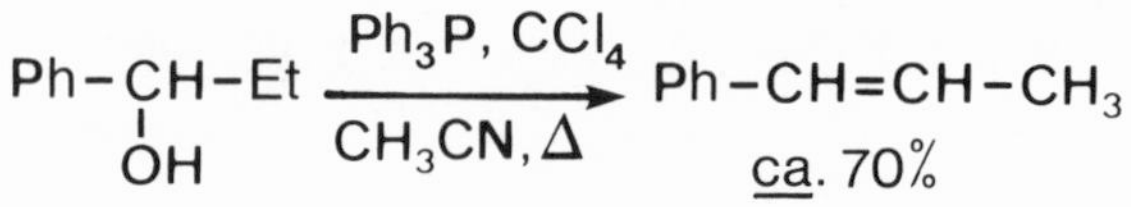

$$\text{Ph-CH-Et} \xrightarrow[\text{CH}_3\text{CN},\Delta]{\text{Ph}_3\text{P, CCl}_4} \text{Ph-CH=CH-CH}_3$$

OH   <u>ca.</u> 70%

The arylation of ketones can be achieved using triaryl-bismuth carbonate (D. H. R. Barton, *et al.*, Chem. Comm., 1980, 246.). Thus quininone is α-arylated in good yield. The reaction is thought to proceed *via* the enol and ethyl acetoacetate can be transformed into either the mono- or diphenyl derivatives using triphenylbismuth carbonate. 2-Naphthol gives 1-phenyl-2-naphthol in 76% yield. Potassium enolates are also phenylated. For example, cholestan-3-one is converted into 2,2-diphenyl-cholestan-3-one (64%) and 4,4-dimethylcholest-5-en-3-one gives the highly hindered ketone, 4,4-dimethyl-2,2-diphenylchlolest-5-en-3-one, in 80% yield.

Pentaphenylbismuth has been used to phenylate 2-alkyl phenols (D. H. R. Barton, *et al.*, Chem. Comm., 1980, 827). Thus 2,6-dimethylphenol gives 2,6-dimethyl-6-phenylcyclohexa-2,4-dienone exclusively in 72% yield. This reaction should be compared with the reaction of 4-methoxyphenyl-lead triacetate with mesitol which affords a mixture of the 2- and 4-aryl-cyclohexadienones in a 4:1 ratio (H. C. Bell, J. T. Pinhey, and S. Sternhell, Austral. J. Chem., 1979, 32, 1551).

8. *Oxidation Reactions*

Ozone is adsorbed on silica gel at -78° and affords concentrations of <u>ca.</u> 4.5% by weight (Z. Cohen, E. Keinan, Y. Mazur, and T. H. Varkony, J. org. Chem., 1975, 40, 2142) and the oxidation of arenes to carboxylic acids proceeds in high yield using the reagent (H. Klein and A. Steinmetz, Tetrahedron Letters, 1975, 4249).

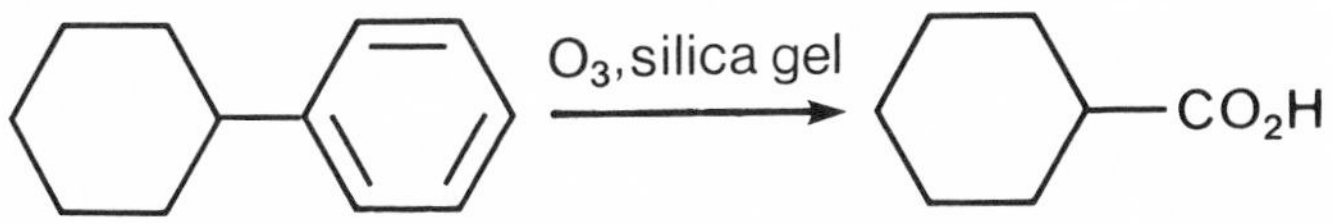

The oxidation of alkylbenzenes with purple benzene (potassium permanganate in benzene containing dicyclohexyl-18-crown-6) proceeds in high yield. Toluene affords benzoic acid quantitatively (D. J. Sam and H. E. Simmons, J. Amer. chem. Soc., 1972, 94, 4024). The selective oxidation of methyl groups in arenes can be achieved using oxygen in the presence of cobalt (II) ions. The reaction involves the conversion of Co(II) to Co(III) which is promoted by the presence of ethyl methylketone. An electron transfer mechanism is proposed (A. Onopchenko, J. G. D. Schulz, and P. Seekircher, J. org. Chem., 1972, 37, 1414). p-Cymene is oxidised to p-isopropylbenzoic acid in 90% yield. The oxidation of γ-phenylbutyric acid with ammonium persulphate in water leads to 4-phenyl-γ-butyrolactone in high yield. The reaction is believed to involve two oxidation steps, first to the benzylic radical and then to the benzylic cation (A. Clerici, F. Minisci, and O. Porta, Tetrahedron Letters, 1974, 4183).

The oxidation of 4-substituted-2,6-xylenols with a variety of reagents, including silver oxide, gives o-quinone methides which either trimerise, capture another hydroxy-compound to form benzyl ethers, or form chroman derivatives as a result of Diels-Alder reactions (D. A. Bolon, J. org. Chem., 1970, 35, 715; 3666).

The oxidation of phenols to o-quinones continues to be a desirable objective. A large number of methods is available when a substituent is present at the para-position. 3,4-Xylenol gives a 50% yield of 4,5-dimethyl-o-benzoquinone using Fremy's salt (H. -J. Teuber, Org. Synth., 1972, 52, 88). The oxidation of 4-aryl-2,6-di-t-butylphenols with oxygen in

the presence of a cobalt (II) complex, followed by decomposition of the resulting hydroperoxides using toluene-*p*-sulphonic
acid, gave 5-aryl-3-t-butyl-*o*-benzoquinones in good yield
(A. Nishinaga, *et al.*, Synthesis, 1977, 270). Potassium
ferricyanide in methanol at pH7 has been used for the
oxidation of 2-amino-3,6-di-t-butylphenol (H. Brockmann and
F. Seela, Ber., 1971, 104, 2751), and the oxidation of
catechol to *o*-benzoquinone has been reported using succinimido
dimethylsulphonium fluoroborate in methylene chloride at
low temperature, followed by the addition of triethylamine
(J. P. Marino and A. Schwartz, Chem. Comm., 1974, 812).

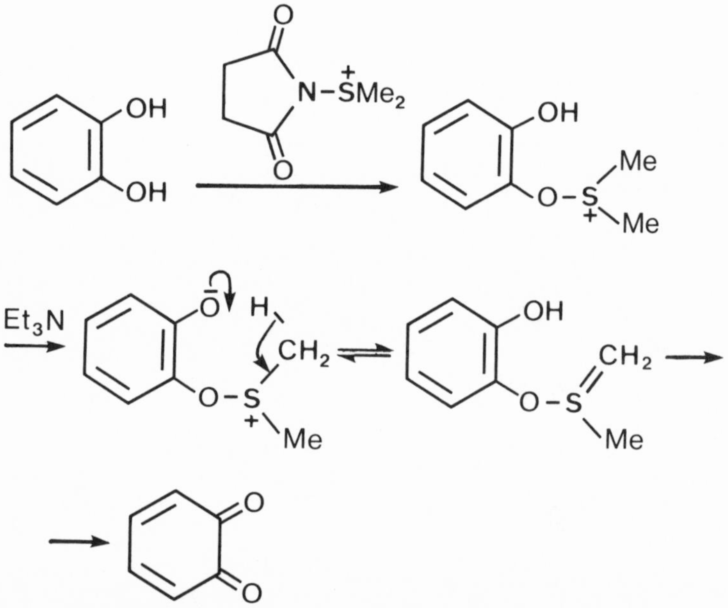

The selective introduction of an oxygen function, *ortho*
to a phenolic group has been achieved using diphenylseleninic
anhydride (D. H. R. Barton, *et al.*, Chem. Comm., 1976, 985).
Yields are generally in the range 60-70% in rapid reactions
(15 min in THF at 50°) and the *ortho*-quinone is normally
obtained even in the absence of a substituent at the 4-position.

Quinols and catechols are both oxidised to the corresponding quinones in quantitative yields using *N*-chloro-succinimide and DMSO followed by the addition of triethylamine (J. P. Marino and A. Schwartz, Chem. Comm., 1974, 812).

Chapter 3

HALOGEN DERIVATIVES OF BENZENE AND ITS HOMOLOGUES

W.J. FEAST

Introduction

For the convenience of the reader the sections and sub-
sections of this chapter have the same titles and numbering
as those used in Chapter 3, Volume IIIA of the Second Edition.
Not all the topics covered in the Second Edition have been the
subject of significant advance, consequently sub-section
numbering is not always sequential. Tabulations of physical
properties and descriptions of individual compounds have been
omitted. Some new sub-sections have been included under 1(b).
Section 3 has been omitted, the data which could have made up
this very small section being included elsewhere.

The decade since the manuscript for the Second Edition
was prepared has been one of phrenetic activity in the field
reviewed, new journals have been spawned and papers published
at an apparently ever increasing rate, this makes the working
chemist's and the reviewer's task of identifying the
significant items difficult. Two developments act in part as
antidotes to the trend described above; thus, computer
searching of the literature has become available to many and
a number of regular surveys of the literature of particular
fields have appeared. Some of these literature compilations
and reviews promised to be of lasting value but unfortunately
lack of a sufficiently large market and rapidly increasing
publishing costs have resulted in several ventures being
abandoned after only a few issues. Of particular relevance
to the topic of this chapter are: "Advances in Fluorine
Chemistry", Butterworth, London, 1, 1960 through to 7, 1973;
"Fluorine Chemistry Reviews", Dekker, New York, 1, 1967
through to 8, 1976; the Chemical Society Specialist
Periodical Reports "Fluorocarbon and Related Chemistry", 1,
1971; 2, 1974 and 3, 1976; and "Aromatic and Heteroaromatic

Chemistry", 1, 1973 and annually to 7, 1979;   and "Aromatic
Compounds" in the MTP International Review of Science,
Organic Chemistry Series One, 1973, Volume 3, and Series Two,
1976, Volume 3. All these review or survey publications have
suspended publication at the time of writing, nevertheless
they will remain important points of access to the literature
of the period. Several other useful and well established
reviews and literature compilations continue to be published,
a listing of relevant publications will be found in the
opening pages of Chapter 3, Volume IIIA and Chapter 3 of the
Supplement to Volume IA/B of the Second Edition of this work.
      A number of new books containing information relevant to
the content of this chapter have also appeared. R.D. Chambers
and S.R. James provide in Chapter 3, Volume 1 of "Comprehensive Organic Chemistry", Pergamon, 1979 a general review
of the chemistry of halogeno compounds. The second (revised)
edition of a well established text has appeared ("Chemistry
of Organic Fluorine Compounds", M. Hudlicky, Ellis Horwood
Ltd., Chichester, 1976) and the chemistry and applications of
fluorinated aromatic compounds receive attention in "Fluorine
in Organic Chemistry", R.D. Chambers, Wiley, New York, 1973
and "Organofluorine Chemicals and their Industrial
Applications", Editor R.E. Banks, Ellis Horwood Ltd.,
Chichester, 1979. "Polychloroaromatic Compounds", Edited by
H. Suschitzky, Plenum Press, London, 1974 has two particularly
useful chapters, "Aromatic and Alkaromatic Chlorocarbons",
M. Ballester and S.O. Olivella (Chapter 1, 196 pages) and
"Polychloroaromatics and Heteroaromatics of Industrial
Importance", M.B. Green (Chapter 4, 72 pages).

1. Nuclear halogen derivatives

(a) Methods of preparation

     (i) Replacement of aromatic hydrogen by halogen
      The majority of papers relating to this topic have been
concerned with the mechanisms by which aromatic halogenations
occur. Some of the more important results are referred to
below but generally space limitations have precluded detailed
discussion of these investigations of mechanisms and the
reader is referred to the appropriate annual literature
reviews, particularly "Organic Reaction Mechanisms"
Interscience, New York, 1966 et seq., and to publications by
P.B.D. de la Mare (Accounts of Chemical Research, 1974, 7,
361; "Electrophilic Halogenation", Cambridge University

Press, 1976; and J. chem. Soc. Perkin II, 1977, 106) and R. Taylor in "Comprehensive Chemical Kinetics", Vol. 13, Elsevier, Amsterdam, 1972.

It is now established that direct fluorination of aromatic compounds is possible under carefully controlled conditions (V. Grakauskas, J. org. Chem., 1969, $\underline{34}$, 2835; and 1970, $\underline{35}$, 723). For example toluene reacts with diluted fluorine in acetonitrile at -35°C to give predominantly *ortho*- and *para*-fluorotoluene, suggesting an electrophilic process. Fluorinations with xenon difluoride have received further attention (M.J. Shaw, H.H. Hyman and R. Filler, J. org. Chem., 1971, 36, 2917; and B. Sket and M. Zupan, *ibid*, 1978, $\underline{43}$, 835); for example, *ortho*-xylene gives a mixture of 3- and 4-fluoro-1,2-dimethylbenzenes. However xenon difluoride is not a particularly convenient reagent and it has been suggested that the product of interaction between xenon hexafluoride and graphite offers practical advantages while effecting similar reactions (H. Seliget *et al.*, J. Amer. chem. Soc., 1976, $\underline{98}$, 1601). Fluorinations with xenon difluoride generally require catalysis by hydrogen fluoride and cation radicals are implicated as intermediates in the process, as they have also been in fluorinations with high valence metal fluorides (W.K.R. Musgrave *et al.*, J. chem. Soc. Perkin I, 1974, 114) and some other halogenation processes (see below). Perhaps the most promising development in the direct aromatic fluorination field is the demonstration that alkali-metal tetrafluorocobaltates(III), $MCoF_4$, react in a controlled manner with aromatic substrates to give fluoro-aromatic products, the spent fluorinating agent being re-generated by reaction with fluorine (A.J. Edwards, R.G. Plevey, I.J. Sallomi and J.C. Tatlow, Chem. Comm., 1972, 1028). This approach offers inter alia the possibility of a one step route from benzene to a variety of fluorinated benzenes including hexafluorobenzene.

The intermediacy of cation radicals has been suggested by Ledwith to account for the observation that a mixture of sodium persulphate, hydrochloric acid and copper(II) chloride is an effective reagent for nuclear chlorination of aromatics (benzene 60% and toluene 80% reaction). The reaction is believed to proceed via an initial electron transfer process giving a cation radical which subsequently reacts with chloride. (A. Ledwith and P.J. Russel, Chem. Comm., 1974, 291 and 959, and J. chem. Soc. Perkin II, 1975, 1503).

$$CH_3C_6H_5 \xrightarrow[CuCl_2]{CH_3CN/H_2O/LiCl} \quad \left[\text{ring with } + \text{ and } CH_3, \; H, \; Cl\cdots CuCl\right] \xrightarrow{Cl^-} CH_3C_6H_4Cl$$

Other examples of this apparent nucleophilic substitution of hydrogen by halogen have appeared; for example, benzene can be converted to chloro-, bromo- or iodo-benzene using cobalt(III) trifluoroacetate and the appropriate lithium halide, again an intermediate cation radical is proposed (M.E. Kurz and G.W. Hage, J. org. Chem., 1977, 42, 4080). Explanations based on cation radicals have been disputed and, for example, it has been suggested that chlorination by sodium persulphate/lithium chloride mixtures in aqueous acetonitrile with or without added copper(II) chloride actually involves molecular chlorine (L. Eberson and L.G. Wistrand, Ann. 1976, 1777).

Various new reagents and modified procedures have been described for the halogenation of aromatics. Chlorination in the presence of silica goes readily at room temperature (C. Yaraslavsky, Tetrahedron Letters, 1974, 3395). Reaction is achieved using a solution of chlorine in carbon tetrachloride, there is no side chain chlorination and the enhanced reactivity is attributed to the polar environment provided by the silica. Brominations and iodinations proceed with very high positional selectivity when a 1:1 mixture of the halogen and antimony pentachloride in carbon tetrachloride is used as reagent. Halogenobenzenes give the *para* bromo or iodo derivatives (99%), the solvent has a significant effect, for example, use of 1,2-dichloroethane results in a loss of selectivity (S. Vemura, A. Onoe and M. Okano, Bull. chem. Soc. Japan, 1974, 47, 147). High positional selectivity for molecular bromination also results from the use of sulphur dioxide as solvent (J.P. Causelier, Bull. Soc. chim. France, 1971, 1785; and 1972, 762). Specific *ortho* bromination of toluene, ethylbenzene and chloro- and bromo-benzenes can be achieved by initial alkylation (isobutene/sulphuric acid) followed by bromination (Br_2/Fe/CCl_4) and final removal of the t-butyl substituent using transalkylation catalysed by aluminium chloride (D. Meidar and Y. Halpern, J. appl. Chem. and Biotechnol., 1976, 26, 590). α,α,α-Tribromoacetophenone and aluminium bromide effect bromination of aromatics and a reactive entity of the form $Br_3\bar{A}lOC(C_6H_5)=CBr_2,Br^+$ has been

suggested (V.A. Smrchek, V.F. Traven and B.I. Stepanov, J.
org. Chem. (U.S.S.R.), 1972, $\underline{8}$, 1810; and 1973, $\underline{9}$, 585).
Thallation as a route to aryl halides was mentioned briefly in
the Second Edition, this procedure has been developed
considerably and aryl fluorides, chlorides, bromides and
iodides have all been prepared. Thallation of appropriate
aromatics with thallium(III) trifluoroacetate gives a
product of the form $ArTl(O_2CCF_3)_2$ which may be reacted with
potassium fluoride followed by boron trifluoride or with
potassium iodide to give respectively aryl fluorides or
iodides (A. McKillop et $al.$, J. org. Chem., 1977, $\underline{42}$, 362;
and Org. Synth., 1976, $\underline{55}$, 70). Thallium(III) chloride
tetrahydrate in carbon tetrachloride, and chlorine in the
presence of other thallium salts are effective electrophilic
chlorinating agents (S. Vemura, K. Sohma and M. Okano, Bull.
chem. Soc. Japan, 1972, $\underline{45}$, 860); and thallium(III) bromide
and bromine in the presence of thallium(III) or (I) salts are
effective brominating agents (McKillop et al., J. org. Chem.,
1972, $\underline{37}$, 88; and S. Vemura et al., Bull. chem. Soc. Japan,
1971, $\underline{44}$, 2490). Chlorine monofluoride reacts with benzene
and toluene in carbon tetrachloride to give the expected
products of electrophilic chlorination (G.P. Gambaretto and
M. Napoli, J. fluorine Chem., 1976, $\underline{7}$, 569). Photochemical
bromination of halogenobenzenes results predominantly in $meta$
substitution (P. Gouverneur and J.P. Soumillion, Tetrahedron
Letters, 1976, 133); with dihalogenobenzenes $ortho$ substitution becomes important and for chloro compounds $ipso$
attack, that is replacement of the chlorine, becomes
significant. Trialkylphosphates have been recommended as
solvents for molecular chlorinations and brominations, such
solvents are particularly useful in cases where the substrate
or product is particularly sensitive to the hydrogen halide
produced as a by-product since it is immediately converted to
an alkyl halide (D.E. Pearson et $al.$, Synthesis, 1976, 621).
Reaction of aromatics with a mixture of sulphuric and nitric
acids results exclusively in iodination (A.R. Butler and A.P.
Sanderson, J. chem. Soc. (B), 1971, 2264), HNO_2I^+ was identified as the electrophile on the basis of kinetic studies.
Bromination of $ortho$-xylene and mesitylene in acetic acid is
catalysed by hydrobromic acid (E.P. Yesodharan, J. Rajaram
and J.C. Kuriacose, Int. J. chem. Kinetics, 1976, $\underline{8}$, 277),
possibly the acid facilitates the loss of bromide ion from
the initial ArH,Br_2 complex to give the Wheland intermediate. Titanium tetrahalides in the presence of peroxy-

acids are effective agents for chlorination, bromination and
iodination, hypohalous acids are proposed as intermediates
(G.K. Chipp and J.S. Grossert, Canad. J. Chem., 1972, $\underline{50}$,
1233). Reactions involving hypohalous acids are one of
several classes of halogenation reaction which have been
the subject of detailed mechanistic investigation (J.H. Ridd
et al., Chem. Comm., 1971, 130 and J. chem. Soc. Perkin II,
1973, 1321; and C.G. Swain and D.R. Crist, J. Amer. chem. Soc.,
1972, $\underline{94}$, 3195). These investigations have resulted in a re-
evaluation of ideas on electrophilic halogenation and the
long favoured electrophilic species Cl^+ and Br^+ are no longer
held to play a significant role in these particular reactions.
The mechanistic scheme shown below provides a satisfactory
rationalization of the evidence in the case of brominations.
The preferred route in the pre-equilibrium leading to the

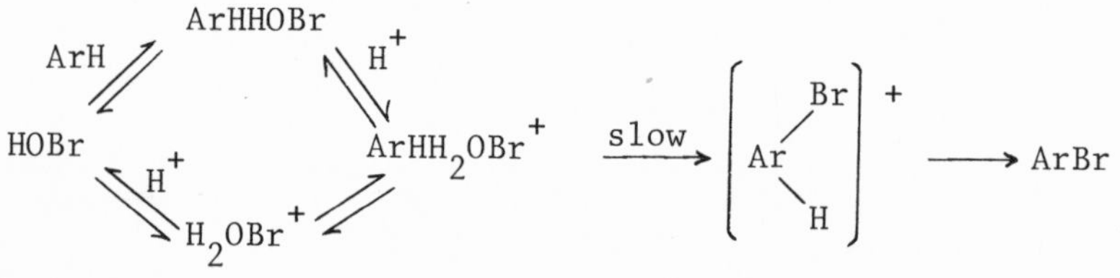

$ArHH_2OBr^+$ complex depending on the relative acidity of the
medium and basicity of the aromatic substrate.

(iv) Replacement of an amino group by halogen
    The Sandmeyer reaction for the synthesis of aryl halides
has been improved; stable aryl diazonium tetrafluoroborates
are reacted first with potassium acetate in the presence of
18-crown-6 and then with bromotrichloromethane (or iodotri-
chloromethane) to give the aryl halide in excellent yield
(S.H. Korzeniowski, G.W. Gokel, Tetrahedron Letters, 1977,
3519). The reactions of aryl amines with an alkyl nitrite
and copper(II) halide is a good route to high yields of aryl
chlorides or bromides (M.P. Doyle, B. Siegfried and J.F.
Dellaria, J. org. Chem., 1977, $\underline{42}$, 3494). A novel route to
aryl iodides from aryl amines has been reported (A.R.
Katritzky, N.F. Eweiss and P.-L. Nie, J. chem. Soc. Perkin I,
1979, 433), the method is summarized below.

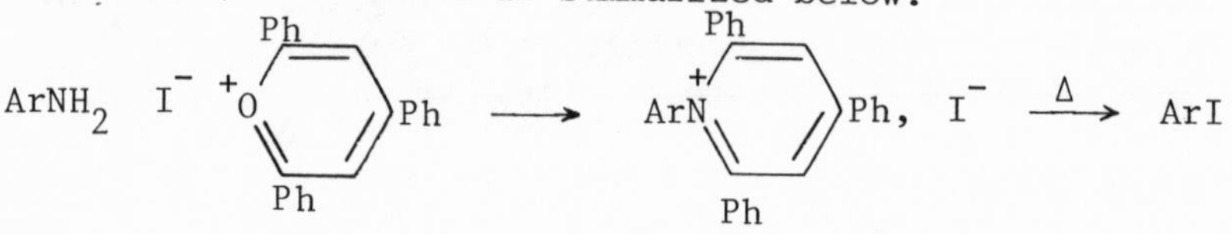

Although the method works well for iodides its generality is uncertain, for example it is applicable to the synthesis of primary alkyl or benzyl fluorides but not aryl fluorides (Katritzky *et al.*, Chem. Comm., 1979, 238).

(vi) Replacement of one halogen by another

The method of halogen exchange has been used for the synthesis of a number of polyfluoroaromatics, some further investigations of the variables involved in the reaction have been published. Thus, some limitations on the usefulness of potassium fluoride in aprotic solvents for converting poly-chloroaromatics to polyfluoroaromatics have been pointed out (R. Bolton, S.M. Kazeroonian and J.P.B. Sandall, J. fluorine Chem., 1976, $\underline{8}$, 471); in boiling sulpholane KF exchange of the chlorines in C_6Cl_6 and $1,4-C_6Cl_4(H)NO_2$ proceeds satisfactorily whereas the reactions of C_6HCl_5 and $1,4-C_6Cl_4(H)F$ do not. Some control over the outcome of exchange fluorination of polychloroaromatics with potassium fluoride or mixtures of it with caesium fluoride in dimethylsulphone is possible (G.C. Finger, D.R. Dickerson and R.H. Shiley, J. fluorine Chem., 1972, $\underline{1}$, 415, and 1972/3, $\underline{2}$, 19). The gas phase reaction of halogenobenzenes with fluorine atoms generated by a radiofrequency discharge results in significant proportion of *ipso* attack in addition to replacement of hydrogen (A.H. Vasek and L.C. Sams, J. fluorine Chem., 1973, $\underline{2}$, 257; and 1974, $\underline{3}$, 397).

(vii) Perhalogenation

Perchlorination has been dealt with at length by Ballester and Olivella (*loc. cit*) this detailed review is particularly welcome since it contains a substantial amount of material previously either unpublished or difficult to obtain access to. A novel perchlorination result has been reported (A.V. Foukin et al., Izvest. akad. Nauk. S.S.S.R., Ser. Khim., 1977, 2388), reaction of benzene with chlorine fluorosulphate at $-100°C$ in a Freon gave C_6Cl_6 (77%). Perbromination and periodination of benzenoid aromatics may be achieved via initial permercuration. Reaction of the aromatic with molten mercuric trifluoroacetate results in per-mercuration, the mercury atoms may then be replaced by bromines by reaction with potassium tribromide (G.B. Deacon and G.J. Farquharson, Austral. J. Chem., 1976, 29, 627) or by iodines by reaction with potassium triiodide (*idem, ibid*, 1977, $\underline{30}$, 170; and L.M. Yagupol'skii, Zhur. org. Khim., 1976, $\underline{12}$, 916). This route to perbromo and periodo compounds appears to offer some convenience and generality.

(xi)   Trimerization of alkynes

Ballester (*loc. cit*) has described the thermal telomerization of perchlorophenylacetylene which at ca. 190°C
gives, inter alia, perchloro-1,2,3- and -1,2,4-triphenyl-
benzenes. Roedig and co-workers have published a number of
examples of the dimerization and telomerization of poly-
unsaturated halogeno compounds which are conveniently
included in this section. For example, the dimerization of
perhalogenovinylacetylenes can lead to either aromatic or
cyclooctatetraene derivatives (A. Roedig *et al.*, Ber.,
1971, 104, 3378 and Ann. 1972, 755, 123).

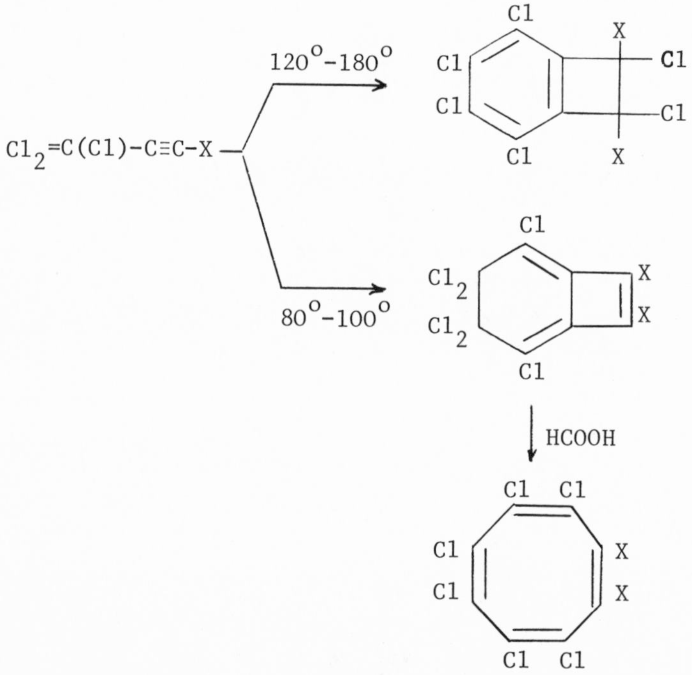

(b)   Properties and reactions

(ii)   Dehalogenation

Aryl halides are conveniently reduced by the homogeneous
reagent system $C_6H_5CCl_3py_3$-$NaBH_4$ (C.J. Love and F.J. McQuillin,

J. chem. Soc. Perkin I, 1973, 2509). Aryl iodides, but not
bromides or chlorides, are reduced by sodium hydride (R.B.
Nelson and G.W. Gribble, J. org. Chem., 1974, 39, 1425).
Electrochemical reduction of halogenoaromatics can result in
the replacement of halogen by hydrogen; however, the reaction
is frequently complex and the overall outcome depends on a
subtle interaction of several factors (K.H. Houser, D.E.
Bartak, and M.D. Hawley, J. Amer. chem. Soc., 1973, 95,
6033). For example the three isomers of fluorobenzonitrile
give different products on electrochemical reduction.

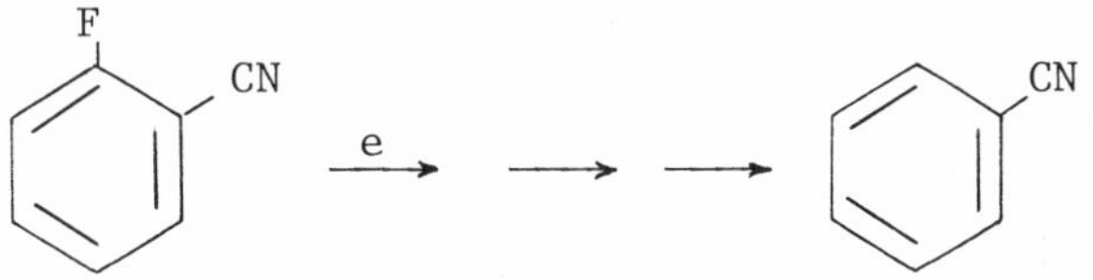

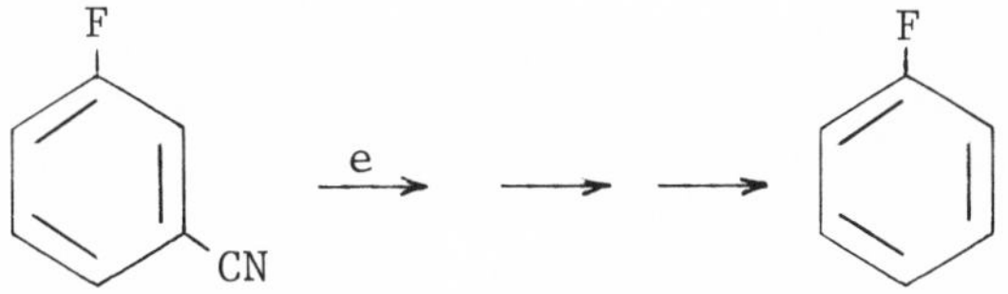

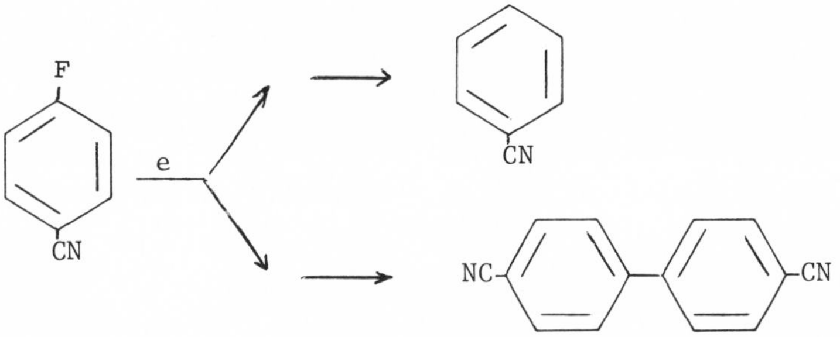

Halogen atoms may also be used as positional protective groups
in the synthesis of polysubstituted aromatics, the halogen
atom being removed at the end of the synthetic sequence by
selective reduction. Considerable control over the course of
the last step has been obtained by careful manipulation of
reaction conditions, as exemplified on the following page.
The use of halogens as positional protective groups in
aromatic synthesis has been reviewed (M. Tashiro, Synthesis,

150

1979 (12), 926).

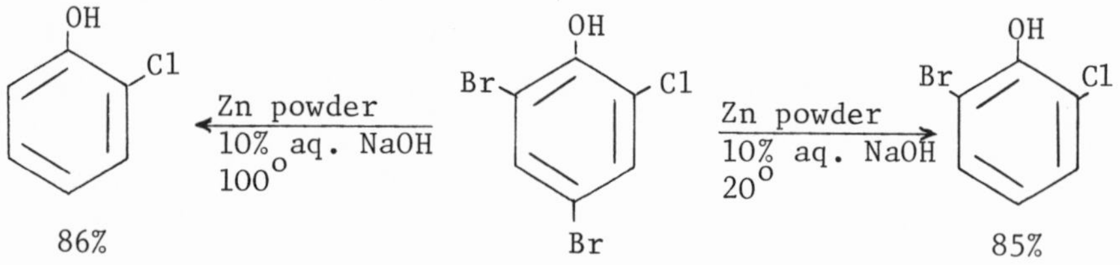

(iii) Electrophilic reaction

The intense interest in electrophilic aromatic substitution has been referenced earlier (1a(i)). Electrophilic substitutions in polyfluorobenzenes have been reviewed (V.M. Vlasov and G.G. Yakobson, Russ. chem. Rev., 1974, 43, 781). A number of examples of electrophilic addition to halogeno-aromatics have appeared, for example pentafluorotoluene gives a 1,4-adduct on reaction with a mixture of nitric and hydrofluoric acids (V.D. Shteingarts et al., Izvest. sibirsk. Otdel. Akad. Nauk S.S.S.R., Ser. Khim. Nauk, 1974, 117; Chem. Abs., 1975, 82, 86237k).

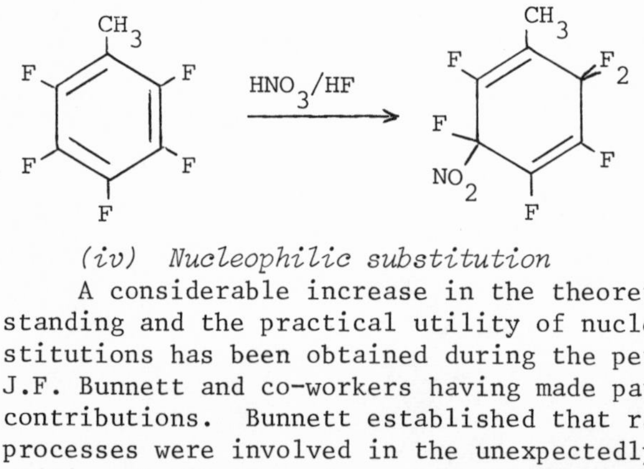

(iv) Nucleophilic substitution

A considerable increase in the theoretical understanding and the practical utility of nucleophilic substitutions has been obtained during the period reviewed, J.F. Bunnett and co-workers having made particularly notable contributions. Bunnett established that radical chain processes were involved in the unexpectedly easy nucleophilic substitution of unactivated aryl halides. In these reactions the first step is electron transfer to the aryl halide to give an anion radical, the transfer being accomplished by an electron donor, or an electrochemical or photochemical process. For example, some of the chain propagation steps in the photochemically promoted reaction between thioethoxide ($C_2H_5S^-$) and iodobenzene are shown below, the reaction is a typical chain process and other

$$C_6H_5I + e \longrightarrow C_6H_5I^{\overline{\cdot}}$$

$$C_6H_5I^{\overline{\cdot}} \longrightarrow C_6H_5{\cdot} + I^-$$

$$C_6H_5{\cdot} + C_2H_5S^- \longrightarrow C_6H_5SC_2H_5^{\overline{\cdot}}$$

$$C_6H_5SC_2H_5^{\overline{\cdot}} + C_6H_5I \longrightarrow C_6H_5SC_2H_5 + C_6H_5I^{\overline{\cdot}} \quad \text{etc., etc.}$$

termination reactions lead to diphenylthioether. The topic
has been reviewed (Bunnett, J. chem. Educ., 1974, 51, 312),
the subject of nucleophilic aromatic substitution forms a
substantial component of the annual literature review "Organic
Reaction Mechanisms", Wiley-Interscience, New York.

The reaction of metal alkyls with halogeno compounds
continues to be a fruitful area of work and the mode of
reaction observed is often very sensitive to the reaction
conditions, particularly to the temperature. For example,
the outcome of the reaction between tertiary butyl lithium
and 1,3,5-trifluorobenzene changes completely as the
temperature is lowered (R.D. Howells and H. Gilman, Tetrahedron Letters, 1974, 1319). Distinct differences in

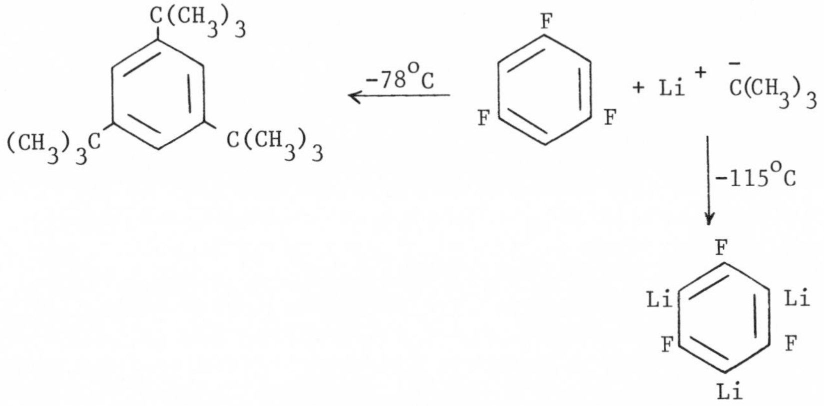

reactivity towards halogen-metal exchange at -100°C lead to
the reactivity series $ArCH_2Br$ > $ArBr$ > $ArCH_2CH_2Br$ >
$ArCH_2CH_2Cl$, this selectivity at low temperatures means that
in some cases substituents normally expected to be reactive
can be tolerated, and this leads to interesting possibilities

of synthesis (W.E. Parham, L.D. Jones, and Y.A. Sayed, J. org. Chem., 1976, 41, 1184), as shown below.

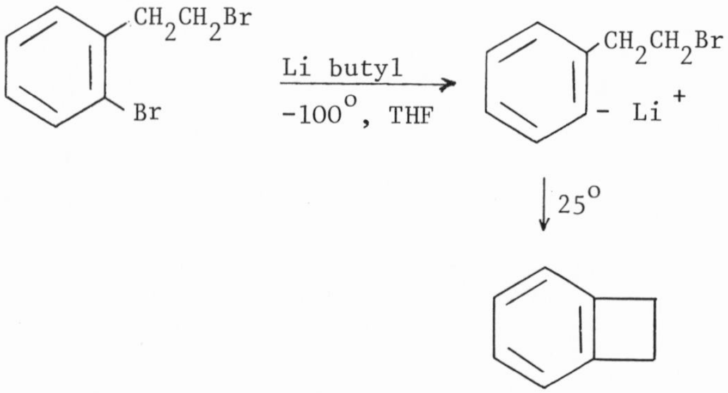

Copper catalysis of nucleophilic substitution has been a subject of mechanistic investigation and many synthetic applications continue to be added to the established lists; for example, reaction of aryl halides with CF_3SCu allows the introduction of the CF_3S^- group (D.C. Remy *et al.*, J. org. Chem., 1976, 41, 1644). The synthetic value of copper catalysis is well established; other metals may also be of value for example, catalysis of the nucleophilic reactions of aryl halides with nickel compounds is established (R. Cramer and D.R. Coulson, J. org. Chem., 1975, 40, 2267 and D.G. Morrell and J.K. Kochi, J. Amer. chem. Soc., 1975, 97, 7262). Palladium salts also can play a part in the synthetic chemists armoury, for example, replacement of bromide by cyanide is achieved even for unactivated aryl bromides using potassium cyanide and palladium(II) salts (K. Takagi et al., Bull. chem. Soc. Japan, 1975, 48, 3298).

Studies of nucleophilic substitution in polyfluoro- aromatics continue to be published in relatively large numbers. Earlier differences of approach over the rationalization of results in this field appeared to have been resolved (W.K.R. Musgrave *et al.*, Chem. Comm., 1974, 239), the matter has been elaborated elsewhere (see for example Chambers and James *loc. cit.* for a summary).

(v) Photochemical reactions

Many photochemical reactions have been described, for example, reaction of aryl fluorides with secondary amines gives both conventional and cine-substitution (D. Bryce-Smith,

A. Gilbert and S. Krestonosich, Chem. Comm., 1976, 405),
para-difluorobenzene giving the *para*-adduct with diethyl-
amine.

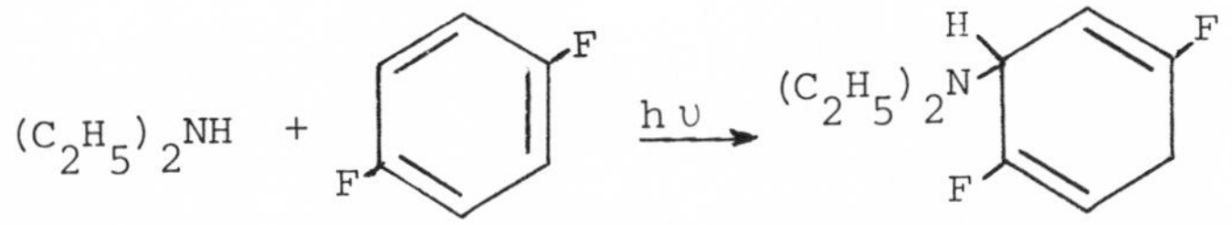

R.H. Schuler, J. Amer. chem. Soc., 1972, $\underline{94}$, 1056).

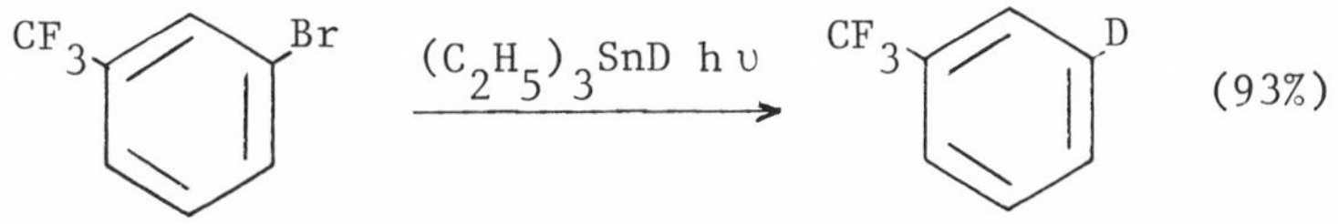

A useful review of photosubstitution of aromatics has
appeared (J. Cornelisse and E. Havinga, Chem. Rev., 1975,
$\underline{75}$, 353). The photoisomerizations of fluorinated aromatics
referred to in the second edition have been considerably
extended; this area has been reviewed by Bryce-Smith and
Gilbert ("Photochemical Reactions of Aromatic Compounds" in
MTP International Review of Science, Organic Chemistry Series
One, 1973, Volume 3, Butterworths; and Tetrahedron, 1976,
$\underline{32}$, 1309; the topic also features prominantly in various
annual literature surveys).

(vi) *Isomerization reactions*

Halogen substituents in benzenoid aromatics change their
relative positions under a variety of experimental conditions. Thus, *meta*-dichlorobenzene is isomerized to a
mixture of *ortho* and *para* isomers when passed over aluminium
chloride and magnesium sulphate at 160° (J. Suwinski and
W. Zielinski, Roczniki Chem., 1977, $\underline{51}$, 455). Perhaps the
best known isomerization is the "halogen dance" described by
Bunnett (Accounts Chem. Res., 1972, $\underline{5}$, 139). Treatment of
1,2,4-tribromobenzene with strong base in liquid ammonia
gives 1,3,5-tribromobenzene via a series of exchanges of the
kind shown on the following page

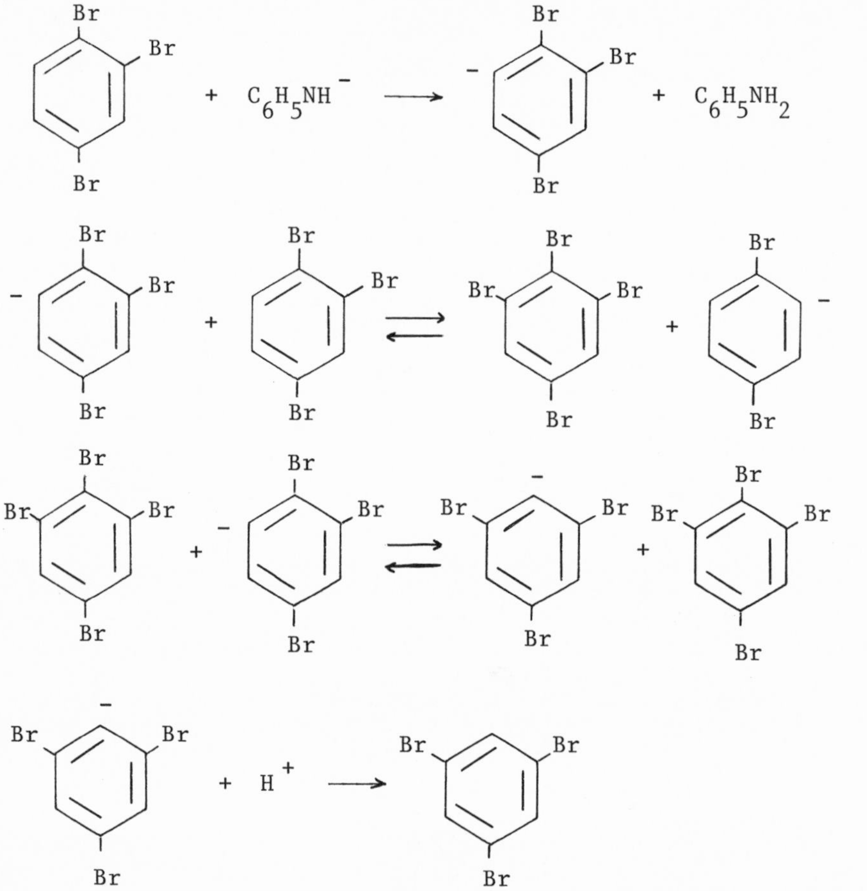

154

The isomerization halogenated bicyclo[2,2,0]hexa-2,5-dienes to halogenobenzenes has been the subject of both theoretical and experimental investigations.

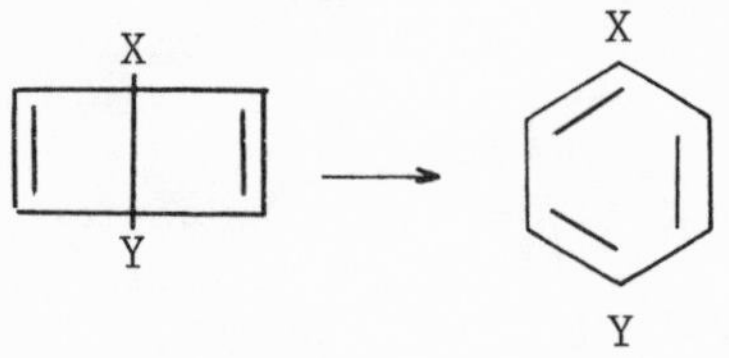

For example, in the isomerization - for Y = H, X = H, Cl, F the relative rates are 1, 90, 370, whereas for X = Y = Cl a dramatic decrease in rate is observed (R. Breslow, J.

Napierski and A.H. Schmidt, J. Amer. chem. Soc., 1972, 94, 5906). Pentafluorobicyclo[2,2,0]hexa-2,5-dienes carrying substituents at position 1 isomerize to pentafluorobenzene derivatives fastest when the substituent is electron withdrawing (E. Ratajczak, Chem. Abs., 1973, 79, 136300n; see also M.J.S. Dewar and S. Kirschner, Chem. Comm., 1975, 463).

(vii) Radical reactions

There have been several studies of the generation and reactivity of halogenoaromatic radicals and the reactions of halogenobenzenes with radicals. Haszeldine and co-workers have shown that the pentafluorophenyl radical can be generated either by photolysis of pentafluorophenyliodide (J. chem. Soc. Perkin I, 1974, 1715) or by hypochlorite oxidation of pentafluorophenylhydrazine (*ibid*, 1974, 1740); the radical displays substantial electrophilic character. Photolysis of trifluoromethyl iodide in benzene or halogenobenzenes gives benzotrifluorides (J.M. Birchall *et al.*, J. chem. Soc. Perkin II, 1975, 435). The thermal reaction between aryl halides and alkenes carrying electron withdrawing substituents also presumably involves radicals; for example, heating a mixture of iodobenzene and methyl acrylate at 150° gives methyl cinnamate (M. Julia and M. Duteil, Bull. Soc. chim. France, 1973, 2790).

Radical attack on perfluoroaromatics and some polyfluoroaromatics gives rise to intermediates which would have to eliminate a fluorine atom to re-aromatize, this process does not occur as readily as loss of hydrogen and consequently the intermediate radicals may undergo other reactions, for example dimerization. Yakobson (Zhur. org.

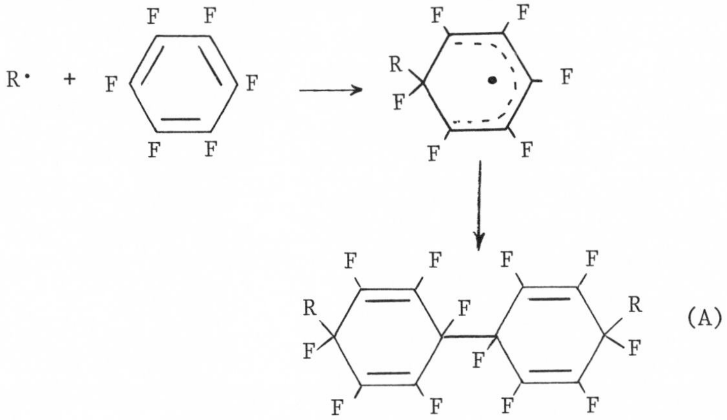

Khim., 1974, 10, 791 and 797; and 1975, 11, 215) and
Williams (J. fluorine Chem., 1974, 4, 347 and 355; and 1975,
5, 61) and their co-workers have reported several examples
of this type of reaction. Direct fluorination of hexafluorobenzene with fluorine gives the dimer A (R = F) (I.J.
Hotchkiss, R. Stephens and J.C. Tatlow, J. fluorine Chem.,
1975, 6, 135). The reactions of polyfluoroaromatic compounds
with aryl and alkyl radicals have been reviewed (L.S. Kobrina,
Russ. Chem. Rev., 1977, 46, 348).

2. Alkylbenzenes with halogen in the side chain

   (a) Methods of preparation
   Many of the methods listed in the Second Edition have
found use in the synthesis of novel compounds, in some cases
reaction procedures have been modified. Thus, for example,
sulphur tetrafluoride/hydrogen fluoride mixtures have been
successfully used in the synthesis of poly(trifluoromethyl)
benzenes from the corresponding polycarboxylic acids (L.M.
Yagupol'skii *et al.*, see Chem. Abs., 1978, 88, 190255).
Chloromethylation has received considerable attention (see,
for example, Specialist Periodical Reports on Aromatic and
Heteroaromatic Chemistry, *loc. cit.*) probably because of the
use of aryl chloromethyl groups as initial anchor points
for polymer bound syntheses of biopolymers.
   Free radicals have been used in a variety of syntheses.
Benzyl radicals add to fluorinated olefins to give poly-
fluoroalkylarenes (H. Kimoto, H. Muramatsu and K. Inukai,
 Chem. Letters, 1974, 791).

$$C_6H_5CH_2 \cdot \; + \; CF_2{=}CFCF_3 \; \longrightarrow \; C_6H_5CH_2CF_2CHF_3$$

Photochemically generated perfluoroalkyl radicals react with
benzene to give perfluoroalkylbenzenes (M.Y. Turkina, I.P.
Gragerov and S.L. Dobychin, Zhur. org. Khim., 1971, 7, 1541;
see also Birchall *loc. cit.*, section 1b(vii)). Likewise,
perfluoroalkyl radicals generated by thermal decomposition of
perfluoroheptylcopper react to give substitution products
with a variety of aromatic substrates (P.L. Coe and N.E.
Milner, J. fluorine Chem., 1972, 2, 167), the observed
substitution pattern suggests that the perfluoroalkyl
radical is highly electrophilic (cf. Haszeldine's work on
$C_6F_5 \cdot$ cited in section 1b(vii)).
   In the field of polyfluoroaromatic compounds Yakobson
and co-workers have made extensive use of the pyrolysis of

polyfluoroaromatics in the presence of known precursors of difluorocarbene; for example, they have shown that co-pyrolysis of tetrafluorophthalic anhydride and tetrafluoro-ethylene yields perfluorobenzobicyclobutene and hexafluoro-benzene can be converted to perfluorotoluene (G.G. Yakobson et al., Izvest. Akad. Nauk. S.S.S.R., Ser. Khim., 1973, 2827 and Zhur. org. Khim., 1975, 11, 2123).

In polymethylbenzenes side chain chlorination may result from initial electrophilic attack at the nucleus (E. Baciocchi *et al.*, Tetrahedron Letters, 1975, 2265, 2268).

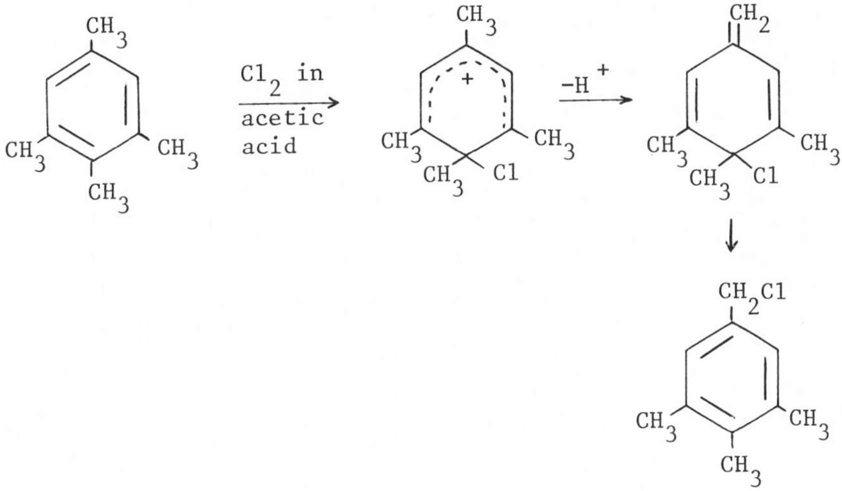

The interpretation put forward by these authors disputes the intermediacy of cation radicals advocated previously (J.K. Kochi, Tetrahedron Letters, 1974, 4305). Tetrakis(trifluoro-methyl)cyclopentadienone reacts with acetylene at room temperature with elimination of carbon monoxide to give the expected benzene derivative (S. Szilagyi, J.A. Ross, and D.M. Lemal, J. Amer. chem. Soc., 1975, 97, 5586). Another novel

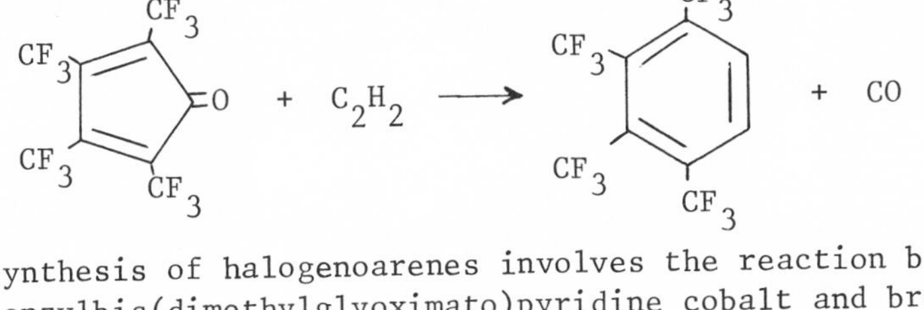

synthesis of halogenoarenes involves the reaction between benzylbis(dimethylglyoximato)pyridine cobalt and bromotrichloromethane in chloroform solution to give β,β,β-tri-chloroethylbenzene and benzyl chloride (T. Tunabiki, B. Dass Gupta, and M.D. Johnson, Chem. Comm., 1977, 653); this reaction being the first example of homolytic displacement by an alkyl group at saturated carbon.

(b) Properties and reactions

(i) Benzylic halides
Benzylic halides are versatile intermediates in synthesis and many examples of their use have been published. For example: they can be methylated with lithium dimethylcopper (G.H. Posner and D.J. Brunelle, Tetrahedron Letters, 1972); some can be coupled in the presence of iron(II) chloride tetrahydrate in dimethyl sulphoxide (T. Shirafuji, Y. Yamamoto and H. Nozaki, Bull. Chem. Soc. Japan, 1971, <u>44</u>, 1994), e.g. $(C_6H_5)_2CCl_2 \longrightarrow (C_6H_5)_2C=C(C_6H_5)_2$.

Two new routes to aldehydes involve *either* the use of methyl thiomethylsulphoxide in the presence of sodium hydride in tetrahydrofuran (K. Ogura and G. Tsuchihashi, Tetrahedron Letters, 1971, 3151).

$$C_6H_5CH_2Br \; + \; CH_3SOCH_2SCH_3 \xrightarrow{\text{NaH-THF}} \xrightarrow{\text{H}^+} C_6H_5CH_2CHO,$$

or mercury(I) nitrate (A. Mckillop and M.E. Ford, Synthetic Comm., 1974, <u>4</u>, 45).

$$C_6H_5CH_2Br + Hg_2(NO_3)_2 \longrightarrow C_6H_5CH_2ONO_2 \xrightarrow{\text{OH}^-} C_6H_5CHO;$$

Finally benzylic bromides can be converted to deuterides with high isotopic purity via a trimethylstannane derivative (W.A. Asomaning, C. Eaborn and D.R.M. Walton, J. chem. Soc. Perkin I, 1973, 137).

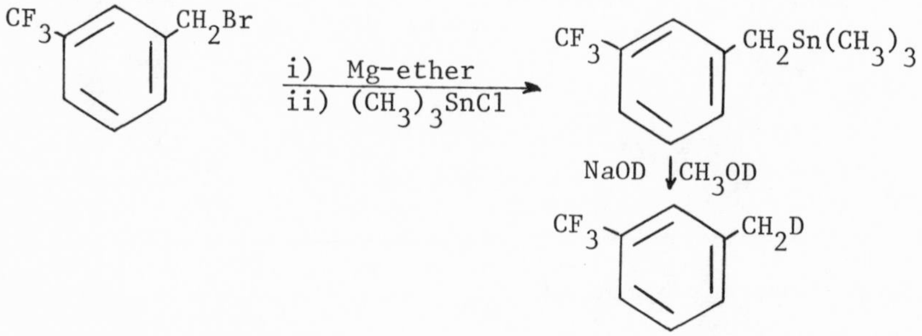

(iv) Compounds with more than one side chain

Dehalogenation of bis-benzylic halides provides some interesting chemistry. Thus 1,3-bis(chloromethyl)-4-methyl-benzene, on reaction with sodium under Wurtz coupling conditions, gives isomeric mixtures of metacyclophanes (C. Glotzmann *et al.*, Tetrahedron Letters, 1975, 675). α,α'-Dibromo-ortho xylene can be debrominated by a complex of copper with t-butylisocyanide to give the much investigated ortho xylylene as an intermediate which may be trapped via Diels-Alder reactions with olefins, the analogous reaction with α,α,α',α'-tetrabromo-ortho xylene provides a trapable dibromo-ortho xylylene (Y. Ito, K. Yonezawa and T. Saegusa, J. org. Chem., 1974, <u>39</u>, 2769).

Hexakis(trifluoromethyl)benzene, as reported in the Second Edition, is subject to photochemical valence bond isomerization to give the Dewar, prismane and benzvalene isomers. The final member of the first complete set of benzene valence isomers, perfluorohexamethylbicyclopropenyl, has now been prepared by coupling of 2-iodotris(trifluoromethyl)cyclopropene (M.W. Grayston and D.M. Lemal, J. Amer. chem. Soc., 1976, <u>98</u>, 1278), it is thermally isomerized to

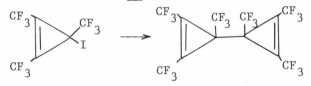

hexakis(trifluoromethyl)benzene. Several papers concerning the isomerization of polyperfluoroalkylbenzenes have been published by Haszeldine and co-workers (Chem. Comm., 1975, 323; J. chem. Soc. Perkin I, 1974, 1736; 1975, 2005 and 2010); somewhat surprisingly it was shown that the equilibrium between hexakis(pentafluoroethyl)benzene and its Dewar benzene analogue favoured the Dewar benzene above 551K.

Chapter 4

NUCLEAR HYDROXY DERIVATIVES OF BENZENE AND ITS HOMOLOGUES

J.H.P. TYMAN

1. Monohydric Phenols

Introduction

Accounts of the chemistry of phenols have appeared in 'Chemistry of the Hydroxyl Group', ed. S. Patai, Interscience, New York, 1971, D.A. Whiting, 'Phenols', in 'Comprehensive Organic Chemistry' ed. by Sir D.H.R. Barton and W.D. Ollis, Pergamon, Oxford 1979 vol. 1, p.707). Summaries of naturally-occurring phenolic compounds are available in the 'Handbook of Naturally Occurring Compounds', T.K. Devon and A.I.Scott, Academic Press, New York, 1975; K.Nakanishi, T.Goto, S.Ito, S. Naton and S. Nozoe, Natural Products Chemistry, Academic Press, New York, 1974.

The natural occurring compounds arising biogenetically from shikimic acid have been reviewed (B. Ganem, Tetrahedron, 1978, *34*,3353; E. Haslam, 'The Shikimate Pathway', Butterworths, London, 1974).

Properties

Spectroscopic properties of phenols have received a lot of attention particularly with regard to ^{1}H NMR and ^{13}C NMR studies. ^{13}C NMR data for phenolic compounds have become available and several publications of general interest have appeared: J.B. Stothers, Carbon-13 NMR Spectroscopy, Academic Press, London, 1972; G.C. Levy and G.L. Nelson, Carbon-13 NMR for Organic Chemists, Wiley Interscience, New York, 1972; L. Crombie, G.W. Kilbee and D.A. Whiting, J. chem. Soc., Perkin 1, 1975, 1497.

Solvent-induced alterations in chemical shifts are well established and can be rationalised in certain cases (A.Pelter *et al.*,) Tetrahedron, 1971, 27, 1625). Lanthanide shifts for monohydric phenols have been analysed (D. Molho *et al.*,

Tetrahedron Letters, 1976, 2825).   Very small couplings have
been observed between aromatic protons and methyl or methylene substituents (F. Scheinmann *et al*., J. chem. Soc., (C),
1971, 3389) and between aromatic protons *ortho* to hydroxyl
groups (N.L. Allinger *et al*., J. Amer, chem. Soc., 1970,
92,5200).

The infrared absorption of phenols has been reviewed.
(C.H. Rochester in 'Chemistry of the Hydroxyl Group', ed.
S. Patai, Interscience, New York, 1971, pt. 1, p. 327).

The mass spectra of phenolic compounds of various types
have been referred to (R.C. Cooks, in 'The Chemistry of the
Hydroxyl Group,ed. S. Patai, Interscience, 1971, ch. 19).

A wide variety of techniques is now available for the
separation of phenols.

Gas liquid chromatographic, thin layer chromatographic,
column chromatographic and high performance liquid chromatographic investigations on phenolic materials have been
summarised (J. Chromatog., Subject index 1970-1975, *5l-ll0*:
266-268:   1976, *lll-l20*:106; 1977, *l2l-l30*:67-68; *l3l-l40*:
69-70; 1978;*l4l-l60*: 454).   The normal phase and reversed
phase HPLC separation of alkylphenols has been described
(J.F. Shabron, R.J. Hurtabise and H.F. Silver, Analyt. Chem.,
1977, *49*,832; D.J. Pietzryk and Chi-Hong Chu, *ibid*.,p. 860).
Analytical procedures for long chain phenols based on TLC,
GLC, and mass spectrometry have been summarised (J.H.P.Tyman,
Chem. Soc. Rev. 1979, *4*,499).   Detection and determination
of the phenolic hydroxyl group has been discussed (S.Siggia,
J.C. Hanne and T.R. Stengle, in 'The Chemistry of the Hydroxyl
group', ed. S. Patai, Interscience, 1971, ch. 6).

The acidity and inter-and intra-molecular hydrogen
bonding of phenols in relation to that of alcohols has been
described in detail (C.H. Rochester, in 'The Chemistry of the
Hydroxyl Group'. ed. S. Patai, Interscience, 1971, ch.7).

The ambident nature of the phenols has been the subject
of several investigations, notably by P. Giacommello and co-
workers.   To avoid solvation problems gas phase reactions
of the t-butyl cation with phenol and with anisole have been
carried out with the finding that the competition between
n-type and π-type nucleophilic centres lies heavily in
favour of the former leading to predominant formation of a
t-butylated oxonium ion (P. Giacommello *et al*., Chem. Comm.,
1976).

Studies have been concerned with the equilibrium
protonation of 1,3,5-trinitrobenzene-phenoxide complex and
are considered to provide the first example of the phenoxide

ion acting as ambident nucleophile (E. Buncel, A. Jonczyk, and J.G.K. Webb, Canad. J. Chem., 1975, 53,3761).

The exchange reactions of ^{18}O labelled phenols have been described (S. Oae and S. Tamagaki, in 'The Chemistry of the Hydroxyl Group', ed. S. Patai, Interscience, 1971, ch. 15).

(a) Preparation

No basically new methods for the general synthesis of phenols have been introduced in the last decade. There has, however, been considerable interest in radiolytic and photo-lytic hydroxylation, acyloxylation, benzoyloxylation and in ι electrochemical procedures. Biological hydroxylations have received some attention.

Enzymic hydroxylation of aromatic rings can be involved with aromatic compounds derived from the shikimate biogenetic pathway (E. Haslam, 'The Shikimate Pathway', Butterworths, London,1974). Cinnamic acid can be hydroxylated to 4-hydroxy, 3,4-dihydroxy and 3,4,5-trihydroxycinnamic acid. A compilation of plant monooxygenases, phenolases, and micro-organisms has been made (S.K. Erickson et al., "Chemistry of the Hydroxyl Group", ed. S. Patai, Interscience, New York, 1971, vol. 2. p 776). In vivo and in vitro hydroxylation can be accompanied by proton rearrangements involving arene oxides (G.J. Kasperek and T.C. Bruice, J. Amer. chem. Soc., 1972, 94,108; G.J. Kasperek et al., ibid., 7876; K.B.Sharphen and T.C. Flood, ibid., 1971, 93,2316).
Chemical hydroxylating reagents are usually based on hydrogen peroxide, a transition metal catalyst and a redox buffer. The chemistry of Fenton's, Hamilton's, and Uden-friend's reagent has been reviewed (D.I. Melelitsa, Russ.chem. Rev., 1971, 40,563; D.F. Sangster in 'Chemistry of the Hydroxyl Group', ed. S. Patai, Interscience, New York, 1971, vol. 1 p. 163). Their synthetic utility is, however, limited and generally low yields result. Phenol has been obtained

in 57% yield from benzene with hydrogen peroxide, with iron
II and copper II ions (K. Omura and T. Matsuura, Tetrahedron,
1968, 24,3475; C. Walling and R.A. Johnson, J. Amer. chem.
Soc., 1975, 97,363).   Peroxydisulphate in the presence of a
suitable metal ion oxidant, particularly copper II causes ring
hydroxylation of benzene derivatives (C. Walling. D.M.Camaioni
and Sung Soo Kim, J. Amer. chem. Soc., 1978, 100,4814).
Hydroxylation reactions of benzene, alkylbenzenes and halogenobenzenes with hydrogen peroxide in superacidic media at
low temperature have given high yields of monohydroxylated
products.   The products are protonated thus becoming deactivated against further reaction.   1,2-Shifts of methyl substituents occur suggesting the intermediacy of benzene oxides
(G.A. Olah and R. Ohnishi, J. org. Chem., 1978, 43,865).

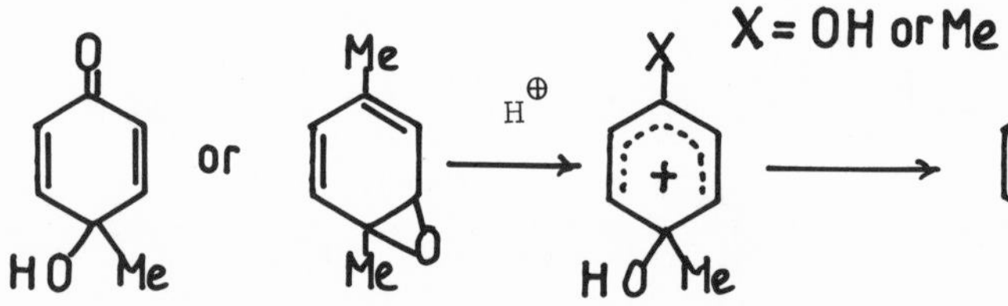

Trifluoroacetoxylation of a number of substituted benzenes
has been achieved anodically as shown in Table 1 (L.Eberson
et al., Acta. chem. Scand., (B), 1975, 29,715)

(Table 1)

Anodic Trifluoroacetoxylation of Substituted Benzenes (PhX)

X	O(%)	m(%)	p(%)	current yield(%)
H	–	–	–	27
CO$_2$Me	51	34	15	65
NO$_2$	22	59	19	60
CF$_3$	35	47	18	31
COMe	54	32	14	67
COPh	70	18	12	21
CN	45	30	25	10

The first step in the reaction is electron transfer from the
aromatic compound.   The initially formed trifluoroacetates
are readily hydrolysed during work-up to give the phenols.
The optimised anodic oxidation of iodobenzene in the presence
of benzene gives 95% yields of diphenyliodonium salts which
can be hydrolysed to phenol and iodobenzene (H. Hoffelner,

H.W. Lorch, and H. Wendt,Electronoalyt Chem. interfacial
Electrochem., 1975, *66*,183). Hydroxylated alkyl benzenes
have been obtained by attack of oxygen (3P) atoms generated
radiolytically from carbon dioxide (A. Hori *et al.*, Chem. Comm.,
1978, 16).

Aryl and aroyl peroxides, arenesulphonyl peroxides
(E.M. Levi, P. Kovacic and J.F. Gormisch, Tetrahedron, 1970,
26,4536) hydroperoxides,peroxy acids and peroxyesters have all
been utlised for hydroxylation and acyloxylation. By heating an alkylbenzene (methyl, ethyl, isopropyl, or isobutyl)
with tert-butylhydroperoxide, Me_3CO_2H, in the presence of
copper (I) chloride or copper naphthenate and decomposing the
resultant t-butyl α- phenylalkylperoxide by heating in the
presence of an acid catalyst while continuously removing the
volatile reaction products (carbonyl compounds and alcohols)
phenol is produced in high yield. The peroxide from ethylbenzene heated at 70-5° in the presence of toluene-p-sulphonic
acid while nitrogen was passed through the mixture gave a 77%
yield of phenol (based on the peroxide consumed) (B.J.Suberoff,
T.S. Simmons, Lummus Co., USP 4,006,194, Chem. Abs., 1977,
86,139619).

Di-isopropylperoxy dicarbonate $(Pr^1OCO_2)_2$ with cupric
chloride gives fair yields with little side reaction.
Toluene gives 85% of the tolylisopropyl carbonate (*o*,57,*m*,15,
p,28) mesitylene a quantitative yield of 2,4,6-trimethyl-
phenylisopropylcarbonate, biphenyl an 86% yield (*o*,47,*m*,4,*p*,49),
and naphthalene an 89% yield (α,93,β8) (P. Kovacic, C.G.Reid,
and M.E. Kurz, J. org. Chem., 1969, *34*,3302). Electrophilic
acyloxylation takes place in the presence of aluminium chloride,
e.g. toluene gives o/p products (∿1:2) in 52% yield while
anisole gives o/p products (∿1:4) in 76% yield. Acyloxylation may also be achieved with lead tetraacetate or tetra-
trifluoroacetate in moderate yields (R.A. McClelland, R.O.C.
Norman, and C.B. Thomas, J. chem. Soc., Perkin I, 1972, 562:
S. Sternell *et al.*, Tetrahedron Letters, 1972, 1763).

Catalytic conversion of benzoic acid into phenol in 76%
yield occurs with steam/air over copper (II) oxide, zirconia
and potassium oxide on alumina during 3 hours at 280-310°C
and 6-7 atm. pressure (Lummus Co., Ger.Pat 2844195).

Many of the methods such as alkali fusion of sulphonates,
hydrothermal decomposition of aryldiazonium salts and acidic
rearrangement of cumene hydroperoxide for the preparation of
phenol are applicable to the alkyl phenols. However, in
certain cases special procedures have been applied. 3-Alkyl-
phenols have been prepared by sulphonation of the alkyl-

benzene, selective hydrolysis of the undesired sulphonate
(4- and 2-isomers) and treatment of the non-hydrolysed alkyl-
benzene sulphonic acid with molten alkali.   (L.Gerd,Koppers
Co. Inc., Ger. Pat., 2,362,884; Chem. Abs., 1974, $8l$,189294).

2-Methyl and 2,6-dimethylphenol have been obtained by the
interaction of phenol and methanol (1:6 molar proportion) with
a catalyst composed by weight of ferric oxide (50), zinc oxide
(45) and chromic oxide (5) in a fluidised or fixed bed at 370°
(K. Shibata, T. Kiyoura and J. Kitagawe, Mitsui-Toatzu Chemicals Co.Ltd., Jap. Pat. 72, 07 020; Chem. Abs., 1972,
76,140184).

Cyclohexadienones are a particular source of alkyl
phenols.   Hydrogenolytic cleavage occurs in hexane solution
with 5% Pd-C of the trimethyl compound shown to give 2,4,6-
trimethyl phenol (B. Miller and L. Lewis, J. Org. Chem., 1974,
39,2605).

2,5-Dimethylcyclohexane-1,4-dione with hot concentrated
hydrobromic acid or hydrochloric acid gives 2,5-dimethylphenol
in 80-85% yield (C.G. Ras, S. Rengaraju, and M.V. Bhatt, Chem.
Comm., 1974, 584).   Cyclohexa-2,4-dienones,which are discussed in detail later,can undergo acid-catalysed rearrangements to give 2,3- and 2,6-di-substituted phenols (B. Miller,
J. Amer. chem. Soc., 1970, 92, 432, 6246, 6252).   With organolithium and organomagnesium reagents the acetate illustrated
gives mixtures of ethers and conjugate addition products.
When the reagent is tertiary or benzylic the latter product
may result exclusively (A.J. Lin and A.C. Sartorelli, J.org.
Chem., 1973, 38,813).

The tetraphenylcyclone shown gives by way of an adduct,
followed by refluxing in bromobenzene for 24 hours, 2,3,4,5-
tetraphenylphenol in 90% yield (E.A. Harrison Jr., Org. Prep.

Int., 1975, 7,71).

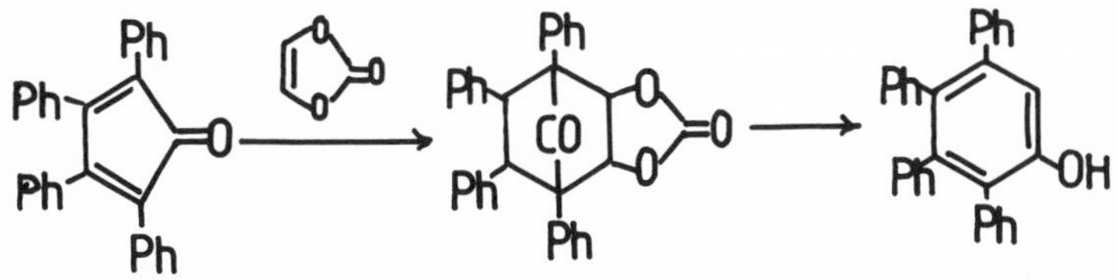

3-Arylphenols can be obtained by treatment of 2-aryl-6-oxocyclohex-1-enyl acetic esters with sodium hydride in dimethylformamide at 100° (D. Nasipuri, A. Bhattacharyya and B.G. Hazra, Chem. Comm., 1971, 660). 4-Phenylphenol has been obtained by way of a phenylepoxyoxepin (F. Bouvelle-Wargnier, M. Vincent and J. Chuck, Chem. Comm., 1979, 584).

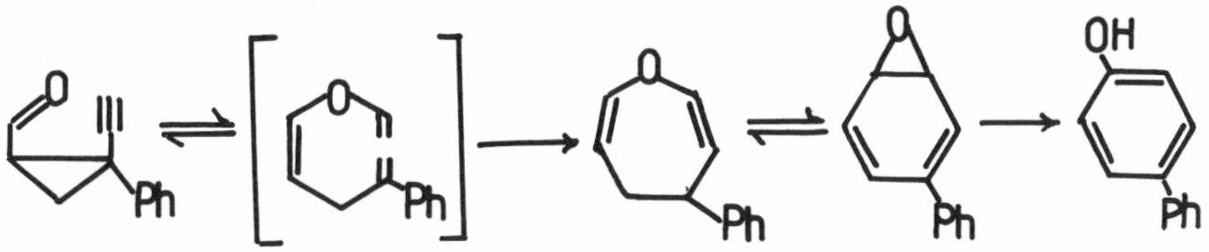

Alkylphenols may be obtained from azasulphonium salts (from N-chlorosuccinimide and dialkylsulphide) in the following way after final desulphurisation (P.D. Gassman and D.R. Amich, Tetrahedron Letters, 1974, 889).

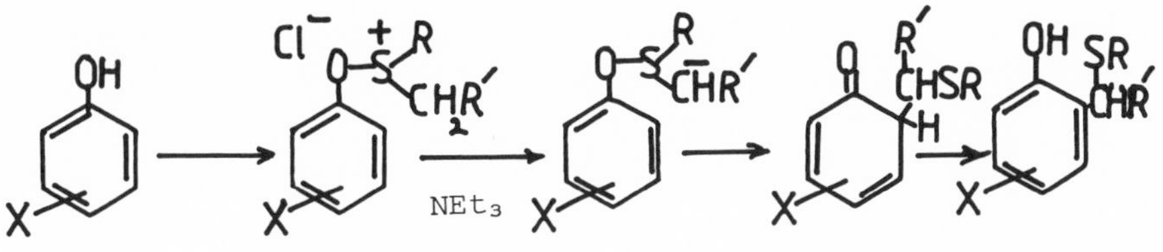

Aryl magnesium bromides give excellent yields of phenols upon reaction with molybdenum peroxide in hexamethylphosphoric triamide/pyridine (N.J. Lewis, S.Y. Gabke, and M.R. De La Mater, J. org. Chem., 1977, 42,1479).

A procedure for the formation of phenols from diazonium salts, considered to be the method of choice, which avoids the usual highly acidic conditions consists in adding copper (I) oxide to the dilute solution of the diazonium salt and a large excess of copper (II) nitrate (T. Cohen, A.G. Dietz Jr., and J.R. Misec, J. org. Chem., 1977, 42,2053).

The Claisen rearrangement, discussed in a subsequent section on phenolic ethers, can be used in certain cases for the synthesis of alkenyl and alkyl phenols.

A number of naturally occurring phenols particularly of the long chain group have been synthesised including the 8',(Z)-monoene, and the 8',(Z), 11'(Z)-diene constituents of cardanol which comprises the four material shown (J.H.P.Tyman and J. Caplin, Chem. and Ind., 1973, 40, 953) by the routes given,

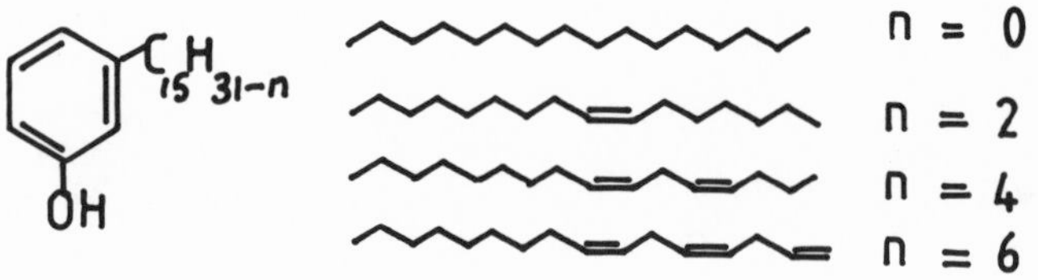

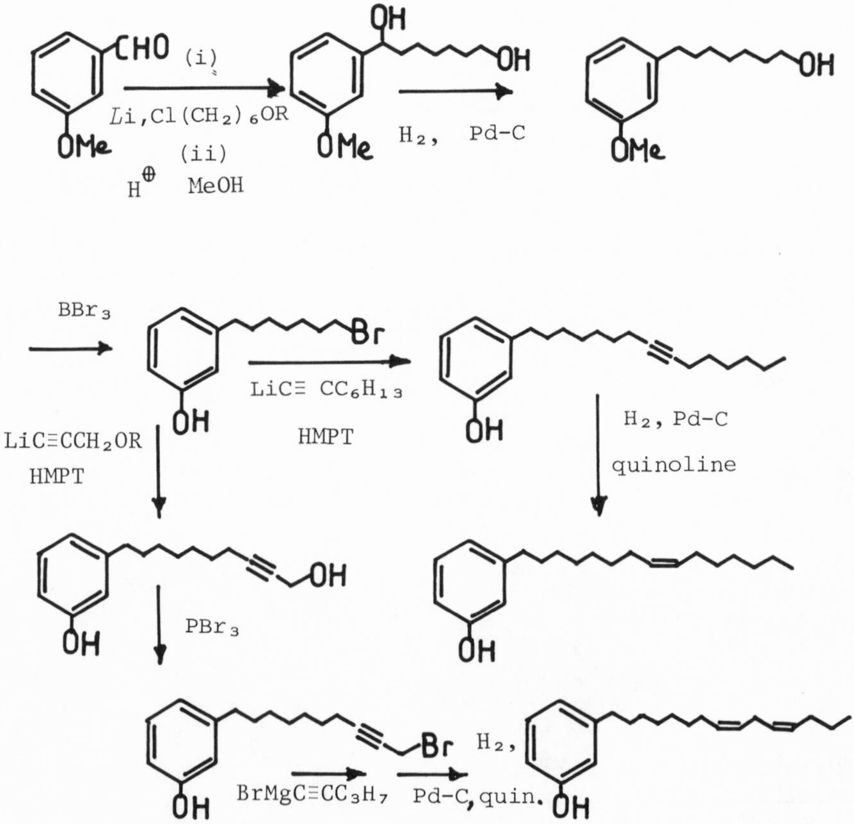

The chemistry of non-isoprenoid long chain phenols has been reviewed (J.H.P. Tyman, Chem. Soc. Rev., 1979, 8, 499).

The natural phenols of *Anacardium occidentale* (cashew nut)
contain the phenolic acid anacardic acid, comprising four
constituents, cardol (see dihydric phenols) and smaller amounts
of cardanol and 2-methylcardol.   In the hot processing
(J.H.P. Tyman, Chem. and Ind., 1980, 57), of the cashew nut
decarboxylation occurs and anacardic acid becomes converted
into cardanol.   The mixture of resulting phenols has important industrial uses.

(b)Reactions of the Hydroxy Group

(i) Esterification
        Phenols can be conveniently esterified directly with
carboxylic acids by the use as a catalyst of boric and sulphuric acids (W.C. Lowrance, Tetrahedron Letters, 1971, 3453)
together with azeotropic removal of water.   Methods involving trifluoroacetic anhydride, trifluoromethane sulphonic
anhydride, polyphosphoric acid or dicyclohexylcarbo-diimide
catalysts are superior to the Fischer-Speier (or sulphuric
acid) method.   The use of the highly active acylation catalysts 4-dimethylaminopyridine and 4-pyrrolidinylpyridine has
been reviewed (G. Hölfe, W. Steglich and H. Vorbrüggen, Angew.
Chem. internat. Edn., 1978, *17*,569).
        The chemistry of the inorganic esters of phenol such as
the phosphates, borates and sulphates has been reviewed
(P. Salomaa, A. Kankaaperä and K. Pihlajd, in 'Chemistry of
the Hydroxyl Group' ed. S. Patai, Interscience, New York, 1971,
vol. 1, p.481).   5-iso-Pentadecylresorcinol-1,3-disulphate has
been isolated from a streptomyces source (H. Umezawa *et al.*,
J. Antibiot., 1971, *24*,870).   Aryloxyphosphonium chlorides
are formed nearly quantitatively from the interaction of
phenols, triphenylphosphine and carbon tetrachloride in acetonitrile at 0° (R. Appel, K. Warning and K.-D. Ziehn, Ann., 1975,
406).   Phenol interacts with phosphorus oxychloride as shown
in the presence of magnesium oxide for a short period (H.Stoye,
E. Ger. P. 134104).

$$C_6H_5OH \ + \ POCl_3 \ \xrightarrow{\text{MgO,75°,15min.}} \ C_6H_5O\overset{\overset{\textstyle O}{\|}}{P}Cl_2 \ + \ HCl$$

Phenylbenzenesulphonate undergoes photochemical rearrangement
into *o*- and *p*-hydroxyphenylsulphone (Y. Ogata, K. Takagi, and
S. Yamada, J.chem. Soc., Perkin II, 1977, 1629).

(ii)   Replacement

Hindered esters can be selectively cleaved in the presence of phenolic ethers by boron trichloride in dichloromethane (P.S. Manchard, Chem. Comm., 1971, 667).   Phenyl 2-
chlorobenzoate (R=*o*-ClC$_6$H$_4$) can undergo a preferential acylation reaction with nitromethane in dimethylsulphoxide containing potassium t-butoxide to give the β-acylnitromethane
(G.F. Field and W.J. Zally, Synthesis, 1979, 295).

$$CH_2NO_2 + RCO_2Ph \longrightarrow RCOCH_2NO_2 + PhOH$$

The replacement of a phenolic hydroxyl by an amino group
is an unusual reaction.   Aryl diethyl phosphates can be
reductively cleaved by potassium with potassamide in anhydrous
ammonia to give the primary arylamine (R.A. Rossi and J.F.
Bunnett, J. org. Chem., 1972, *36*,3570).

$$ArOH \longrightarrow ArOPO(OEt)_2 \xrightarrow{K} \left[ArOPO(OEt)_2\right]^{\overline{\cdot}} \longrightarrow Ar^{\cdot}$$

$$+ (EtO)_2PO_2^- \xrightarrow{NH_2^-} \left[ArNH_2\right]^{\overline{\cdot}} \longrightarrow ArNH_2$$

The rearrangement of 2-aryloxy-2-methylpropanamides into
N-aryl-2-hydroxy-2-methylpropanamides in hexamethylphosphoric
triamide is another route for the conversion of phenols into
anilines (R. Bayles *et al.*, Synthesis, 1977, 33).
Phenols can be converted into thiols through the dialkylthio-
carbamate and thence into the corresponding sulphonic acid
(J.E. Cooper and J.M. Paul,J. org. Chem., 1970, *35*,2046).

A useful addition to synthesis is an efficient procedure
for replacement of hydroxyl group by hydrogen and several new
methods have appeared.   The catalytic reductive removal of
phenolic hydroxy groups in the form of their iso-urea or
carbonimidic ester derivatives can be carried out at ambient
temperature (E. Vowinkel and C. Wolff, Ber., 1974, *107*,907,
1379;E.Vowinkel and I. Buthe, *ibid.*, 1353).   A complementary
method is the smooth cleavage of the potassium aryl sulphate
derivative by catalytic hydrogenation (W. Lonsky, H. Traitler,
and K. Kratze, J. chem. Soc., Perkin I, 1975, 169).
Sulphonic esters have also been used in catalytic hydrogenation with palladium-carbon at moderate temperature and

pressure (K. Clauss and H. Jensen, Angew. Chem. internat. Edn., 1973, *12*,913). The hydrogenolysis of various derivatives is much less drastic than the direct replacement of the phenolic hydroxyl and methoxyl groups by lithium aluminium hydride at 350° (Th. Severin and I. Ipach, Synthesis, 1973, 796). The aryloxytetrazole formed from the phenol and 5-chloro-1-phenyl-tetrazole has been another useful derivative for enabling regiospecific removal of the phenolic hydroxy group (e.g. in a recent morphine synthesis, L. Maat *et al.*, Rec. trav. Chim., 1976, *95*,24). Another derivative for this type of reaction is the carbamate (J.D. Weaver and E.J. Eisenbraum, Chem. and Ind., 1973, 187).

$$\text{ArOH} \longrightarrow \text{ArOCONHPh} \xrightarrow[\text{Pd-C}]{\text{H}_2,} \text{ArH}$$

(iii) Protection

The need to protect the phenolic hydroxyl group during synthesis frequently arises and for this, conversion into the methyl ether has been widely used. The benzyl ether can be readily prepared by refluxing the phenol with benzyl chloride in acetone solution containing potassium iodide and potassium carbonate and cleavage can be effected either with trifluoro-acetic acid at ambient temperature or by hydrogenolysis over a palladium catalyst. Selective blocking procedures have been described (L. Crombie, P.J. Ham, and D.A. Whiting, Phytochem., 1973, *12*,703). The difficulty of the removal of the methyl group in certain instances with existing reagents has led to the examination of numerous alternative ways of protection. The trimethylsilyl group has been widely used in organic synthesis and also for volatilising phenolic materials for gas liquid chromatography. Trimethylsilylation can be effected with trimethyl silyl chloride, hexamethyldisilazane ($\text{Me}_3\text{SiNHSiMe}_3$), bis-(trimethylsilylacetamide) or bis-(trimethylsilyltrifluoroacetamide) usually with a tertiary amine (e.g. pyridine) present. Silyl ethers are readily hydrolysed but under anhydrous conditions they generally resist nucleophilic attack. n-Butyldimethylsilyl ethers are slower to cleave upon hydrolysis. The trimethylsilyl group is valuable for protecting the hydroxyl groups in halogeno phenols during coupling reactions (F.D. King and D.R.M.Walton, Synthesis, 1976, 40). Methoxymethyl, introduced by way of chloromethyl methyl ether, unfortunately a highly carcino-genic material, is a convenient protective group which is readily removed by dilute acid (D.G. Roux *et al.*, Tetrahedron

Letters, 1976, 1033; Y.P. Yardley and H. Fletcher III, Synthesis, 1976, 244).   The β-methoxy ethoxymethyl group has
also been used (E.J. Corey, J.-L. Gras and P. Ulrich, Tetrahedron Letters, 1976, 809).   As an alternative to the benzyl
ethers, 9-anthrylmethyl ethers have been used as blocking
groups for phenols and other acidic compounds.   Ready cleavage occurs on brief treatment with sodium thiomethoxide in
hexamethylphosphoric triamide at ambient temperature (N. Korn-
blum and A. Scott, J. Amer.chem. Soc., 1974, *96*,590).   The
use of the phenacyl group, readily removable by zinc and
acetic acid, for both phenols and acids has been reported
(J.B. Hendrickson and C. Kandall, Tetrahedron Letters, 1970,
343).   For the protection of phenols (and alcohols) the *o*-
nitro benzoyl group is valuable (D.H.R. Barton, I.H. Coates
and P.G. Sammes, J. chem. Soc., Perkin I, 1973, 599).   The
recent uses of positional protective groups in selective
substitution reactions to form phenols, diphenylmethanes and
biphenyls have been reviewed (M. Tashio and G. Fukata, Org.
Prep. Proced. internat., 1976, *8*,51) and the protection of
phenols in general has been discussed (J.F.W. McOmie in
'Protective Groups in Organic Chemistry', Plenum, London 1973).
The loss in yield consequent upon insertion and removal of
the protective group may mitigate against its use and in
certain cases it can be dispensed with.

(C)   Reactions of the Aromatic Ring

(i)   Protonation and Deuteriation
      Ring protonation, deuteriation of phenols and hydrogen
abstraction are of practical and theoretical interest in
phenol chemistry.   O-Deuteriophenols with high isotopic purity have been prepared by the cleavage of trimethylsilyl derivatives with O-deuteriomethanol (A.R. Barsindale, C.Eaborn
and D.R.M. Walton, J. chem. Soc., (C) 1970, 1577). In the gaseous phase ring protonation of phenol has been calculated to
be more favourable than O-protonation and a deuterium exchange
experiment supports this (D.J. De Frees, R.T. McIver, and
W.J. Hehre, J. Amer. chem. Soc.,1977, *99*,3853).   Hydrogen
abstraction from phenols has been correlated with quantum
mechanical indices and over a wide range of reactivities is
more complex than indicated by a simple Hammett relationship
(I. Luskovits, J. Kardos, and M. Simonyi, Tetrahedron Letters,
1974, 2685).   A sufficient proportion of the protonated
phenol (or the ether) can be found in 'superacid' solution
to permit spectroscopic examination.   p-Protonation of simple

phenols (and their ethers) was mainly observed with some int-
erconversion between the stereoisomers shown (G.A.Olah and
Y.K. Mo, J. Amer. chem. Soc., 1972, *94*,5341).

0-Protonation and di-protonation have also been observed.

C-Protonation provides a carbon electrophile and this has been
employed in synthesis (J.-P. Gerson and J.-C. Jacquessy, Chem.
Comm., 1976, 652)   Ring protonation may also lead to retro
Friedel-Crafts reactions (U. Svanholm and V.D.Parker, J.chem.
Soc., Perkin I, 1973, 562; T. Watanabe, Org. prep. Proceed.
internat., 1974, *6*,107,117).

(ii) Electrophilic Substitution

     Electrophilic substitution reactions of phenols have been
reviewed (D.A.R. Haffer and J. Vaughan, in 'Chemistry of the
Hydroxyl Group', ed. S. Patai, Interscience, New York, 1971,
p.393)   and a general review has appeared (R. Taylor, 'MTP
International Review of Science, Organic Chemistry, Series
One', Butterworths, London, 1973, vol. 3, p.2.).   They
occur more readily in phenols than in phenyl ethers, explicable by an inductomeric effect of the hydroxyl bond electrons
on conjugation in the transition state.   Hydrogen bonding to
the solvent has also been emphasised as an important factor.
Nitration and halogenation of phenol in organic solvents are
irreversible reactions with a constant *o/p* ratio in the former
and a lower ratio in the latter.   Sulphonation and Friedel-
Crafts alkylation are reversible with *o*-products predominating
at low temperature and *p*-products at high temperatures in the
former.   In Friedel-Crafts alkylation the *o/p* ratios are
different with kinetic and thermodynamic control.   Numerous

examples of electrophilic substitution are detailed in ensuing sections on derivatives of monohydric and polyhydric phenols.

(d) Phenol Oxidation

(i) Oxidative Coupling

Oxidation of phenols and in particular oxidative coupling has been widely studied with regard to new reagents, reaction techniques and synthetic applications.   O-C Coupling can occur as opposed to C-C coupling.   Oxidation with high potential oxidants in acidic media is likely to encourage phenoxonium species and O-C coupling, while alkaline conditions with lower potential oxidants lead to phenoxyl radicals and C-C coupling. Phenoxonium cations may arise by either one electron oxidation of phenoxyl radicals or by disproportionation, and their int-ermediacy in coupling has been suggested (W.A. Waters, J.chem. Soc., (B), 1971, 2026; also D.G. Hewitt, J.chem. Soc., (C) 1971, 1950).

$$2 \, Ar^{\bullet} \xrightarrow{\text{1 e}^{-} \text{ transfer}} ArO^{-} \; + \; ArO^{+}$$

The extent of the involvement of cations in system with one electron oxidants however remains uncertain although with two electron oxidants (dichlorodicyanoquinone) experimental evidence has been forthcoming.   Inorganic oxidants may give rise to reactive intermediates consisting of organometallic species or outer sphere complexes rather than radicals or cations.   2,2'-Dihydroxy-4-methoxychalcone is oxidised by alkaline ferricyanide to give an insoluble complex from which the (O-C) product, 2'-hydroxy-6-methoxy aurone is formed on acidification without further oxidation.

'Pummerer's' ketone (the structural reassignment of which led in 1963 to a one-step synthesis of the lichen metabolite, usnic acid by Barton et al.,can be obtained from p-cresol, by a number of different oxidants.   This system continues to be worked upon (P.L.Pauson et al., J. chem. Res., (S), 1977, 12 et seq.;L.R. Mahoney and S.A. Weiner, J. Amer. chem. Soc.,

1972, *94*,585, 1412). The product distribution of Pummerer's ketone shown (formed by *o/p*, C-C coupling), the *o,o*-product and a minor product of O-C coupling is independent of pH and the redox potential of the oxidant but dependent on the temperature,

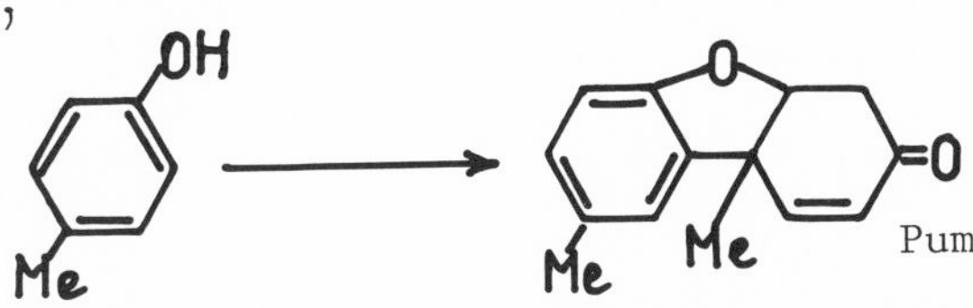

Pummerer's ketone

indicating the reversibility of the coupling and the irreversibility of the ensuing enolisation. The kinetically favoured *o/p* coupling product dissociates readily in acetonitrile leading to dimeric products containing 63% of Pummerer's ketone; but 0% in benzene containing silver carbonate on celite (Fétizon reagent).

A number of new oxidising reagents have been introduced which results in greater specifity and less complexity. Aqueous iron(III)chloride can be replaced by the iron complex $[Fe(DMF)_3Cl_2]^+[FeCl_4]^-$. In dimethylformamide (DMF) solution this reagent effects the following oxidation in 39% yield ($R^1=R^2=H$) (S. Tobinaga and E.Kotani, J. Amer.chem. Soc., 1972, *94*,309), reminiscent of the synthesis of erysodienone in the erythrina alkaloid series (A.I. Scott *et al.*,Chem. Comm., 1966,142). Other synthetic applications of oxidative coupling have been studied, the first involving the protection of a t-amine in the coupling process by means of a borane complex (M.A.Schwarz,B.F.Rose and B. Vishavajgala, J. Amer. chem. Soc., 1973, *95*,612; S.M.Kupchan and A.J.Liepa, *ibid.*, 4062).

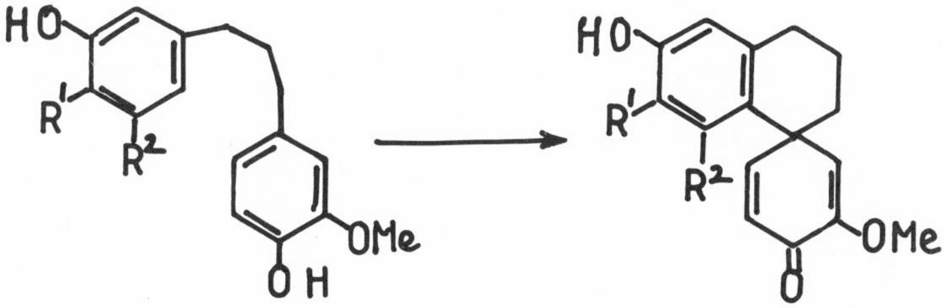

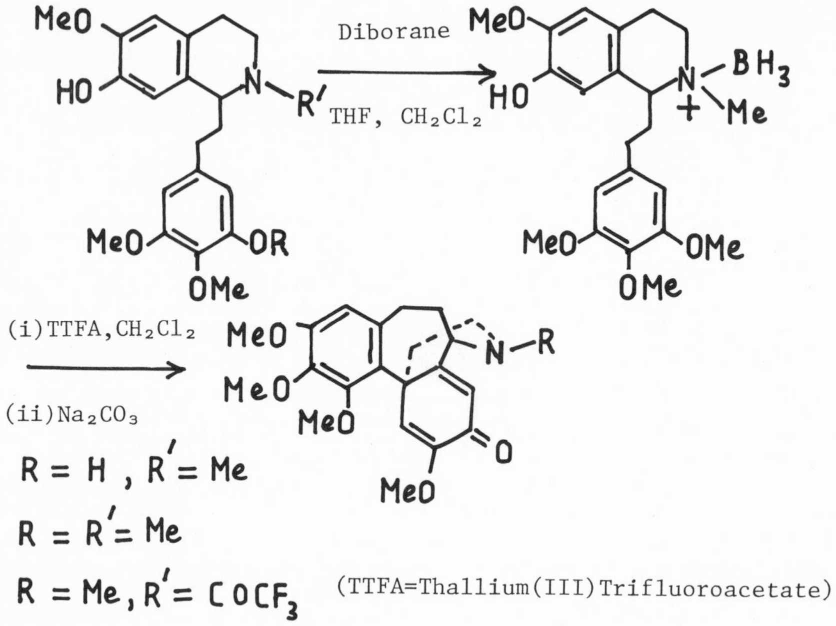
Diborane
THF, CH_2Cl_2
(i)TTFA,CH_2Cl_2
(ii)Na_2CO_3
R = H , R' = Me
R = R' = Me
R = Me, R' = COCF$_3$
(TTFA=Thallium(III)Trifluoroacetate)

Manganese(III)acetylacetonate has been used in organic solvents for various O-C couplings of oximes leading to novel heterocyclic syntheses with quinone by-product formation (A.R. Forrester, R.H. Thomson and S. -O.Woo, J. chem. Soc., Perkin I, 1975, 2340).

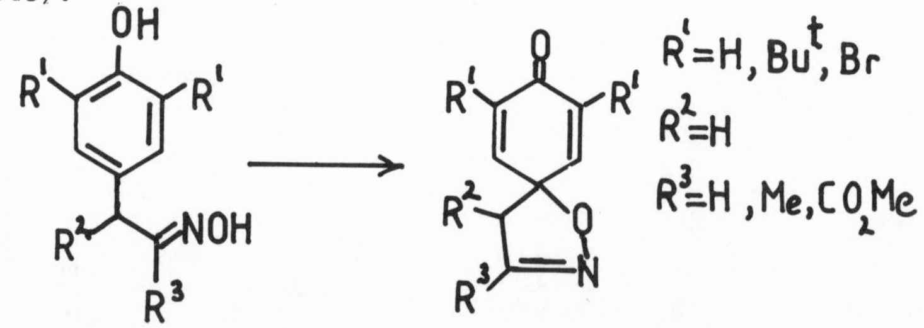
R^1=H, Bu^t, Br
R^2=H
R^3=H , Me, CO$_2$Me

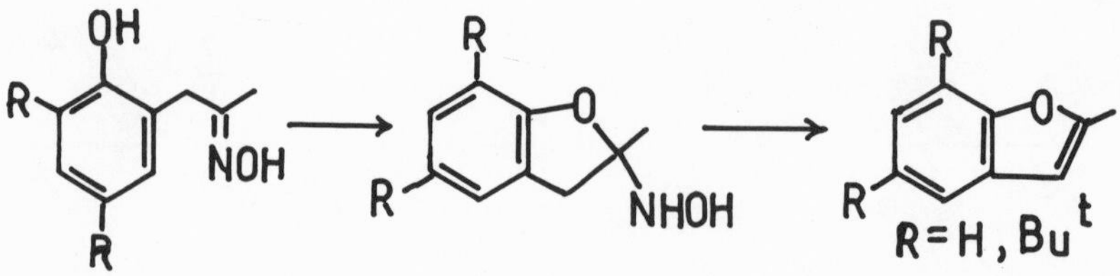
R=H , Bu^t

Heterogeneous as opposed to homogeneous oxidative C-C coupling can be effected by silver carbonate on celite (Fétizon reagent) (V. Balogh, M.Fétizon and M. Golfier, J.Org. Chem., 1971, 36,1339).

Oxidative coupling of conjugated olefinic C atoms as in propenylphenol and in coniferyl alcohol (relevant to lignin formation biosynthetically) has been investigated in various model system (K.V. Sackanen and A.F.A. Wallis, J. chem. Soc., Perkin I, 1973, 1869, 1873; J.M.Haskin, K.Weinges and R. Spänig, 'Oxidative Coupling of Phenols', Arnold, London 1968.

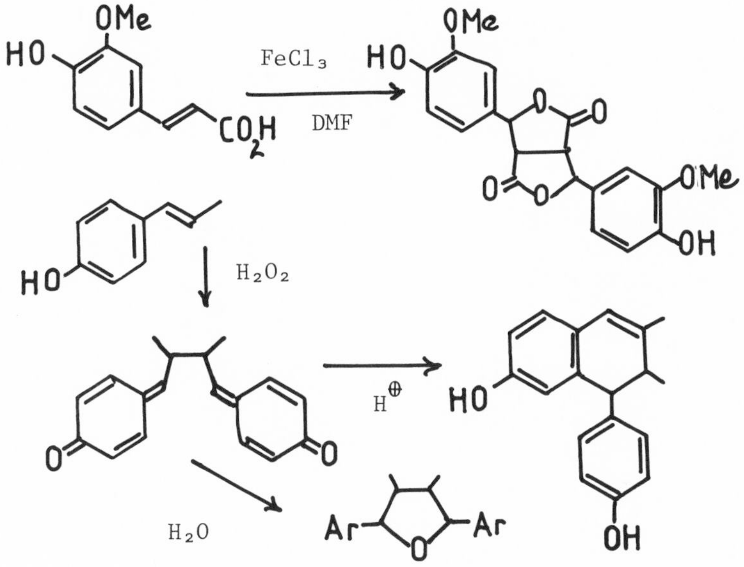

A related coupling involving a phenolic tolan shown gives a cyclobutane derivative (S. Hanff and A. Rieker, Tetrahedron Letters, 1972, 1451).

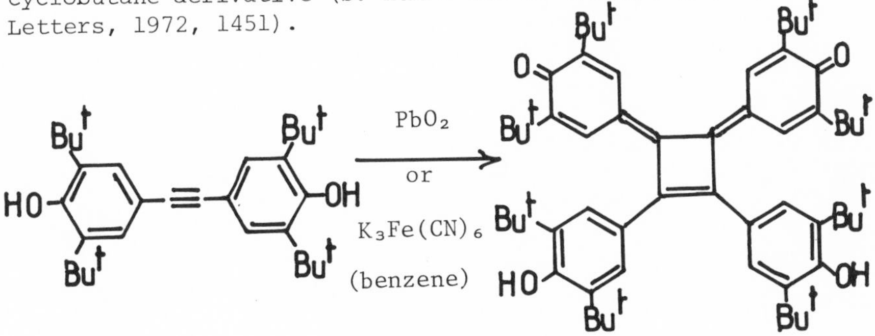

The stilbene quinone shown (R=t-Bu), a known coloured oxidation product of 2,6-di-t-butyl-4-methylphenol, undergoes acid-catalysed hydration to give the bis-(phenyl)acetaldehyde (L. Taimr and J. Popíšil, Tetrahedron Letters, 1972, 4279). This may in the reviewer's opinion have some value in the avoidance of yellow discolouration encountered in the use of 2,6-di-t-butyl-4-methylphenol (Topanol O) as an anti-oxidant.

Advances in oxidation techniques have been achieved by conducting the coupling electrochemically through anodic oxidation. This avoids side reactions and further oxidation of products which can occur in conventional chemical oxidation. The stability of cations in trifluoroacetic acid and trifluoroacetic acid-fluorosulphonic acid permits the electrochemical procedure to be stopped at the cation radical stage (U.Svanholm et al., Tetrahedron Letters, 1972, 2271). Intermolecular and intramolecular C-C and C-O couplings have been described (J.M. Bobbitt and R.C. Hallecher, Chem. Comm., 1971, 543; J.R. Falck, L.L. Miller and F.R. Sternitz, Tetrahedron, 1974, 30,931; J.M.Bobbitt et al., J.org.Chem., 1975, 40,2924). Useful references on this technique have been collected by K. Korinek and T.F.W. McKillop (Ann. Report, 1971, 68B,315).

Photochemical oxidative couplings have been employed in many different synthetic applications (T. Kametani and K. Fukomoto, Accounts chem. Res., 1972, 5,212) and related reactions have been reviewed (H. Dürr, 'Methoden der Organischen Chemie, Photochemie I', Houben-Weyl, George Thieme Verlag, Stuttgart, 1975, vol. 4/5a, p. 642).

(ii) Other Oxidations

Much interest has been shown in reactions leading to quinol and quinone formation by a wide variety of reagents.

The oxidation of 2-benzylphenol with 2,3-dicyanobenzoquinone in methanol gives 2-benzyl-1,4-benzoquinone (J.M.Singh and A.B. Turner, J. chem. Soc., Perkin I, 1972, 2296) and not 2-hydroxybenzophenone as previously reported.   Quinols are formed via a C-methyl migration from 4-alkylphenols by oxidation with thallium perchlorate (Y. Yamada *et al.*, Chem. Comm., 1974, 661; Y. Yamada and K. Hosaka, Synthesis, 1977, 53).

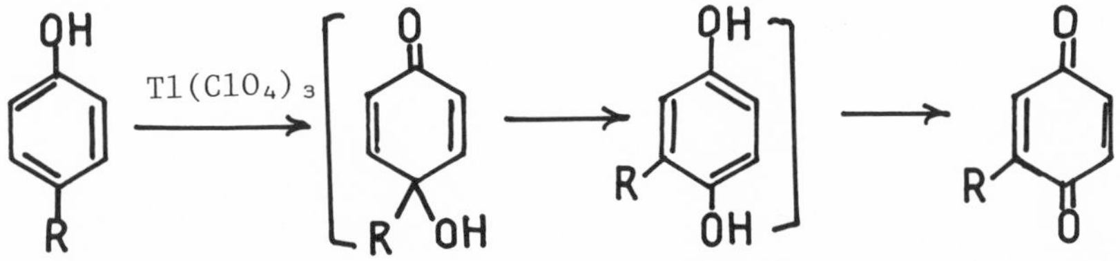

By contrast thallium (III) nitrate (TTN) in methanol results in a 4-alkyl, 4-alkoxycyclohexa-2,5-dienone (A.McKillop *et al.*, J. org. Chem., 1976, *41*,282), discussed in a subsequent section.

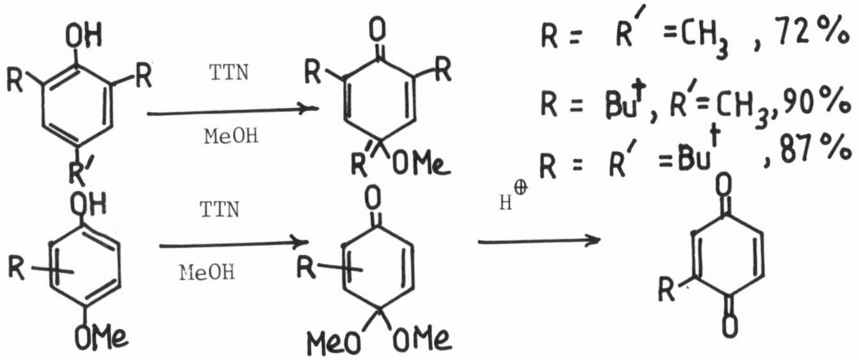

The reaction is postulated to involve *ipso* thallation followed by nucleophilic displacement of thallium. The oxidation of 4-alkylphenols with silver oxide giving the quinone methide is discussed in a later section.   In the oxidation of phenol by peracetic acid, *o*- and *p*-hydroxylation first occurs, *o*-dihydroxybenzene then being oxidised to hexa-2,4-dienoic acid while *p*-dihydroxybenzene furnishes 1,4-benzoquinone (R.A.G. Marshall and R. Naylor, J. chem. Soc.,Perkin II, 1974, 1242). *o*-Quinones are produced from phenols including those with unblocked *p*-positions by oxidation with diphenylselenic anhydride (D.H.R. Barton *et al.*, Chem. Comm., 1976, 985).

The oxidation of 2,6-disubstituted phenols typically by potassium nitrosodisulphonate (Fremy's salt), the Teuber reaction, has been reviewed (H. Zimmer, D.C. Laskin and S.W.Horgan

Chem. Review, 1971, *7l*, 229). A phase transfer procedure with Fremy's salt has been described (G.L.Olson, *et al.*, J. org. Chem., 1980, *45*,803). The synthesis of 2,6-di-t-butyl-1,4-benzoquinone has been described from 2,6-di-t-butyl-phenol (B.A. Carlson and H.C. Brown, Org. Synth., 1978, *58*,24). By contrast 2,6-di-t-butyl-4-iodophenol stirred with potassium t-butoxide in t-butanol for fourteen hours under nitrogen gives a 95% yield of 3,3',5,5'-tetra-t-butyl-4,4'-diphenoquinone (P. Bartholmei and P. Boldt, Angew. Chem., 1975, *87*,39). Under oxidative conditions 2,6-di-t-butyl-4-(*p*-methoxyphenyl) phenol gives by oxygenation in the presence of potassium t-butoxide in t-butanol the 'cyclone', 3-*p*-methoxyphenyl-2,5-di-t-butyl-2,4-cyclopentadiene in 97% yield by the route shown (A. Nishinaga *et al.*, Angew Chem., 1976, *88*,154; cf. Synthesis, 1976, 604).

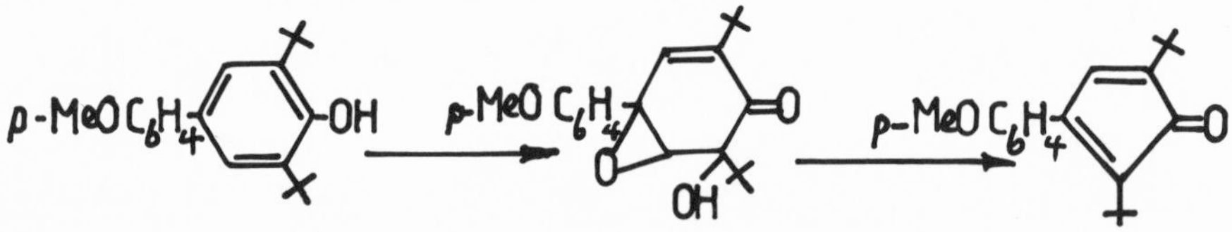

2,4-di-t-butyl-4-methylphenol when oxidised with manganese dioxide in methanol gives the acetal shown (C.M.Orlando, J. org. Chem., 1970, *35*,3714).

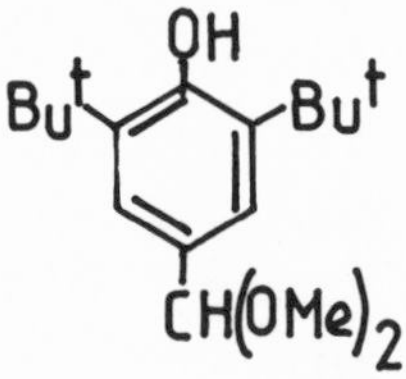

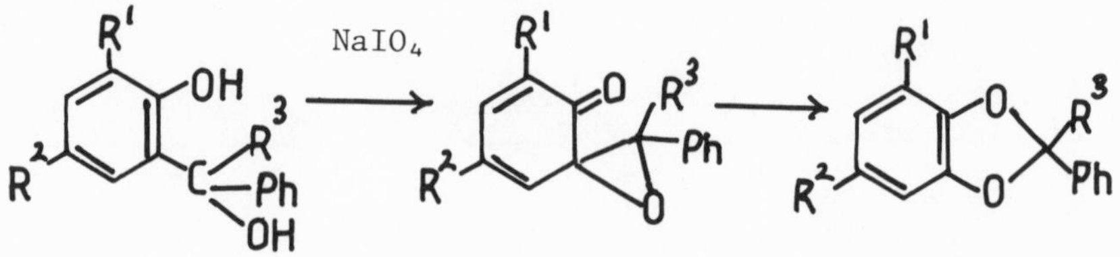

Periodate oxidation of 2-hydroxyalkylphenols, in particular those bearing one or two phenyl groups, gives methylenedioxy derivatives (H.Becker, T. Bremholtz and E. Adler, Tetrahedron Letters, 1972, 4205; H. Becker and T. Bremholz, *ibid.*, 1973, 197) by way of a cyclohexadienone intermediate.

Anodic hydroxylation of 4-propenylphenol at a lead dioxide electrode in aqueous sulphuric acid gives a cyclohexadienone in good yield although phenol yields 1,4-benzoquinone
under similar conditons (A. Nilsson, A. Ronlán and V.D.Parker,
J. chem. Soc., Perkin I, 1973, 2337).

Anodic methoxylation of phenols has been investigated
for the synthesis of quinones, quinone hemiacetals, 4-alkyl-
4-methoxycyclohexa-2,5-dienes and 2-alkyl-2-methoxycyclo-hexa-
3,5-dienes (A. Nilsson *et al.*, J. chem. Soc., Perkin I, 1978,
696). The relative ease of oxidation of phenols and their
ethers has been studied electrochemically (O.Hammerich, V.D.
Parker, and A. Ronlán Acta chem. Scand., 1976, *B30*,89) and
the former are more easily oxidised (U. Palmquist, A. Ronlán,
and V.D. Parker, *ibid*, 1974, *B28*,267). Reaction pathways
have been examined in the direct autoxidation of phenol with
sodium methoxide in hexamethylphosphoric triamide solution and
the electron spin resonance spectra of semiquinones have been
observed (W.T. Dixon, P.M.Kok, and D. Murphy, Tetrahedron
Letters, 1976, 623). The oxidation of di and polyhydric
phenols is discussed subsequently.

(iii) Cyclohexadienone formation
The oxidation of phenols as well as certain electrophilic
substitution reactions often involves 1,2- or 1,4-cyclohexa-
dienone formation. The autoxidation of the anti-oxidant
2,6-di-t-butyl-4-methylphenol and 4-alkyl analogues is catalysed by cobalt complexes and gives 2,6-di-t-butyl-4-n-alkyl,
4-hydroxycyclohexa-2,5-dienone in high yield by way of the
corresponding hydroperoxide (T. Matsuura, K. Watanabe, and
A. Nishinaga, Chem. Comm., 1970, 227). The hydroperoxides
of the 2,5-dienones are obtained from phenols by oxidation
with Ce(IV) oxide-hydrogen peroxide (D.H.R. Barton, P.D.Magnus,
and J.C. Quinney, J. chem. Soc., Perkin I, 1975, 1610). Both
2-hydroxy and 4-hydroxycyclohexadienones are obtained from

phenols with diphenylselenic anhydride, although phenolate anions give only the former (D.H.R. Barton, P.D. Magnus, and M.N. Rosenfeld, Chem. Comm., 1975, 301). 2,6-Dimethylphenol in acetic acid with sodium bismuthate gives chiefly (by 2 electron oxidation) 2-acetoxy-2,6-dimethylcyclohexa-3,5-dien-one (E. Kon and E.McNee, J. org. Chem., 1976, 4*l*,1646). The well-known Wesseley oxidation can be extended to acryloxylat-ion and the product from 2,4,6-trimethylphenol undergoes a thermal intramolecular Diels-Alder addition (D.J.Bichan and P. Yates,Canad. J. Chem.,1975, 53,2054).

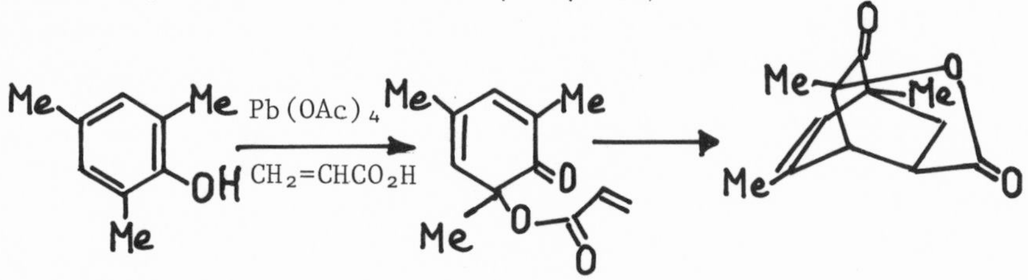

Cyclohexadienones are capable of various reactions as shown in the formation of 2-methylresorcinol diacetate from an *o*-dienoneacetate and in the reversible rearrangement of an *o*-dienonebenzoate (D.H.R. Barton, P.D. Magnus and M.J.Pear-son, J. chem. Soc., (C), 1971, 2231).

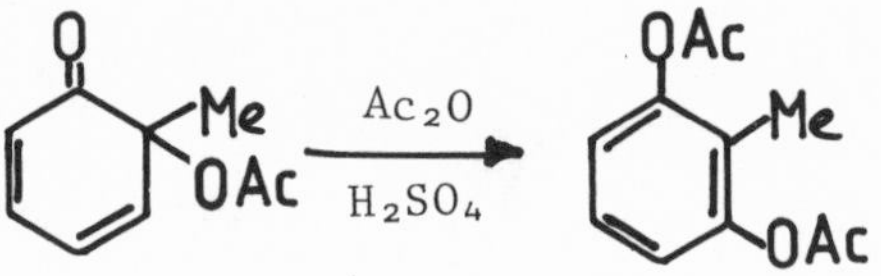

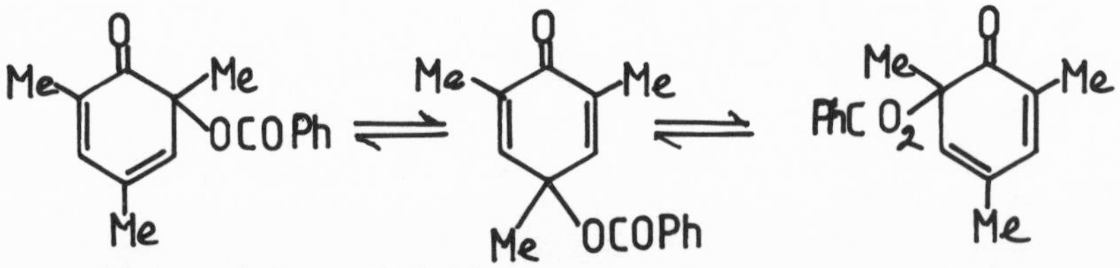

The cyclohexa-2,4-dienones produced by allylation of sodium 2,6-dimethylphenoxide and 2,4,6-trimethylphenoxide undergo different thermal reactions as shown (C.P.Falshaw, S.A. Lane and W.D. Ollis, Chem. Comm., 1973, 491).

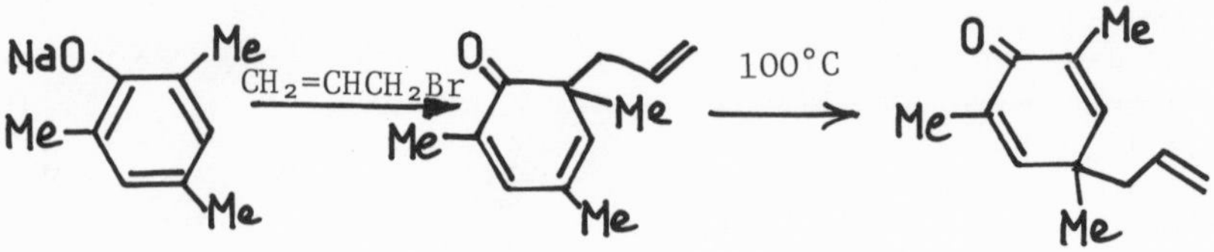

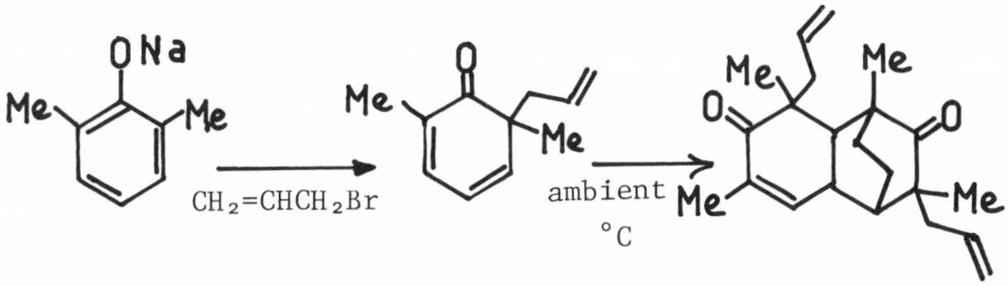

When salicyl alcohols are oxidised with sodium periodate, spiro-epoxycyclohexa-2,4-dienones are formed which are stable if a bulky group or halogen substituent is present in the nucleus. Otherwise they undergo Diels-Alder dimerisation. On photolysis they give salicyl aldehydes in high yield (H.-D. Becker, T. Bremholtz and E. Adler, Tetrahedron Letters, 1972, 4205).

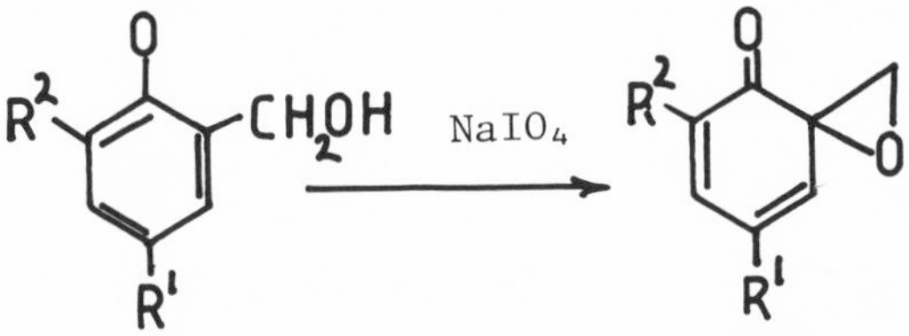

Thallium(III)perchlorate with 3,4-alkylenephenols has been reported to yield 4-hydroxy-3,4-alkylenecyclohexa-2,5-dienones.

The many rearrangements undergone by cyclohexadienones have been reviewed (B. Miller, Accounts chem. Res., 1975, *8*, 245; M.J. Perkins and P. Ward, 'Mechanisms of Molecular Migrations', ed. B.S. Thyagarajan, Wiley Interscience, 1971, vol. 4, p.55; H.J. Shine, MTP International Review of Science, Organic Chemistry, Series One, Butterworths, London, 1973, vol. 3, p.65; R.S. Ward, Chem. in Brit., 1973, *9*,444).

In the anodic intramolecular coupling of a series of α, ω-bis(hydroxyaryl) alkanes, cyclohexadienone formation takes place in one ring in the usual way but cyclisation is dependent on the length of the alkylene chain (V.D. Parker *et al.*, J. Amer. chem. Soc., 1976, *98*,2571)(Scheme 1). Methyldihydroxydiphenyl ethers arise from certain 2,5-dienones by 1,3- rather than 1,4- addition of methylmagnesium iodide (I.G.C. Coutts and M. Hamblin, Chem. Comm., 1976, 58).

Scheme 1

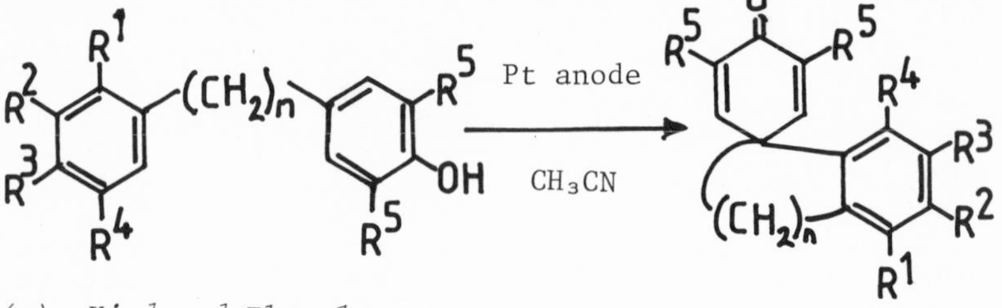

(e) Hindered Phenols

Many of the reactions described have involved phenols
bearing a t-butyl group, often referred to as hindered phenols
when the substituent is *ortho* to the hydroxyl group, due to
their masked phenolic character (e.g. acidity and alkali
solubility). Their participation as antioxidants in prevent-
ion of autoxidation in many different systems is to a large
extent associated with the formation of cyclohexa-2,5-dien-
ones.

Oxidation of 4-aryl-2,6-di-t-butylphenols with oxygen in
the presence of a cobalt (II) complex followed by decomposition of the intermediate hydroperoxide with toluene-p-sulphonic acid gives 5-aryl-3-t-butyl-o-benzoquinones in good yield
(T. Matsuura *et al.*, Synthesis, 1977, 270). In the *trans* t-
butylation of arenes with 2,4-di-t-butylanisole, the t-butyl
group next to the methoxyl is transferred (K. Shimada, Bull.
chem. Soc., Japan, 1976, *49*,1375).

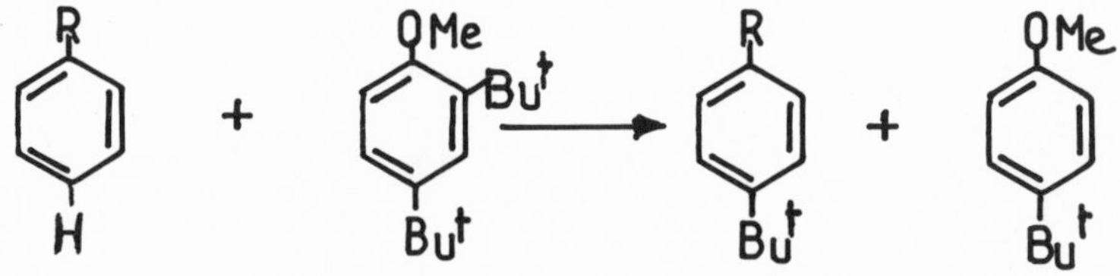

Radical formation from the t-butyl group can occur as shown
in the following photolytic reaction of the quinone derived
from 2,6-di-t-butylphenol (R.H. Thompson *et al.*, Chem. Comm.,
1973, 844).

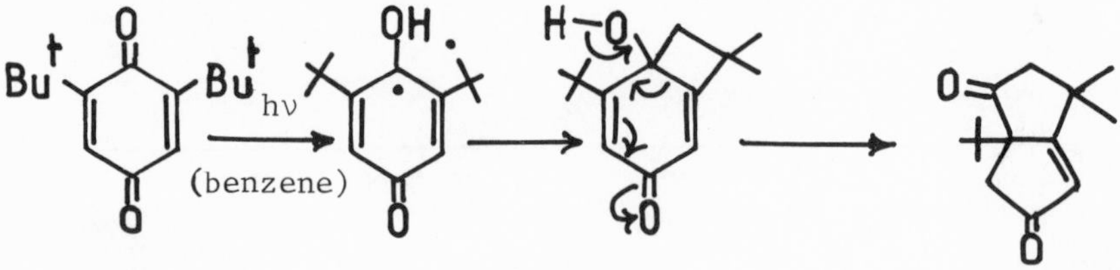

A modified reducing agent produced from lithium aluminium hydride with two moles of 2,6-di-t-butylphenol followed
by one mole of neopentyl alcohol exhibits remarkable stereoselectivity (99%) for the reduction of 3,3,5-trimethylcyclo-
hexanone to the *trans* axial alcohol (H. Haubenstock, J. org.
Chem., 1975, *40*,926). Remarkably, in the gas phase the t-
butyl cation reacts with phenol and with anisole to give predominantly the t-butylated oxonium ion (P. Giacomello *et al.*,
J. Amer. chem. Soc., 1977, *98*,4101, 5022). Through the
different ease of removal of a t-butyl group, the *m*- and *p*-
cresols may be separated by way of their di-t-butyl derivatives (L. Crombie *et al.*, J. chem. Soc.,(C), 1971, 3634).
2-t-Butylphenol (and a wide variety of substituted phenols)
may be converted into the magnesiobromo derivative with ethyl
magnesium bromide which reacts with benzaldehyde (and substituted benzaldehydes) to give the triarylmethane compound shown
in nearly quantitative yield (G. Casiraghi *et al.*, J. chem.
Soc., Perkin I, 1974, 2077). The reaction proceeds by way of
the diarylcarbinol).

With cinnamldehyde the first stage of the reaction is followed
by cyclisation rather than a second substitution.

By the reaction of a hindered phenol with trichlorocycl-
opropenium tetrachloroaluminate followed by oxidation of the
bis-(hydroxylaryl) cyclopropenone formed, diquinocyclopropan-
ones are produced (R. West and D.C. Zecher, J. Amer. chem.
Soc., 1970, *92*,155, 161; R. West *et al.*, J. org. Chem., 1975,
40,2295).

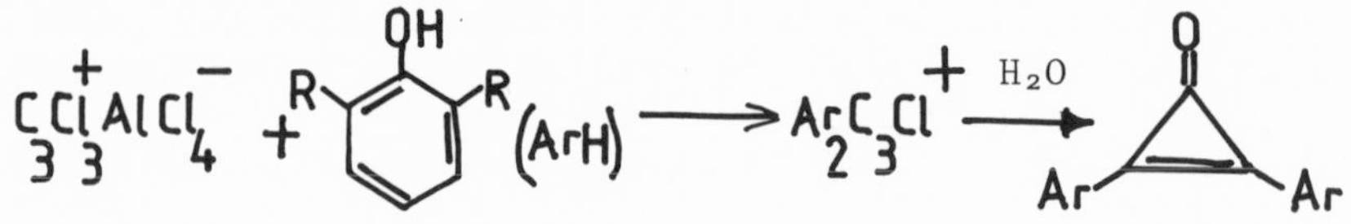

$$\overset{+}{\underset{3}{C}}\overset{-}{Cl_3}\overset{-}{AlCl_4} + \overset{OH}{\underset{(ArH)}{R\text{⬡}R}} \longrightarrow Ar_2\overset{+}{\underset{3}{C}}Cl \xrightarrow{H_2O} Ar\text{△}Ar$$

The hindered base potassium 2,6-di-t-butylphenoxide has been used in the synthesis of α-chloronitriles and of dialkyl-nitriles from trialkylboranes and dichloroacetonitrile (H. Nambu and H.C. Brown. J. Amer. chem. Soc.,1970, 92,5790, 1761).

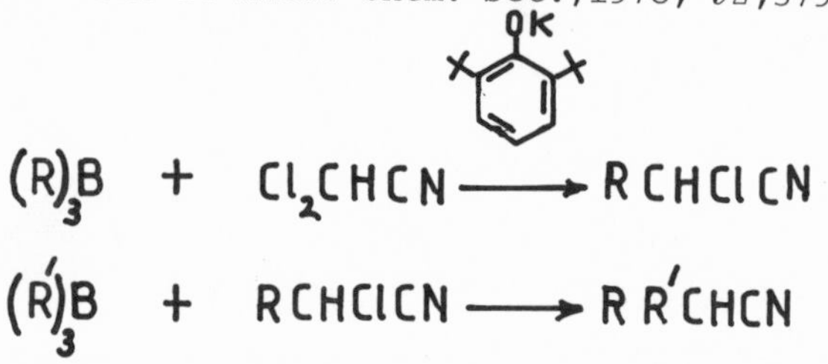

$$(R)_3B + Cl_2CHCN \longrightarrow R CHCl CN$$

$$(R')_3B + RCHClCN \longrightarrow R R'CHCN$$

Lithium salts of hindered phenols have been used in the synthesis of certain 2-alkylcyclohexanones (E.J. Corey and R.H.K. Chen. Tetrahedron Letters, 1973, 3877).

2-t-Butyl-4-methylphenol with aqueous (40%) glyoxal in glacial acetic containing concentrated hydrochloric acid gives upon refluxing for sixteen hours a 90% yield of 5-methyl-7-t-butyl-2(3H)benzofuranone (R.W. Layer, J. heterocyclic Chem., 1975, 12,1067).

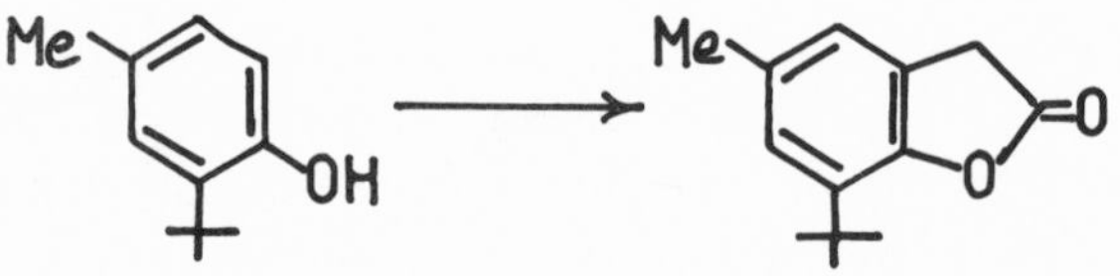

The lack of polarity in hindered phenols and mono-t-butyl methoxy analogues results in unusual TLC and GLC chromatographic properties (J.H.P. Tyman and A.J. Matthews, Chem. and Ind., 1977, 740).

(f) Reactions of Alkyl Phenols

Although the t-butyl phenols undergo some of the reactions of the methyl and polymethyl phenols, the latter as a group can also be used to synthesis cyclic compounds which can themselves undergo useful subsequent transformations.

2,5-Dimethylphenol with oxalyl chloride furnishes, after

lithium aluminium hydride reduction followed by oxidation with periodate, 2-(1,2-dihydroxyethyl)-3,6-dimethylphenol in 85% yield. Thus a phenolic aldehyde can be produced which due to steric hindrance would otherwise be difficult to obtain (D.J. Zwanenburg and W.A. Reynen, Synthesis, 1972, 624).

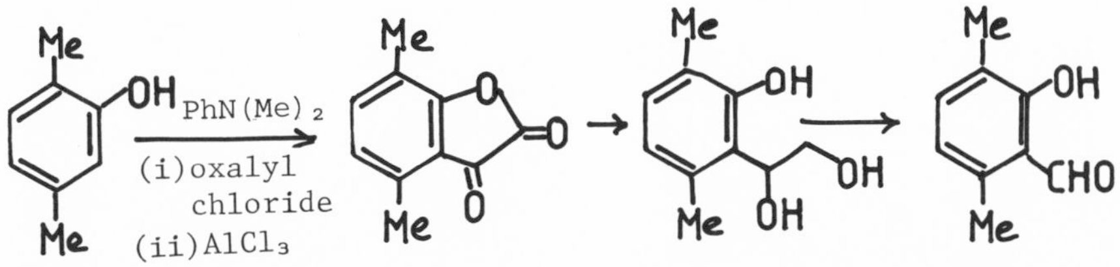

p-Chlorophenol and phenylacetylene react in the presence of phenoxymagnesium bromide (from phenol and ethylmagnesium bromide) in refluxing xylene to give an 87% yield of the o-vinylated phenol (C. Casiraghi et al., Synthesis, 1977, 122), a potential source of other cyclic products.

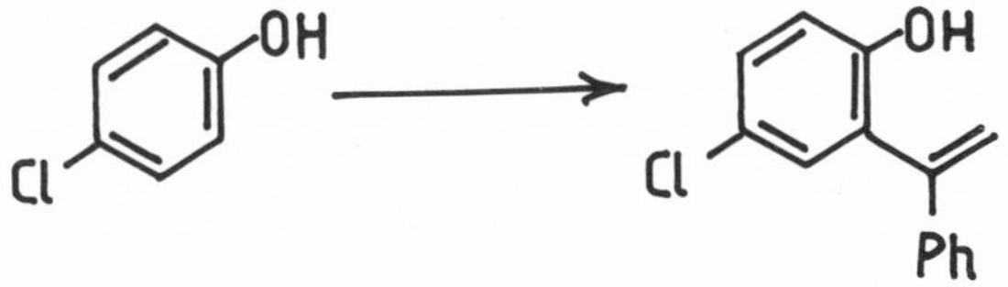

The pent-4-enylphenol shown undergoes thermal rearrangement possibly through the agency of cyclohexa-3,5- and 2,5-dienones (C. Moreau, F. Rouessac and J.M. Conia, Tetrahedron Letters, 1970, 3527).

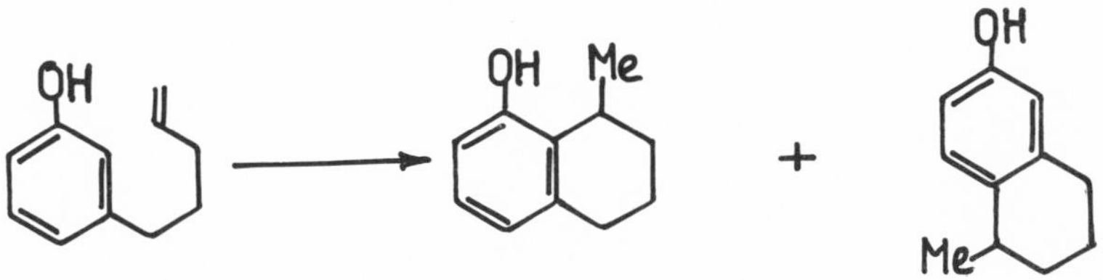

7-Hydroxyindenones can be synthesised in 72% yield from a phenol and dichloromaleic anhydride added to a paste of aluminium chloride and sodium chloride at 140-145° and then briefly heated to 175° (G. Roberge and P. Brassard, Synth. Comm., 1979, 9,129). Phenol and methyl acrylate with alum-

inium chloride during ten hours at 25° followed by brief
warming furnish a 57% yield of dihydrocoumarin (V.B. Reddy,
Chem. and Ind., 1976, 414) ordinarily troublesome to prepare
(G.R. Brown, J. Grundy and J.H.P. Tyman, J. chem. Soc., Perkin
1, 1981, 336.

Methylthiomethylation of phenols in the *o*-position may be
effected (R=H), or in the *p*-position (R=alkyl) with dimethyl
sulphoxide (DMSO) and dicyclohexylcarbodi-imide (DCC) followed by acidic treatment as shown (P. Claus, Monatsh, 1971, *102*,
913, 1072;   J.P. Marino, K.E. Pfitzner and R.A. Olofson, Tetrahedron,1971, *27*,4181;   R.A. Olofson and J.P. Marino, *ibid.*,
4195).   Such products can be desulphurised to yield alkyl
phenols.

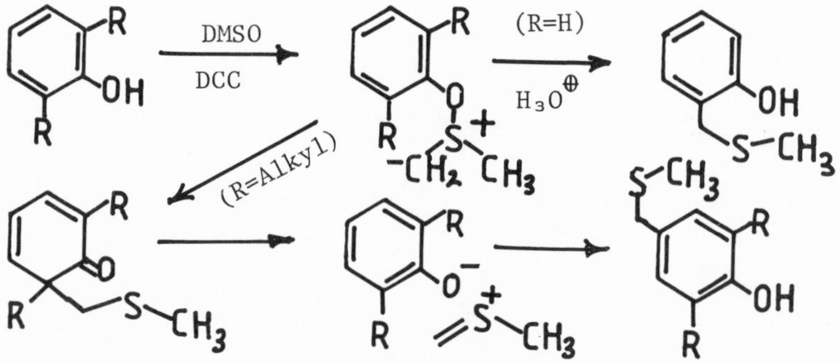

Phenolic iso-chroman-3-ones can be obtained by the *o*-hy-
droxymethylation of homoisovanillic acid in the presence of
benzeneboronic acid and paraformaldehyde in refluxing benzene
with azeotropic removal of water during twenty four hours
followed by water treatment at 100°C to cleave the organoboron
compound (W. Nagata *et al.*, Chem. pharm. Bull., Japan, 1975,
23,2867).

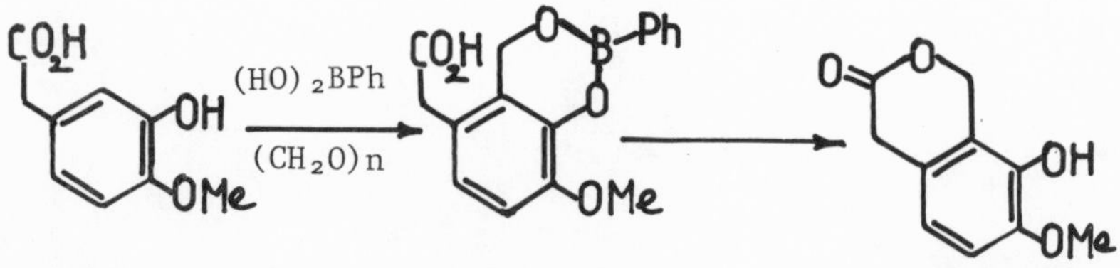

2,6-Dimethylphenol has been used for deriving a borane
for the subsequent synthesis of certain ketones(B.A. Carlson
and H.C. Brown Org. Synth., 1978, *58*,24).

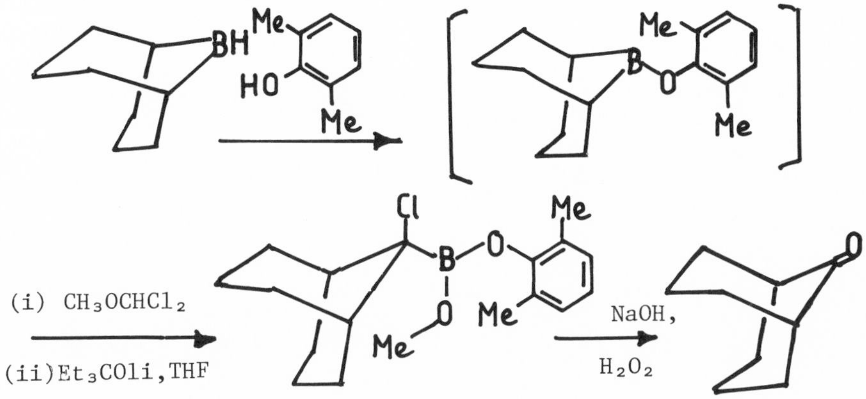

(i) CH_3OCHCl_2

(ii) Et_3COli, THF

Friedel Crafts benzylation of 2,6-dimethylphenol under a variety of conditions gave an unusually high proportion (40%) of the *m*-product and in the case of 2,6-dimethylanisole or isopropyl 2,6-dimethylphenyl ether the *m*-benzyl derivative is the main product. It is considered that the novel products arise by direct *m*-attack since all other reasonable mechanisms have been eliminated (M.P. McLaughlin, V. Creedon, and B.Miller, Tetrahedron Letters, 1978, 3537)

Pentamethylphenol, dibromomethane and aluminium bromide heated at 95°C during four hours followed by treatment at -50° with methanolic sodium methoxide gives 2,3,4,6,9-pentamethyl-tropone (N.F. Salakhutdinov *et al.*, Zh. org.Khim., 1979, *15*, 1468).

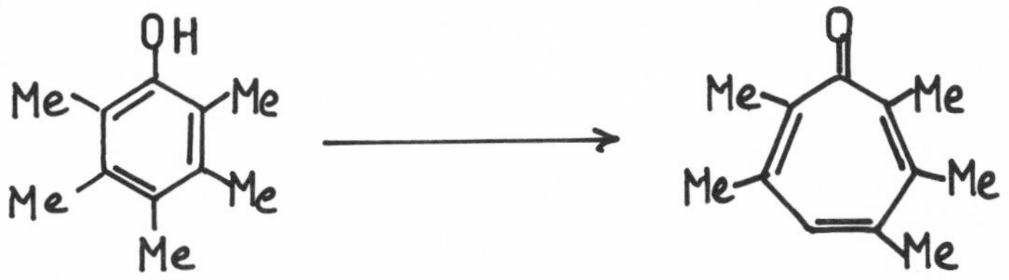

Alkylphenols, benzeneboronic acid, formaldehyde and a little propanoic acid refluxed in benzene for ten hours and the product refluxed for two hours in benzene containing propylene glycol give the *o*-hydroxymethyl derivative shown (W.Nagata, K. Okada and T. Aoki, Synthesis, 1979, 365).

(g) Quinone Methides

An aspect of phenolic chemistry that has received increasing attention during the past decade is the formation of
ortho-and *para*-quinone methides and their rearrangements and
additive reactions.   On account of their instability they
are generated in situ and used without isolation.   They may
be prepared by photoenolisation of 2-acylphenols (P.Heinrich,
'Methoden der Organische Chemie, Photochimie II', Houben-Weyl,
Georg Thieme Verlag, Stuttgart, 1975, vol.4,5b, p.795),through
thermally allowed phototropic shifts,by elimination from phenolic Mannich bases, from the oxidation of 4-alkylphenols in
carbon tetrachloride with silver oxide (L.K. Dyall and S.Win-
stein, J. Amer. chem. Soc., 1972, *94*,2196), by other procedures such as pyrolysis of salicyl alcohols and from Wittig
reactions with quinone reactants (G.L.McIntosh and O.L.Chap-
man, Chem. Comm., 1971; 771, G. Cardillo, L. Merlini and S.
Servi, *Ann. Chim.* (Rome), 1970, *60,654)*.

Carpanone has been synthesised from a quinone methide
(O.L. Chapman *et al.*, J. Amer. chem. Soc., 1971, *93*,6696).

Oxidation of appropriate phenols can be achieved through
high potential quinone (DDQ) dehydrogenation and the 2-(3'-
methylbut-2'-enyl)phenol shown gives, by way of the quinone
methide the 2,2-dimethylchromen in the final electrocyclisat-
ion (G. Cardillo, R. Cricchio and L. Merlini, Tetrahedron,
1968, *24*,4825). (Scheme 2).

An orange-red crystalline *o*-quinone methide of unusual
stability has been isolated from the silver oxide oxidation
of 2-cinnamyl-4,5-methylenedioxy phenol and the corresponding

compound from 2-(4'-methoxybenzyl)-4,5-methylenedioxyphenol rapidly dimerises in polar solvents (L. Jurd, Tetrahedron, 1977, *33*,163) (Scheme 3).

Scheme 2

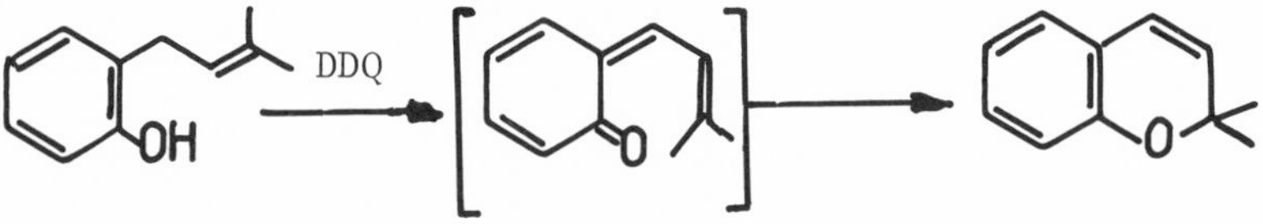

Scheme 3

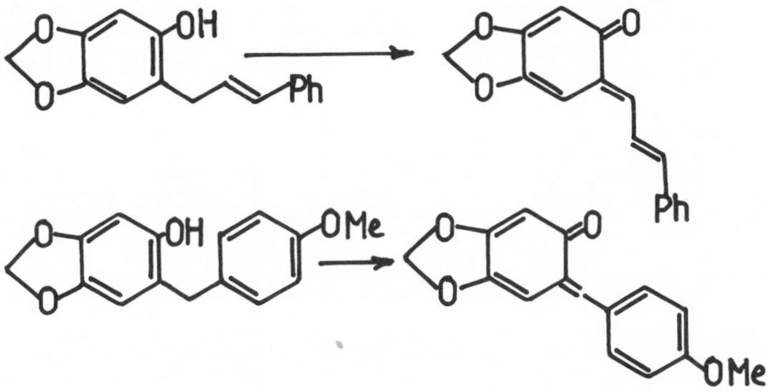

The stable *o*-quinone methide shown below has been prepared by the reaction of 2,4-dimethylphenol with two equivalents of 1,3-benzodithiolylium tetrafluoroborate in acetonitrile and subsequent addition of triethylamine (K. Yamashita *et al.*, Chem. Letters, 1977, 789).

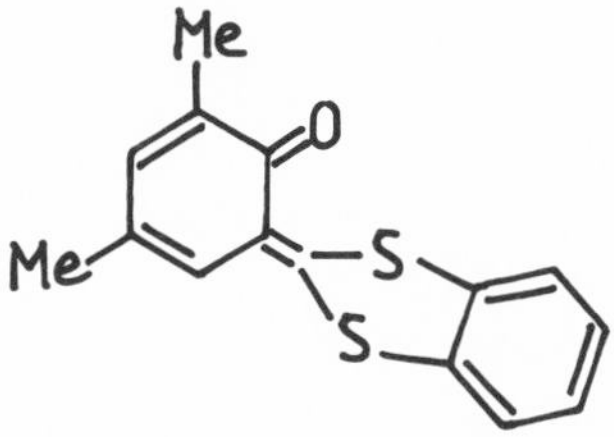

By contrast quantitative isomerisation of the hindered quinone methide shown occurs on neutral alumina (D. Braun and B. Meier, Angew. Chem. internat. Edn., 1971, *10*,566).

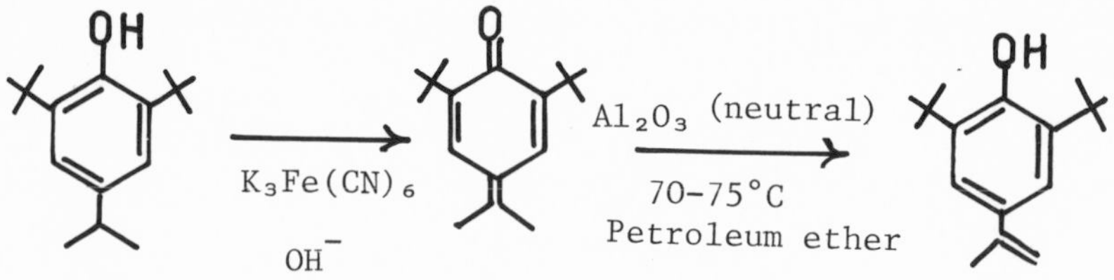

The vinylic *o*-quinone methide shown below forms, in concentrated solution in an aprotic solvent or in the solid state, a dimer by Diels-Alder addition of the terminal double bond of one molecule to the side chain diene system of a second molecule. Products of nucleophilic addition have been observed and with water or methanol, 1,4 and 1,6-addition (J.A. Hemingson, Tetrahedron Letters, 1977, 616 2967). The related methylene compound has also been studied (G. Leary *et al.*, J. chem. Soc., Perkin II, 1977, 1737).

The mechanistic pathways of certain reactions have been clarified by deuterium labelling. 2-Vinylphenols at 143° undergo reversible, 1,5-hydrogen shifts to give *o*-quinone methides and the same process leads to rapid (E,Z) isomerisation of 2-(prop-1'-enyl)phenols (H.-J. Hauser, Helv.,1977, *60*,2007).

The thermally allowed prototropic shift illustrated, results in a 2-alkylchromene (E.E. Schweizer, D.M. Crouse, and D.L. Dalrymple, Chem. Comm., 1969 354).

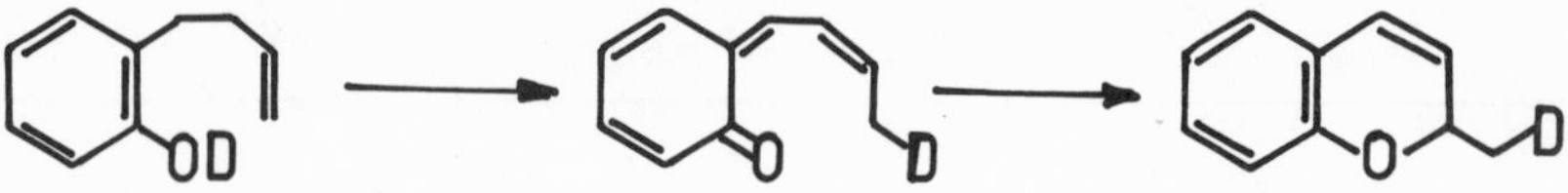

Gas phase pyrolysis has been used to generate materials closely related to the o-quinone methides. Thus phenyl propargyl ether at 450°C gives benzocyclobutene (31%) and β-indanone (26%) by the following probable route (W.S. Trahanowsky and P.W. Muller, J. Amer. chem. Soc., 1972, 94,5911).

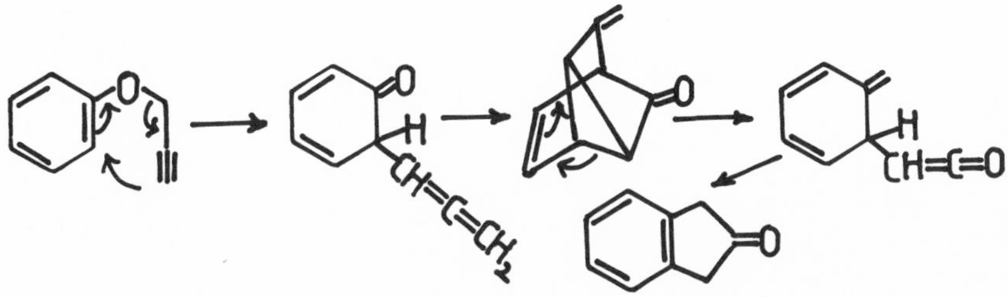

The o-quinone methide illustrated is an intermediate in the photochemical reactions of duroquinone and 2,4,5-trimethylhomogentisic lactone (D. Creed, Chem. Comm., 1976, 121).

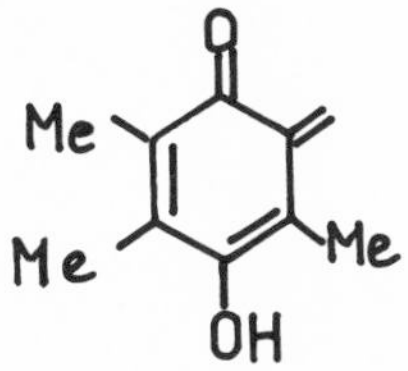

The thione methide shown results from the irradiation of 3H-benzo(b)thiophene-2-one. The structure was proved by isolation of the adduct with N-phenylmaleimide (G. Jacqmain et al., Tetrahedron Letters, 1973, 3655).

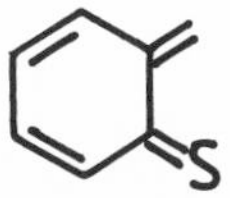

The 3,5-dialkyl-4-hydroxybenzyl derivatives shown (X=NR$_2$, Cl or SCSNR$_2$) are valuable reagents for the 4-hydroxybenzylation of activated methylenic carbanions, a reaction believed to proceed by way of the p-quinone methide (A. Schmidt and H. Brunetti, Helv. 1976, 59,522).

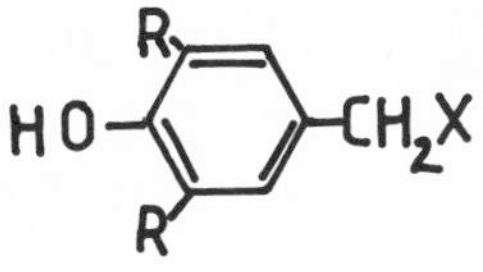

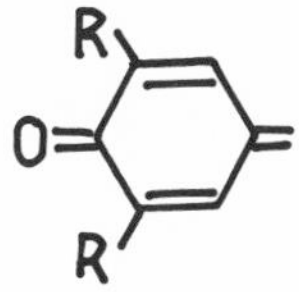

(h) Phenolic Aldehydes, Ketones and Acids

A review of acylation by the Vilsmeier-Haack-Arnold reaction contains reference to phenolic materials (C. Jurz, Adv. org. Chem., 1976, 9,225(.

o-Phenolic aldehydes have been synthesised by the interaction of aryloxymagnesium halides with ethyl orthoformate followed by acidic hydrolysis (C. Casiraghi et al., J. chem. Soc., Perkin I, 1974, 2077).

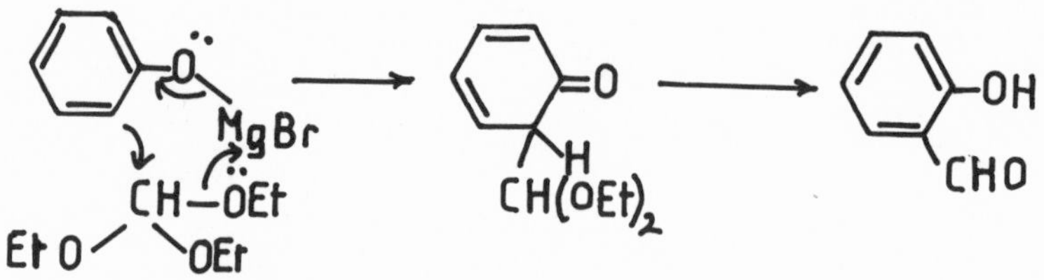

Selective formylation of phenols with 3-methoxyprop-2-enyl formate (J.W. Scheeren and H. Bats, Rec. trav. Chim., 1979, 98, 324) in ethereal solution during eight hours at ambient temperature in the presence of toluene-4-sulphonic acid refers to O not C-attack. o-Formylation of p-substituted phenols by the use of an azathiasulphonium salt has been described (P.D. Gassman and D.R. Amich, Tetrahedron Letters, 1974, 3463) and proceeds as shown.

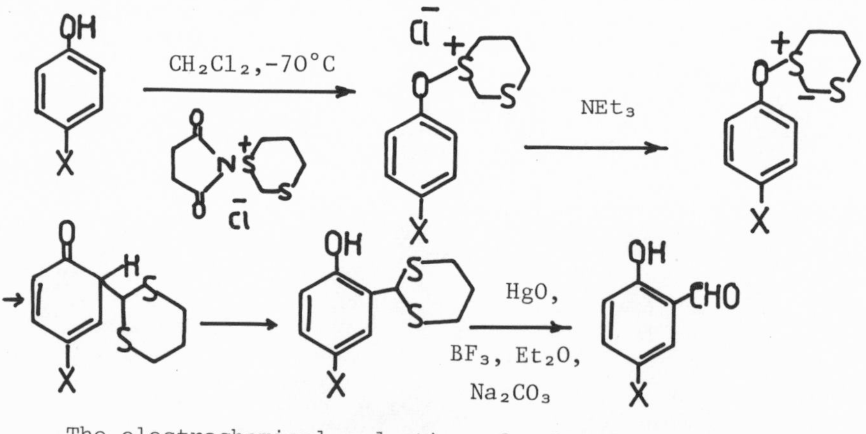

The electrochemical reduction of salicylic acid to salicylaldehyde has been carried out in the presence of borate ion (J.Ch. Hofmann, P.M. Robertson, and N. Ibi, Tetrahedron Letters, 1972, 3433). The preparation of the sterically hindered

3,6-dimethylsalicylic aldehyde from 2,5-dimethylphenol has been referred to earlier. *o*- and *p*-Hydroxyaldehydes are obtained by photochemical interaction of phenol, N,N-diethyl-aniline and chloroform in methanol solution (K. Hirao and O. Yonemitsu, Chem. Comm., 1972, 812).

Deuterium and tritium labelling studies have shown that the formyl proton in salicylaldehyde produced by the Reimer-Tiemann reaction is not derived from the aryl residue and accordingly protonation of the anion illustrated must occur before re-aromatisation (D.S. Kemp, J. org. Chem., 1971, *36*,202).

The Smiles and related rearrangements of aromatic systems, some involving phenols have been reviewed (W.E.Truce, E.M. Kreider and W.W. Brand, Org. Reactions, 1970, *18*, 99).

A considerable amount of work has been carried out on the Fries rearrangement and certain acid-catalysed reactions have been shown to be reversible (F. Effenberg, H. Klenk, and P.L.Reiber, Angew. Chem. internat. Edn., 1973, *12*,795). In its conventional form where hydrogen chloride is freely evolved the reaction is acid-catalysed and although considered to be intramolecular, it must in part be intermolecular since a mixture of *p*-cresyl benzoate and 4-methyl-2-chloro-phenyl acetate gave all four possible *o*-hydroxyketones (M.J.S. Dewar and L.S. Hart, Tetrahedron, 1970, *26*,973).

The *p*-product in Fries reactions has been suggested to form by deacylation-reacylation. The *o*-product which is probably thermodynamically preferred, may arise intramolecularly and irreversibly since a chelate complex is produced. Hydrogen fluoride has been recommended as a catalyst (J.R. Norell, J. org. Chem., 1973, *38*,1924). A synthetic application leading to a Vitamin K analogue and involving an unsaturated acid chloride has been described (N.I.Burckner

and N.L. Bauld, J. org. Chem., 1972, *37*,2359), although the
yields were inferior to the conventional Friedel-Crafts alkylation approach.

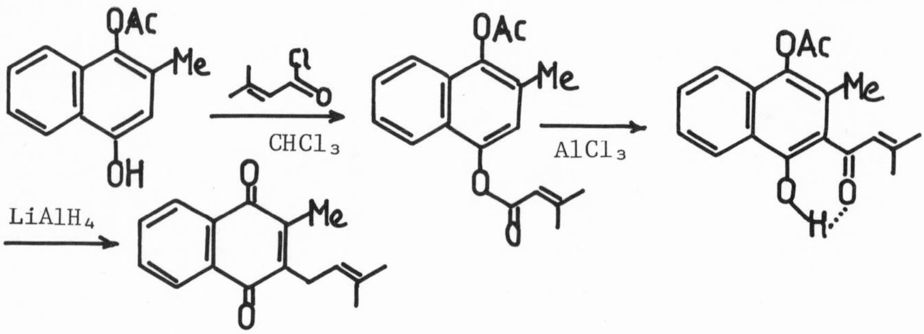

The Fries reaction may be carried out photochemically
(D. Bryce-Smith and A. Gilbert, in 'MTP International Review
of Science, Organic Chemistry, Series One', Butterworths,
London, 1973, vol.3, p.131).   The primary product involves
singlet radicals (solvated) which recombine to give the rearranged product (C.E. Kalmur and D.M. Hercules, Tetrahedron
Letters, 1972, 1515;   J.W. Meyer and G.S. Hammond, J. Amer.
chem. Soc., 1972, *94*,2219;   W. Adam, J.A. Sanabia,and H. Fischer, J. org. Chem., 1973, *38*,2591).   Although not without
side reactions it has been used in one instance (V.T. Ramak-
rishnan and J. Kogan, J. org. Chem., 1970, *35*,2898) as an
alternative in the flavanoid series to the Algar-Flynn-Oyamda
reaction which effects, by way of alkaline peroxide, the
oxidation shown.

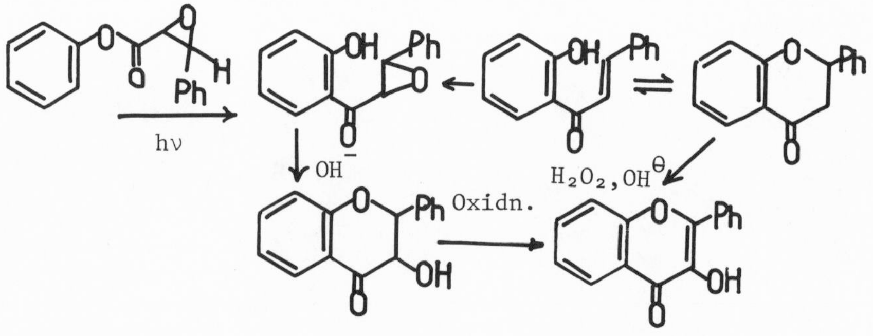

Photolysis at low temperature of benzene oxide, and poly-

cyclic oxides results in formation of phenol and a ketene
(B. Witkop *et al.*, J. Amer.chem. Soc., 1974, *96*, 5579).

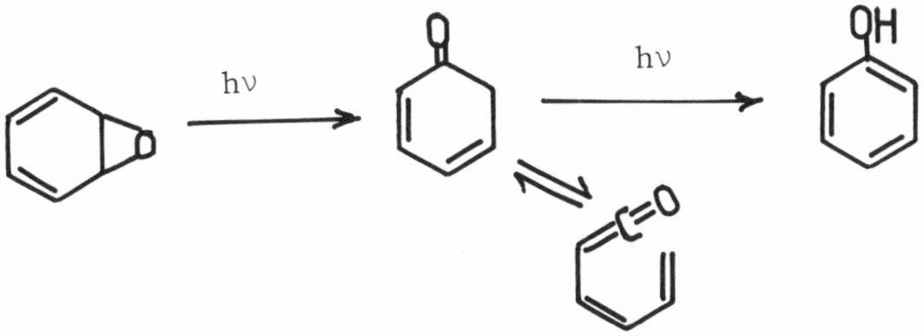

Salicylic acid methyl ether can be converted into a 2-alkylbenzoic acid with alkyllithium or Grignard reagent by way of the oxazoline derivative, (A.I.Meyers, R. Gabel, and E.D. Mikelieh, J. Org. Chem., 1978, *43,*1372), and offers an alternative route to earlier procedures, (J.H.P. Tyman and A.A.Durrani, Chem. and Ind., 1971, 934; 1973, 762; J. chem Soc., Perkin I, 1979, 2079).

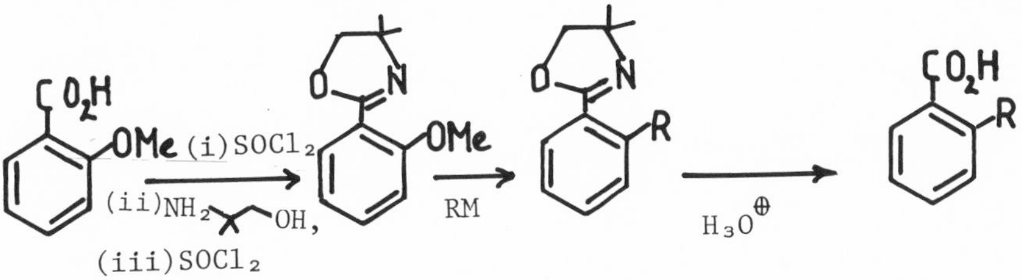

Phenolic esters have been prepared by the Diels-Alder reaction from 2,5-dialkylfurans and dimethyl acetylenedicarboxylate (DMAD) (P. Vogel, B. Willhelm, and H. Prinzbach, Helv., 1969, *52,* 584.

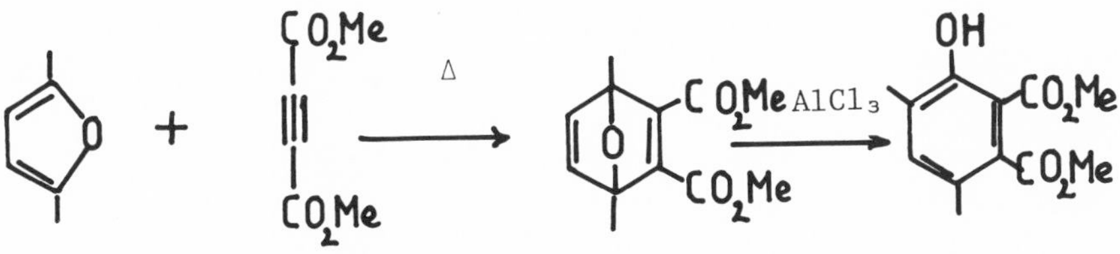

Terephthalates have been prepared by related routes (R. M. Acheson *et al.*, J. Chem. Soc., Perkin I, 1974, 1177).

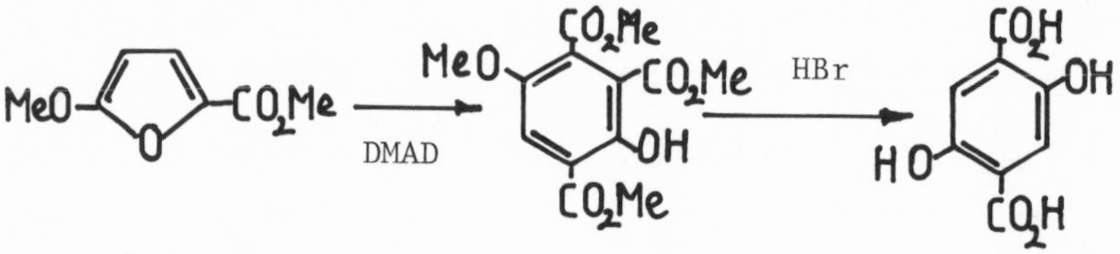

2-(p-Acetylaminophenyl furan reacts in a similar way to give dimethyl 3-hydroxy-6-(p-acetylaminophenyl) phthalate (A.F. Oleinik *et al.*, Khim Geterotsikl. Soldin, 1979, 17; Chem. Abs. 1979, *90*, 151897).

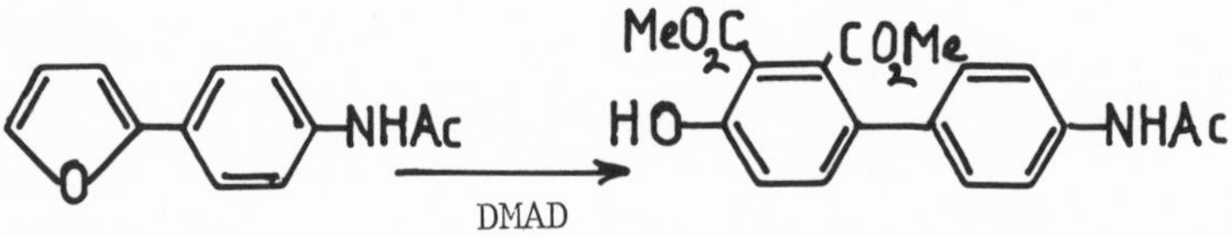

Dimethyl 4-hydroxyphthalate and a variety of other structures have been prepared by a related Diels-Alder reaction by way of the trimethylsilyl ether of an enol of the readily available *trans* 4-methoxybutene-2-one as shown in which 1,4-elimination of methanol occurs during the reaction (S. Danischevsky and T. Kitahara, J. Chem. Soc., 1974

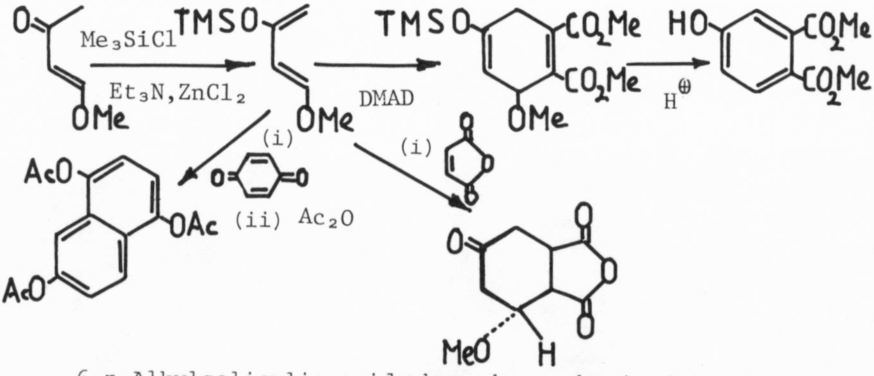

6-n-Alkylsalicylic acids have been obtained by the thermal rearrangement of basic copper salts of 2-n-alkyl benzoic acids (A.A. Durrani and J.H.P. Tyman, J. Chem. Soc.,Perkin I, 1979, 2069) or better from *o*- or *m*-fluoroanisole with n-alkyllithium followed by carbonation and demethylation (A.A.Durrani and J.H.P. Tyman, J.chem.Soc., Perkin I, 1979, 2079).

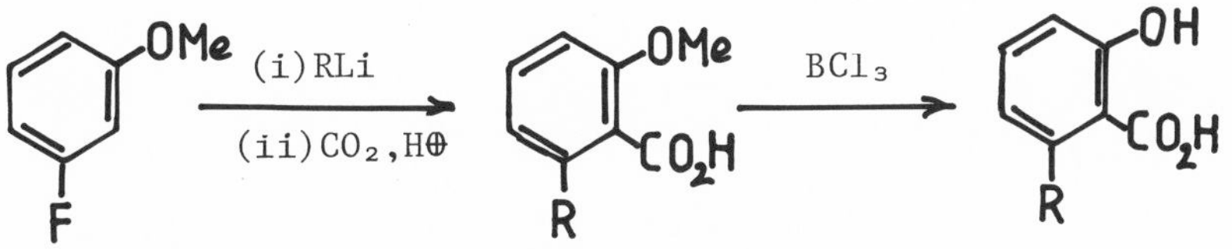

The synthesis of gingkolic acid, 6-(8'(Z)-pentadecenyl)-salicylic acid, by a related reaction has been reported (J.H.P. Tyman, J. org. Chem., 1976, $4l$,894). This material is identical to the monoene constituent of anacardic acid the principal component of natural cashew nut shell liquid from *Anacardium occidentale*.

(i) Phenolic Ethers

(i) Preparation
Several novel procedures for aryl ether formation have been described both with respect to the methylating agent and mode of formation of the nucleophile.

The important technique of phase transfer catalysis has been applied to the synthesis of phenolic ethers leading to a simple procedure for alkylation of the phenoxide ion with an alkyl halide or sulphate in a two phase dichloromethane/water system with a quaternary ammonium hydroxide as the effective reagent for transport of the phenoxide ion between the two phases (A. McKillop, J.-C. Fiaud, and R.P. Hug, Tetrahedron, 1974, 30, 1379). In the following scheme, $Q^{\oplus} = (R)_4N^{\oplus}$,

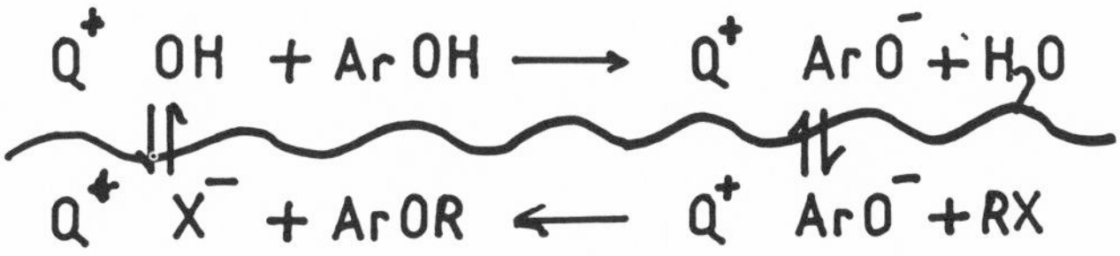

A convenient synthesis of alkylaryl ethers involves interaction of a phenol and an alcohol in the presence of triphenylphosphine and diethyl azodicarboxylate (A.K. Bose *et al.*, J. chem. Soc., Perkin. I, 1975, 46). Unactivated aryl halides react with sodium methoxide in hexamethylphosphorictriamide to give unusually good yields of the methyl ether (J.E. Shaw, D.C. Kunerth and S.B. Swanson, J. org. Chem., 1976,

$4l$,732). Sodium hydride in tetrahydrofuran at ambient temperature has been used as a reagent in the methylation of
highly hindered phenols (B.A. Strocknoff and N.L. Benoiton,
Tetrahedron Letters, 1973, 21). The oxonium salt shown,
O-methyldibenzofuranium fluoroborate obtained by heating 2-
methoxydiphenyl-2'-diazonium fluoroborate in benzene is an
extremely powerful Meerwein-type methylating agent (H.Heaney
et al., Chem. Comm., 1972, 315).

Phenol is converted into anisole by heating it with
dimethylcarbonate and a catalytic amount of 4-dimethylamino-
pyridine for ten hours at 180° C (BASF, Eur.P.62). Methylenation of catechol can be effected by its reaction with
dihalogenomethanes in dimethylformamide containing potassium
or caesium fluoride (J.H. Clark, H.L. Holland, and J.M.Miller,
Tetrahedron Letters, 1976, 3361). Intramolecular transeth-
erification of 2-methoxy-5-nitro-N,N-bis(β-hydroxyethyl)anil-
ine can be effected (ICI, Brit. Pat. 1539509) by the phase
transfer technique.

Mechanistic studies (S. Rajan and P. Sridaran, Tetrahedron Letters, 1977, 2177) have revealed an $S_{RN}1$ reaction

(radical chain reaction) of halogenobenzenes with phenoxide ion
in aqueous t-butanol in the presence of sodium amalgam. In
the slow reaction of diazomethane with certain phenols, catalysed by methanol, boron trifluoride or fluoroboric acid,
O-methylation is not the only reaction pathway (J.St.Pyrek
and O. Achmatowicz Jr., Tetrahedron Letters, 1970, 2651).
(Scheme 4).

Diazomethane in etheral solution at 0° reacts with 6-
pentadecyl salicyclic acid to give the methyl ester without
methyl ether formation (J.H.P.Tyman, J.Chromatog., 1975,

lll., 285).

Scheme 4

R=H,n-C₃H₇,allyl

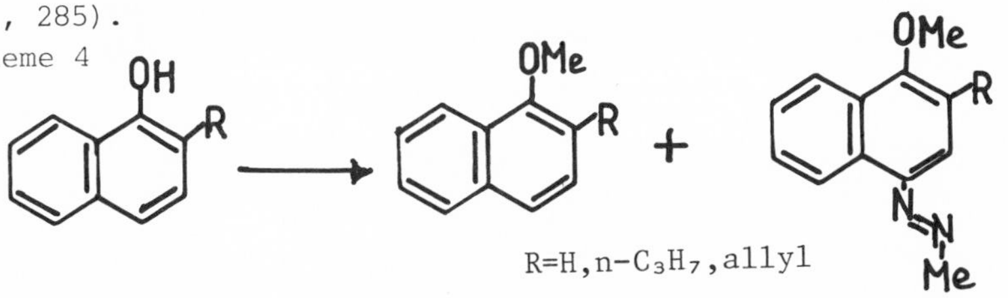

(ii) Dealkylation, Alkylation and Coupling

The removal of the alkyl group, usually methyl, can be effected in a number of ways, by for example refluxing with sodium thiomethoxide in dimethylformamide (G.I. Feutrill and R.N. Mirrington, Tetrahedron Letters, 1970, 1327). Partial demethylation of orcinol dimethyl ether can be effected (R.N. Mirrington and G.I. Feutrill. Org. Synth., 1974, 53,93).

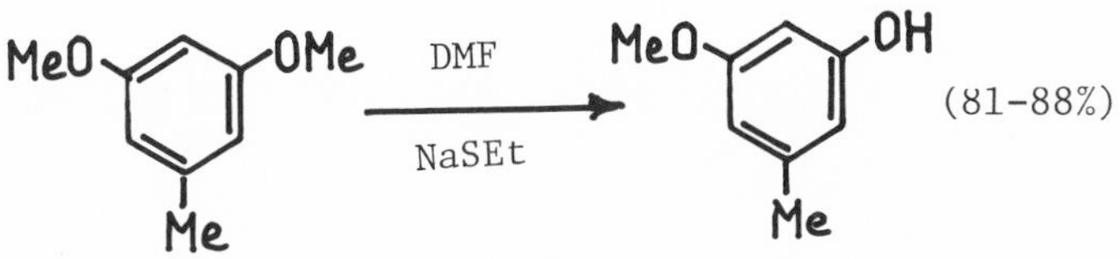

(81-88%)

Boron triodide in dichloromethane has been used for the demethylation shown (R.C. Ronald and J.M. Lansinger, Chem. Comm., 1979, 124).

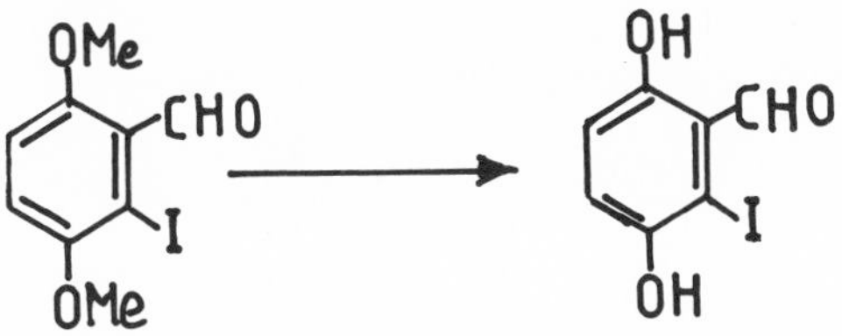

Boron tribromide behaves similarly to the iodide whereas boron trichloride can effect partial demethylation with sel⁻ective removal of methyl adjacent to a carbonyl-containing group (A.A. Durrani and J.H.P. Tyman, Tetrahedron Letters, 1973, 4839). More drastic reagents can cause side reactions and with pyridine hydrochloride, C-demethylation as well as O-demethylation can occur (J.H.P. Tyman, Chem. Comm., 1972, 914).

Electrolytic cleavage of the benzyl ether group can be selectively achieved in a molecule having an aryl ether group (S.M. Weinreb *et al.*, J. org. Chem., 1975, 40,1356). Tritylfluoroborate in dichloromethane at ambient temperature

has been employed for abstracting an hydride ion to achieve
debenzylation (D.H.R. Barton *et al.*, Chem. Comm., 1971, 861).

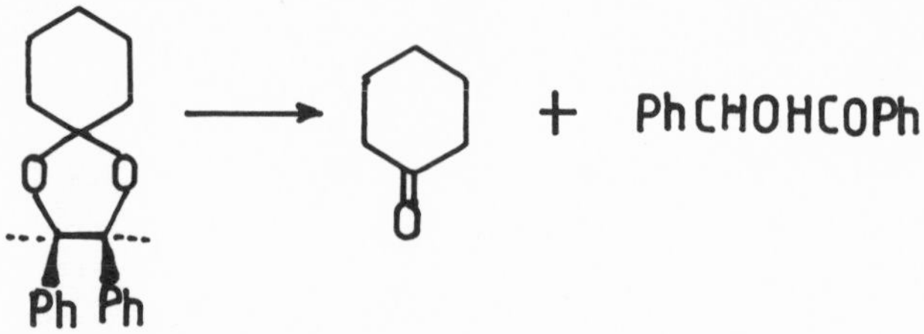

+ PhCHOHCOPh

Alkylating reagents such as the oxonium salt shown can
be prepared from anisole in the presence of a superacid salt
(G.A. Olah and E.G. Melby, J. Amer. chem. Soc., 1973, *95*,4791;
J.T. Adams *et al.*, *ibid.*, 6027).

$$PhOMe + MeO_2C Cl + Ag SbF_6 \longrightarrow Ph\overset{+}{O}Me_2 SbF_6^- + AgCl$$

Nitromethane reacts with anisole in acetic acid with
manganese (III) acetate present, probably by way of nitromet-
hyl radicals, to give mono substitution products in 77% yield,
consisting of a mixture of isomers (*o*, 71, *m* 5, *p*,24%) (M.E.
King and T.R. Chen, J. org. Chem., 1978, *43*, 239).

Anisole can be phenylated by benzenediazonium fluoroborate in trifluoromethanol (P. Burn and H. Zollinger, Helv.,
1973, *56*,3204). Resorcinol dimethyl ether can be substituted
by way of the dienyl cation shown (T.G. Bonner, K.A.Holden
and P. Powell, J. organometal. Chem., 1974, *77*, C37).

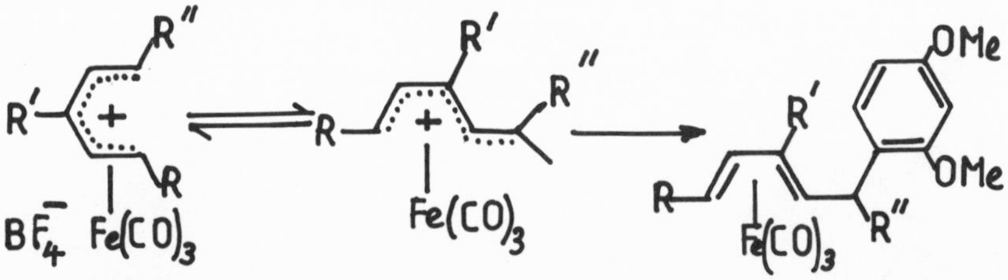

The dianion of phenoxyacetic acid has been used in conjunction with 9-borabicyclo[3,3,1]nonane to extend in high
yield a terminal alkene by two carbon atoms (S. Hara, K. Kish-
imuro, and A. Suzuki, Tetrahedron Letters, 1978, 2891).
(Scheme 5).

The dienes obtained from phenolic methyl ethers by the
Birch reduction, in the form of their iron tricarbonyl adducts

react with the triphenylmethyl cation to afford stabilised
ions which can undergo nucleophilic substitution, the dihydro
compounds produced being capable of rearomatisation (A.J.Bir-
ch *et al.*, J. chem. Soc., Perkin I, 1973, 1882).(Scheme 6).
Scheme 5

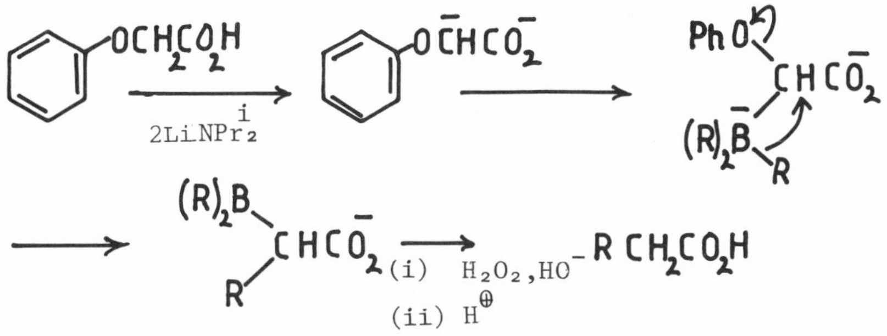

Scheme 6

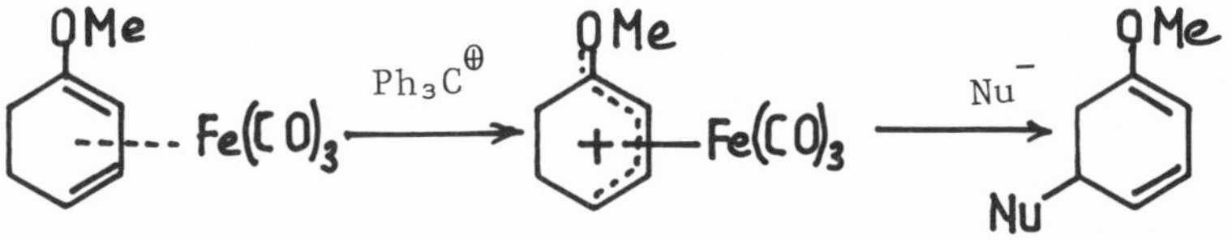

Cycloalkylation of a m-alkylanisole with stannic chloride in benzene has been described (P.A.Gueco and R.S.Finkel-
hor, Tetrahedron Letters, 1974 527).

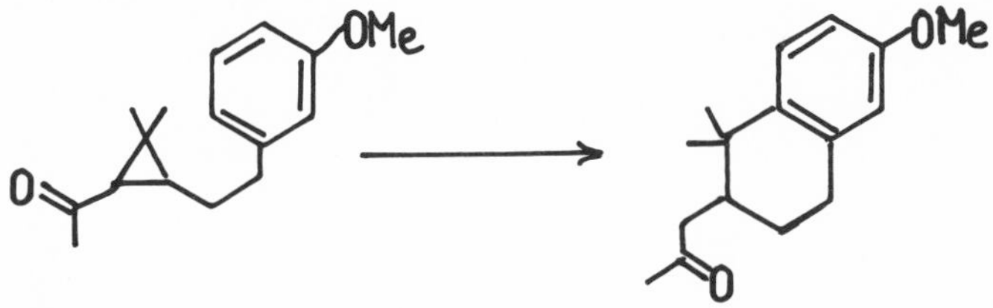

Although phenolic ethers are less easily oxidised than
the parent phenols (U.Palmquist, A.Ronlán and V.D.Parker,
Acta. chem. Scand., 1974, *28*(B), 267), their electrochemical
oxidation has been described (M.Sainsbury and J.Wyatt, J.chem.
Soc., Perkin I, 1976, 661). Anodic oxidation of alkyl anisoles and of hydroquinone ethers, in methanol solution, leads
by way of the cation radical to nuclear methoxylated products
and with p-alkyl anisoles to benzyl ethers or benzaldehyde
dimethyl acetals (A.Nilsson *et al.*, J.chem.Soc., Perkin I,
1978, 708). 2,5-Dimethylanisole undergoes oxidative coup-

ling by treatment with thallium(III) trifluoroacetate in trifluoroacetic acid to give 2,2',5,5'-tetramethyl-4,4'-dimethoxydiphenyl probably by a radical cation mechanism (A.McKillop, A.G. Turrell and E.C. Taylor, J. org. Chem., 1977, 42,764). Reductive coupling of 4-methoxy-2,6-dimethylacetophenone takes place with titanium(III) trichloride-lithium aluminium hydride (McMurry reagent) to give the hindered stilbene illustrated (A.L. Baumstark, E.H.J.Bechara and M.J. Semigran, Tetrahedron Letters, 1976, 3265).

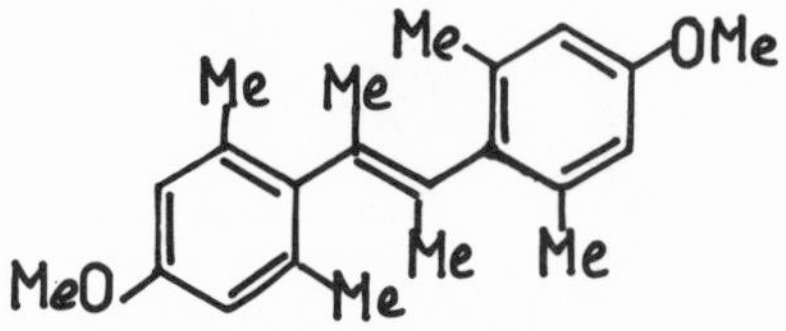

(iii) Cyclisation and Rearrangement

Cyclisations of appropriately substituted phenolic ethers have been effected.   The base-catalysed reaction of the aryloxyacetone shown has been employed for a synthesis of psoralene (J.K.MacLeod and B.R. Worth, Tetrahedron Letters, 1972, 237, 241).

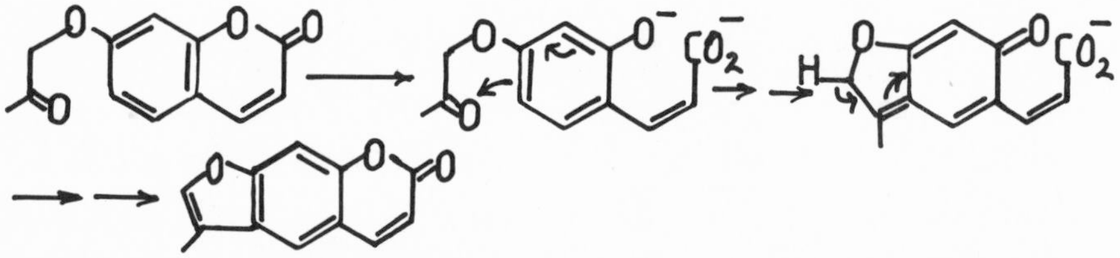

The unsymmetrical bis(methoxyphenyl)alkane shown undergoes isocyclic ring closure, partial demethylation and dearomisation in a 'superacidic' medium (J.-P. Gerson and J.-C. Jacquery, Chem. Comm., 1976, 652).

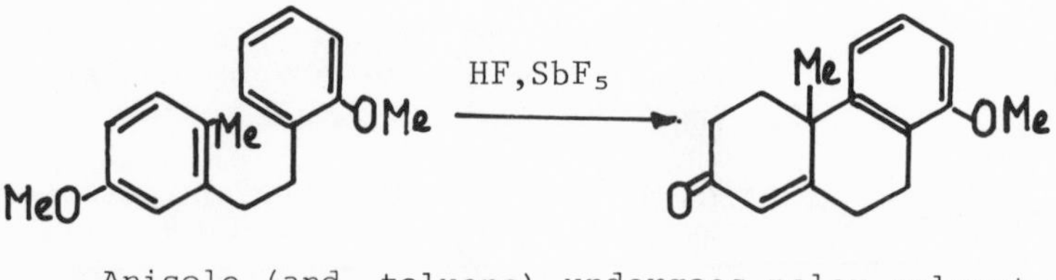

Anisole (and toluene) undergoes polar solvent-promoted 1,3-photocyclo addition with ethyl vinyl ether indicating that it is not a homopolar process as previously thought, and this type of procedure has been used as a route to the bicyclo-[3,2,1]octen-8-one system (A.Gilbert and G.Taylor,Chem.Comm., 1977, 242; J.A.Orrs and R. Srinavasan, J.org. Chem., 1977 *42*, 1321).

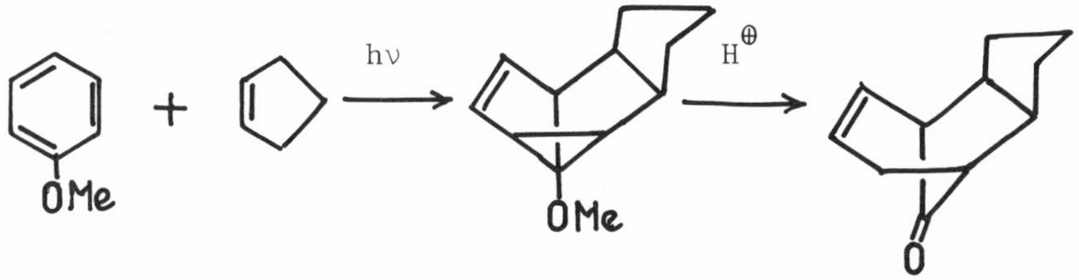

The photoaddition of maleimide to anisole has been studied (D. Bryce-Smith, A. Gilbert and B. Halton, J.chem.Soc., Perkin I, 1978, 1172) and found to give a number of complex products.

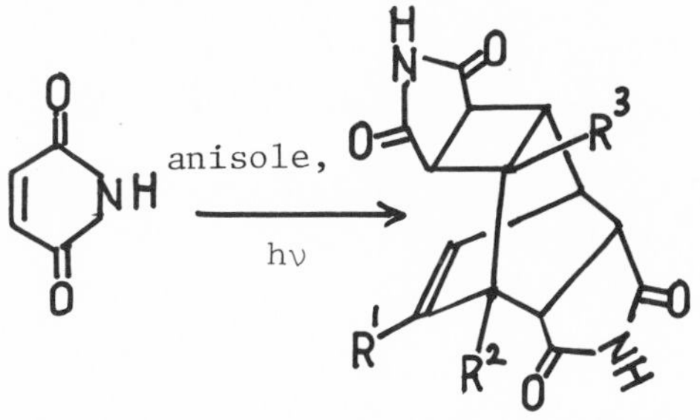

$$R^2 = R^3 = H \ , \ R' = OMe$$
$$R^3 = OMe, R' \quad R^2 = H$$
$$R^3 = R' = H \ , \ R^2 = OMe$$

Irradiation of a mixture of anisole and acrylonitrile in methanol and in acetonitrile gives respectively o- and p-cyanoethylated products and [2+2] cycloadducts (M.Ohashi, Y. Tanaka and S. Yamada, Chem. Comm., 1976, 800). (Scheme 7)

Preparative and mechanistic aspects of the Claisen allyl ether rearrangement have attracted continued interest. This reaction and the Cope rearrangement have been reviewed (S.J. Rhoads and N.R. Raulins, Org. Reactions, 1975, *22*,1). The o-Claisen rearrangement with benzenoid (and naphthalenic) systems has been extensively studied (N. Sarcevic, T.Zsindely and H. Schmid, Helv., 1973, *54*, 1457; 1975, *58*, 6110). The rearrangement in hot diethylaniline of the conjugated ether shown is an example of a 5s,5s sigmatropic rearrangement which has been shown to involve a ten-membered transition state (Gy. Frater and H. Schmid, Helv., 1970,*53*,269) (Scheme 8)

Scheme 7

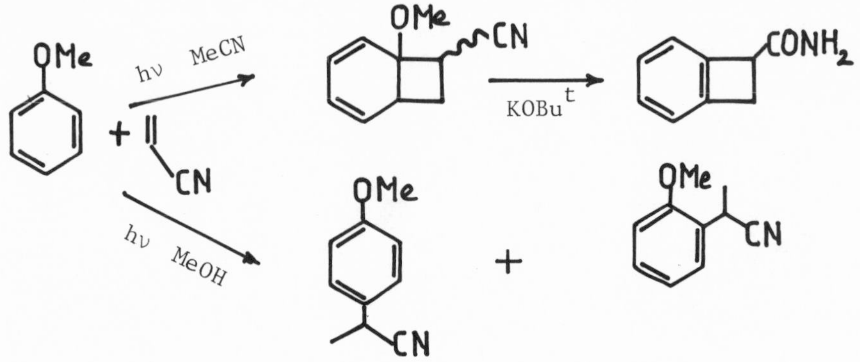

Scheme 8

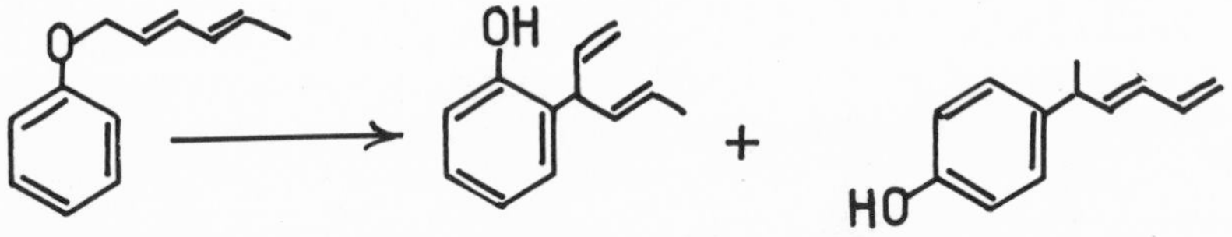

The allyl ether indicated undergoes an unusual Claisen rearrangement under thermal (or catalytic conditions) to give the *m*-allylic substance shown by an intermolecular mechanism (E.K. Aleksandrova, Zhur. org. Khim., 1974, *10*,825; Chem.Abs., 1974, *81*,25254)

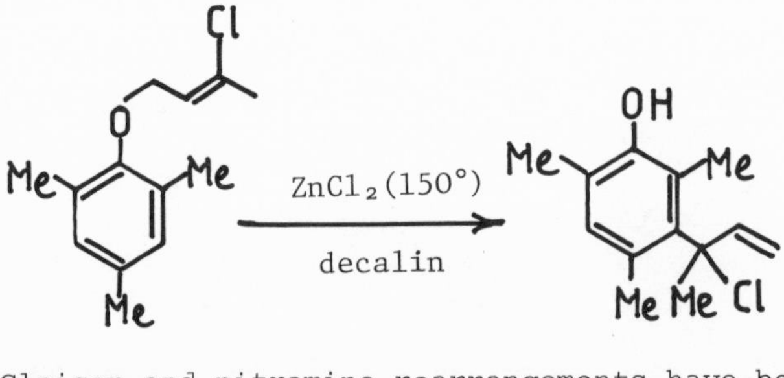

The Claisen and nitramine rearrangements have been compared with respect to the products from allyl *m*-chlorophenyl ether and *N*-methyl, *N*-nitro-*m*-chloroaniline (W.N. White and J.R. Klink), J. org. Chem., 1977, *42*,166). There is only a superficial resemblance and the product distribution is different.

Allyl phenyl ethers undergo *o*-Claisen rearrangement on dissolution in trifluoroacetic acid at ambient temperature

(U. Svanholm and V.D. Parker,Chem. Comm., 1972, 645). Large
salt effects and isotope effects confirm the existence of a
polar mechanism and transition state and indicate that proton
transfer is involved in the rate-determining step (U. Svanholm
and V.D.Parker, J.chem. Soc., Perkin 11, 1974, 169). Acetic a
dride in the presence of an acetate effectively trapped the
acetate giving an improved yield of the thermodynamically less
stable conventional Claisen rearrangement products of γ-sub-
stituted allyl aryl ethers (D.S. Karanlosky and Y. Kishi,
J. org. Chem., 1976,41,3026).

(77%)   (23%)

Allyl 2-dimethylaminophenyl ether by Claisen rearrangement
gives as one of the products, 2-dimethylamino-4-allyl phenol
(W.D. Ollis *et al.*, Chem. Comm., 1973, 651, 653, 654) by a
series of 2,3 and 3,3-rearrangements. Cinnamyl phenyl allyl
ether in the Claisen rearrangement gives both *o*- and *p*-subs-
tituted products. The ratio is dependent on the acidity of
the solvent with no abrupt change of rate, contrary to previous reports, in liquid-crystal solvents on proceeding from
the nematic to the isotropic phase. Alignment of the molecule in the liquid crystal is not considered important for
the formation of the transition state (M.J.S. Dewar and B.
Nahlovsky, J. Amer. chem. Soc., 1974, *96*,460). Phenyl
propargyl ether upon gas phase pyrolysis at 460°C yields
2-indanone (26%) and benzocylobutene (31%) (W.S. Trahanowsky
and P.W. Mullen, J. Amer. chem. Soc., 1972, *94*,5911).

(iv)   Electrophilic and Nucleophilic Substitution
       The chlorination of anisole by hypochlorous acid, catal-
ysed by cyclodextrins such as dodecamethyl-α-cyclodextrin or
an O-alkylated polymer from α-cyclodextrin and epichlorhydrin
has been shown to give exclusive *p*-substitution (R. Breslow,
H. Kohn, and B. Siegel, Tetrahedron Letters, 1976, 1645).

Electrochemically-generated iodine, probably as $CH_3C \equiv N - \overset{\oplus}{I}$
in acetonitrile, is a powerful electrophile and anisole added
to the preformed reactant gives *p*-iodoanisole in 90% yield
(L.L. Miller and B.F. Watkins, J. Amer. chem. Soc., 1976,

98,1515).

Mechanistic studies of the nucleophilic substitution reactions of polynitro phenolic ethers have continued to engage attention.   The ^{1}H NMR spectrum of the σ-complex in the reaction of 2,4,6-trinitroanisole with n-butylamine (not in large excess) to give N-(n-butyl)picramide has been obtained by a flow technique (C.A. Fyfe *et al.*, Chem. Comm., 1977, 335). Kinetic and equilibrium data have been reported for σ-complexes resulting from sodium ethoxide and a series of 6-substituted-2,4-dinitrophenetoles (M.R. Crampton, J.chem.Soc., Perkin II, 1977, 1442).

Catalysis by the piperidine in its reaction in benzene with 2,4-dinitrophenyl 4-nitrophenyl ether in the reaction shown has been discussed in terms of concerted proton transfer and expulsion of the leaving group (D. Spinelli, G.Consiglio, and R. Noto, J. chem. Soc., Perkin II, 1977, 1316)

In the uncatalysed reaction of morpholine with 2,4-dinitrophenylphenyl ether in dipolar aprotic solvents, evidence suggests that fast deprotonation and slow expulsion of the leaving group does not occur (D. Ayediran *et al.*, J. chem. Soc., Perkin II, 1977, 597).   Shortcomings of the concerted mechanism with media of low dielectric constant have been discussed and an alternative suggested involving aggregate formation in which synchronous bond-making and breaking does not occur. Similarly the base-catalysed and general acid-catalysed mechanism for substitution in activated aromatic compounds by amines in protic solvents involving fast deprotonation and rate-limiting catalysed expulsion of the leaving group, is not considered a significant pathway (C.F. Bernasconi, R.H. De Rossi, P. Schmid, J. Amer. chem. Soc., 1977, *99*,4090).   Rather the mechanism is believed to involve a rate-limiting deprotonation of the substance shown when the expulsion of the

leaving group (e.g. $\overset{\ominus}{O}$ Ph) is relatively fast or a rapid equilibrium deprotonation succeeded by a rate-limiting non-catal-
ysed expulsion in the case of a very poor leaving group (e.g.
X = OMe). References of interest with respect to phenolic
ethers are given in a review of nucleophilic aromatic substitution of nitro compounds (J.R. Beck, Tetrahedron, 1978, *34*,
2057).

(j) Halogenophenols

A variety of new reagents for halogenation, often specifically, have been described. Monochlorination of phenols
(or anilines) can be readily effected with trichloroisocyan-
uric acid (E.C. Juenge, D.A. Beal and W.P. Duncan, J. org.
Chem., 1970, *35*,719). Chlorination at 15° with sulphuryl
chloride of 2-methylphenol containing catalytic amounts of
aluminium chloride and diphenyl sulphide leads to 2-methyl-4-
chlorophenol (W.D. Watson, Tetrahedron Letters, 1976, 2591).
Chlorine or t-butylhypochlorite in carbon tetrachloride
gives high yields of o-chlorophenols apparently by an ionic
mechanism (W.D. Watson, J. org. Chem., 1974, *39*,1160). 2,6-
Dichlorophenol is produced (86% yield) from the bis-phenol shown
by addition of aluminium chloride to a benzene solution during
five hours at 50°C (M. Tashiro, H. Watanabe and K. Oe, Org.
prep. Proc. Int., 1975, *7*,189).

Cyclohexadienones frequently arise in halogenations involving
alkylphenols. The reaction of twenty-six different phenols
in the solid state with gaseous chlorine gives in each case
high yields of substitution products often with isolatable
cyclohexadienones as intermediates (R. Lamarfine and R. Perrin,

J. org. Chem., 1974, *39*,1744). 4-Methylphenol in dichloromethane added to antimony pentachloride at -70° to -50° gives
4-chlorocyclohexa-2,5-dienone in 92% yield (A.Nilsson,A. Ronlán and V.D. Parker, Tetrahedron Letters, 1975, 1077). 3,4-
Dimethylphenol, the methyl ether and the acetate upon poly-
chlorination give the products shown by way of a cyclohexadienone (in case of the phenol) (P.B.D. de La Mare and B.N.B.
Hannan, Chem. Comm., 1971, 1324). Abnormal paths in the halogenation of aryl acetates probably by way of *ipso*-substitut-
ion have been described (P.B.D. de La Mare, B.N.B. Hannan, and
N.J. Isaacs, J. chem. Soc., Perkin II, 1976, 1389).

With 3,4-dimethylphenol and a molar proportion of chlorine
the corresponding 4-chlorocylohexadienone is formed. It is
not necessary for all *o*- and *p*- positions to be blocked for
cyclohexadienones to be produced. Certain halogenocyclohex-
adienones can themselves be halogenating agents. The bromo
compound shown derived from phenol is a selective mono brominating reagent for primary, secondary and tertiary aromatic
amines (V. Calo *et al.*, J. chem. Soc., (C), 1971, 3652).
Aniline yields p-bromoaniline (96%) and o-bromoaniline (4%)

Tribromophenol with bromine through the intermediacy of the
same dienone has been used for the monobromination of 3-trif-
luoromethyldimethylaniline to give the 4-bromo product in
82-90% yield (G.J. Fox and G. Hallas, Org. Synth. 1976,
55,20).
Anomalous high proportions of *p*-products in the bromination of

phenols have been shown to arise from 1,2 or 1,4-electrophilic addition of hydrogen bromide (V. Calo, J. chem. Soc., Perkin II, 1974, 1189, 1192) the 1,2 compound rearranging to give the 1,4-product which then affords the phenol.   An *ipso*-substituted cyclohexadienone is formed in the bromination of 3,4-dimethylphenol and possible pathways for related species have been discussed (P.B.D. de la Mare, N.J. Isaacs and P.D. McIntyre, Tetrahedron Letters, 1976, 4835).   2,6-di-t-Butyl phenol gives the related 4-bromocyclohex-2,5-dienone which slowly enolises to the 4-bromo product.   *o*-Bromination of phenols is favoured under Ti(IV) catalysis (M.V. Sargent, J. chem. Soc., Perkin I, 1973, 340).   *p*-Bromination of *o*-chlorophenol occurs in 98% yield in 1,1,1,-trichloroethane solution containing triethylamine hydrochloride (Ciba-Geigy A.G., Brit. Pat. 2006189).

Treatment of a phenol with thallium (I) acetate and iodine causes selective *o*-iodination (R.C. Cambie *et al.*, J.chem. Soc., Perkin II, 1976, 1161).

Another procedure for *o*-iodination employs bis-trifluoroacetoxythallation with decomposition of the intermediate by potassium iodide followed by heating (A. McKillop *et al.*, J. Amer. chem. Soc., 1971, *93*, 4841, 4845).

$$ArH \;+\; Tl(OCOCF_3)_3 \longrightarrow Ar(Tl\,OCOCF_3)_2$$

$$Ar\,Tl\,I_2 \longrightarrow ArI \;+\; Tl\,I$$

In the gas phase reactions of fluorine-substituted aromatics with hydroxide and alkoxide ions, reaction pathways have been interpreted as involving loss from the Meisenheimer complex of either a fluoride ion to give the phenol or ether, or loss of an alkyl fluoride or hydrogen fluoride to give the phenoxide ion (S.M.J. Briscese and J.M. Riveroo, J. Amer.chem. Soc., 1975, *97*,230).

Halogenated phenols have uses as bactericidal and fungicidal compounds and 2,2'-dihydroxy compounds in which the rings can be joined by a wide variety of bridging groups (e.g. "hexachlorophene") and atoms have enhanced properties in this direction although they have fallen out of favour in recent years.

2,5-Dichlorophenol has been identified in the defensive secretions of grasshoppers and is thought to stem from 2,4-dichlorophenoxy acetic acid as a metabolite (J. Meinwald *et al.*, Science, 1971, *172*,277).

2,6-Dichlorophenol is the sex pheromone of the female tick,

Amblyomma americanum (.R.S.Berger, Science, 1972, *177*, 704).
Phenol itself has been identified as a sex pheromone of the
grass grub beetle, *Costelytra zealandica.*

The chemistry of some halogenation materials of marine
origin with phenolic structure has been investigated.
Laurintenol from *Laurencia intermedia* has been converted into
aplysin which occurs in *Aplysin kurodai* and has itself been
synthesised (Y. Hirata *et al.*, Tetrahedron 1969, 25, 3509).

(k) *Nitrophenols*

Aspects of work on nitration, covering phenols, have
been reviewed (S.R. Hartshorn and K. Schofield, Prog. org.
Chem., 1973, *8*, 278).
The mechanism of nitration of phenol and of anisole has been
discussed.   Passage of oxygen through a solution of 3-methyl
-4-nitrophenol in aqueous acetonitrile at 35° containing concentrated nitric acid gives 3-methyl-4-nitrosophenol (Chemin-
ova A/S, Belg. P. 872530).   4-Nitro-2,6-diphenylphenol can
be synthesised from 5-nitropyrimidine and dibenzylketone with
potassium ethoxide in ethanol as shown (P. Barczynski and H.C.
van der Plas, Rec. trav. Chim., 1978, *97*, 256).   In general
the course of the reaction depends on the nature of the ketone and the base and can lead to the formation of nitropyrid-
ines, nitrophenols and pyridones by addition of the anion of
the ketone to C6 and C2 followed by appropriate cleavage and
cyclisation.

The ambident nitrite ion bonds to aromatic carbon by way
of its nitrogen or oxygen atom (N-attack or O-attack) but the
end product of reaction with a 2-halogeno, 4-halogeno or 2,4
dihalogenonitrobenzene is always the nitrophenoxide owing to
the reactivity of the nitro intermediate which can sometimes
be isolated (T.J. Broxton, D.M. Muir and A.J. Baker, J. org.
Chem., 1975, *40*, 2037, 3230; D.H. Rosenblatt, W.H.Goodwin and

R.D. Goodin, J. Amer. chem. Soc.,1973, *95*,2133).

The practical hydrolysis of nitrohalogenobenzenes to nitrophenols has been described (ICI, Brit. P. 1238372). Trifluoromethylnitrochlorobenzenes can be selectively converted into phenols by means of powdered sodium hydroxide without hydrolysis of the trifluoromethyl group (R.L. Jacobs, J. org. Chem., 1971, *36*,242).

Nitrosation of phenols (and naphthols) can be brought about photochemically in neutral solution in the presence of sodium nitrite (F.D. Salva and G.R. Olin, J. Amer. chem. Soc., 1975, *97*,5631).

The mechanism of nitrosation of anisole (and phenol) has been studied. Ether cleavage occurs with formation of p-nitrosophenol (B.C. Challis and A.J. Lawson, J. chem. Soc., (B), 1971, 776).

A study of the nitration of anisole (and related compounds) by nitric acid in aqueous sulphuric acid has indicated that the unprotonated species is involved. The *o/p* ratio in the products varies considerably and this is considered to be explicable in terms of the combined effects of hydrogen bonding and a diffusion-controlled reaction. With o- and p-methylanisole, *ipso*-substitution at C-Me is important and in the case of the latter this leads to 4-methyl-2-nitrophenol as a major product by subsquent water attack and loss of methoxyl group (K. Schofield *et al.*, J. chem. Soc., Perkin II, 1977, 248). In acetic anhydride, the nitration of *p*-methylanisole (and some other *p*-substituted toluenes) involves the formation of 4-methyl-4-nitrocyclohexa-2,5-dienone (M.P. Hartshorn *et al.*, Austral. J. chem., 1977, *30*,113).

o-Nitrophenols formed by the aromatisation of 4-methyl-4-nitrocyclohexa-2,5-dienones in a wide variety of solvents are believed to arise by a formal 1,3-shift of the nitrogroup and suggests the existence of a radical dissociation-recombination process (C.E. Bame and P.C. Myce, J. Amer. chem. Soc., 1978, *100*,973; R.G. Coombes and J.G. Goldring, Tetrahedron Letters, 1978, 3583).(Scheme 9).

In the nitration of O-acyl derivatives of phenol o- and p-products are formed exclusively when the acyl group is acetyl or methanesulphonyl whereas if it is phosphoryl a considerable proportion of the *m*-product is obtained (M. Attira *et al.*, Chem. Comm., 1976, 466). The spiro compound shown has been formed by the intramolecular trapping of a Wheland intermediate, in the nitration of 3,4,5-trimethylphenyl iso-valerate (M. Shinoda and H. Suzuki, Chem. Comm., 1977, 479). (Scheme 10).

214

Scheme 9

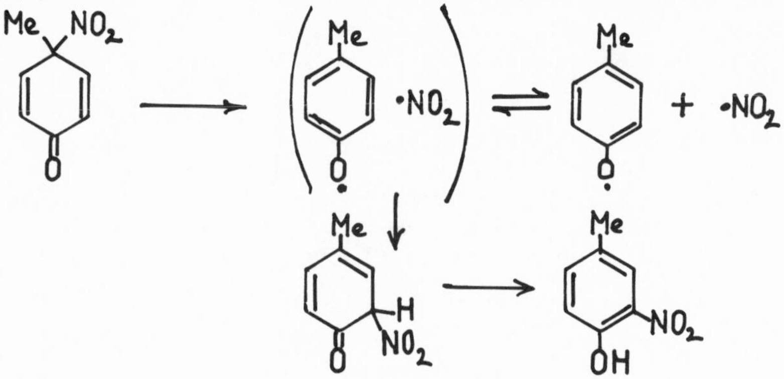

Scheme 10

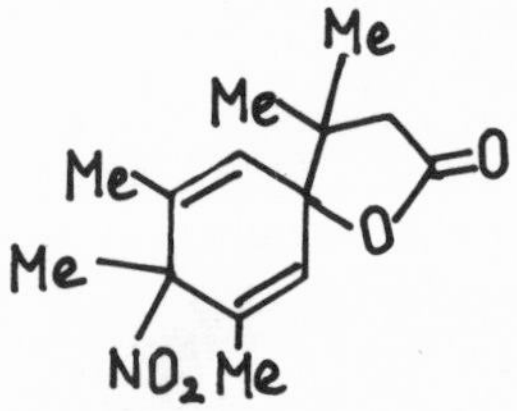

Methyl 4-hydroxybenzoate (in ethanol) has been nitrated
by warming it for a short time with an aqueous solution of
mercuric nitrate and potassium nitrite.   Reduction *in situ*
of the product with dilute ammonium sulphide in the presence
of hydrogen sulphide gives methyl 3-amino-4-hydroxybenzoate
(J. Lazer, L. Mod, and E. Vinkler, Acta chim. acad. Sci. Hung.,
1979, *99*,149).

The addition of various anions to polynitrophenyl ethers
is known to give coloured adducts known as 'Meisenheimer complexes'.   Methyl picrate reacts with methoxide ion as shown,
kinetic control giving initially the 1,3-dimethoxy compound
which under thermodynamic control is slowly converted into the
more stable 1,1-adduct.   With cyanide ion the 1,3-adduct is
preferred kinetically and thermodynamically (J.W. Larsen, J.H.
Fendler, and E.J. Fendler, J. Amer. chem. Soc., 1969, *9?*,5903);
A.R. Norris, Canad. J. Chem., 1969, *47,* 2895).

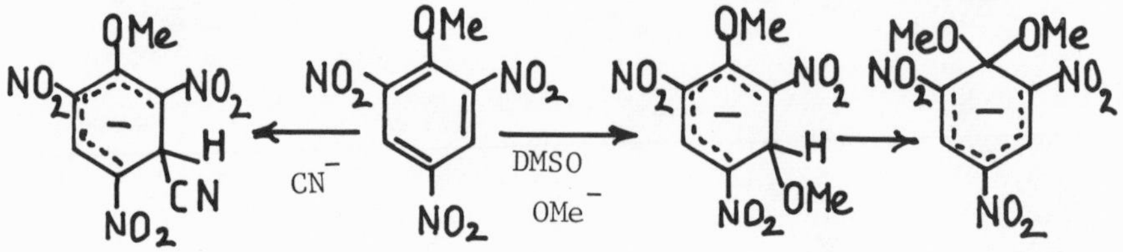

The chemistry involved generally in the interaction of anions with polynitroaryl compounds has been reviewed (M.R. Crampton, Adv. phys. org. Chem., 1969, *7*,211; E. Buncel, A.R. Norris and K.E. Russell, Quart. Review., 1968, *22*,123). 2,4,6-Trinitro-3-hydroxybenzoic acid reacts with hydroxide ion in the presence of acetone to give a dark red trianion adduct which with another hydroxide ion gives a tetranion as shown whereas picric acid gives monocyclic and bicyclic anions (K.A. Kova, Arch. Pharm., 1974, *307*,100).

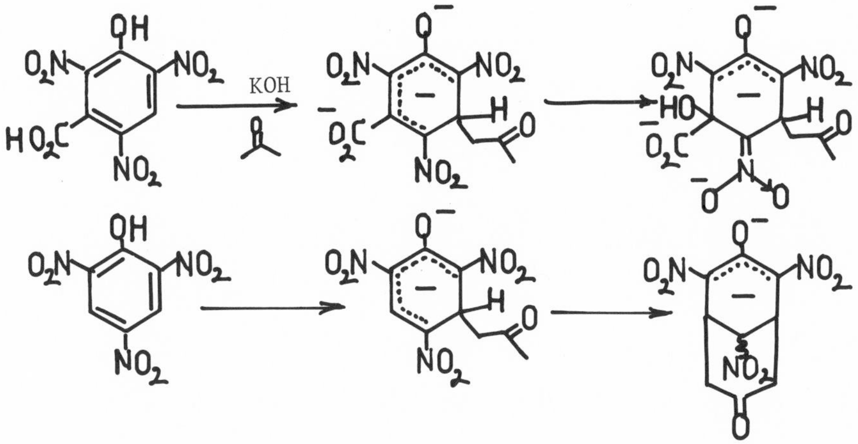

(1) Amino phenols

o-Acylamino phenols can be synthesised from o-hydroxyaldehydes and o-hydroxyketones by interaction of the sodium salts with chloramine as shown (R.A. Crocker and P. Kovacic, Chem. Comm., 1973, 716).

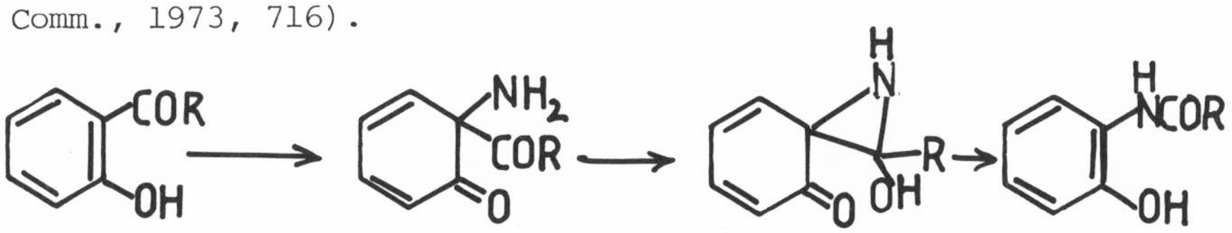

o-Ylides of the type shown prepared from *o*-dimethylamino phenol are isolable and upon heating rearrange to the benzyl ether whereas the corresponding *p*-compounds are unstable above 0° (W.D. Ollis *et al.*, Chem. Comm., 1973, 651). (Scheme 11).

m-Amino phenols can be prepared indirectly by Wesseley oxidation and reaction of the acetoxycyclohexadienone with an amine as shown. (Scheme 12).

Scheme 11

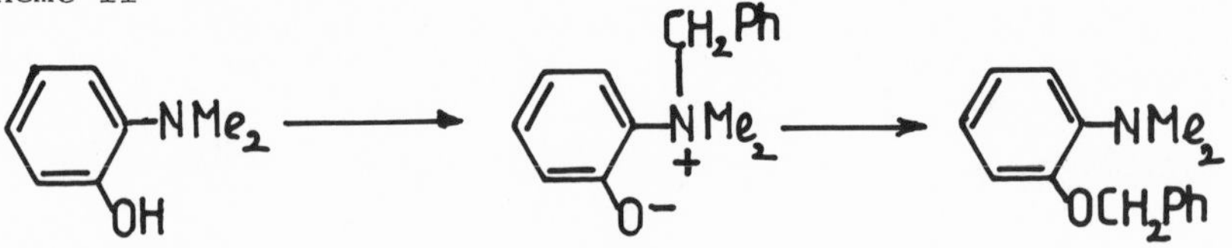

Scheme 12

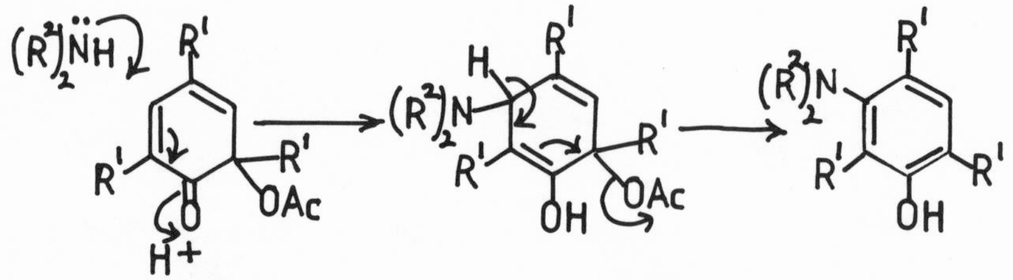

In the reaction of 3-aminophenol with benzenesulphonyl chloride selective reaction to give a benzenesulphonate can be achieved (Casella.Farb.Maink. Belg.P 866275).

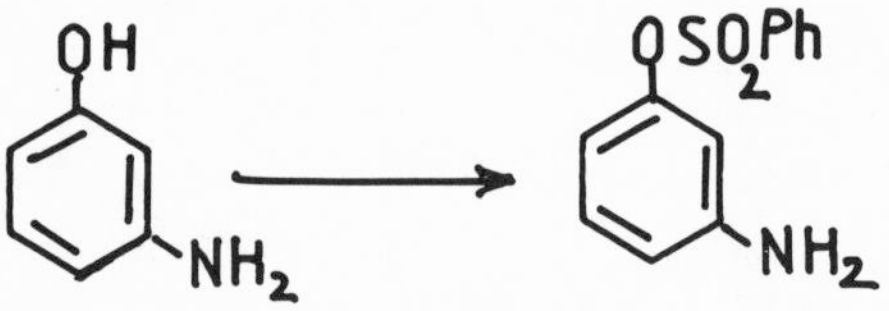

(m)   Azo compounds and Diazonium salts

The union between the positive diazonium cation and a phenolic anion has been shown to involve, in part at least, a radical ion pair intermediate from observation of ^{15}N CIDNP signals in the ^{15}N NMR spectrum (K.A. Bilevich *et al.*, Chem. Comm., 1972, 1058).

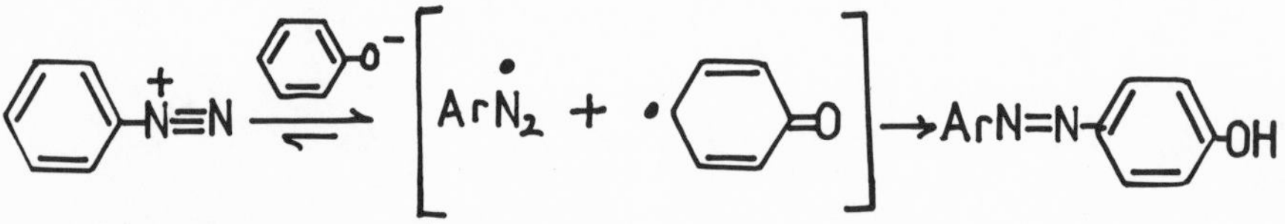

Electrophilic displacement of the diazonium cation from *o*-arylazophenols has been observed and it appears to be a better leaving group than $Cl^{\oplus}$, $Br^{\oplus}$, or $NO^{\oplus}$ (N.J. Burce, J. chem. Soc., Perkin I, 1974, 942). Results from the diazocoupling reactions of 4-substituted phenolate anions suggest that the methoxy group activates to electrophilic attack the position

meta to it in contrast to the behaviour expected from its σ
value (J. Kulič, M. Titz, and M. Večeřa, Coll. Czech. chem.
Comm., 1975, *40*,405).

Diazotisation of tetrafluoroanthranilic acid with nitro-
sylsulphuric acid in acetic acid proceeds with replacement of
fluorine to give the very stable diazonium salt shown (C.W.
Rees *et al.*, J. chem. Soc., Perkin I, 1972, 1315).

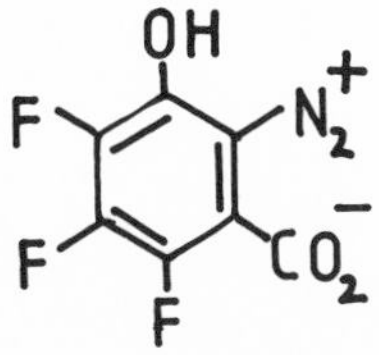

The diazotisation of 2-aminotetrafluorophenol in 70% aqueous sulphuric acid gives 1-diazotetrafluorobenzene oxide in
quantitative yield.   With benzonitrile, the 2-phenylbenzox-
azole shown was formed (R.N. Hazeldine *et al.*, J. Chem. Soc.,
(C), 1971, 562).

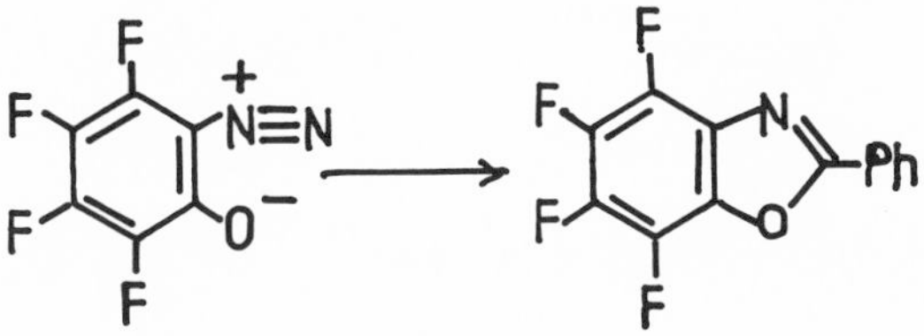

Studies have continued to appear on the Wallach rearrangement.   Phenylnaphthyl (and naphthylnaphthyl)azoxybenzenes
give the product with the hydroxyl group on the naphthyl ring
regardless of the initial position of the oxygen of the azoxy
group (A. Dolenko and E. Buncel, Canad. J. Chem., 1974, *52*,-
623).   A novel Wallach rearrangement product from 4,4'-dial-
kylazoxybenzenes has been investigated and found to yield,
together with other materials, 2-alkyl-5-(4-alkylphenylazo)
phenol   as shown.

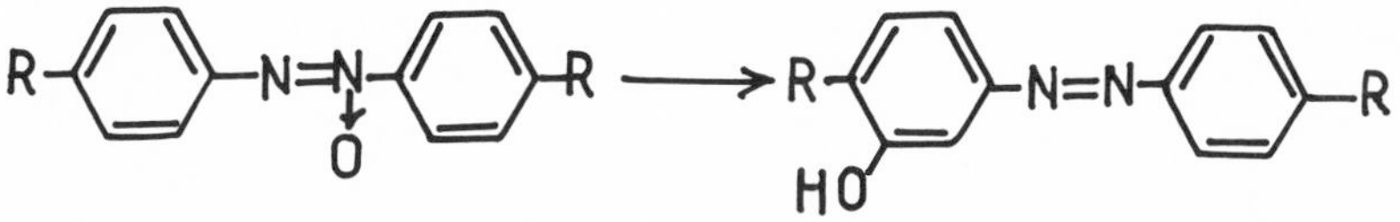

The thermal Wallach rearrangements of azoxybenzenes
and that of the 1:1 complexes with antimony pentachloride in
nitrobenzene have been studied and found to give *o*-hydroxy-

azobenzenes exclusively (I. Skimao and H. Hashidzume, Bull. chem. Soc., Japan, 1976, *49*,754; J. Yamamoto *et al*., Chem. Letters, 1976, 261). The rearrangement of azoxyarenes has been carried out photochemically to yield *o*-hydroxyazo compounds (Y. Katsuhara, Tetrahedron Letters, 1973, 1323). o-Nitrobenzanilide has been photochemically transformed to the *o*-hydroxyazo compound shown by way of the carbozyazoxybenzene (B.C. Gunn and M.F.G. Stevens, Chem. Comm., 1972, 835).

(n) Sulphonation and Related Reactions

Aspects of the mechanism of sulphonation relevant to phenols have been given (B.V. Smith, Org. React. Mech., 1975,275; 1976, 301. Synthesis of sulphonic acids generally has been described (G.C. Barrett in Organic Compounds of Sulphur, Selenium and Tellurium, ed. D.H. Reid, The Chemical Society, London, 1970, vol.1, ch.2, vol.3, ch.1).

In the reaction of phenol with fluorosulphonic acid and excess hydrogen fluoride, 4,4'-dihydroxydiphenylsulphone is formed in 87% yield, the reaction being commenced at - 100° and completed at 75° over six hours at 5.95 ats. (W. Ger. P., 2,804,080).

2. DIHYDRIC PHENOLS

(i) Preparations

A mixture of the monomethyl derivatives of catechol (40%) and hydroquinone (55%) can be obtained in 70% yields, by the hydroxylation of anisole with hydrogen peroxide in the presence of aluminium chloride (M.E. Kurz and G.J. Johnson, J.org. Chem.,1971, *36*,3184). The interaction of allyltrimethylsilane with various *p*-benzoquinones in the presence of titanium tetrachloride in dichloromethane gave allyl-substituted hydroquinones (A. Hosomi and H. Sakurai, Tetrahedron Letters, 1977, 4041). (Scheme 13).

The interaction of π-allyl nickel bromide complexes with *p*-quinones usually leading to 2-allylhydroquinones has been demonstrated to involve unstable allyl-quinol intermediates which rearrange to the final product (L.S. Hegedur and B.R. Evans, J. Amer. chem. Soc., 1978, *100*,3461). The photoaddition of p-benzoquinone to cyclooctatetraene in non-acidic solvents gives the spiro adduct (together with the peroxide in presence of oxygen) which upon warming in acetic acid rearranges to a mixture of the two hydroquinone products shown (E.J. Gardner *et al.*, J. Amer. chem. Soc., 1973, *95*,1693).(Scheme 14).

Scheme 13

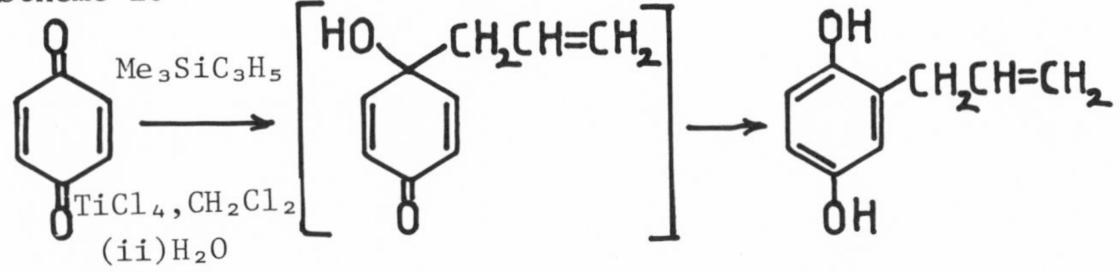

Scheme 14

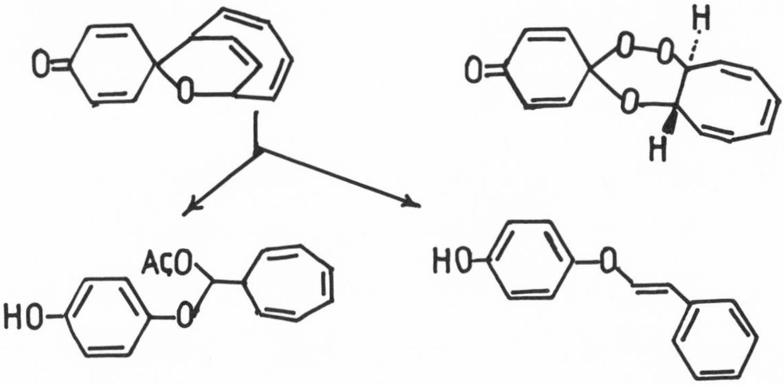

Palladium (II) acetate with anisole gives *m*-acetoxyanisole by way of the intermediate shown (L. Eberson and L. Gomez-Gonzales, Chem. Comm., 1971, 263).

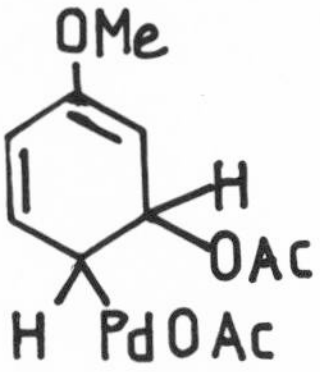

Cyclopropenones have been used for a synthesis of 2-alkylresorcinols by reaction with a Grignard reagent (R. Breslow, M. Oda and J. Pecorano, Tetrahedron Letters, 1972, 4415,4419).

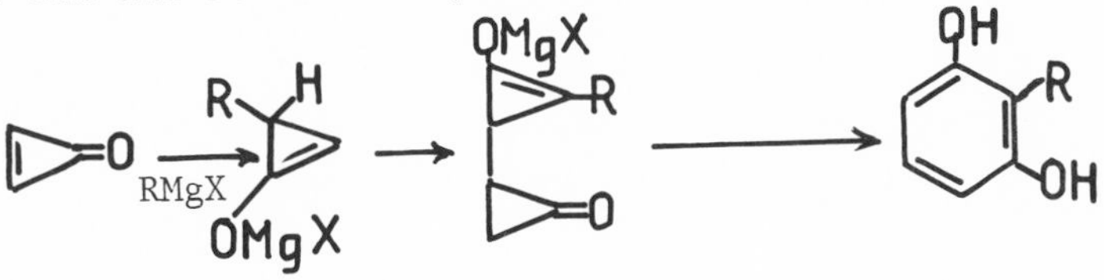

With organolithium reagents only polymeric products were isolated.

A route to 5-substituted resorcinols based on the Michael reaction has been described (A.A. Jaxa-Chamiec, P.G. Sammes and P.D. Kennewell, Chem. Comm., 1978 118).

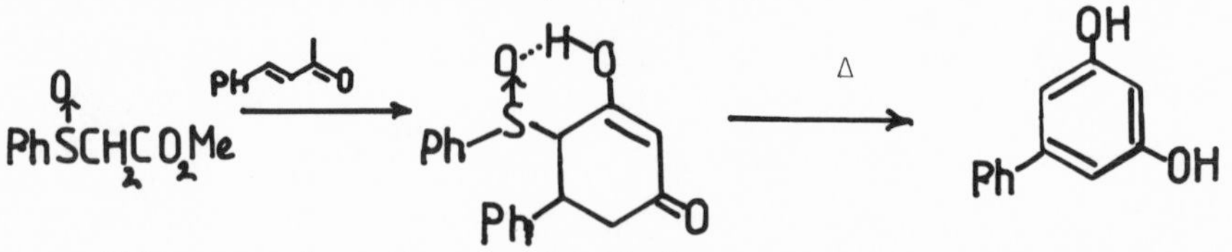

Other novel routes to 5-alkylresorcinols have been described. Birch reduction of 3,5-dimethoxybenzoic acid followed by alkylation and oxidative decarboxylation provides a general route (A.J. Birch and J. Slobbe, Tetrahedron Letters, 1976, 2079)

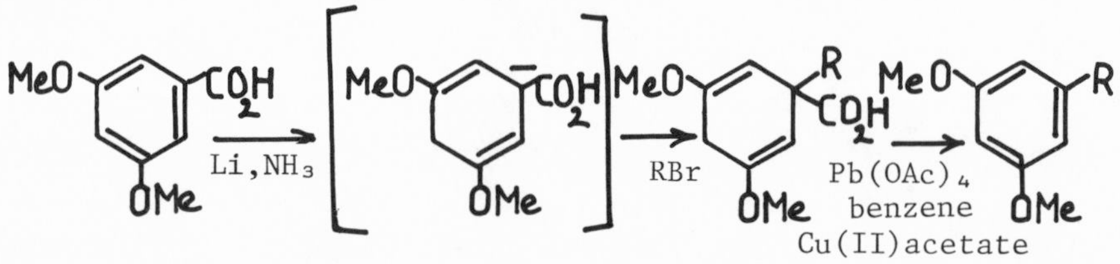

Olivetol (5-n-amylresorcinol) has been synthesised from 3,5-dimethoxybenzaldehyde by way of a Claisen condensation and further stages (H.G. Krishnamurthy and J.S. Prasad, Tetrahedron Letters, 1975, 2511).

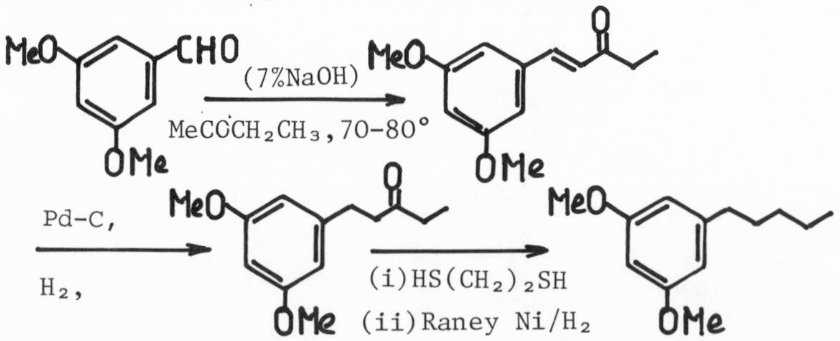

The interaction of 3,5-dimethoxyfluorobenzene with alkyllithium followed by water and demethylation with boron tribromide gives a general route to 5-n-alkylresorcinols (A.A. Durrani and J.H.P. Tyman, Tetrahedron Letters, 1973, 4839).

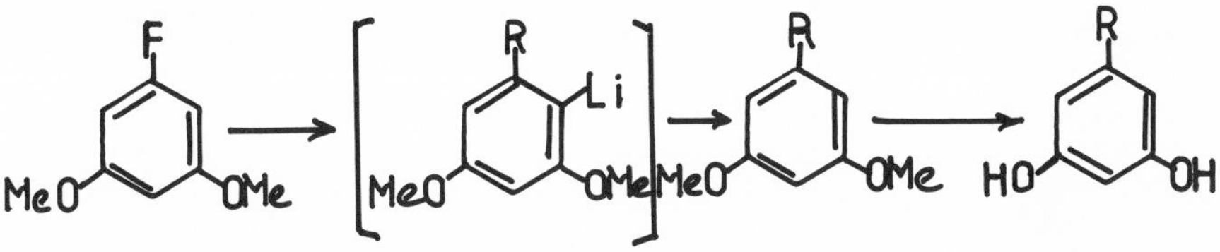

4,6-Diphenylresorcinol monomethyl ether has been prepared by the route shown (W. Reid, W. Kuhn and A.H. Schmidt, Angew. Chem. internat. Edn., 1971, *10*,736), reaction at a lower temperature giving a different product.

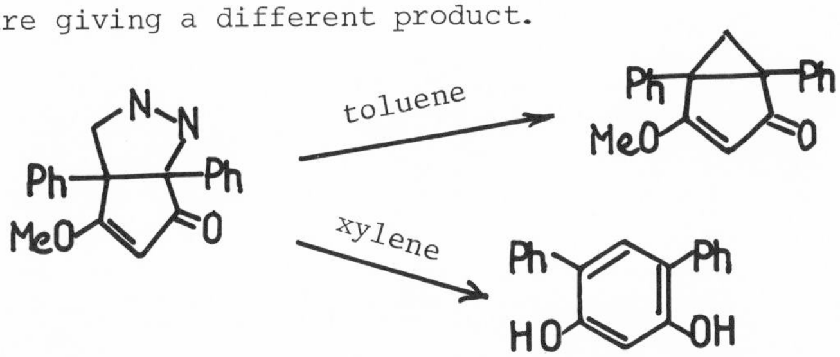

Resorcinol has been synthesised by the hydrolysis of m-phenylenediamine with aqueous ammonium bisulphate (Kopper Co. Inc., Ger. P. 2,459,149; Chem. Abs., 1976, *85*,124609). 2-Methyl and 4-methylresorcinol can also be prepared from the corresponding toluenediamines (USP 3,933,925; Chem. Abs., 1974, *84*,164412). An improved work-up for the alkali fusion process from benzene-1,3-disulphonic has been described (Mitsui Toatsu Chemicals Inc., Jap. P. 74 40,455; Chem. Abs., 1975, *82*,139633).

The conversion of o-hydroxymethylphenols into catechol acetals has been referred to (under cyclohexadienone chemistry). Benzoyloxylation with benzoyl peroxide introduces the benzoyloxy group mainly *ortho* to an existing hydroxyl but *para*-products can be formed by [3,3] migration of the acyloxy group around the ring periphery of the dienone intermediate (D.H.R. Barton, P.D. Magnus and M.J. Pearson, Chem. Comm., 1969, 550). The mechanism shown accounts for the *o*-substitution.

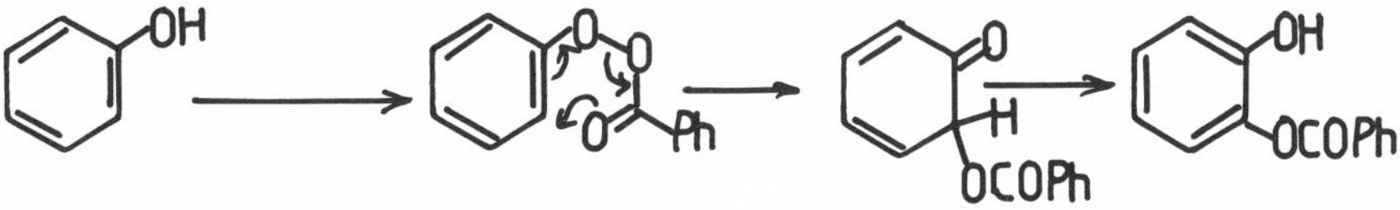

The preparation of 2-acetoxyhydroquinones by way of the

Thiele-Winter reaction has been reviewed (J.F. McOmie and J.M. Blatchly, Org. Reactions, 1972, *19*,1199). The reaction has been applied to a series of methoxyphenyl, hydroxyphenyl and aryl-substituted p-benzoquinones. The acetoxylation of 2-phenyl and 2-(4-substituted phenyl) quinones occurs mainly or exclusively *para* to the aryl group, that of 2-methoxy-3-phenylquinones occurs *para* to the phenyl group and in 2-hydroxy-5-phenyl, 2-methoxy-5-phenyl, 2-hydroxy-6-phenyl and 2-methoxy-6-phenylbenzoquinones, acetoxylation occurs *ortho* to the phenyl group (J.F. McOmie *et al.*, J. chem. Soc., Perkin I, 1975, 309). The anodic oxidation pathways of 2-substituted hydroquinones, catechols and p-substituted phenols have been investigated and lead to 3-hydroxylation when an electron-withdrawing substituent is present (L. Papouchado, G. Petrie and R.N. Adams, J. electroanalyt. Chem. interfacial Electrochem., 1972, *38*,389).

Nitromesitylene when heated in an autoclave with 10% aqueous phosphoric acid at 250° gives trimethylhydroquinone and 2,6-dimethylhydroquinone.

Several syntheses of orsellinic acids (or the dimethyl ethers) and 5-alkylresorcinols (by ready decarboxylation) have appeared, some of which are based on the biosynthetic 'polyketide' route. These have been reviewed (T. Money, Chem. Review, 1970, *70*,553). Extensive work has been carried out on the synthesis of a wide variety of poly-β-keto acids (T.M.Harris *et al.*, J. Amer. chem. Soc., 1973, *95*,6865). Not all these procedures have proved straight-forward and in certain cases pyrones (masked poly-β-ketones) have been prepared which are then cyclised to orsellinic acids or their derivatives. Methylated triacetic lactone has been converted to ethyl dimethylsellinate by the following route (D.A. Griffin and J. Staunton, Chem. Comm., 1975, 675).(Scheme 15).

Improvements in the synthesis of orsellinic acid have been made in the route based on the Michael addition reaction (G.M. Gaucher and M.G. Shepherd, Biochem. Prep., 1971, *13*,70). (Scheme 16).

The structure of the product of bromination is dependent on the molar ratio of bromine to the keto ester and to the time of bromination (M.V.Sargent and P.Vogel, J. chem. Soc., Perkin I, 1975, 1986).

Scheme 15

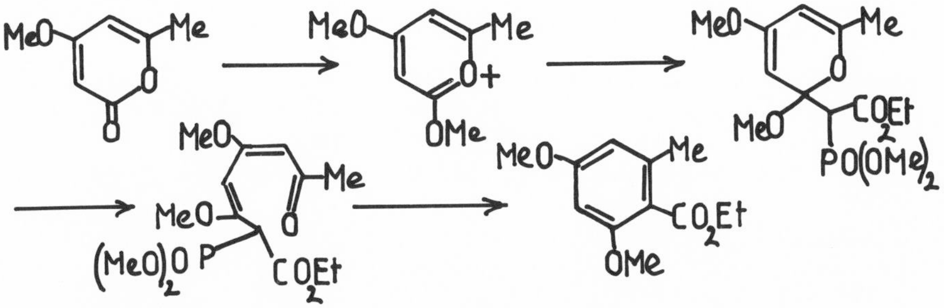

Scheme 16

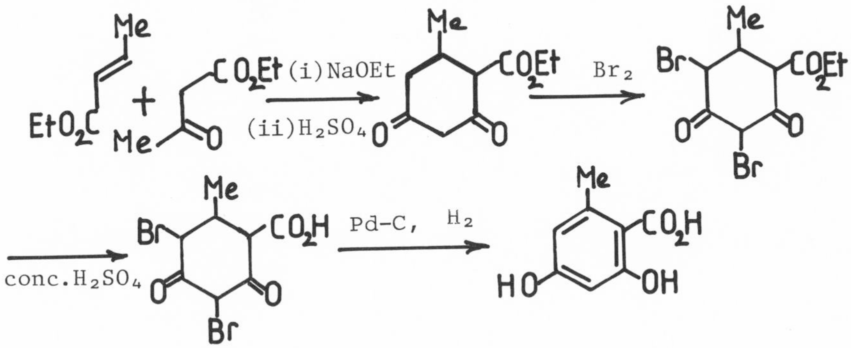

The Gatterman reaction on orcinol followed by oxidation has been used (E.M. Gruneiro and E.J. Gross, A. Assoc. Quim. Argrent. 1971, 59:259; Chem. Abs., 1972, *76*,59151). Carbonation of the lithium intermediate shown followed by demethylation with aluminium chloride (A.A. Durrani and J.H.P. Tyman, J. Chem. Soc., Perkin I, 1980, 1658, has given alternative routes to homolgous orsellinic acids.

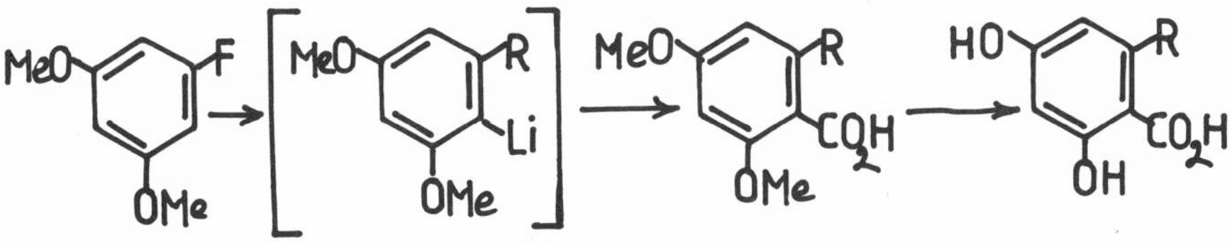

Dimethoxy and methoxyhydroxybenzoic acid derivatives

have been obtained by Diels-Alder reactions of a silylated
dienol as shown (S. Danischefsky, R.K. Singh, and R.B. Gamill,
J. org. Chem., 1978, *43*, 379; S. Danischefsky, M.P. Prisbylla
and S. Hiner, J. Amer. chem. Soc., 1978, *100*,2918).

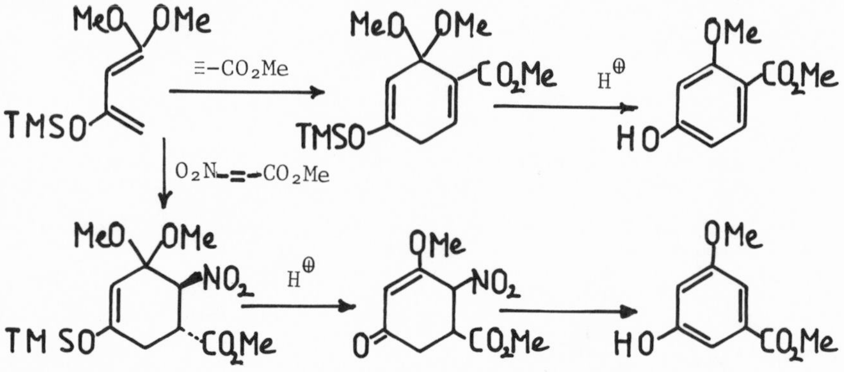

(ii) Reactions

Considerable attention has been devoted to the preparation of quinones from dihydric phenols.   Monoethers of hydroquinone (and of catechol) hindered towards electrophilic substitution are efficiently cleaved by nitrous acid to the quinones and alcohols as shown (D.H.R. Barton, P.G. Gordon and
D.G. Hewitt, J.chem. Soc., (C), 1971, 1206).

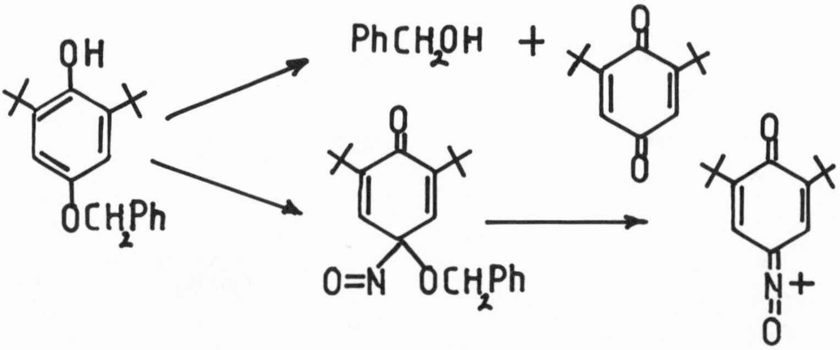

Oxidative decarbonylation of 2,4,6-tri-t-butylresorcinol
to give 2,3,5-tri-t-butylcyclopentadienone is believed to proceed by way of a singlet *m*-quinone (W.H. Starnes, D.A. Plank
and J.C. Floyd, J. org. Chem., 1975, *40*,1124).

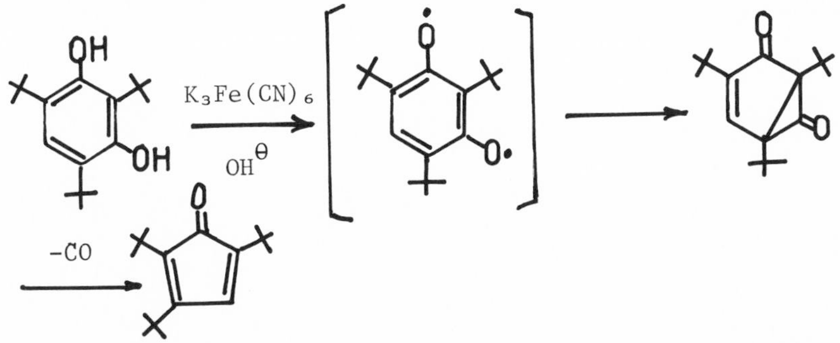

Oxidative amination of hydroquinones with sodium iodate in the presence of the appropriate amine gives 2,5-dialkylamino-1,4-benzoquinone in good yield (W. Schäfen and A. Agnado, Angew. Chem. internat. Edn., 1971, 10,405). Oxidative amination of 2-methylhydroquinone with sodium iodate in the presence of aniline gave mainly 5-anilino-2-methylhydroquinone, the novel 6-anilino isomer and the known 3,6-dianilino compound (S.C. Srivastava and U. Hornemann, ibid., 1976, 15,109). 2,3,4,5-Tetramethylhydroquinone is converted into the quinone by treatment in the dark with a flavin N-oxide (W.H. Rastetter et al., J. Amer. chem. Soc., 1979, 101,2228). 2-Formylhydroquinone diacetate gives the quinone diacetate shown in quantitative yield with silver (II) oxide (D.V. Rao, H. Ulrich and A.A.R. Sayigh, J. org. Chem., 1975, 40, 2548).

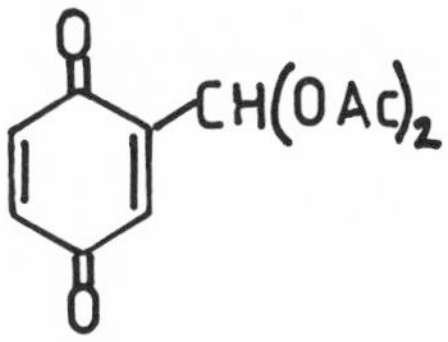

The complex from N-chlorsuccinimide and triethylamine is an efficient oxidising agent for the conversion of catechols and hydroquinones into quinones (H.D. Durst, M.P. Mack and F. Wudl, J. org. Chem., 1975, 40,268). The N-dimethylthiosuccinimide shown is a mild oxidising agent for the conversion of catechols to o-benzoquinones (J.P. Marino and A. Schwartz, Chem. Comm., 1974, 812). (Scheme 17). o-Quinones can be obtained by selective oxidation with phenylseleninic anhydride ortho to the hydroxyl group (D.H.R. Barton et al., Chem. Comm., 1976, 985). The oxidation of 3-pentad-

ecylcatechol to the *o*-quinone and its reaction to form diphen-
yls and other products has been examined (A.V. Balint J.R.
Dawson and C.R. Dawson, Anal. Biochem., 1975, *66*, 340).
(Scheme 18.)
Scheme 17

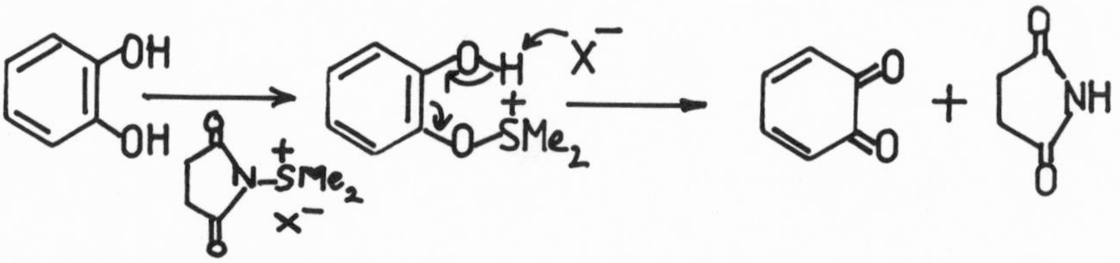

Scheme 18

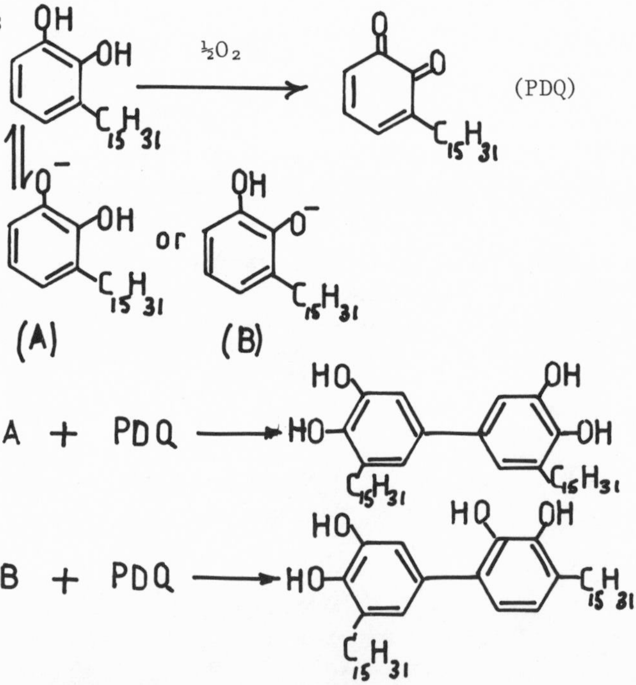

(PDQ)

(A)   (B)

A + PDQ ⟶

B + PDQ ⟶

Catechol oxidised by lead dioxide in methanol containing
sodium methoxide gives, by a series of nucleophilic substitutions, 4,5-dimethoxy-o-benzoquinone.   Synthetic uses of nucleophilic reactions leading to the isoflavone shown have been
described based on the use of this and related quinones (C.A.
Weber-Schilling and H.-W. Wanzlick, Ber., 1971, *104*, 1518)

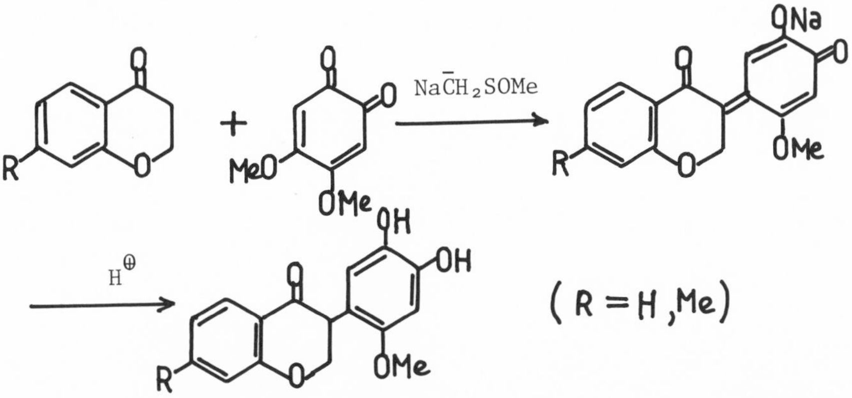

The electrochemical production of o-quinones in the presence of 1,3-dicarbonyl compounds provides in high yields a convenient route to certain heterocycles as shown (Z. Grujic, I. Tabakovic and M. Tikovonik, Tetrahedron Letters, 1976, 4823).

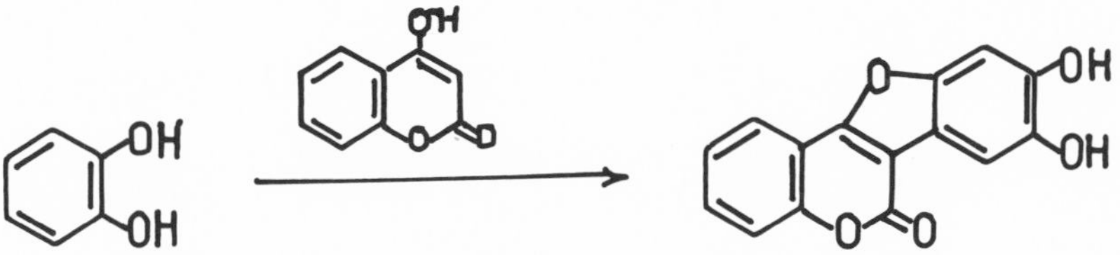

The hydroquinone ether shown can be anodically oxidised in methanolic potassium hydroxide and the product by treatment with ethereal toluene-4-sulphonic acid gives the quinone bisketal (G.C. Fink and D.B. Summers, Trans. electrochem. Soc., 1978, *74*, 325).

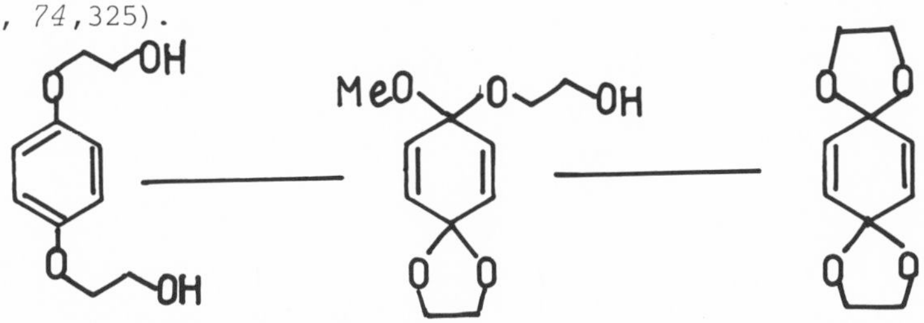

The formation of the monoaryl ether from a certain diaryl compound can be achieved as indicated (Ciba Geigy AG, W. Ger. P. 2847662).

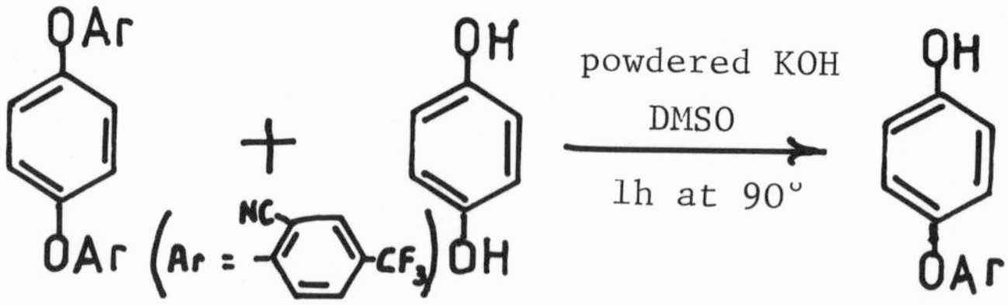

The phenolic quinone, polyporic acid, 2,5-dihydroxy-3,6-diphenyl-1,4-benzoquinone, undergoes oxidative rearrangement in dimethylsulphoxide/acetic anhydride to give a dilactone (R.J. Wilholm and H.W. Moore, J. Amer. chem. Soc.,1972, *94*, 6152).

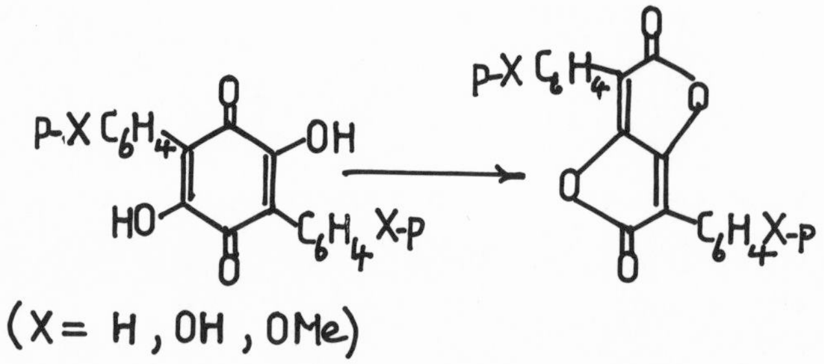

$(X= H , OH , OMe)$

Semi-quinones have been detected by electron spin resonance studies in the oxidation of 1,2 and 1,4-dihydroxyarenes with potassium superoxide and in the former case the products were further oxidised to carboxylic acids (E. Lee-Ruff, A.B.P. Leber and J. Rigaudy, Canad. J. Chem., 1976, *54*,1837, 3303).

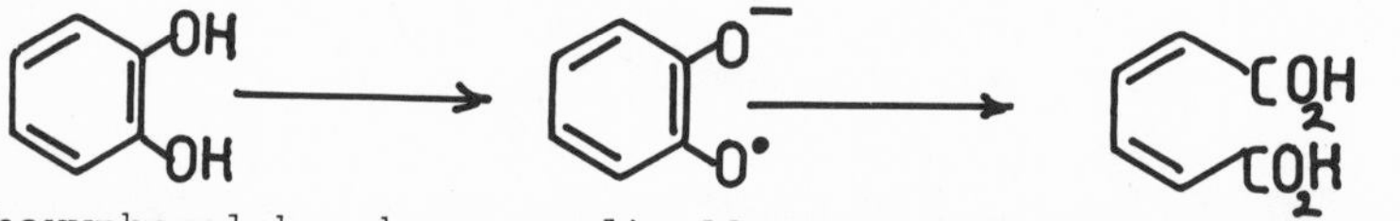

Methoxyphenol has been anodically oxidised in methanol containing lithium perchlorate (amperostatically in an undivided cell) and has given the quinone semi-acetal in 97% yield (A. Nilsson, A. Ronlán and V.D. Parker, Tetrahedron Letters, 1975, 1107). (Scheme 19).

Oxidative ring opening of catechols to monomethyl *cis,* *cis*-muconate in 70-77% yield takes place with oxygen in methanolic pyridine containing copper (I) chloride (J. Tsuji and H. Tokayanagi, J. Amer. chem. Soc., 1974, *96*,7349., Tetrahedron Letters, 1975, 1245). (Scheme 20).

Scheme 19

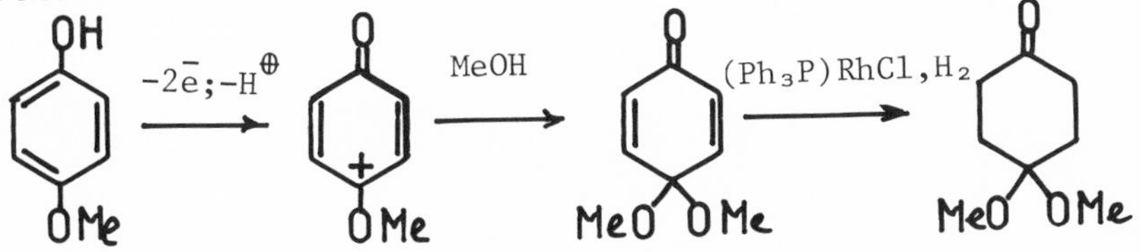

Scheme 20

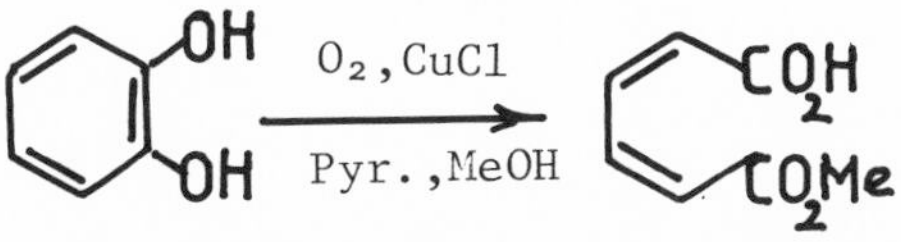

Phenol was similarly oxidised but in poor yield (J. Tsuji and
H. Tokayanagi, Tetrahedron, 1978, *34*,631).   Cyclic esters
of catechols with inorganic acids are known, particularly
those containing either Phosphorus, as in the examples shown
(E. Cherbuliez, Organic Phosphorus Compounds, Ed. G.M. Kosol-
apoff and L. Maier, Wiley Interscience, 1976, Vol.6, Ch.6,
p.211), sulphur atoms or boron atoms (M.A. Ferreria, M. Moir
and R.H. Thompson, J. chem. Soc., Perkin I, 1974, 2429).

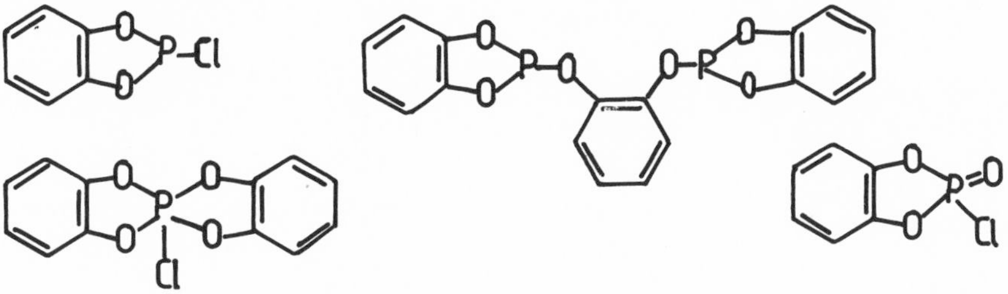

(iii)   Alkylation, Acylation and Cyclic Products
     In the conversion of o-dihydroxybenzenes into the corr-
esponding methylenedioxybenzenes phase transfer catalysis is
valuable (A.P. Bashall and J.F. Collins, Tetrahedron Letters,
1975, 3489).
     1,3-Dimethoxybenzene upon reduction with sodium/ammonia,
interaction of the product with t-butyllithium and alkylation
affords a route to 2-alkylcyclohexan-1,3-diones (E. Piers and
J.R. Grierson, J. org. Chem., 1977, *42*,3755).   (Scheme 21).
     Certain dimethoxybenzyl alcohols undergo ring saturation
and demethoxylation as well as hydrogenolysis of the hydroxyl
group with active palladium carbon catalysts (J.H.P. Tyman,
J. appl. Chem. Biotechnol., 1975, *25*,761).   2-Methyl-1,3-
dimethoxybenzene formylates in the Vilsmeier-Haack reaction
and Baeyer Villiger oxidation of the product with *m*-chloro-

perbenzoic acid followed by hydrolysis yields the phenol
shown (I.M. Godfrey, M.V. Sargent and J.H. Elix, J. chem. Soc.,
Perkin I, 1974, 1353).   (Scheme 22).

Scheme 21

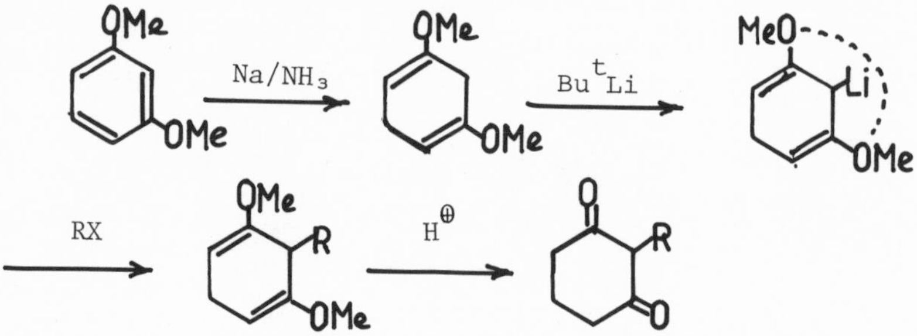

Scheme 22

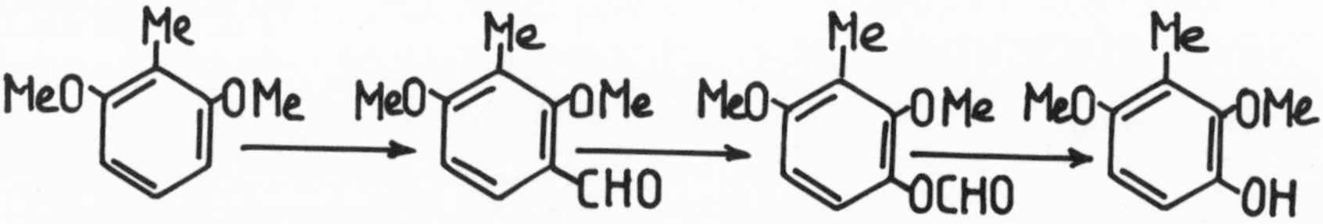

Acylhydroquinone methyl allyl ethers readily rearrange
in dimethylaniline at 170°, by way of sigmatropic shifts in
dienone intermediates as shown (C.P. Falshaw, S.A. Lane, and
W.D. Ollis, Chem. Comm., 1973, 491).

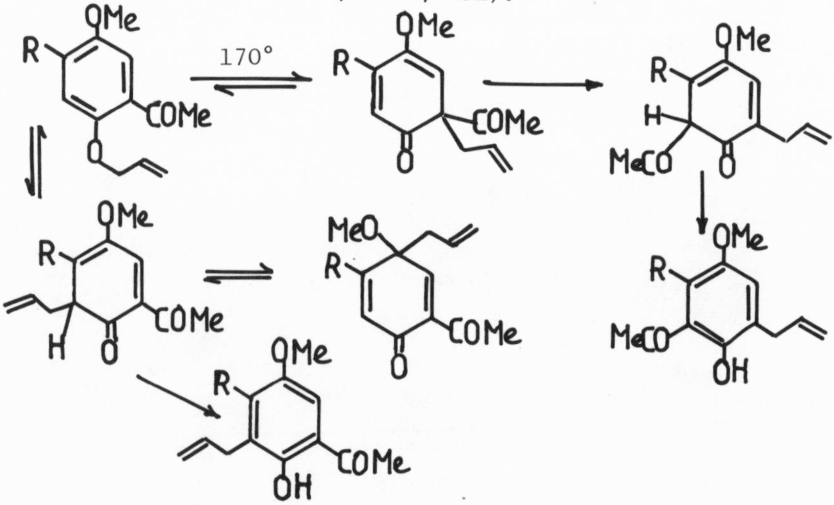

The condensation of hydroquinones and halogenated maleic anhydride yields polyhydroxylated naphthoquinones in excellent yield (R. Huot and P. Brassard, Canad. J. Chem.,1974, 52,838). Resorcinol and 2-amino-4-methylpyrimidine react by 1,3-addition during twenty-four hours in trifluoroacetic acid/benzene (1:2) to give the oxadiaza compound shown as the trifluoroacetate (W.P.K. Girke, Ber., 1979, 112,1).

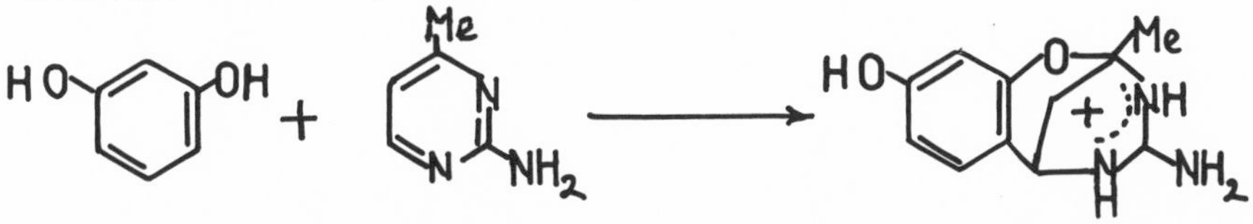

1,3-Dimethoxybenzene reacts with phthalimido-nitrene in benzene to give the 3H-azepin, but in acetic acid the insertion product shown is formed (D.W. Jones, Chem. Comm., 1973, 61).

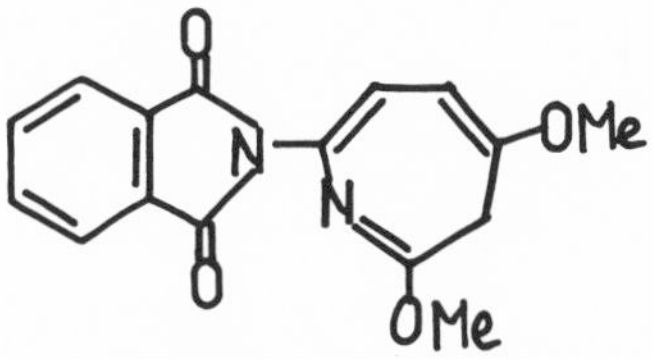

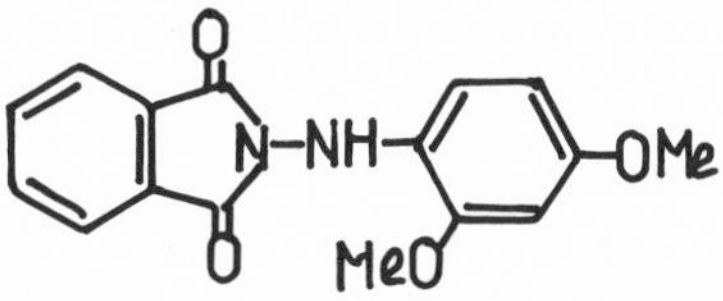

The dicarbethoxy derivative of catechol indicated undergoes Dieckmann cyclisation to give a macrocyclic bis β-keto ester, hydrolysable to a diketone and thence by reduction affords a dibenzo crown compound (G.R. Brown and J.H.P. Tyman, Chem. and Ind., 1970, 436; G.R. Brown, J. Grundy and J.H.P. Tyman, J. Chem. Soc., Perkin I,1981, 336).

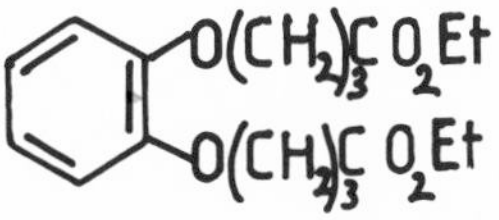

Hydroquinone monomethyl ether reacts with 1,3-dichloro-3-methylbutane in the presence of nickel bis(acetylacetonate) to give the chroman derivative shown (F. Camps *et al.*, Synthesis, 1979, 126). (Scheme 23).

Trimethylhydroquinone, as the monomagnesiobromide derivative, and phytal dimethyl acetal interact to give a chromen, reminiscent of the reaction of cinnamaldehyde with phenoxy magnesiobromide, and hydrogenation affords α-tocophenol (J.O. Asgill, L. Crombie, and D.A. Whiting, Chem. Comm., 1978, 54).

A large volume of work has appeared on the reaction of dihydric phenols with allylic alcohols some serving as biogenetic models. Resorcinol interacts with cinnamyl alcohol in aqueous formic acid to give the products shown (S. Mageswaren, W.D. Ollis, R.J. Roberts and I.O. Sutherland, Tetrahedron Letters, 1969, 2897). (Scheme 24).

Scheme 23

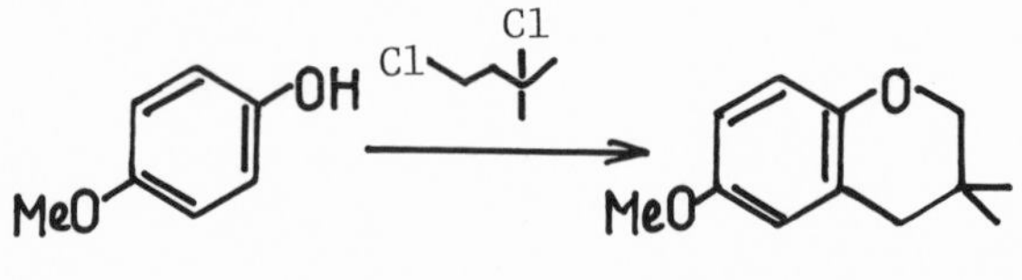

Scheme 24

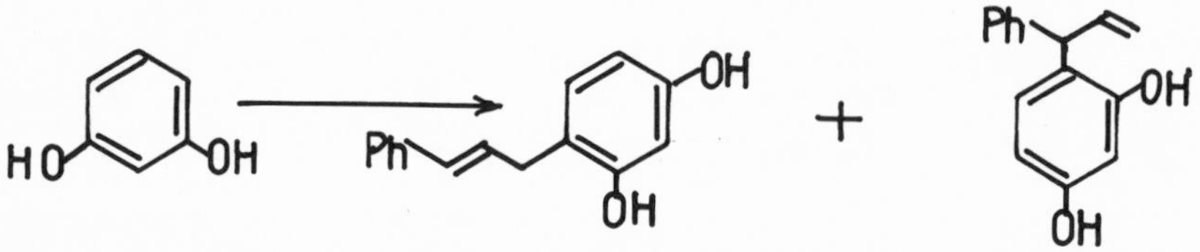

Cinnamyl alcohol with 2-methoxyhydroquinone in aqueous citric acid forms exclusively a similar terminal olefin obtusaquinol which can isomerise, cyclise to obtusafuran or, in acetic acid, transalkylate with pyrogallol (L. Jurd, K. Stevens and G.Manners, Tetrahedron Letters, 1973, *29*, 2347). The reactions were regarded as reversible Friedel-Crafts alkylations involving carbonium ions but the weakly acidic conditions suggest concerted reactions.

Obtusaquinol

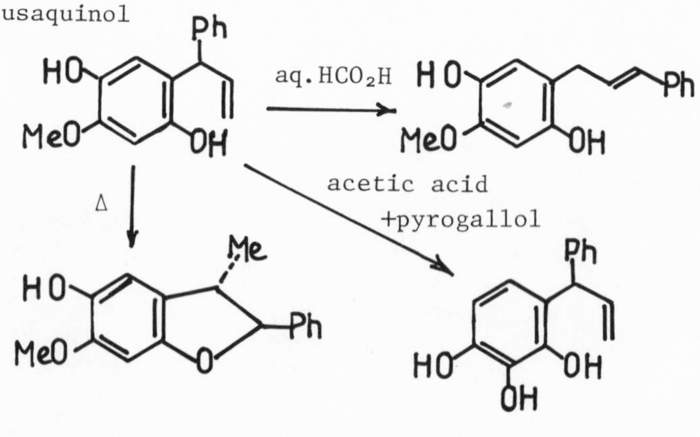

Epoxides can react as indicated (L. Jurd, G. Manners and K. Stevens, Chem. Comm., 1972, 992).

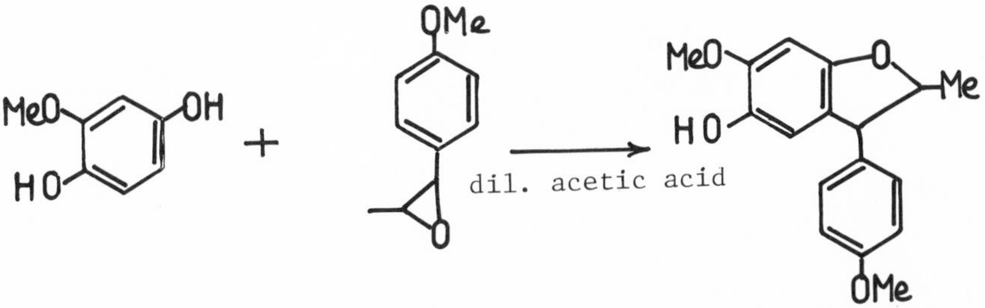

Olivetol with cyclic allyic alcohols in dilute acidic conditions forms isomeric products related to the tetrahydro-cannabinols (B. Cardillo, L. Merlini and S. Servi, Tetrahedron Letters, 1972, 945).

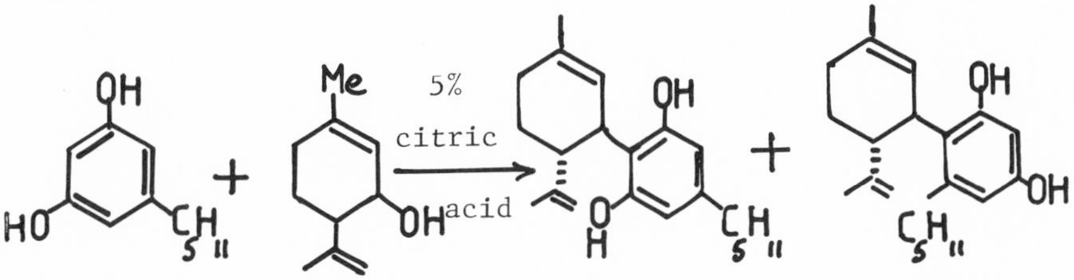

With orcinol and 2-methylbut-3-en-2-ol in dilute citric acid a wide variety of isoprenylation and cyclisation reactions can occur. Apart from the major product formed in 18% yield the four substances shown were isolated (L. Jurd, G. Manners and K. Stevens, Tetrahedron 1972, *28*,2949).

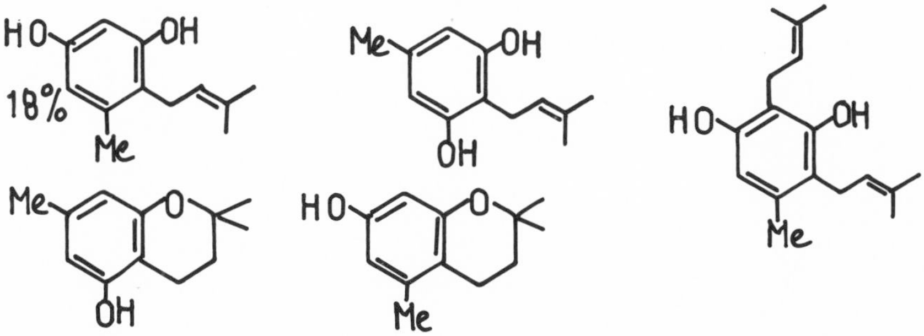

Geraniol reacts with orcinol to give similar and more complex

products and the nerol/orcinol reaction has also been studied
(B.J. Molyneux and L. Jurd, Tetrahedron, 1970, 26,4743).

2,6-Dihydroxy-4-methylfarnesylbenzene, grifolin, an antibiotic metabolite of *Grifola confluens* has been synthesised
by the interaction of farnesyl bromide with orcinol in dioxan
containing silver oxide (J.H.P. Tyman, W.A. Baldwin, and C.J.
Strawson, Chem. and Ind., 1975, 41).

Interaction of olivetol with the carane epoxide shown
gives with boron trifluoride, Δ^1-*trans* tetrahydrocannabinol
and with toluene-4-sulphonic acid the Δ^6- isomer, in both
cases accompanied by other products (R.K.Razdan and G.R. Han-
drick, J. Amer. Chem. Soc., 1970, 92,6061).

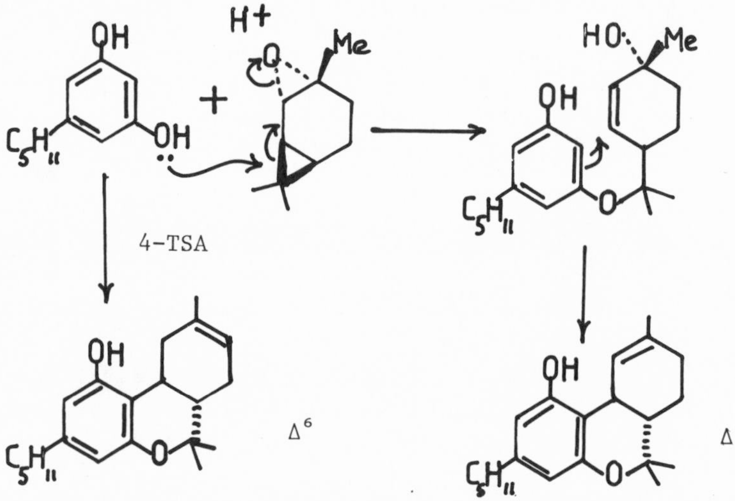

In contrast to the reaction of olivetol and citral under
acidic conditions to give tetrahydrocannabinols, quite different products arise under basic conditions.

Resorcinol and phloroglucinol in their respective dioxo
and trioxo forms show similarities to malonic acid in their
reactivity towards citral in pyridine solution.   Thus 5-n-
amylresorcinol, phloroglucinol (and also 5,7-dihydroxy coum-
arin) react to give tetracyclic structures designated as cit-
rans (L. Crombie and R. Ponsford, J. Chem. Soc (C), 1971, 788,
796, 804).   The reaction pathway, involving a monochromen,is
thought to be as shown (D.G. Clarke, L. Crombie and D.A. Whiting, J. chem. Soc., Perkin I, 1974, 1007; Chem. Comm., 1973,
582; W.M. Bandaranaike, L. Crombie, and D.A. Whiting, J.Chem.
Soc. (C), 1971, 811; L. Crombie *et al.*, J. chem. Soc.,Perkin
I, 1974, 998).

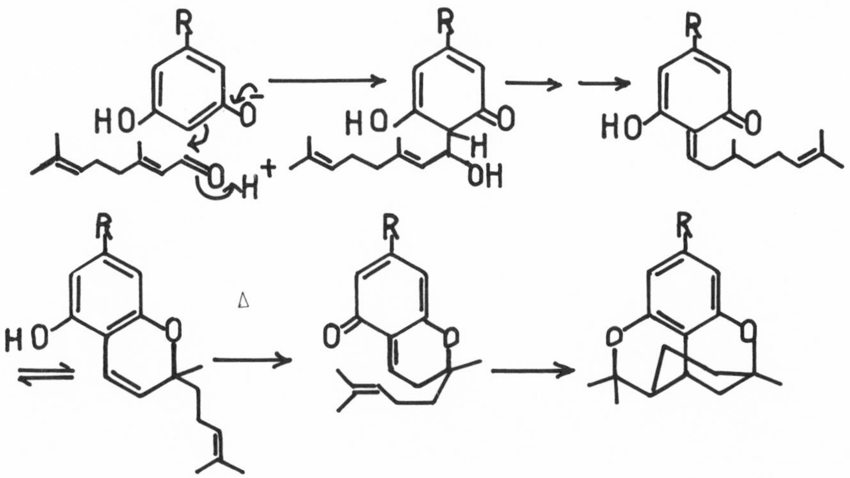

Monochromenylations of phloroacetophenone and phloroglucinald-
ehyde (L. Crombie, D.A. Slack and D.A. Whiting, Chem. Comm.,
1976, 139) give mixtures in which the more acidic product formed by isomerisation, predominates as shown (G.D. Manners and
L. Jurd, Chem. Comm., 1976, 448).

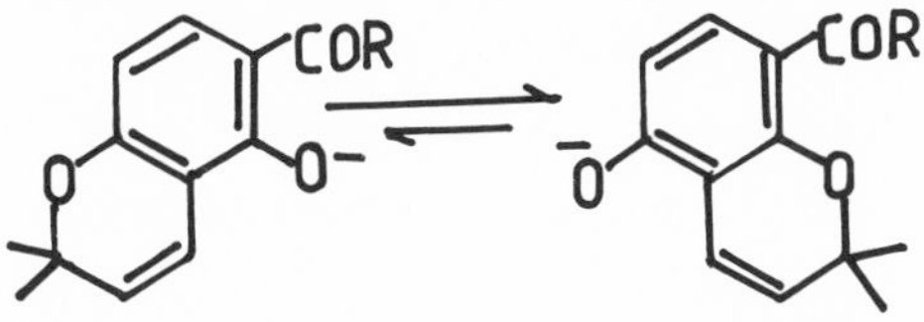

Over the past decade a number of long chain resorcinols,
catechols and hydroxyquinones have been examined.   They include persoonol (1,R=undec-3'(Z)-enyl) from *Persoona elliptica*
(J.R. Cannon and B.W. Metcalf, Austral. J. Chem., 1971, 24,
1925) the dimethyl ether of which was synthesised by the
route shown.   (Scheme 25).
    Cardol, a major component phenol of *Anacardium occident-
ale* comprises four constituents, the saturated (1;R=$C_{15}H_{31}$,n=0,
the 8'(Z)-monoene (1;R=$C_{15}H_{29}$,n=2), the 8'(Z),11'(Z)-diene
(1;R=$C_{15}H_{27}$,n=4) and the 8'(Z), 11'(Z),14'-triene (1;n=C_5H_{25},-
n=3).   The monoene (1;R=$C_{15}H_{29}$,n=2) and the diene dimethyl
ether has been synthesised (J.H.P. Tyman and S.W.D. Odle, Chem.
and Ind., 1975, 88; C.J. Baylis, S.W.D. Odle and J.H.P. Tyman,
J. chem. Soc., Perkin I, 1981, 132).

Scheme 25

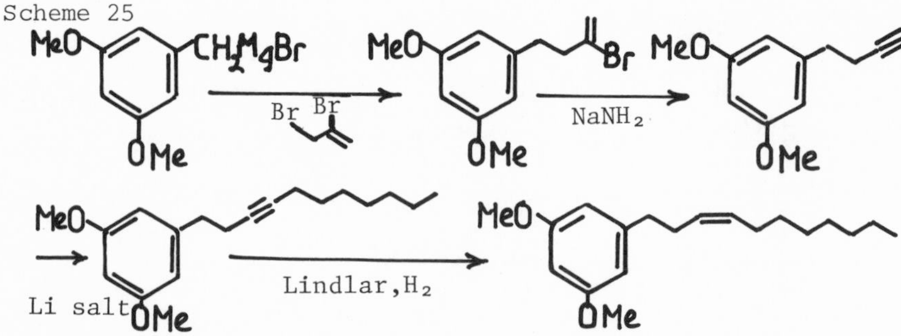

Other products relating to the preceding (in addition to 1; R=$C_{13}H_{27}$ and $C_{15}H_{31}$) have been isolated from the algal source *Cystophora torulosa* (R.P. Gregson *et al.*, Austral. J. chem., 1977, *30*, 2527. They comprise two unsaturated compounds; 1; R=$C_{14}H_{23}$),  and R=$C_{18}H_{31}$

2-Methylcardol (2; R=pentadecyl, R'=methyl) also contains four similarly related constituents and the saturated member has been synthesised (J.H.P. Tyman, Perkin 1, 1973, 1693).

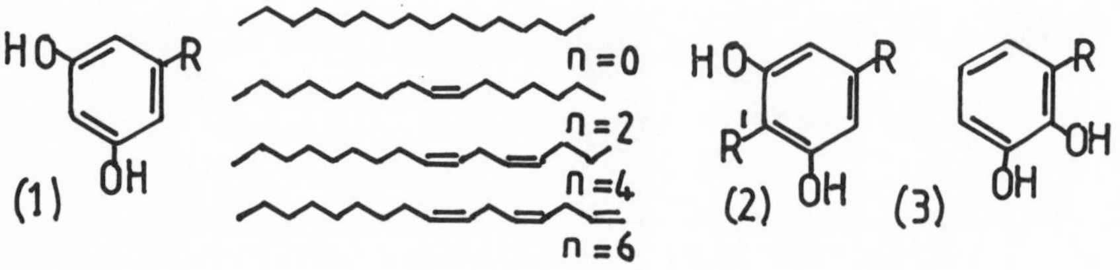

2-n-Propyl-5-n-amyl resorcinol (2; R'=C_3H_7, R=C_5H_{11}) has been isolated from an antibiotic source (T. Kitahara and N. Kando, J. Antibiot., 1975, *28*, 943). The disulphate of a resorcinol (2; R=isopentadecyl, R'=H) has been obtained from a streptomyces source (H. Umegawa *et al.*, J. Antibiot., 1971, 24, 870).

Urushiol, from *Rhus vernicifera*, has four constituents, the saturated (3; R=n-$C_{15}H_{31}$), the related 8'(Z)-monoene (3; R=n-$C_{15}H_{29}$), the 8'(Z),11'(Z)-diene (3; R=$C_{15}H_{27}$), and the 8'-(Z),11'(Z),13'(Z) triene (3; n=$C_{15}H_{25}$).
Rhus toxicodendron contains urushiol comprising the same saturated, monoene and diene constituents but a triene with 8'(Z),11'(Z)14' unsaturation. The monoene (3; R=pentadec-8-(Z) enyl has been synthesised, J.H.P. Tyman and C.H. Khor, Chem. and ind., 1974, *526*) by the route shown, giving a purer product than the original method (B. Loev and C.R. Dawson, J, org. Chem., 1959, *24*, 980).

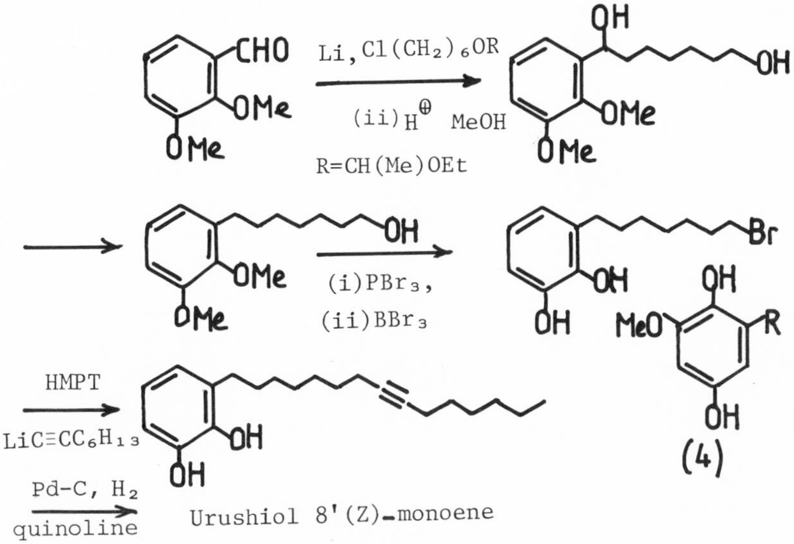

Urushiol 8'(Z)-monoene

Improved syntheses of 4-methyl and 4,5-dimethyl-3-pentadecyc-
atechol through chloromethylation of the dibenzyl ether followed by hydrogenolysis have been described (D.I. Lerner and
C.R. Dawson, J. org. Chem., 1973, *38*,2096).   3-n-Alkylcate-
chols of varying side chain length and branched chain alkyl
groups have been synthesised (C.R. Dawson and A.P. Kurtz, J.
med. Chem., 1971, *14*, 729, 733).   Hydroquinones (4;R=n-pentyl)
include micoridin (G.B. Marini-Bettòlo, Gazz. chem. Ital.,
1971, *41*,101).

3   POLYHYDRIC PHENOLS AND THEIR DERIVATIVES
(i) Natural occurrence
A number of novel polyhydric phenols have been found in
such widely diverse natural sources as higher plants, bacteria,
insects and marine organisms.   Representative members are 1,-
3,4-trihydroxy-2-methylbenzene (versicolin) an antifungal metabolite, and the quinhydrone shown from the ascomycete, *Nec-
tra coryli* (R.W. Richards, J. Antibiot., 1971, *24*,715; M.S.R.
Nair and M. Anchel, Tetrahedron Letters, 1972, 795).
Brominated compounds include the dienone acetal shown from
Verongia fistularis (G.M. Sharma, B. Vig and P.R. Burkholder,
J. org. Chem., 1970, *35*,2823), the iso-oxazoles aerothionin
and homoaerothionin (R.H. Thomson *et al.*, J. chem. Soc.,

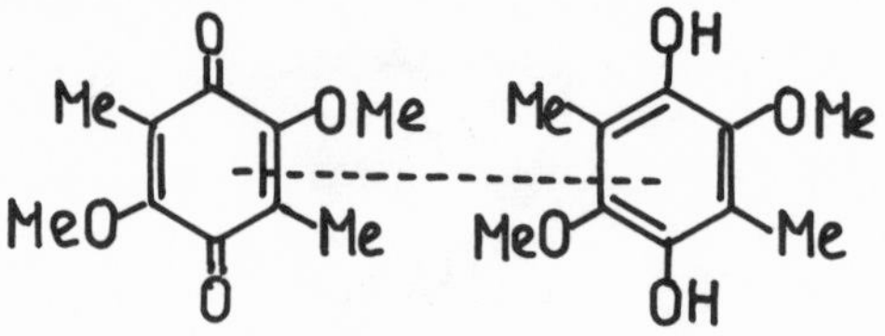

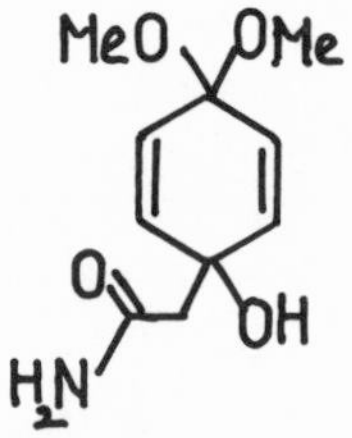

Perkin I, 1972 18) from the sponges *Verongia thiona*
(≡*Aplysina*) and *aerophoba* respectively, and aerophysinin-1,
(E. Fattonisso, L. Minale and G. Sodano, J. chem.Soc.,Perkin 1,
1972, 16) the formula of which is shown.

The phloroglucin monomethyl ether shown has been isolated
from the algal source *Cystophora torulosa* (R.P. Gregson, R,
Kazlauskas, P.T. Murphy and T.J. Wells, Austral. J. chem.,
1977, *30*,2527), and it appears to be the first non-conjugated
tetraene to be detected.

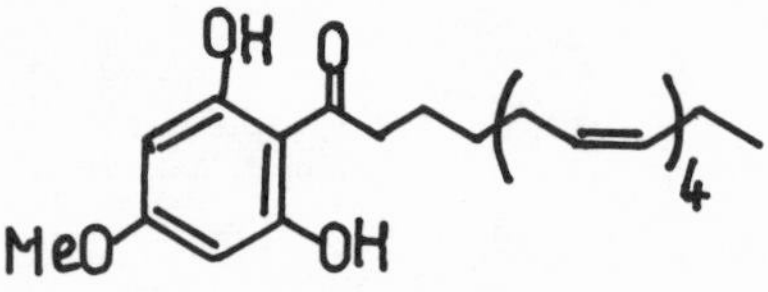

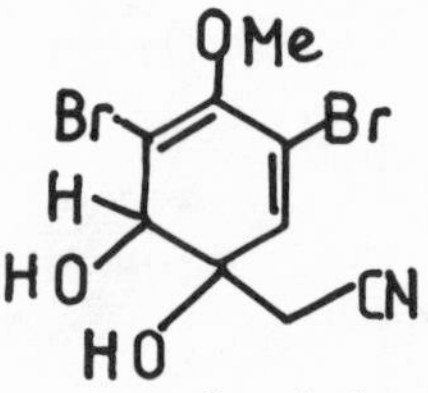

aerophysinin-1

(ii) Reactions

Most of the reactions on this field involve the methyl
or other ethers of the polyhydroxyl phenols. Thus, 2,4,6-
trimethoxytoluene in isobutyric acid saturated with boron
trifluoride gives after methanolysis of the reaction product
the *o*-hydroxyketone shown, a demethylation similar to that
obtained with boron trichloride, (G.P.Schiemerz and U.Schmidt,
Ann. Chim., 1976, 1514).

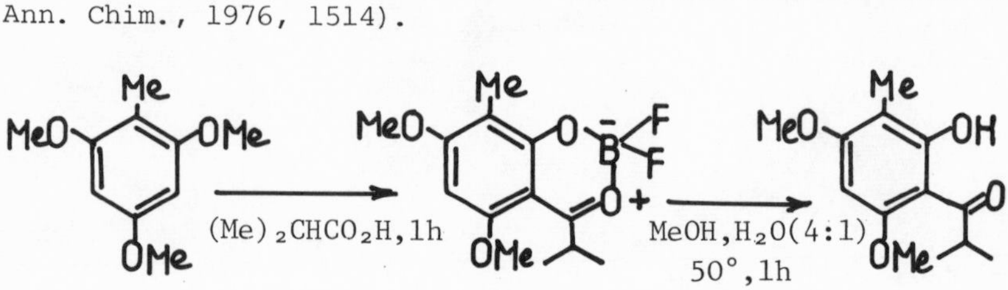

The allyl ether shown (R=H) prepared from phloroacetophenone and propargyl bromide followed by reduction gives upon thermal rearrangement, deoxyhumulones, but with R=Me, only products of C-alkylation are obtained (E. Collins and P. Shannon, J. chem. Soc., Perkin I, 1974, 944).

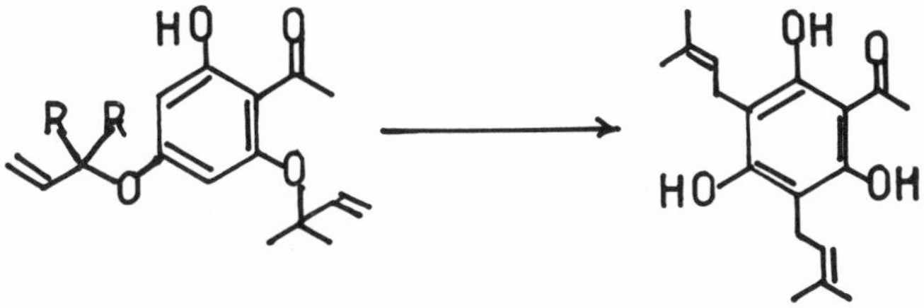

The di-t-butyl-phenolic dimethyl ether shown gives a product by one electron oxidation presumably through secondary radical formation on methoxy carbon (R.C. Eckert, Han-Ming Chang and W.P. Tucker, J. org. Chem., 1974, *39*, 718).

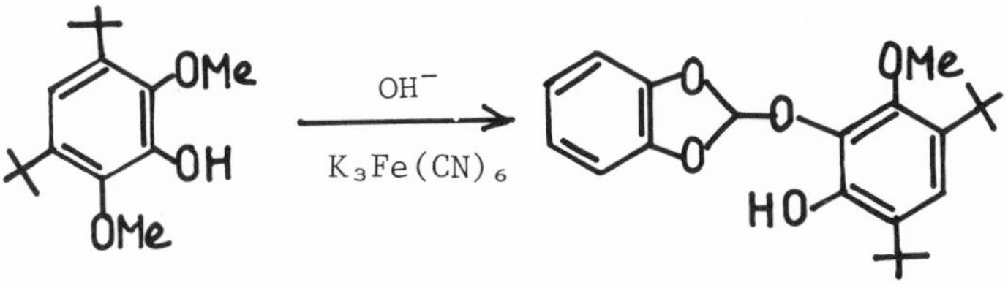

Chapter 5

MONO NUCLEAR HYDROCARBONS CARRYING SUBSTITUENTS ATTACHED
THROUGH SULPHUR AND SELENIUM: THIOPHENOLS, SULPHIDES, ETC.

J.H.P. TYMAN

Introduction

Although no major new methods of preparation of thiophenols have been introduced a vast amount of work has been carried out on the use of thiophenol and thiophenylchloride for
the synthesis of many acyclic and alicyclic compounds.   Reviews covering the chemistry of thiophenols have appeared
('The Chemistry of the Thiol Group', ed. S. Patai, Wiley, New
York, 1974; A. Ohno and S. Oae, in 'Organic Chemistry of Sulphur', ed. S. Oae, Plenum, New York, 1976; G.C. Barrett, in
'Organic Compounds of Sulphur, Selenium, and Tellurium', The
Chemical Society, London, 1970).

Properties

The strength of the S-H bond (339 kJ mol^{-1}) is lower than
that of the O-H bond (462 kJ mol^{-1}) and accounts for much of
the use to which thiophenol has been put in organic chemistry.
Although thiols show reduced hydrogen-bonding compared with
alcohols, thiophenols do form stronger hydrogen bonds to ac-
ceptors than do alkane thiols.   The weaker intermolecular
hydrogen bonding in thiophenol is reflected however in the
b.p. (169°) compared with that of phenol (182°).   Weaker
complexes are formed between thiophenols and NMR shift reagents than those formed with alcohols (H. Yamagawa, T. Kato,
and Y. Kitahara, Tetrahedron Letters, 1973, 2137; H. Duddeck
and W. Dietrich, *ibid.*, 1975, 2925).   Proton transfer rates
to thiophenols are some orders of magnitude less than those
for alcohols and phenols (J.J. Delpuech and D. Nicole, J. chem.
Soc., Perkin II, 1974, 1025.   A significant difference between the thiophenols and phenols is shown in the reaction of
the former with oxygen resulting in the formation of disulphides ArSSAr.   Reactions can often be accounted for in terms
of the homolysis of the SH bond (ArSH→ArS· + H·) which results
in radical scavenging properties.

On the basis of proton acidity criteria the order of acidity is thiophenols > phenols > ethanethiol > methanol, and the reverse for basicity, although by carbon acidity comparison the series is phenol > thiophenol > methanol > ethanethiol.

A study of the effect of resonance interaction of the -SH group in certain p-substituted thiophenols and a comparison of the resultant acidities with those of the corresponding phenols and benzoic acids has been made (G. Maccagnani and G. Mazzanti, in 'Chemistry of the thiol group', ed. S. Patai, Interscience, 1974, ch. 9, p. 424).

The 'valency shell expansion' of sulphur has been used to explain the higher acidity of hydrogen atoms α- to sulphur compared to those α- to oxygen. The higher stability of α-mercaptocarbanions through electron pair acceptor conjugation as shown has been criticised on energetic considerations (S. Wolfe, Acc. chem. Res., 1972, 5, 102).

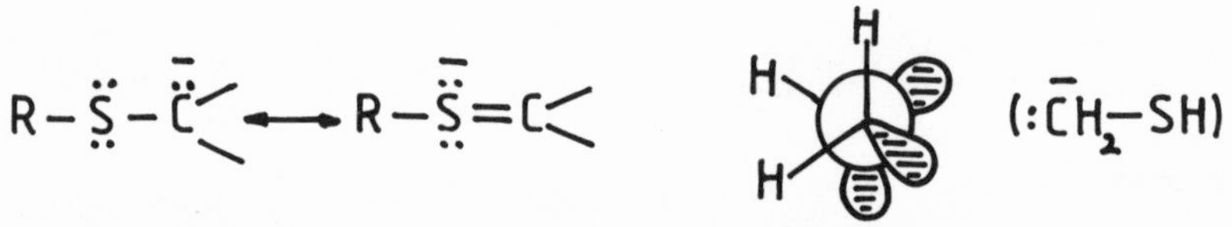

Molecular orbital calculations predict an energy minimum with the conformation illustrated without requiring sulphur d-orbital (valency expansion) contributions.

Variations of molecular geometry resulting from conformational and environmental effects may be considered capable of influencing molecular properties of thiols. An investigation of the effects of various substituents on the acidities of phenols and thiophenols does not afford evidence of the valency shell expansion in sulphur (G. Chuchani and A. Frölich, J. Chem. Soc., (B), 1971, 1417). Valency shell expansion means a significant contribution from the structure shown with sulphur having a negative charge compared with that from the structure with a positive charge on sulphur

A comparison of substituent effects on the proton and carbon basicities of thiophenoxide ions has been made (M.R. Crampton, J. Chem. Soc., (B), 1971, 2112). By comparison with $HO^{\ominus}$, $PhS^{\ominus}$ possesses considerably less basicity and

this is probably related to the lower degree of solvation of
thiolate anions (C.A. Bunton, S.K. Huang and C.H. Paik,
Tetrahedron Letters, 1976, 1445). The effect of substituents
upon the acidity of thiophenol has been discussed(M.R. Crampton,
in 'The Chemistry of Thiol Group', ed. S. Patai, Interscience,
1974, Ch. 8, p. 402).   The mono anion shown is stabilised by
hydrogen bonding and this can explain the higher acidity of
o-thiolbenzoic acid as the free acid; it does not undergo the
alternative S, H interaction shown (N. Mori et. al., Bull.
chem. Soc. Japan, 1971, 44, 1858).   A thiolate anion is a

'soft base' in Pearson's terminology.   The higher nucleophilicity of the thiolate anion (e.g. thiophenoxide) compared
with that of alkoxide anion can be explained in terms of the
higher polarisability of sulphur compared with that of oxygen
and, by higher product stability if this is reflected in the
energy of the transition state.   Thiols such as thiophenol
are very considerably more active than alcohols towards addition to unsaturated systems and basic catalysts are not required (Y. Ogata and A. Kawasaki, J. chem. Soc., Perkin II,
1975, 134; G.E. Lienhard and W.P. Jencks, J. Amer. chem. Soc.,
1966, 88, 3982; M. Friedman, J.F. Calvin and J.S. Wall, ibid.
1965, 87, 3672).
   The spectroscopic properties of aromatic thiols have been
examined in detail.   Characteristic ultraviolet absorption
occurs in the regions 235-240 nm and 265-295 nm and in the
infrared the S-H stretching frequency in the region 2540-2590
cm^{-1} is more easily recognised that that of the C-S bond in
the range 580-760 cm^{-1}.   Infrared studies have been used to
study the degree of delocalisation of the sulphur lone pairs
and their participation in conjugative interactions in aromatic
systems.   The sequence N > O > S has been shown for aliphatic
thiols (P.J. Krueger, J. Jan and H. Wieser, J. mol. Structure,
1970, 5, 375).   From an ir-spectroscopic investigation of o-
aminobenzene thiols the existence of an out-of-plane thiol
group stabilised by 3d-conjugation of sulphur has been postulated (P.J. Krueger, Tetrahedron, 1970, 26, 4753).
   Interaction between the π-orbitals of the benzene ring and
sulphur d-orbitals in thiophenol and p-methylthiophenols has

been studied (G. Di Lonardo and C. Zauli, J. chem. Soc., (A),
1969, 1305). Photoelectron spectral work supports these
findings (D.C. Frost et. al., J. phys. Chem., 1972, 76, 1030).
The spectral properties of aromatic thiols and their general
detection and determination have been discussed (A. Fontana
and C. Toniolo, in 'The Chemistry of the Thiol Group', ed.
S. Patai, Interscience, 1974, Ch. 5; S. Wawzonek in 'Techniques
of Chemistry', vol. 1 Pt. IIA, eds. A. Weissberger and
B.W. Rossiter, Wiley, 1971, p. 50; R.E. Humphrey et al.,
Michrochem. J., 1971, 16, 429; L.C. Gruen and B.S. Harrap,
Analyt. Biochem., 1971, 42, 377).

The mass spectra of thiophenol, and substituted thiophenols have been discussed (C. Lifshitz and Z.V. Zaretskii,
in 'The Chemistry of the Thiol Group, ed. S. Patai,
Interscience, 1974, Ch. 6, p. 330).

Although aliphatic thiols occur widely naturally, aromatic
thiols do not, possibly because of the ease with which they
are oxidised and transformed into other substances.

Methods of Preparation

The synthesis of aromatic thiols has been discussed
(J.L. Wardell, in 'The Chemistry of the Thiol Group', ed.
S. Patai, Interscience, 1974, Ch. 4, 163; G.C. Barrett, in
'Comprehensive Organic Chemistry', ed. Sir D.H.R. Barton and
W.D. Ollis, Pergamon, 1979, vol. 3, part 11, 3).

The main methods are by the reduction of phenylsulphonyl
chlorides, of diphenyldisulphides, from phenols themselves, by
the hydrolysis of O-alkyl S-phenylxanthates, from isothiuronium
salts, through the interaction of organometallic reagents with
sulphur, by the hydrolysis of 2,4-dinitrophenylaryl sulphides,
the nucleophilic substitution of reactive phenyl halides and by
the demethylation of arylthioethers.

1 Reduction of sulphonyl chlorides and reactions of sulphenyl halides

(a) Sulphonyl chlorides

The general route is as shown from the aryl sulphonyl
chloride prepared from chlorosulphonic acid and the hydrocarbon or by the reaction of benzenesulphonic acid or its sodium salt with phosphorus pentachloride or thionyl chloride.

$$ArH + HOSO_2Cl \rightarrow ArSO_2Cl \rightarrow ArSH$$

This route is perhaps the most reliable and high yielding for thiophenol itself (R. Adams and C.S. Marvel, Org. Synth., 1921, 1, 71).

An examination of the reduction of alkylbenzene sulphonyl chlorides by the use of lithium aluminium hydride, zinc and dilute acid, red phosphorus and iodine in acetic acid and tin (II) chloride in hydrochloric acid solution (A. Meacock and J.H.P. Tyman, unpublished work) has shown that the last method is the best. It is a one-pot method proceeding in excellent yield and has apparently only been applied previously to the p-sulphonyl chloride of cinnamic acid (C.G. Overberger, H. Biletch and F.W. Orthung, J. org. Chem., 1959, 24, 289). The use of acetic acid, commonly employed with stannous chloride, is unnecessary.

$$RC_6H_5 \xrightarrow{H_2SO_4/SO_3} RC_6H_4SO_3H \xrightarrow{SOCl_2} AC_6H_4SO_2Cl \xrightarrow[H^{\oplus}]{SnCl_2}$$

$$RC_6H_4SH$$

The reaction of arylsulphonyl chlorides with hexacarbonyl molybdenum in anhydrous tetramethyl urea provides a valuable route for their reduction to disulphides (H. Alper, Angew. Chem. internat. Edn., 1969, 8, 677).

$$ArSO_2Cl \xrightarrow{Mo(CO)_6} ArSSAr \qquad (55-80\%)$$

Pentacarbonyl iron is a useful reagent for the selective preparation of thiosulphate esters from arylsulphonyl chlorides including those containing nitro groups. (H. Alper, Tetrahedron Letters, 1969, 1239).

$$ArSO_2Cl \xrightarrow[CH_3CON(CH_3)_2]{Fe(CO)_5/BF_3.Et_2O} ArSSO_2Ar$$

1,3-Dimercaptobenzene (1,3-benzene dithiol) has been prepared from 1,3-benzenedisulphonic acid by the following route (F. Vögtle, R.G. Lichtenthaler and M. Zuber, Chem. Ber., 1973 106, 719).

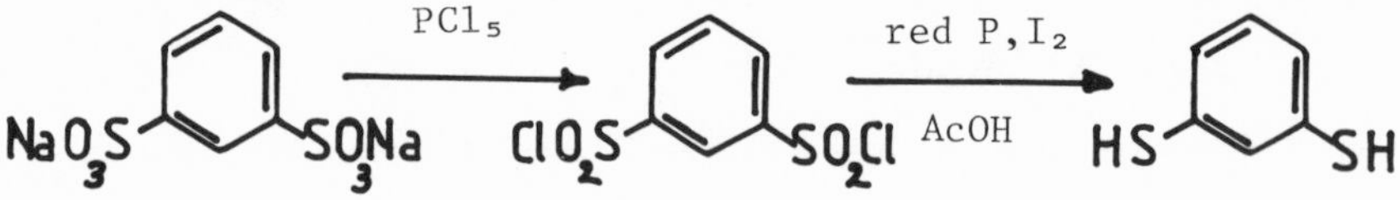

(b) *Sulphenyl chlorides*

This route is based on the reaction of 2,4-dinitrophenyl-sulphenyl chloride with hydrocarbons in the presence of aluminium chloride, followed by alkaline methanolysis of the phenyl 2,4-dinitrophenylsulphide formed as shown (J.L. Wardell, *ibid*, p. 245; N. Kharasch and R. Swidler, J. org. Chem., 1954, 19, 1704; C.M. Buess and N. Kharasch, J. Amer. chem. Soc., 1950, 70, 3529; N. Kharasch, G.I. Gleason and C.M. Buess, *ibid*, 1950, 72, 1796). For the compounds RAr (R=H, Me, Br, Cl) the

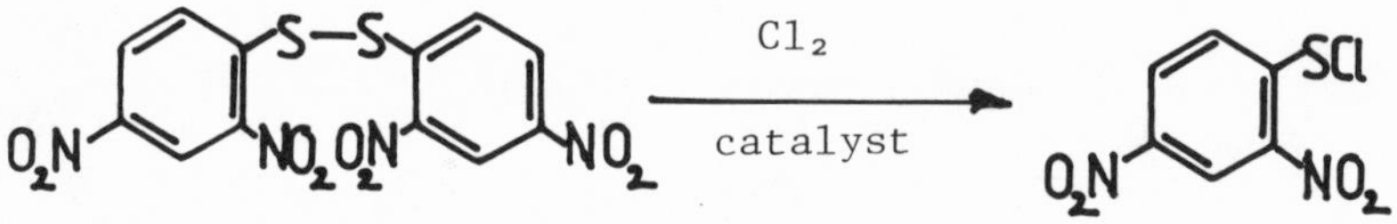

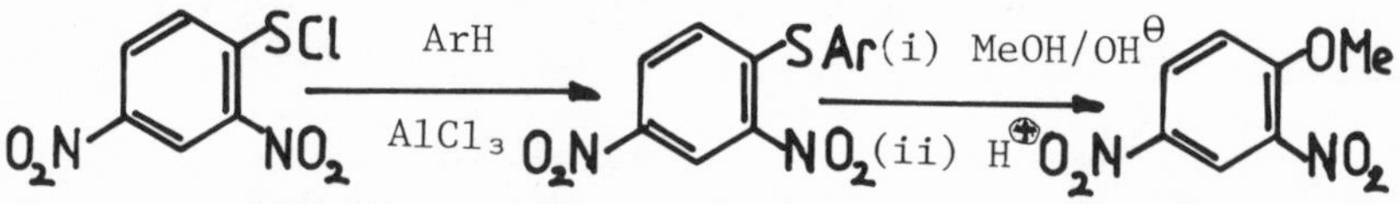

corresponding thiophenols p-RC₆H₄SH are produced in 80, 80, 79 and 76% yield respectively.

2 *Reduction of diphenylsulphides*

Disulphides have often been synthesised in preference to the direct formation of thiols. In this indirect approach the extra step involved must be seen to result in an improved yield compared with the direct route. The disulphide/ polysulphide route involving diazonium salts has none of the explosive hazard connected with the xanthate route (discussed below). The general procedure and a specific example are shown (R. Livingstone *et al*, J. chem. Soc., Perkin I, 1972, 819).

$$Ar\,N{=}N^{+} + S_n^{--} \longrightarrow N_2 + Ar S_n\,Ar \xrightarrow{[H]} Ar\,SH$$

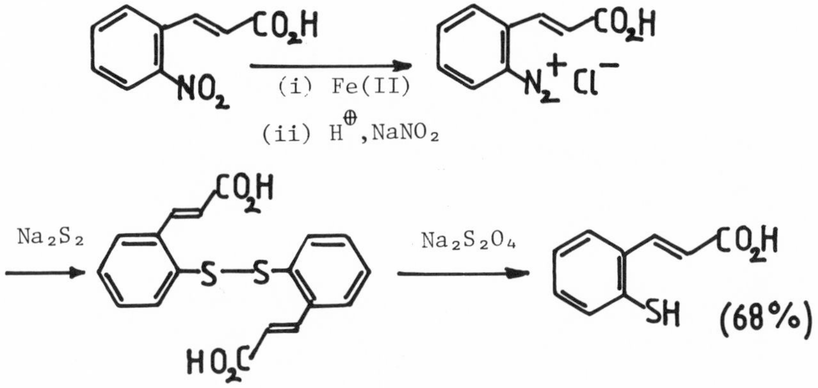

Disulphides are also obtainable from thiocyanates by alkaline hydrolysis (discussed under routes from diazonium salts).

Many reducing agents have been applied for the reduction of disulphides including lithium aluminium hydride, sodium borohydride, zinc and acid or alcohol, tin and acids, hydrogen sulphide and its metal salts, triorganophosphine, glucose and base, hypophosphorous acid, sodium in ammonia, electrolytic reduction and ultraviolet irradiation (J.L. Wardell, in 'The Chemistry of the Thiol Group', ed. S. Patai, Interscience, Ch. 4, p. 221).

3 *From phenols*

Several routes exist for transforming a phenol into the corresponding thiophenol. The best known is the Newman–Kwart rearrangements (M.S. Newman and F.W. Hetzel, Org. Synth., 1971, **51**, 139) typified by the following scheme.

$$\text{ArOH} \xrightarrow[\text{(ii)}\ R_2NCCl]{\text{(i)\ Base}} \overset{\overset{\text{S}}{\|}}{\text{ArOCNR}_2} \xrightarrow{\Delta} \overset{\overset{\text{O}}{\|}}{\text{ArSCNR}_2} \xrightarrow[\text{(ii)}\ H^{\oplus}]{\text{(i)}\ ^{\ominus}\text{OH}} \text{ArSH}$$

o-t-Butylphenol has been converted into *o-t*-butylthiophenol in 65% overall yield with preliminary formation of the dialkyl-thiocarbamate in pyridine/xylene solution (H. Kwart and H. Omura, J. Amer. chem. Soc., 1971, <u>93</u>, 7250).

The second transformation is that of the thioncarbonates to the thiolcarbonate (J.L. Wardell, in 'The Chemistry of the Thiol Group', ed. S. Patai, Interscience, 1974, Ch. 4, p. 201) (the Schonberg rearrangement) followed by hydrolysis, the disadvantage being that only half the phenol used is converted into thiol.

$$2\text{ArOH} \xrightarrow{SCCl_2} (\text{ArO})_2C{=}S \xrightarrow{\text{heat}} \overset{\overset{\text{O}}{\|}}{\text{ArOCSAr}} \xrightarrow{H_2O} \text{ArOH} + \text{ArSH} + CO_2$$

A third procedure, the Kawata-Harano-Taguchi rearrangement of a xanthate, can be effected with aluminium chloride as shown (T. Kawata, T. Harano and T. Taguchi, Chem. Phar. Bull., (Japan), 1973, <u>21</u>, 404) or thermally (Y. Araki, Bull. chem. Soc., Japan, 1970, <u>43</u>, 252).

$$\text{ArOH} \xrightarrow[\text{(ii)\ RX}]{\text{(i)\ }CS_2,KOH} \text{Ar}\overset{O}{\underset{S}{\diagup\diagdown}}C{-}SR \xrightarrow{AlCl_3,CS_2} \text{Ar}\overset{O}{\underset{S}{\diagup\diagdown}}C{-}SR \xrightarrow{^{\ominus}\text{OH}} \text{ArSH}$$

A variation of this is the rearrangement of the thioncarboxylate, (Y. Araki and A. Kaji, *ibid*, 1970, <u>43</u>, 3214).

$$\text{ArOH} \xrightarrow{ClCR}^{\overset{\text{S}}{\|}} \overset{\overset{\text{S}}{\|}}{\text{ArOCR}} \xrightarrow{\text{heat}} \overset{\overset{\text{O}}{\|}}{\text{ArSCR}} \xrightarrow{\text{OH}^{\ominus}} \text{ArSH}$$

Generally the Newman-Kwart procedure is the preferred procedure and the yield at each of the three stages is greater than 80%.

All four thermal rearrangements are considered to be unimolecular processes involving nucleophilic attack by sulphur on the aromatic ring and involve a cyclic transition state. Electron withdrawing substituents facilitate the reaction. In the rearrangement of an unsymmetrical thioncarbonate, sulphur becomes bonded to the ring with the greater electron withdrawing ability.

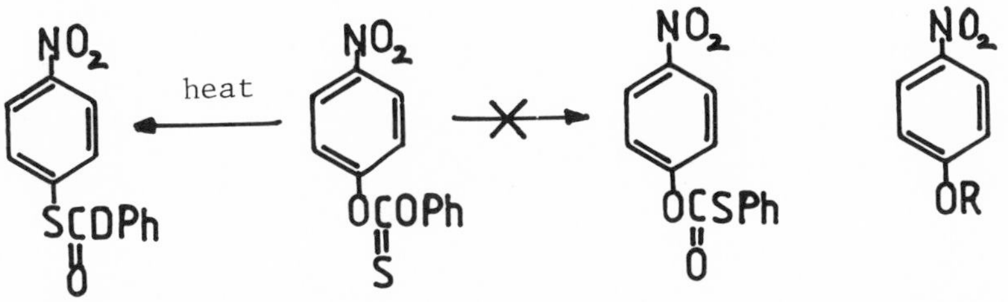

Kinetic studies indicate that the first order rate constants have the following values with the different thione groupings shown (A. Kaji, Y. Araki and K. Miyazaki, Bull. chem. Soc., Japan, 1971, 44, 1393). This probably accounts for the efficiency of the Newman-Kwart method.

R	$-C\!\!-\!\!NMe_2$ $\parallel$ S	$-C\!\!-\!\!Ph$ $\parallel$ S	$-C\!\!-\!\!SPh$ $\parallel$ S	$-C\!\!-\!\!OPh$ $\parallel$ S
Type	Newman-Kwart	thion-carboxylate	xanthate	Schönberg
$K(Sec^{-1})$	1.21×10^{-3}	1.18×10^{-4}	1.06×10^{-4}	2.34×10^{-5}

The conversion of OH into SH has been compared to that of Br into SH and confirms the value of the former reaction (L. Field and P.R. Engelhardt, J. org. Chem., 1970, 35, 3647) providing that the phenolic compound is readily available. (Scheme 1)

4 *From diazonium salts*

Aromatic thiols can be readily produced from (i) *S*-aryl-xanthates, (ii) S-aryl *iso*-thiuronium salts and (iii) thio-cyanates. All three classes of compound are produced from diazonium salts as shown below.

Scheme 1

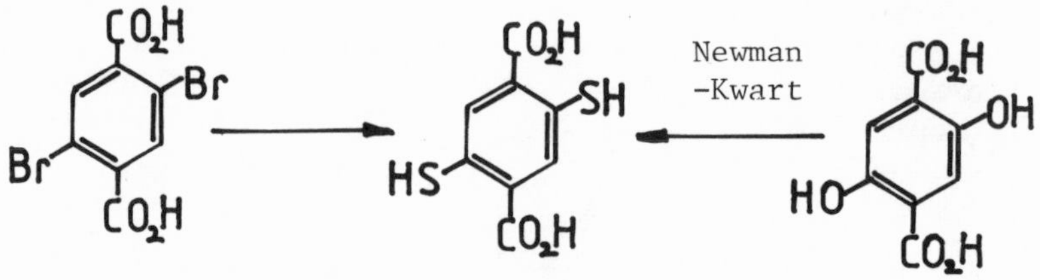

Scheme 2

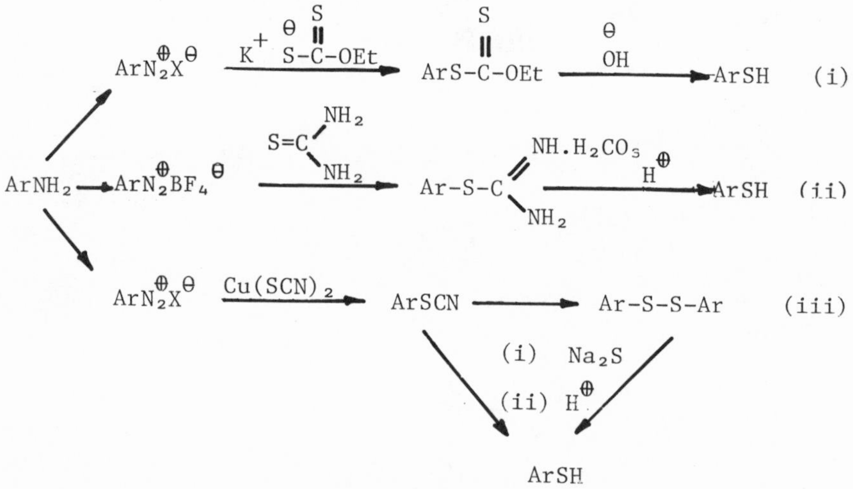

The xanthate route, consisting generally of the reaction between a diazonium salt and an alkyl xanthate, has some dis-advantage in that an explosion can result from cold mixing of potassium ethyl xanthate and a diazonium salt followed by sub-sequent heating (D.S. Tarbell and D.K. Fukushima, Org. Synth., 1947, 27, 81; 1967, 47, 107). The use of a nichrome stirrer or a catalytic amount of nickel ion has been recommended to mini-mise the accumulation of diazonium xanthate. The reaction bet-ween diazonium salts and alkyl xanthate leads also to diaryl dithiocarbonates $(ArS)_2 C=O$ (reducible to thiols by $LiAlH_4$) and alkyl alkylxanthates (J.L. Wardell in 'The Chemistry of the Thiol Group', ed. S. Patai, Interscience, 1974, Ch. 4, 195).

In procedure (ii) for the preparation of thiols, the S-aryl iso-thiuronium salt need not be isolated but can be treated in situ with bicarbonate ion followed by acidification to give yields from 25 to 50% based on the diazonium tetrafluoroborate. (B.V. Kopylova, M.N. Khasanova and R. Kh. Freidlina, Bull. Acad. Sci., USSR, 1970, 582). The alternative alkaline hydrolysis of the iso-thiuronium compound is less effective. In the case of p-chloroaniline a dithiol is produced, as from p-phenylenediamine, on account of the activating influence of the diazonium substituent. o- and m-Chloroaniline behave normally.

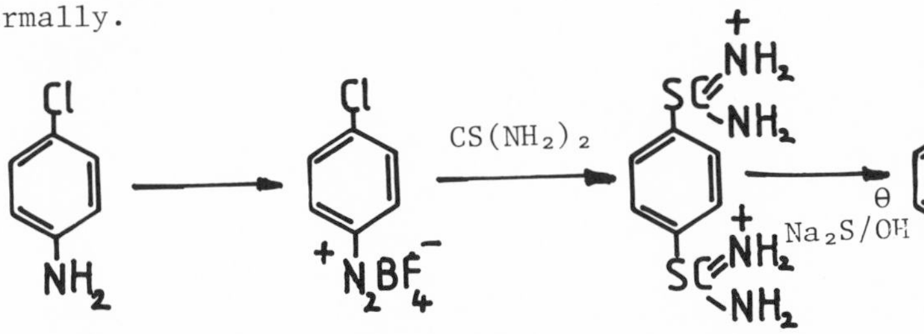

Procedure (iii) as exemplified below is an excellent method of obtaining thiols and has been discussed further (J.L. Wardell, loc. cit.). 2-Amino-4-nitrobenzoic acid similarly gives 2-thiacyano-4-nitrobenzoic acid which by reduction with alkaline sodium sulphide gives the corresponding thiol (overall yield 25-30%).

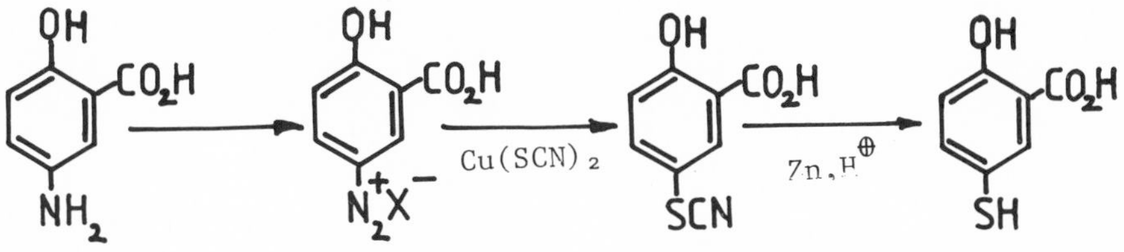

5 Formation from phenols and aromatic amines by thiacyanation

Aromatic compounds having strong electron donating substituents react with thiocyanogen or thiocyanogen halides even in the presence of nitro, halogen and carbalkoxy groups. The formation of aromatic thiols by the reduction of thiocyanates formed by thiacyanation reactions is a useful, if somewhat neglected procedure. The general procedure shown below has

been discussed (J.L. Wardell, *loc. cit.*, p. 234) and thiacyanation reviewed (J.L. Wood, Org. Reactions, 1946, 3, 240).

$$ArH + (NCS)_2 \longrightarrow ArSCN \longrightarrow ArSH$$

Thiocyanogen $(NCS)_2$ can be generated at low temperature in a
number of different ways. Treatment of phenol and ammonium
thiocyanate at -20° to 5° in liquid sulphur dioxide or methanol at 0° with chlorine gave 4-thiocyano-ophenol which upon
refluxing with aqueous sodium hydroxide gives 4-hydroxythio-
phenol (R.J. Laufer, Chem. Abs., 1971, 75, 88310).

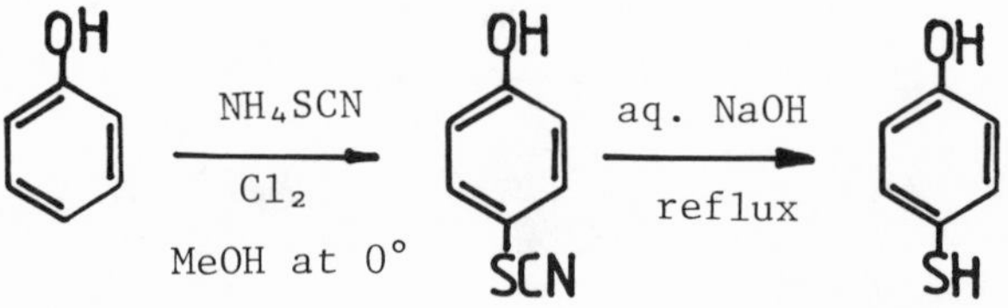

The 2-methyl, 2-t-butyl and 3-methyl derivatives are similarly prepared. Dimethylaniline gave 4-thiocyanodimethylaniline
which upon reductive cleavage with sodium in liquid ammonia
gives 4-dimethylaminothiophenol (72%). 1-Amino-4-thiocyano-
2,3,5,6-tetramehtylbenzene, formed from aminodurene with lead
thiocyanate, bromine and hot aqueous sodium hydroxide, gives
the 4-thiol (22%). The 1-hydroxy analogue gives the corresponding 4-thiol (53%) (T. Wieland and E. Bäuerlein, Chem.
Ber., 1964, 97, 2103; J.L. Wardell, *loc. cit.*). The preceding route can be regarded as essentially one of initial
electrophilic substitution followed by hydrolysis.

6   *Formation from aromatic halides*

(a) *Use of alkali metal sulphides*

The preparation of aromatic thiols from aryl halides
usually only proceeds efficiently if electron withdrawing
groups are present in the nucleus. The routes shown in the
table (J.L. Wardell, *loc. cit.*, p. 182) involve classical
type nucleophilic substitution reactions; aprotic solvents do
not appear to have been generally used.

Halide	Thiol	Method	Yield
p–NO$_2$C$_6$H$_4$Cl	p–NO$_2$C$_6$H$_4$SH	(i) Na$_2$S$_2$ (ii) OH$^\ominus$ (iii) H$^\oplus$	60–65
p–NO$_2$C$_6$H$_4$Cl	p–NH$_2$C$_6$H$_4$SH	(i) Na$_2$S (ii) HOAc	69
o–NO$_2$C$_6$H$_4$Cl	o–NO$_2$C$_6$H$_4$SH	(i) Na$_2$S$_2$ (ii) Zn, HOAc	68
p–SO$_2$MeC$_6$H$_4$Br	p–SO$_2$MeC$_6$H$_4$SH	(i) Na$_2$S$_2$ (ii) glucose/OH$^\ominus$ (iii) H$^\oplus$	27
hexachloro- benzene	pentachloro- thiophenol	(i) NaSH, liq NH$_3$ (ii) H$^\oplus$	96
1,2,3-tri- chlorobenzene	2,3-dichloro- thiophenol	"	20
p–ClC$_6$H$_4$Cl	p–SHC$_6$H$_4$Cl	"	0

The reactions of various halonitrobenzenes with thiolates have been comprehensively reviewed (T.J. De Boer and I.P. Dirkx, Activating effects of the nitro group in aromatic substitutions in 'Chemistry of Nitro, Nitroso Groups', ed. A. Feuer, Interscience, New York, 1969, p. 487).

The reactions of 2,4-dinitrofluorobenzene illustrated occur primarily by way of an addition-elimination mechanism involving a Meisenheimer complex the formation of which is considered to be the rate-determining step, although other variables are the stereochemistry of the entering group, and leaving group effects.

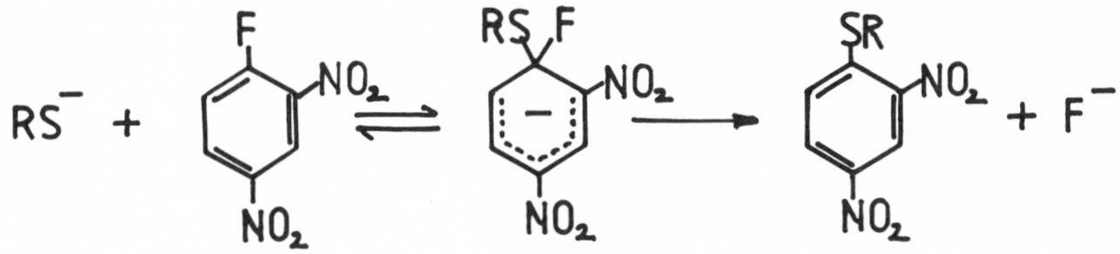

In the use of sodium sulphide followed by acid hydrolysis the formation of amino compounds is a limiting factor but with sodium disulphide in alkaline conditions good yields of the nitrothiol are obtained in a one step process. For the preparation of aminothiols, the di(nitrophenyl)disulphide can be prepared and isolated followed by reduction of both the nitro and the disulphide groups.

With polychloro compounds replacement of both chlorine and nitro groups has been reported to occur in liquid ammonia or methanol (Y. Takikawa and S. Takizawa, Chem. Abs., 1972, $\underline{77}$, 5081). Hexafluorobenzene (like the corresponding polychloro-

benzenes) with the thiolate anion $SR^{\ominus}$ (R=H, Me, Et, Ph, C_6H_5) undergoes nucleophilic substitution giving C_6F_5SR, $C_6F_4(SR)_2$, $C_6F_2(SR)_4$ as the major products (K.R. Langille and M.E. Peach, J. fluorine Chem., 1972, $\underline{1}$, 407).

Unactivated chloro compounds do not readily undergo reaction with sodium sulphide even at high temperature in the presence of aprotic solvent.

Thiophenol has been produced from hydrogen sulphide and halobenzenes in the gas phase with a catalyst composed of activated charcoal and a metal sulphide (K. Mori *et al.*, Japan Pat., 70, 19046; Chem. Abs., 1970, $\underline{73}$, 55182).

(b) *Use of heavy metal alkyl sulphides*

The conversion of aryl bromides to thiophenols by prolonged interaction with a copper (I) alkyl sulphide in boiling quinoline followed by cleavage of the arylalkyl sulphide is a method related to the previous method (a). (R. Adams and A. Ferretti, J. Amer. chem. Soc., 1959, $\underline{81}$, 4939; A. Ferretti, Org. Synth., 1962, $\underline{42}$, 54.   Copper (I) alkyl and aryl sulphides are prepared by the reactions shown.   Copper (I)phenyl sulphide reacts with aryl halides in boiling quinoline during ten hours while copper (I) butyl sulphide only reacts with aryl bromides.

$$Cu_2O + 2PhSH \xrightarrow{\text{EtOH,N}_2, \text{ reflux (3 h)}} (PhSCu)_2 + H_2O$$

$$Cu_2O + 2BuSH \longrightarrow (BuSCu)_2 + H_2O$$

$$ArX + PhSCu \longrightarrow ArSPh + CuX$$

$$PhBr + 2BuSH \longrightarrow BuSPh + CuBr$$

Alkylaryl sulphides are cleaved by sodium in liquid ammonia, lithium in methylamine and other metal/amine systems to give the thiophenol in yields ranging from 70 to 100%. The cleavage reactions are considered to be aromatic S_N2 reactions rather than involving arynes. Unsymmetrical aryl sulphides with a single substituent give in most cases mixtures of thiols but thiophenol is the major product. The reaction has a considerable scope and the yields in the examples shown below are satisfactory, copper (I) ethyl sulphide being used for the sulphide preparation and sodium/liquid ammonia for the cleavage followed by acidification.

Aryl Halide	Sulphide	Thiol
1,4-dibromo-benzene	1,4-di(ethylthio)-benzene (96%)	1,4-benzene-dithiol (98%)
1,2-dibromo-benzene	1,2-di(ethylthio)-benzene (58%)	1,2-benzene-dithiol (72%)
1,3,5-tribromo-benzene	1,3,5-tri(ethylthio)-benzene (35%)	1,3,5-benzene-trithiol (85%)

The nucleophilic substitution reaction of an aryl halide or an aryl sulphonate with the thioacetate ion or thiourea in an aprotic solvent, followed by hydrolytic cleavage or hydrogenolysis of the thiol ester with lithium aluminium hydride (P.A. Bobbio, J. org. Chem., 1961, 26, 3023)(J.L. Wardell,*loc. cit.* p. 189) appears to be a possible route to thiophenols which has not been examined for aryl compounds although it has been investigated with secondary alcohols (J.L. Wardell, *ibid,* p. 208) and ethyl xanthate instead of thiolacetate for the synthesis of optically active thiols (E. Beretta *et al.*

Synthesis, 1974, 425). The interaction of an aryl bromide with copper (I) thiolacetate in hot quinoline would appear to be a potential reaction.

$$ArBr \xrightarrow[\text{Aprotic solvent}]{CH_3COS^{\ominus}Na} ArS\overset{\overset{\displaystyle O}{\|}}{C}CH_3 \xrightarrow[\text{dioxan}]{LiAlH_4} ArSH$$

$$ArOSO_2TS \xrightarrow[\text{aprotic solvent}]{CH_3COS^{\ominus}Na} ArS\overset{\overset{\displaystyle O}{\|}}{C}CH_3$$

$$+ TsSO_3^{\ominus}$$

7 *From Organometallic Compounds*

Related to the preceding routes is the prior conversion of the aryl bromide or hydrocarbon in certain cases, to the aryl lithium with lithium or butyl lithium followed by reaction with sulphur and final acidification (E. Jones and I.M. Moodie, Org. Synth., 1970, 50, 104).

$$ArLi \xrightarrow{S} ArSLi \xrightarrow{H^{\oplus}} ArSH$$

$$C_6H_5Br \xrightarrow{BuLi} C_6H_5Li \xrightarrow{S} C_6H_5SLi \xrightarrow{H^{\oplus}} C_6H_5SH \ (62\%)$$

In general the use of organometallic compounds derived from the more electro positive elements (M) has been extensively applied to the preparation of aromatic thiols (J.L. Wardell, *loc. cit.* p. 213).

$$RM + S \longrightarrow RSM \longrightarrow RSH$$

As an alternative to final acidification, lithium aluminium hydride reduction can be used since in certain cases with Grignard reagents and sulphur, disulphides are formed which would be stable to the final step of the reaction sequence. A number of examples of this reaction are shown below.

PhBr $\xrightarrow[\text{(iii) H}^{\oplus}]{\text{(i) Mg (ii) S}}$ PhSH (30%)

3-t-BuC$_6$H$_4$Br $\xrightarrow[\text{(iii) LiAlH}_4]{\text{(i) Mg (ii) S}}$ 3-t-BuC$_6$H$_4$SH (83%)

4-FC$_6$H$_4$Br $\xrightarrow[\text{(iii) H}^{\oplus}]{\text{(i) Mg (ii) S}}$ 4-FC$_6$H$_4$SH (26%)

4-Me$_2$NC$_6$H$_4$Br $\xrightarrow[\text{(iii) H}^{\oplus}]{\text{(i) Li (ii) S}}$ 4-Me$_2$NC$_6$H$_4$SH (50%)

8 *Other Methods*

The reaction of 4-substituted-1,2-benzoquinones with thiourea under acidic conditions followed by alkaline cleavage provides a route to certain thiophenols (H.-W. Wanzlick, *et al*; Tetrahedron Letters, 1970, 1271; J.L. Wardell, *loc. cit.*, p. 191, 193). 1,4-Benzoquinones react similarly and the thiosulphate ion may replace thiourea. (Scheme 3).

p-Phenylenediamine reacts with thiosulphate ion in the presence of chromate or dichromate to give, after reduction by zinc and acid, 2,5-diaminothiophenol.

The reaction of sulphur monochloride with aromatic amines having a free *ortho* position (the Herz reaction) gives thiazathiolium salts which on alkaline treatment followed by acidification yield *o*-aminothiophenols. The reaction has technical importance in the preparation of sulphur dyes (Scheme 4).

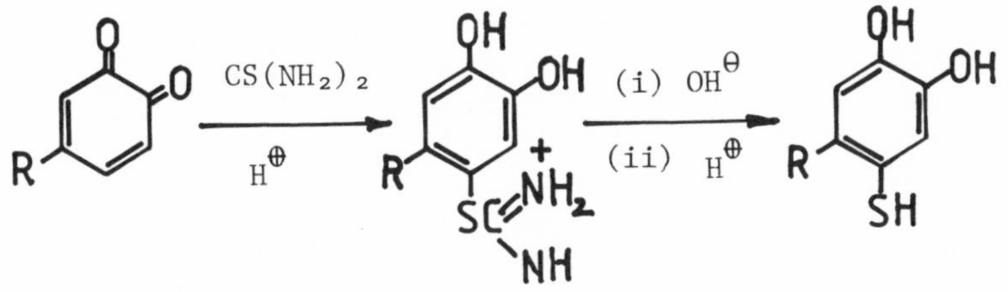

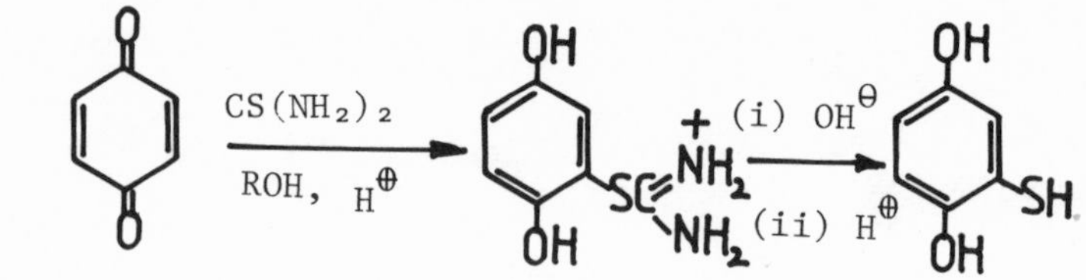

Scheme 3

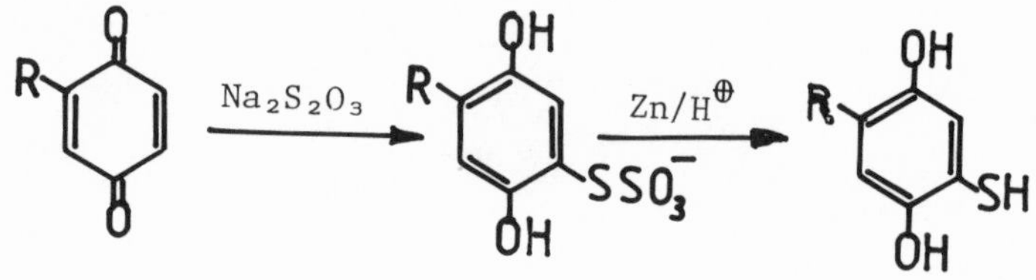

Scheme 4

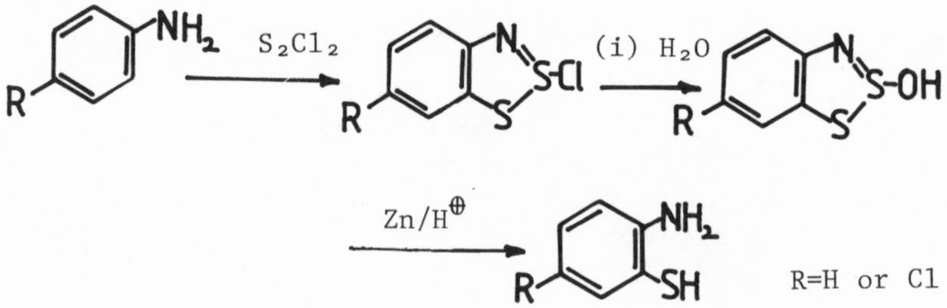

R=H or Cl

Aromatic compounds with other pronounced electron donating groups also react with sulphur monochloride and sulphur dichloride in the presence of hydrogen sulphide to give disulphides reducible to thiophenols. (W. Hahn and K. Goliasch, Belg. Pat. 635,634; E.B. Hotelling *et. al.*, J. org. Chem., 1959, 24, 1598; W.K. Warburton, Chem. Rev., 1957, 57, 1011). (Scheme 4).

Sulphur can react with diphenylsulphone to yield diphenylsulphide and no hydrogen of the aromatic ring is replaced (S. Oae, *et. al.*, Bull. chem. Soc., Japan, 1971, 44, 445). Bromobenzene and certain halogen substituted bromobenzenes react with sulphur to give diphenyl polysulphides which are reducible to thiophenols (S. Oae and Y. Tscuchida, Tetrahedron Letters, 1972, 1283).

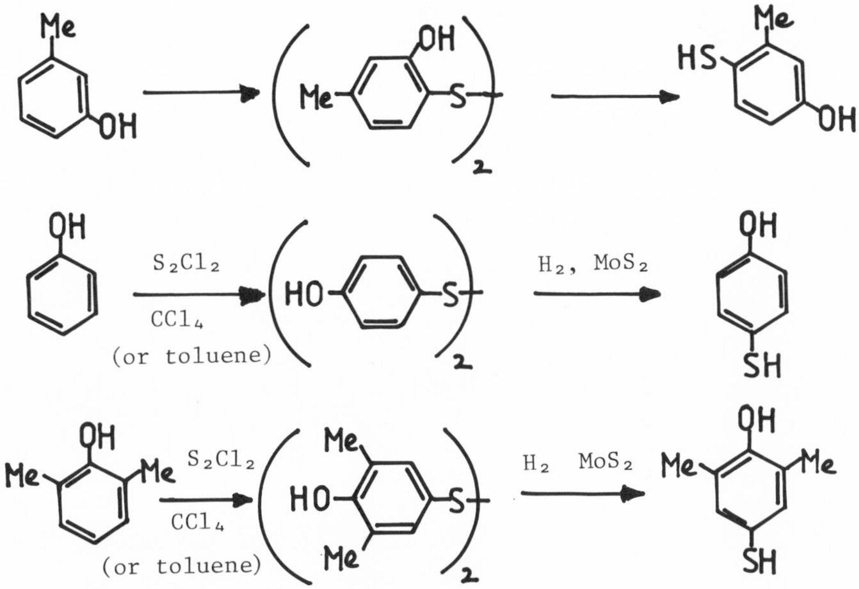

PhBr $\xrightarrow{\text{S}}$ PhS$_x$Ph + S$_2$Br$_2$ $\quad$ (x $\geqq$ 2)

PhS$_x$Ph $\xrightarrow[\text{(ii) H}^{\oplus}]{\text{(i) LiAlH}_4}$ PhSH

The reaction with bromobenzene proceeds without replacement of hydrogen atoms, although this partially occurs with chlorobenzene. Fluorobenzene does not react at all. The reactivity thus decreases with increase in the strength of the carbon–halogen bond.

These results are interesting to compare with the reaction of toluene and sulphur (in the presence of water) whereby benzoic acid is produced by an oxidative process.

It can be seen from the preceding preparations that numerous functional groups in aromatic compounds can be transformed into the thiol group. The carboxyl group has not apparently been converted into the thiol group directly either by catalytic or photochemical means. Xanthate derivative of alkane and arylalkane carboxylic acids are smoothly converted into acyl and xanthate radicals on photolysis and the acyl radicals suffer decarbonylation to give S-alkyl or S-arylalkyl xanthates from which thiols are obtained by hydrolysis. (cf. D.H.R. Barton et al., J. Chem. Soc., 1962, 1967).

$$RCOCl \xrightarrow{\overset{\displaystyle S}{\underset{}{\parallel}} \text{NaSCOEt}} RCO\overset{\displaystyle S}{\underset{}{\parallel}}SCOEt \xrightarrow{h\nu} RCO\cdot + \cdot SCOEt$$

$$RCO\cdot \xrightarrow{h\nu} R\cdot + CO$$

$$R\cdot + \cdot S\overset{\displaystyle S}{\underset{}{\parallel}}COEt \longrightarrow RS\overset{\displaystyle S}{\underset{}{\parallel}}COEt \longrightarrow RSH$$

9 *Conversion of the SMe to the thiol group.*

The synthesis of thiophenols has been realised by the demethylation and methanolysis of aryl methyl thioethers (J.M. Lavanish, Tetrahedron Letters, 1973, 3847).

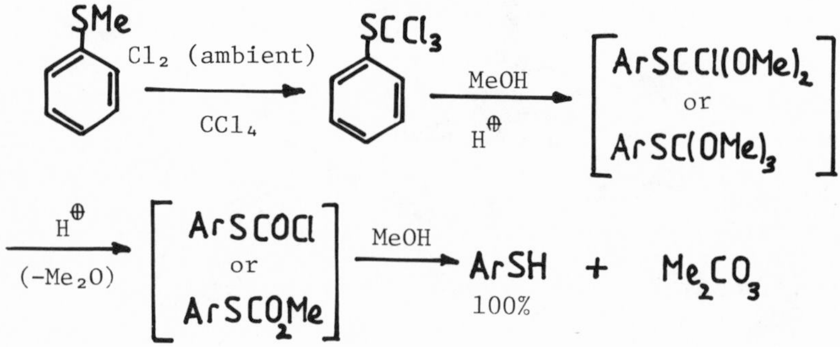

Similarly 4-hydroxythiophenol (*via* the *O*-acetate) is produced in 86% yield (overall).

Reactions of Thiols

(1) *Oxidation*

(a) *With halogens*

Aromatic thiols can be oxidised by a variety of reagents to disulphides and thence to further products dependent on the reaction conditions. The types of substances produced and their inter-connection are shown in the following scheme. The oxidation of thiols has been discussed (G. Capozzi and G. Modena, in 'The Chemistry of the Thiol Group', ed. S. Patai,

Interscience, 1974, Ch. 17; P.C. Jocelyn, 'Biochemistry of the Thiol Group', Academic Press, London, 1972, Ch. 4, p. 94).

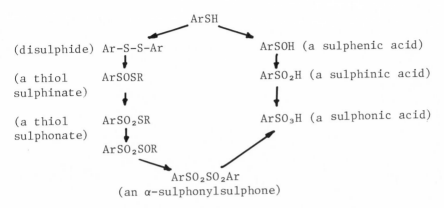

(disulphide) Ar–S–S–Ar

(a thiol sulphinate) ArSOSR

(a thiol sulphonate) ArSO₂SR

ArSO₂SOR

ArSH

ArSOH (a sulphenic acid)

ArSO₂H (a sulphinic acid)

ArSO₃H (a sulphonic acid)

ArSO₂SO₂Ar
(an α-sulphonylsulphone)

The oxidation of aliphatic thiols has a relevance in connection with hydrocarbon sweetening and sewage treatment but in chemical studies aromatic thiols have received almost equally prominent attention. With reference to technological aspects, hydrogen peroxide has been studied as a selective oxidant (W.H. Kibbel, Jr., C.W. Raleigh and J.A. Shepherd, 27th Annual Purdue Industrial Waste Conference, Purdue University, Lafayette, Indiana, USA (1972).

The interaction of chlorine and of bromine with thiols in both aqueous solvents (giving sulphonyl halides or sulphonic acids (C.R. Russ and I.B. Douglass, in 'Sulphur in Organic and Inorganic Chemistry', vol. 1. ed. A. Senning, Dekker, 1971, Ch. 8; P.S. Magee, *ibid.* Ch. 9) and in anhydrous conditions (S.R. Sandler and W. Karo, Organic Functional Group Preparation, vol.12 - III, Academic Press, 1972, Ch. 4) has been studied and the observed reactions are as shown in the following equations.

$$RSH + X_2 + H_2O \longrightarrow RSO_2X + HX$$

$$RSH + X_2 \longrightarrow RSX + HX \quad (1) \quad \text{(sulphenyl halide)}$$

$$RSX + X_2 \rightleftharpoons RSX_3 \quad (2) \quad \text{(an aryl sulphur trihalide)}$$

$$RSH + RSX \longrightarrow RSSR + HX \quad (3)$$

$$RSSR + X_2 \rightleftharpoons 2RSX \quad (4)$$

With aryl sulphur trihalide the equilibrium is shifted to the
left with increasing temperature. With chlorine and bromine
the equilibrium in equation (4) is largely towards the right
whereas for iodine it is towards the left since hardly any
sulphenyl iodides are known (E. Ciuffarin and G. Guaraldi, J.
org. Chem., 1970, 35, 2006; L. Field, J.L. Vanhorne and
L.W. Cunningham, *ibid*., 1970, 35, 3267). On account of the
fact that with iodine reaction (1) proceeds to completion to
the right and equilibrium (4) lies heavily in favour of the
disulphide, thiol can be quantitatively titrated by an iodo-
metric procedure providing pH and any overoxidation are carefully controlled (J.P. Danehy and M.Y. Oester, J. org. Chem.,
1967, 32, 1491).

Chlorine and bromine are believed to interact in aprotic
(but not nucleophilic) solvents with thiols, and sulphides are
thought to behave similarly although no mechanistic studies
have been described (G. Capozzi, G. Melloni and G. Modena,
J. org. Chem., 1970, 35, 1217; E. Kühle, Synthesis, 1970, 561;
G.E. Wilson Jr. and M.G. Huang, J. org. Chem., 1970, 35, 3002;
H. Kwart and H. Omura, J. Amer. chem. Soc., 1971, 93, 7250).

$$R - SH + X_2 \rightleftharpoons R - \overset{\oplus}{\underset{\underset{X}{|}}{S}} - H + X^{\ominus} \longrightarrow RSX + HX$$

Sulphenyl halides (referred to in a subsequent section) readily
undergo nucleophilic attack as shown. However in the case of
water because of the instability of the sulphenic acids, which

$$RSX + NuH \longrightarrow RSNu + HX \quad (\text{NuH can be } R_2NH, HSCN, ROH, RSH \text{ etc})$$

have never been isolated, thiolsulphinate esters are formed
instead. However disproportionation of both the sulphenic
acids and the thiolsulphinates formed finally gives disulphides and thiolsulphonates as shown. (J.L. Kice and
J.P. Cleveland, J. Amer. chem. Soc., 1973, 95, 104; J.L. Kice
and J.P. Cleveland, *ibid*., 109; J.L. Kice, Progr. inorg. Chem.,
1972, 17, 147). (Scheme 5).

In the pyrimidine shown, intramolecular hydrogen bonding
evidently resuls in some stabilisation of the sulphenic acid
(T.C. Bruice and A.B. Sayigh, J. Amer. chem. Soc., 1959, 81,
3416; B.C. Pal *et al*., J. Amer. chem. Soc., 1969, 91, 3634).
(Scheme 6).

Scheme 5

$$RSX + H_2O \longrightarrow [RSOH] + HX \quad \text{(a sulphenic acid)}$$

$$RSOH + RSX \longrightarrow R\underset{\underset{O}{\|}}{S}-S-R + HX$$

$$[2RSOH] \longrightarrow R\underset{\underset{O}{\|}}{S}-S-R + H_2O \quad \text{(thiolsulphinate)}$$

$$2R\underset{\underset{O}{\|}}{S}-S-R \longrightarrow R-S-S-R + RSO_2-SR \quad \text{(thiolsulphonate)}$$

Scheme 6

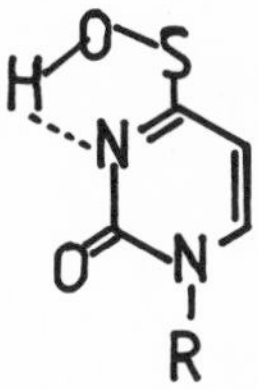

Attempts have been made to establish the formation of the sulphenyl cation from the sulphenyl halide.

$$RSX \rightleftharpoons RS^{\oplus} + X^{\ominus}$$

Dissolution of 2,4-dinitrobenzenesulphenyl chloride in concentrated sulphuric acid showed evidence of the formation of a cationic species (K.C. Malhotra and J.K. Puri, Indian J. Chem., 1971, 9, 1409) but its nature is not certain (E.A. Robinson and S.A.A. Zaidi, Canad. J. Chem., 1968, 46, 3927). Strong interaction has been found between the o-nitro-group and sulphenyl sulphur in methyl o-nitrobenzene sulphenate and in the alkaline rearrangement of 2-nitrobenzenesulphenyl anilides (C. Brown, Chem. Comm., 1969, 100) suggesting that in the case of o-nitrobezene derivatives the alternative cyclic cation,may have formed.

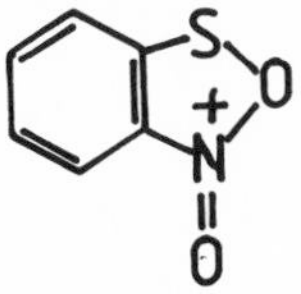

Kinetic evidence indicates that substitution at sulphenyl sulphur is a bimolecular process and the only unimolecular process reported to accord with the equation has been the solvolysis of 2,4,6-trimethoxybenzenethiolarylsulphonate (H. Kloosterziel and J.H. Wevers, 5th Symposium on Organic Sulphur Chemistry, Lund, Sweden, 1972). Nevertheless a dimeric cationic species has been reported for alkyl sulphenyl chlorides and from preliminary evidence seems also to be formed in the aryl series (G. Capozzi, V. Lucchini and G. Modena, Chem. and Ind., (Milan), 1972, _54_, 41; unpublished work).

$$2ArSCl + BF_3 \xrightarrow[\text{FSO}_3\text{H or H}_2\text{SO}_4]{\text{liq. SO}_2} Ar-\overset{+}{\underset{Cl}{S}}-Ar + BF_3Cl^{\ominus}$$

Analogous species are formed from disulphides and sulphenyl chlorides (F. Pietra and D. Vitali, J. chem. Soc., (B), 1970, 623).

$$R-S-S-R + R-S-Cl + BF_3 \xrightarrow[\text{FSO}_3\text{H or H}_2\text{SO}_4]{\text{liq. SO}_2} R-S-\overset{+}{\underset{SR}{S}}-R + BF_3Cl^{\ominus}$$

The over-oxidation of aliphatic disulphides having β-carboxyl groups whereby β-carboxysulphonic acids are finally formed, by way of the initial formation of a sulphenic anhydride, has an analogy in the aromatic series. In the case of 2-thiolbenzoic acid, _o_-sulphobenzoic anhydride can be trapped in the manner shown (L. Field, P.M. Giles Jr. and D.L. Tuleen, J. org. Chem., 1971, _36_, 623).

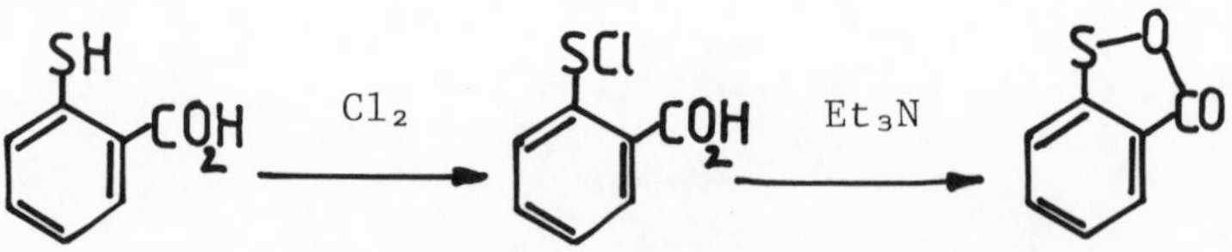

(b) _By dimethyl and other sulphoxides_

Oxidation with dimethyl sulphoxide and tetramethylenesulphoxide (sulpholane) has been reviewed (W.W. Epstein and F.W. Sweat, Chem. Rev., 1967, _67_, 247; C.R. Johnson and J.C. Sharp, Quarterly Report of Sulphur Chemistry, 1969, _4_, 2).

With thiols the reaction takes the following course.

$$2RSH + (Me)_2SO \longrightarrow RSSR + Me_2S + H_2O$$

The rate of oxidation, which is catalysed by aliphatic and aromatic amines, is dependent on the acidity of the thiol and the structure of the sulphoxide. A possible two stage mechanism is shown, like that for other oxidations with sulphoxides (D. Landini $et\ al$, J. Amer. chem. Soc., 1970, $\underline{92}$, 7168) involving a tetra-coordinated sulphur intermediate, stable examples of which have been obtained (I.C. Paul, J.C. Martin and E.F. Perozzi, $ibid$ 1972, $\underline{94}$, 5010).

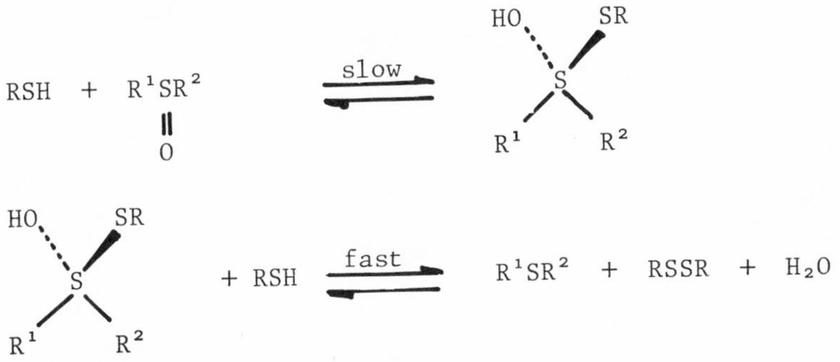

This mechanism does not explain the role of acidic and basic catalysts and it may be more complicated than as illustrated in the fast stage (G. Modena, Int. J. sulphur Chem., 1972(C), $\underline{7}$, 95).

(c) *Use of diethyl azodicarboxylate*

Diethyl azodicarboxylate oxidises thiols to disulphides at ambient temperature in the dark.

$$EtO_2CN=NCO_2Et + RSH \longrightarrow RSSR + EtO_2CNHNHCO_2Et$$

Benzene thiol gives a 90% yield in the absence of solvent after 24 hours. 2-Aminobenzenethiol gives 67% disulphide in refluxing benzene during 4 hours. The reaction is catalysed by triphenyl phosphine (N. Yoneda, K. Suzuki and Y. Nitta, J. org. Chem., 1967, $\underline{32}$, 727) K. Kato and O. Mitsunobu, $ibid$. 1970, $\underline{35}$, 4227).

(d) *By N-halo reagents and related comounds*

N-Chlorosuccinimide, *N*-bromosuccinimide, *N*- chlorobenzo-triazole, and 'positive halogen compounds' such as dichloro-iodobenzene interact with thiols to give sulphenyl halides or disulphides dependent on the reagent/reactant ratio (S.R. Sandler and W. Karo, 'Organic Functional Group Preparation', vol. 12-III, Academic Press, 1972, ch. 4). 2,4,4,6-Tetra-

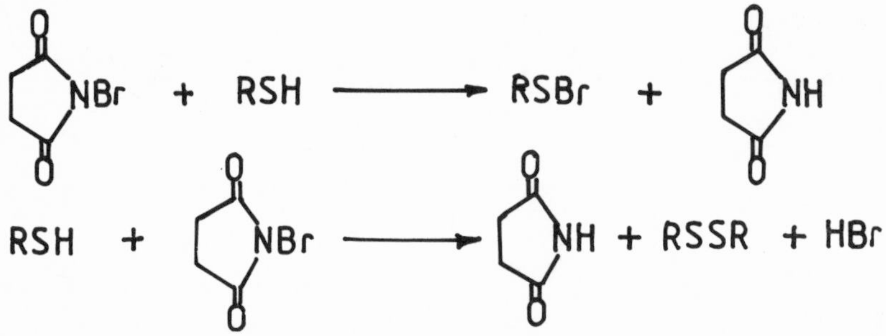

bromo-cyclohexa-2,5-dienone has been used to convert benzene-thiols to the corresponding sulphenyl bromides (V. Calo *et al.* J. sulphur Chem., 1971, $\underline{1}$, 130).

(e) *Use of metal ions and oxides*

Transition metal ions or oxides oxidise thiols. Many of these metal ions exert a catalytic effect on the autoxidation of thiols. Iron (III) cyanides oxidise thiols to disul-phides in the absence of air and the reaction has considerable technological relevance with regard to synthetic rubber.

$$2RSH + 2Fe^{3+} \longrightarrow RSSR + 2Fe^{++} + 2H^{\oplus}$$

Oxidations in alkaline and acidic media have been examined and reveal a complex situation. Several possible mechanisms have been proposed for the various conditions (R.C. Kapoor, R.K. Chohan and B.F. Sinha, J. phys. Chem., 1971, $\underline{75}$, 2036).

Other heavy metal ions in higher oxidation states give disulphides from thiols and possible ionic and radical path-ways have been discussed (G. Capozzi and G. Modena, in 'The Chemistry of the Thiol Groups', ed. S. Patai, Interscience,

1974, ch. 17, p. 803). The use of electron spin resonance has been found valuable. In the oxidation of thiols with manganese (III) acetylacetonate, the disulphide is considered to arise from the interaction of a sulphenium ion with the thiol. The participation of thiyl radicals (RS·) was eliminated from consideration (by the absence of addition products when an alkene was present) and it appears implicit that oxidation of the thiyl radical to the sulphenium ion is faster than its dimerisation (T. Nakaya, H. Arabori and M. Imoto, Bull. chem. Soc. Japan, 1970, $\underline{43}$, 1888).

$$RSH \longrightarrow RS\cdot$$

$$RS\cdot \longrightarrow RS^{\oplus}$$

$$RSH + RS^{\oplus} \longrightarrow RSSR + H^{\oplus}$$

(f) *By molecular oxygen*

The ready oxidation of thiols by air, the influence of metal ion catalysis, the effect of ultraviolet light and the acceleration of their autoxidation by weak and strong bases are of considerable importance in petroleum and biological chemistry. This type of oxidation results in the transformation of unpleasant odorous materials into odourless stable disulphides. Extensive work has been carried out on aliphatic thiols. In a comparative study (C.F. Cullis *et al*. J. appl. Chem., 1968, $\underline{18}$, 1330) aromatic thiols were almost, with the exception of t-butyl thiol, the less reactive towards oxidation by molecular oxygen. The product is the disulphide but in the presence of a large proportion of a strong base, with a prolonged reaction time or in aprotic solvents, sulphonic acid appears (D.P. Danehy and W.E. Hunter, J. org. Chem., 1967, $\underline{32}$, 2047) possibly by the following mechanism involving disproportionation of sulphenate ion.

$$RS-SR + OH^{\ominus} \rightleftharpoons RSOH + RS^{\ominus}$$

$$RSOH + OH^{\ominus} \rightleftharpoons RSO^{\ominus} + H_2O$$

$$3RSO^{\ominus} \longrightarrow RSO_3^{\ominus} + 2RS^{\ominus}$$

In the case of thiophenol, co-catalytic effects by anthraquinone-1-sulphenic acid, *t*-butyl hydroperoxide and by phenylbenzenethiol sulphonate have been found. Many aspects of the

oxidative mechanism are uncertain and discussions have centred
on the involvement of thiyl radicals and possibly radical ions
(G. Capozzi and G. Modena, 'Oxidation of Thiols' in 'The
Chemistry of the Thiol Groups', ed. S. Patai, Interscience,
1974, ch. 17, p. 816).   Thiyl radicals dimerise extremely
rapidly (M.J. Hoffmann and E. Hayon, J. Amer. chem. Soc., 1972,
94, 7950) which may explain why no solvent interaction products
have been isolated.   It has been observed that thiyl radicals
generated in aqueous solution by flash photolysis give rise
to the radical ion shown.

$$RS\cdot \;+\; RS^{\ominus} \;\rightleftharpoons\; RS\text{-}\overset{\cdot}{S}\text{-}R$$

Similar species have been observed in the reaction of various
disulphides with hydrated electrons.   In the reduction of

$$RSSR \;+\; \varepsilon \;\longrightarrow\; RS\text{-}\overset{\cdot}{S}\text{-}R$$

naphthalene 1,8-disulphide, a long-life radical ion is observed (A. Zweig and A.K. Hoffmann, J. org. Chem., 1965, 30,
3997; G. Gaspari and S. Granzow, J. phys. Chem., 1970, 74,
836),   the esr spectrum of which indicates that the electron
is on the sulphur atoms. Similar disulphide radical ions may
indeed be reaction intermediates.

Amines function as catalysts in the oxidation of thiols
probably by converting them to the more susceptible anion.
Aromatic thiols, being more acidic and dissociated, are oxidised faster than arylalkane and alkane thiols (G. Capozzi
and G. Modena, *ibid.* p. 817).

Heavy metal salts added to basic aqueous solutions of
thiols enhance the rate of oxidation (to disulphide only)
(C.F.Cullis *et al.* J. appl. Chem., 1968, 18, 335).

$$4RSH \;+\; O_2 \longrightarrow 2RSSR \;+\; 2H_2O$$

Aromatic thiols such as thiophenol show a low rate of oxidation compared with n-alkane thiols.

Hydroquinone and p-phenylenediamine derivatives in basic
conditions can serve as catalysts in the autoxidation of
thiols to disulphides. Flavin derivatives oxidise thiols to

disulphides in the absence of oxygen giving the reduced form
of the dye which is then oxidised with regeneration of the
original oxidant (M.J. Gibian and D.V. Winkelman, Tetrahedron
Letters, 1969, 3901).

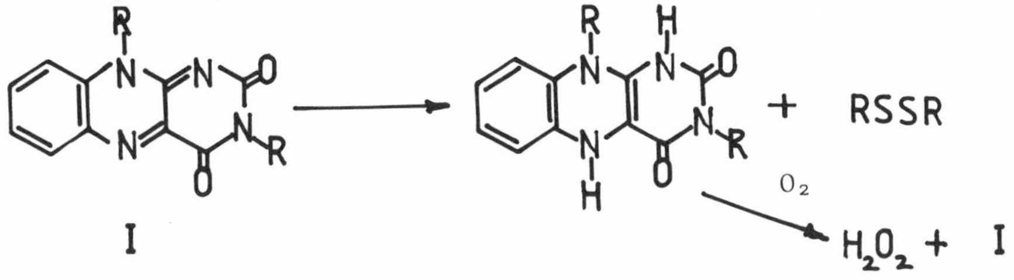

$$I \longrightarrow \quad + \quad RSSR$$

$$\searrow^{O_2} \quad H_2O_2 + \quad I$$

The co-oxidation of thiols in the presence of alkenes
has been widely studied.   Hydroperoxysulphides are formed
which then rearrange to give β-sulphinyl alcohols.   The transfer of oxygen is thought to be intramolecular and evidence
of this has been obtained in the study of the thiophenol/
indene system in the presence of oxygen (H.H. Szmant and
J.J. Rigau, Tetrahedron Letters, 1967, 3337; H.H. Szmant and
J.J. Rigau, J. org. Chem., 1972, 37, 447).   Three of the four
possible stereoisomeric 2-phenylsulphinyl-1-indanols are
formed.   The general reaction scheme and the formulae of the
indanols are shown below.

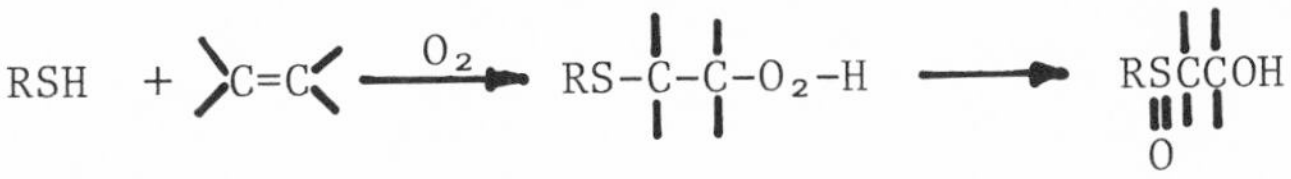

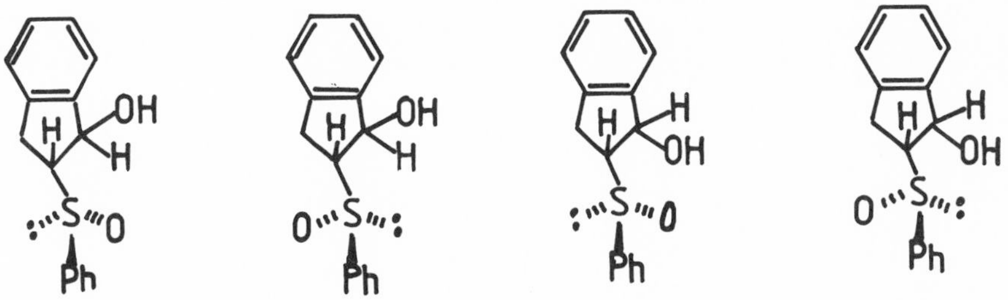

$$RSH \; + \; \text{>}C=C\text{<} \xrightarrow{O_2} RS\text{-}\overset{|}{C}\text{-}\overset{|}{C}\text{-}O_2\text{-}H \longrightarrow RS\overset{||}{C}\overset{||}{C}OH$$

The results show that in benzene solution a 5.4:1 _trans/
cis_ mixture of hydroperoxides is formed and hence the co-oxidation is stereoselective rather than stereospecific as
suggested earlier (A.A. Oswald and T.J. Wallace, in 'Organic
Sulphur Compounds', vol. 2, ed. N. Kharasch, Pergamon, ch. 8).
Thiols are unstable to light and undergo photolytic oxi-

dation to the disulphide. The photochemistry of thiols including aromatic thiols has been reviewed (A.R. Knight, in
'The Chemistry of the Thiol Group', ed. S. Patai, Interscience,
1974, ch. 10, p. 455).

$$2RSH \xrightarrow{h\nu} RSSR + H_2$$

Similar processes occur in aqueous solution as well as in
other solvent solutions. The photolysis of thiols has been
used as a source of hydrogen atoms (W.A. Pryor and
J.P. Stanley, J. Amer. chem. Soc., 1971, <u>93</u>, 1412). In the
majority of cases the sulphur-hydrogen bond breaks resulting
in thiyl radicals which can then undergo dimerisation and
other reactions. Carbon-sulphur bond fission occurs with
shorter wavelength light (D.M. Graham and B.K.T. Sie, Canad.
J. Chem., 1971, <u>49</u>, 3895). The removal of odour resultant
upon photolytic oxidation of thiols has some practical signifi
-cance.

(2) *Substitution Reactions*

Most attempts to effect electrophilic substitution reactions of aryl thiols give transformation products of the SH
group which are generally more sensitive. Attempts to
nitrate or brominate aryl thiols give the disulphide which
may then undergo some nuclear substitution. It has been seen
that disulphides may be formed from thiophenols with *N*-
halosuccinimides whereas phenols under similar circumstances
give halophenols. Diazonium salts with thiophenols yield
diazosulphides rather than benzeneazothiophenols.Thiophenols
interact with t-alcohols in the presence of acids to give
sulphides instead of nuclear substitution products. If *m*-
methoxythiophenol is acetylated in the presence of aluminium
chloride 2-acetyl-5-methoxythiophenyl acetate is obtained due
to the strongly activating methoxyl group. The lack of
nuclear acylation with acyl chlorides is explained in terms
of the involvement of either of the intermediates shown
respectively. (Scheme 7).

For the preceding reasons the -SH groups must be protected prior to the majority of attempted electrophilic substitution reactions. Protective groups for thiols have been listed
in general, certain of which are relevant to aromatic thiols
(Y. Wolman, in 'Chemistry of the Thiol Group', ed. S. Patai,

Interscience, 1974, Ch. 4. p. 669).
Scheme 7

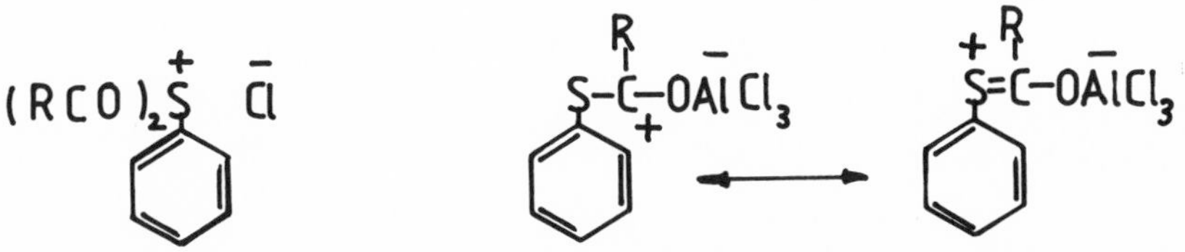

A useful general method for the protection of aromatic
thiols is by interaction with 3-nitrobenzylidene acetophenone.
The products can be acetylated, brominated and nitrated and
the protective group then quantitatively removed (G. Maccagnani
and G. Mazzanti, in 'Directing and Activating Effects',
'Chemistry of the Thiol Group', ed. S. Patai, Interscience,
1974, Ch. 9, p. 432; G.C. Barrett, in 'Organic Sulphur Compounds', Comprehensive Organic Chemistry, Sir D.H.R. Barton
and W.D. Ollis, Pergamon, 1978, vol. 1, p. 17).

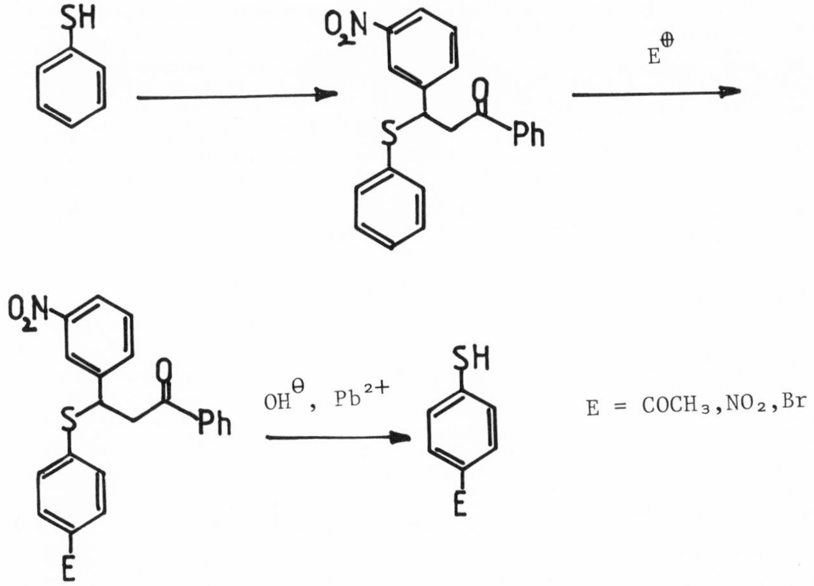

$E = COCH_3, NO_2, Br$

Other protecting groups such as carboxymethyl, acetonyl and
cyanomethyl are generally less effective.

Protodesilylation, the acid-catalysed solvolytic cleavage of the aryl-silicon bond in aromatic compounds ArSiR$_3$, of aryl thiols has been investigated (F. P. Bailey and R. Taylor, J. chem. Soc., (B), 1971, 1446). Protodesilylation is an electrophilic substitution in which a solvated proton is the attacking species (C. Eaborn and R.W. Bott, in 'Organo Metallic Compounds of the Group IV Elements', ed. A.G. MacDiarmid, Dekker, 1969, vol. I, p. 407). Attempted application of a hyperconjugation theory (C.K. Ingold, 'Structure and Mechanism in Organic Chemistry', 2nd Edn., Cornell University Press, Ithaca and London, 1969, p. 111) to the sulphur series failed to explain the results for thiophenols and thioanisoles (F.P. Bailey and R. Taylor, J. chem. Soc., (B), 1971, 1446) although in other thiols and sulphides it has been considered that the hyperconjugative ability of the C-S bond is nearly equal to that of the C-H bond. The former plays no determining role in the conformation of allylmethyl sulphide. (W. Schäfer and A. Schweig, Chem. Comm., 1972, 824; Tetrahedron Letters, 1972, 5205).

Friedel-Crafts alkylation has been discussed (G. Maccagnani and G. Mazzanti, *loc. cit.* p. 434). Other alkylating procedures have been described (C.L. Zundell and L. Choron, Ger. Pat., 1518, 460; Chem. Abs., 1971, 74,141253), but the properties of the products not described. Thiophenols with a t-alcohol or thiol in the presence of aluminium chloride give p-t-alkyl substituted thiophenols although with strong acids sulphides are formed.

Tritylation of aromatic thiols and aminosubstituted benzene thiols gives trityl arylsulphides and not nuclear substitution products as with phenols (G. Chuchani and K.S. Heckmann, J. chem. Soc., (C), 1969, 1436). The attack of the trityl cation on sulphur is considered as evidence for the expansion of its valency shell, according to the following mechanism (M.L. Eberhardt and G. Chuchani, Tetrahedron, 1970, 26, 955). (Scheme 8).

Thiophenol under the conditions of the Reimer-Tiemann reaction yields phenyl orthothioformate (HC(SPh)$_3$, a substance which has uses in enabling a number of interesting nucleophilic reactions to be effected.

The Friedel-Crafts alkylation as opposed to acylation with aromatic thiols has been discussed (G. Maccagnani and G. Mazzanti, *loc. cit.* p. 434).

Scheme 8

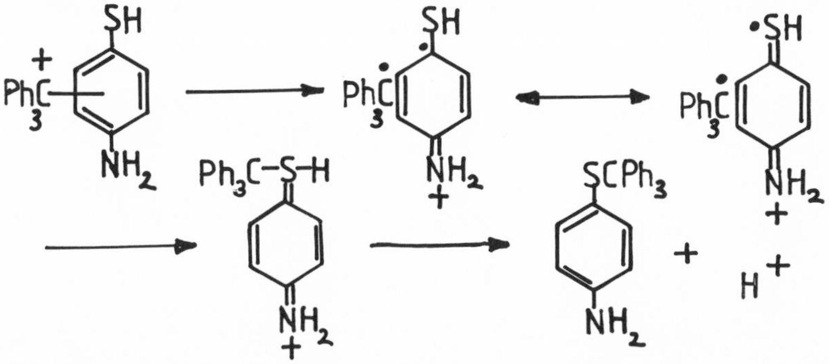

(3) *Rearrangements*

Thiols in the form of derivatives in contrast to phenols undergo a prolific number of rearrangements some examples of which are described in this section. Rearrangements involving thioethers are discussed subsequently.

A thermal rearrangement involving migration of the trimethylsilyl (and trimethylgermanyl) group from sulphur to aromatic carbon has been described (A.R. Bassindale and D.R.M. Walton, J. organometall., Chem., 1970, 25, 389).

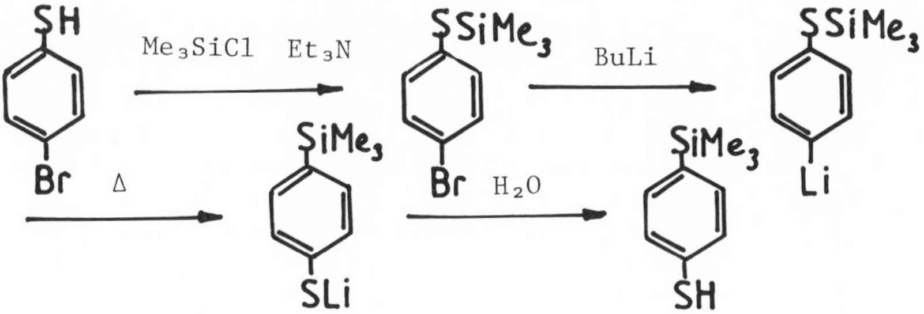

Intramolecular migration of an alkyl group occurs in the following reaction of methyl 2-mercapto benzoate with base or with a primary amine RNH_2 to give the amide. This is reminiscent of the dealkylation of ethers by thiol salts referred to in Chapter 4. It has been explained through an intramolecular S_Ni mechanism involving a six-membered transition state (J.C. Grivas and K.C. Navada, J. org. Chem., 1971, 36, 1520). (Scheme 9).

The Claisen rearrangement in the thiophenol series has been shown to occur. It is catalysed by nucleophiles and it is suggested that nucleophilic attack at the allylic carbon triggers a concerted process (H. Kwart and J.L. Schwartz, J. org. Chem., 1974, **39**, 1575).

Scheme 9

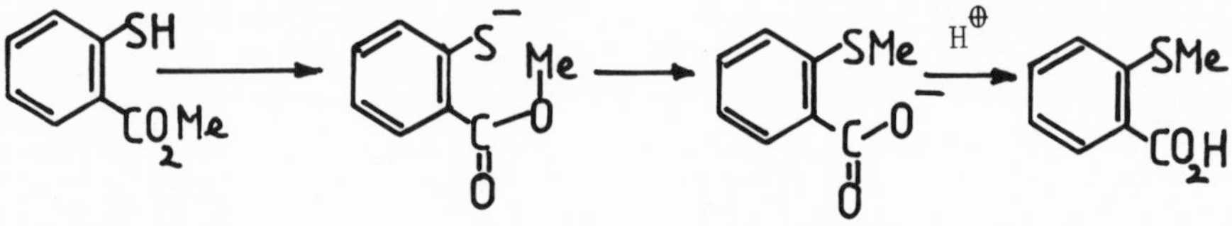

Aryl migration occurs in the Smiles rearrangement (W.M. Truce, E.M. Kreider and W.W. Brand, Org. Reactions, 1970, **18**, 99) in which the thiol group can be involved either as the nucleophile in the conversion of mercaptodiarylethers to hydroxyaryldisulphides or as the displaced group, as shown in the following equations respectively.

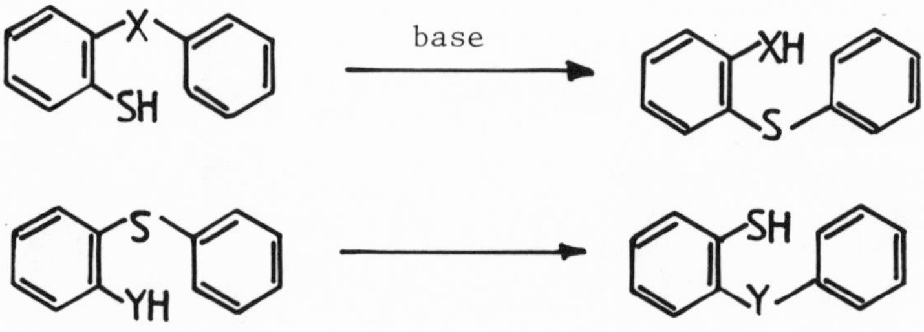

X=O, Y=NHAc

Reactions of the second type afford an easy synthetic route to phenothiazines of pharmaceutical interest as illustrated in the following example which also shows the course of the reaction sequence,

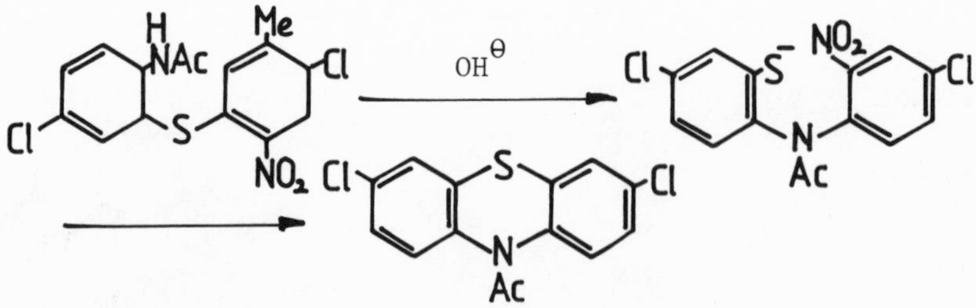

The rearrangement of 2-amino-2'-nitro-5'-methyldiphenyl sulphide at 190° or in dimethylacetamide gives the phenothiazine shown, contrary to previous reports, and a dibenzothiophenone (F.A. Davis and R.B. Wetzel, Tetrahedron Letters, 1969, 4483).

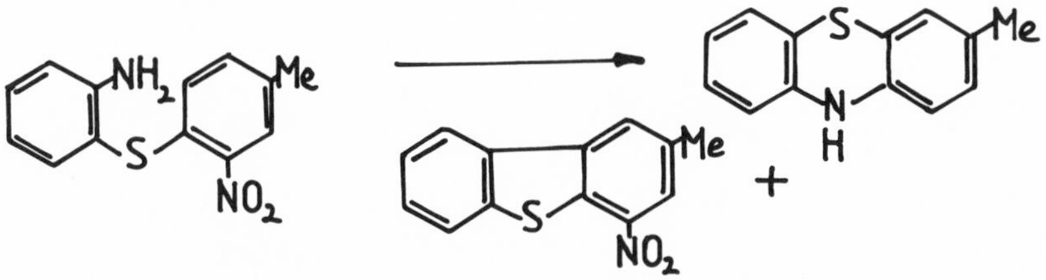

The Smiles rearrangement has been effected photochemically (K. Matsui, *et.al.*, Tetrahedron Letters, 1970, 1467).

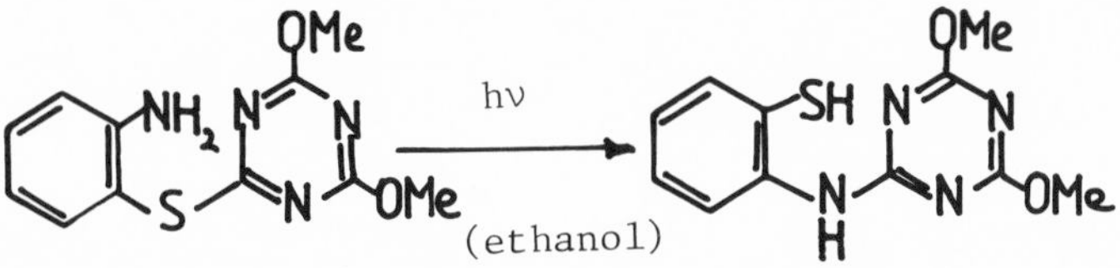

Rearrangements of the Newmann or Schönberg type involving conversion of *O*-thioacyl to *S*-acyl compounds have already been referred to in the section concerned with the synthesis of thiols. An intramolecular four-membered cyclic mechanism appears to be involved in the thermal rearrangement of thione-carbamates to thiolcarbamates from the effect of substituents, kinetic results and experiments with a mixture of two thione-carbamates (K. Miyazaki, Tetrahedron Letters, 1968, 2793).

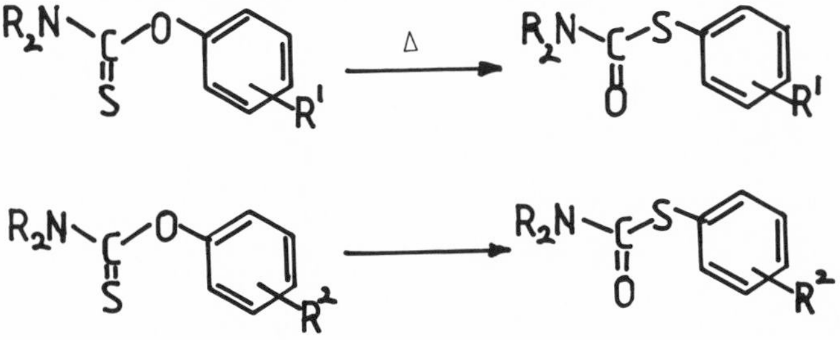

Rate enhancement due to the steric influence of hindered rotation in *o*-substituted compounds is found and ascribed to the increased nucleophilicity of the polarised form shown.

Electron donating groups R, in the order shown, also promoted the rate.

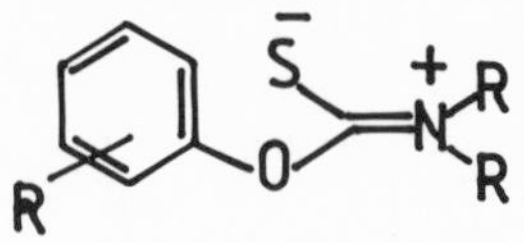

$$R = Bu^i > Pr^n > Et > Me$$

Synthesis of sulphonic acids is achieved by making use of the rearrangement products, followed by appropriate steps, (J.E. Cooper and J.M. Paul, J. org. Chem., 1970, 35, 2046).

Diaryl xanthates rearrange by way of a four membered cyclic transition state to diaryldithiocarbonates and a similar pathway accounts for the conversion of arylthiobenzoates into arylthiolbenzoates (Y. Araki, Bull. chem. Soc. Japan, 1970, 43, 252; Y. Araki and A. Kaji, *ibid.*, 1970, 43, 3214).

Certain substituted 2-alkylbenzene thiols undergo ready cyclisation or in certain cases exist in tautomeric equilibrium with cyclic products as shown in the following examples. In the first example (H. Kwart and J.G. George, Chem. Comm., 1970, 433) the propargylic thiol is converted into the benzomethylthiophen by hot quinoline.

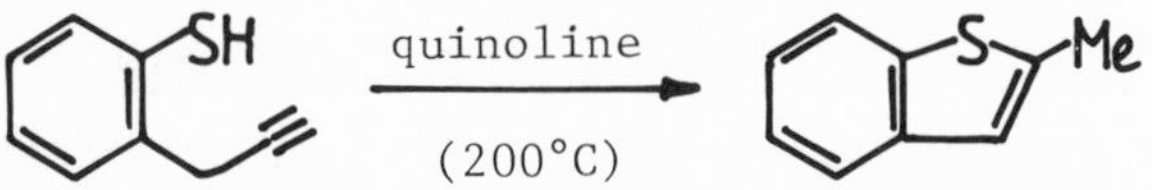

o-Cyanomethylbenzenethiol (o-mercaptobenzylcyanide) has been shown to exist only in a cyclic form (G.W. Stacy, F.W. Villaescusa and T.E. Wollner, J. org. Chem., 1965, 30, 4074).

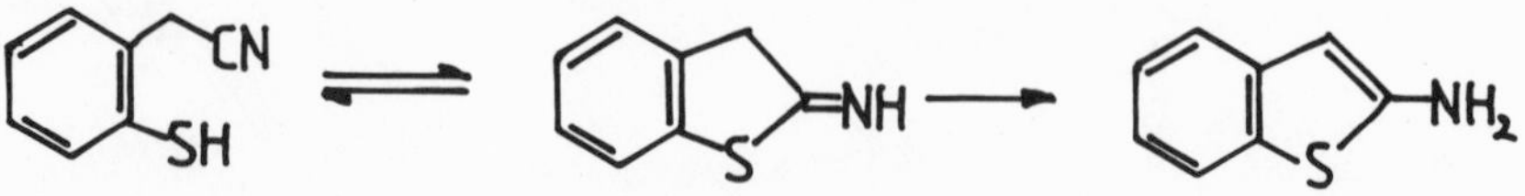

In the naphthalenic compound shown the imino form is stable because of the instability of the amino compound (G.W. Stacy, *et.al.*, *ibid.* 1964, 29, 607).

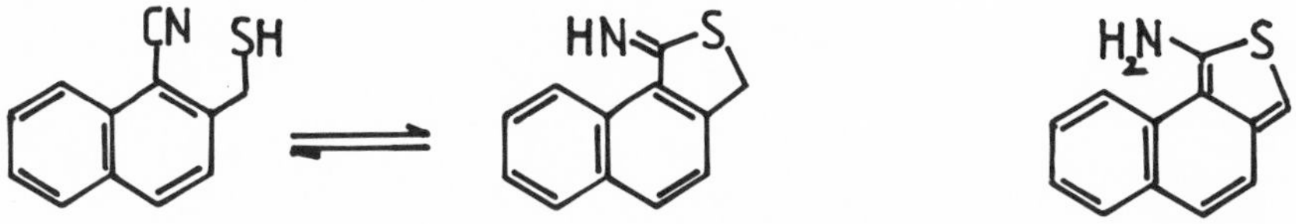

In the case of the six-memberd ring compounds shown below the amino compound (X=CO₂Et,CN) predominates.

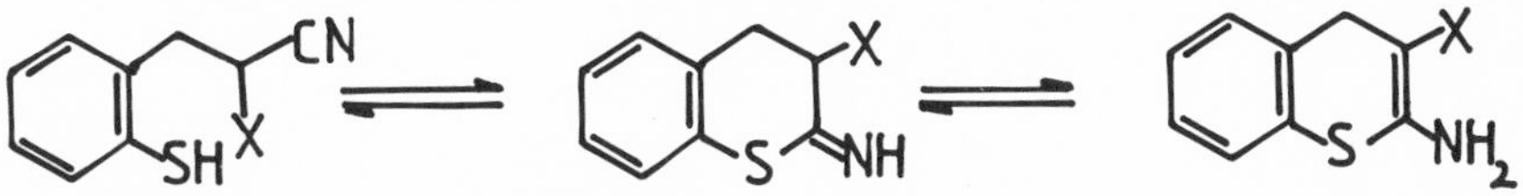

With X=H, the existence of the imino/open chain form could not be proved conclusively (G.W. Stacy, D.L. Eck and T.E. Wollner, J. org. Chem., 1970, 35, 3495). The above examples are without parallel in phenol chemistry and demonstrate the nucleophilic capability of the -SH group, a subject discussed at length in an ensuing section.

Photochemical rearrangement of the *o*-vinyl phenol shown, α-(*o*-hydroxybenzylidene)-γ-butyrothiolactone, takes place to give 3-(2-mercaptoethyl)coumarin, by attack of the phenolic OH upon the activated C=O group (H. Zimmer, *et.al.*, Tetrahedron Letters, 1968, 5435).

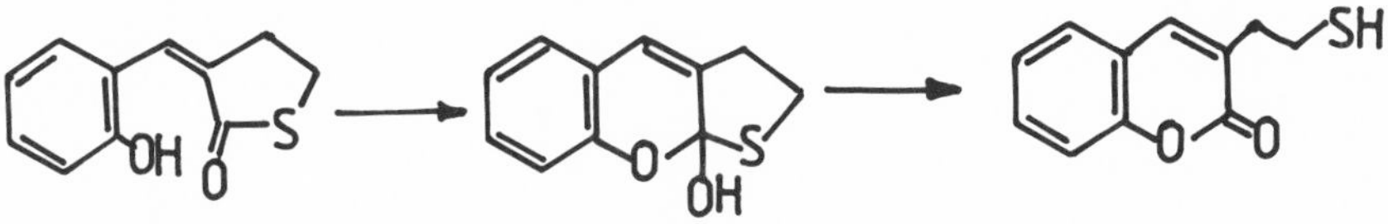

Numerous examples of rearrangements involving thiols related to the penicillins have been described (S.Wolfe *et.al.*, Canad. J. Chem., 1968, 46, 2549), and suggest to the author

that in the benzenoid series the following might occur to give
a final structure capable of quinone methide type reactions
and other synthetic uses.

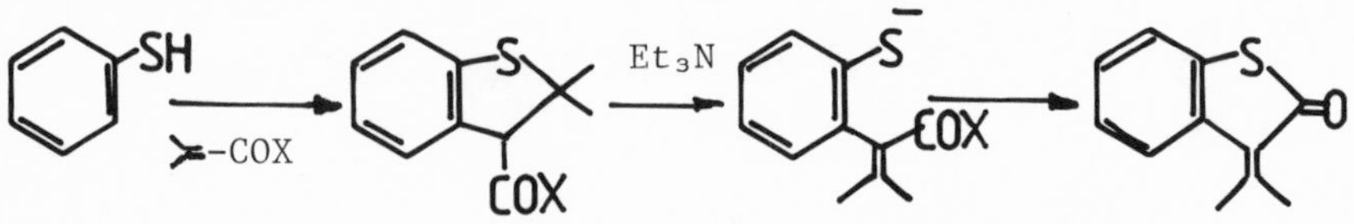

The thiol ester group is a source of a few rearrangements.
Although the conventional Fries rearrangement of thiol esters
is unsuccessful (D.S. Tarbell and A.H. Herz, J. Amer. chem.
Soc., 1953, 53, 1668) it has been effected photochemically
(E.L. Loveridge, B.R. Beck and J.S. Bradshaw, J. org. Chem.,
1971, 36, 221) to give the products shown, albeit in low yield.

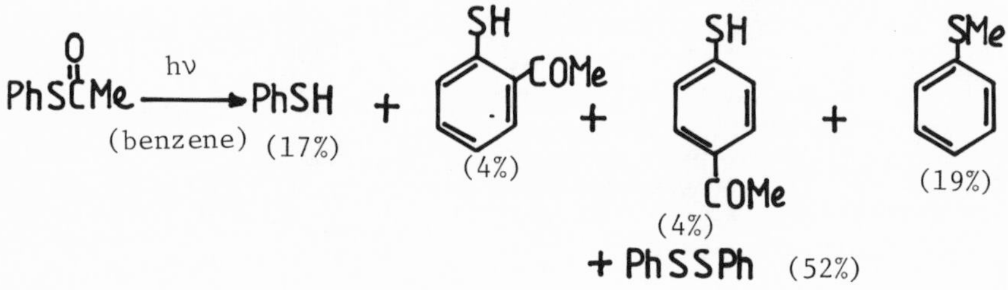

The proportion of thiophenol increases with reaction time.
The initial reaction is the cleavage of the S-CO bond followed
by hydrogen abstraction and other processes. In the reported
photolysis of the p-tolyl analogue in cyclohexane (J.R.Brunwell
Chem. Comm., 1969, 1437) the formation of the thiol was not
mentioned.

In the acidolytic cleavage of the epoxide, phenyl α-methyl-
trans-thiolglycidate, by means of boron trifluoride etherate
the enol tautomer (presumably stabilised by hydrogen bonding)
is obtained, the whole thiol ester group having migrated,
(J. Wenple, J. Amer. chem. Soc., 1970, 92, 6694; S.P. Singh
and J. Kagan, *ibid.*, 1969, 91, 6198).(Scheme 10).

(4) *General reactions and catalytic uses*

(Scheme 10)

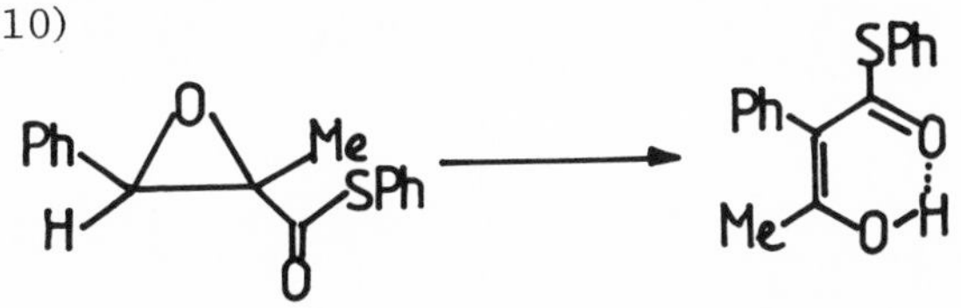

In the reaction of diazonium cations, $XC_6H_4N_2^{\oplus}$, with thiophenoxide, the kinetics indicate that the *syn*-diazothioether is formed rapidly and is followed by *syn-anti* isomerisation. With *p*-nitro and *p*-cyanobenzene diazonium chlorides the two stages can be distinguished. In the case of benzene diazonium salts, and the *p*-Me and *p*-OMe compounds first-order kinetics are observed with thiophenoxide. The rate-determining step in these cases is considered to be probably the slow formation of the *syn*-diazothioether followed by rapid isomerisation to the *anti* compound (C.D. Ritchie and P.O.I. Virtanen, J. Amer. chem. Soc., 1972, 94, 1589).

Unlike phenols, thiophenols have a number of general uses, apart from their special application in organic synthesis, as in certain organic chemical procedures, shown in the following examples.

p-Methoxydiazonium fluoroborate treated in water/pentane with thiophenol yields anisole (T.Shono,Y. Matsumara and K. Tsubata, Chem. Letters, 1979, 9, 1051).

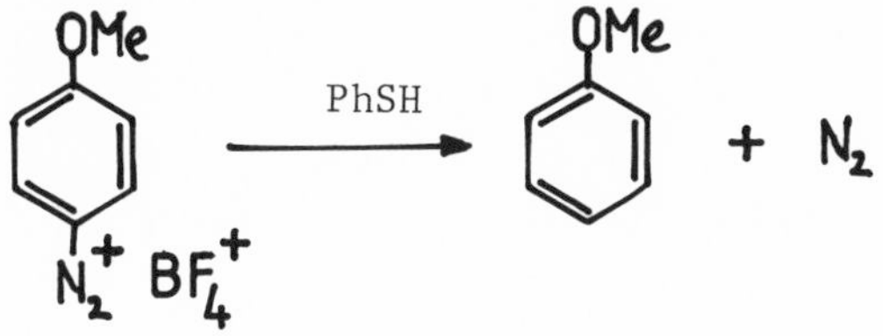

The isomerisation of *cis* to *trans*-alkenes can be effected with thiophenol (C.A. Hendrick *et. al.*, J. org. Chem., 1975, 40, 1). Thiophenol in conjunction with azo bis-isobutyronitrile effects an enrichment of reaction mixtures in the less abundant *trans*-isomers. (U.T. Bhalerao and H. Rapoport, J. Amer. chem. Soc., 1971, 93, 4835).

In the sigmatropic rearrangement of allylic sulphoxides to sulphenates (P.A. Grieco, Chem. Comm., 1972, 702; D.J. Abbott and C.J.M. Stirling, J. chem. Soc., (C), 1969, 818), sodium thiophenoxide can be recovered unchanged.

Although desulphurisation of thiols can be effected class-
ically with Raney nickel, photoreduction occurs smoothly in
triethyl phosphite (G.C. Barrett, *loc. cit.*, p. 16).

$$PhSH \ + \ (EtO)_3P \longrightarrow PhH \ + \ (EtO)_3P=S$$

Displacement of the -SH group by NH_2 in aza-heterocyclic systems can be effected with formamidine acetate.

(5) *Nucleophilic reactions*

In this section three main groups of reaction are discussed, substitution, dealkylation and addition reactions. Examples within the scope of carbon chemistry have been considered
and the extensive reactions of the compounds of transition and
other elements, excluded. Excellent coverage of these aspects
has been given (M.E. Peach, 'Thiols as Nucleophiles' in 'The
Chemistry of the Thiol Group' ed. S. Patai, Interscience, New
York, 1974, ch. 16, p. 721). Aspects of substitution reactions
have been described in the preparative section, addition reactions are much involved in the synthetic uses of thiols, discussed at length in a subsequent section, and in the present
context the interest lies in obtaining a sulphur-containing
product. Dealkylation reactions have a subsidiary value in
the preparation of thiophenolic ethers.

Many unusual nucleophilic substitution reactions without
parallel in the oxygen series or by comparison with nucleophilic attack on carbon have been listed involving formation
of special compounds which may have some synthetic uses
(G.C. Barrett, 'Thiols' in 'Organic Sulphur Compounds', *loc.*
cit.,) Aromatic sulphenyl halides are discussed under
sulphides.

The preparation of salts of thiols is often a necessary
step to their usage in nucleophilic reactions and in organic
synthesis. The alkali metal salts being ionic can be prepared in a variety of ways and in the case of aqueous procedures the anhydrous salt can be obtained by azeotropic removal
of water (H.E. Jones, Chem. Abs., 1970, 72, 54727; 1972, 77,
74822). With other solvents, salts and solvated salts can be
isolated (C.R. Lucas and M.E. Peach, Inorg. nucl. Chem. Lett.,
1969, 5, 73; Canad. J. Chem., 1970, 48, 1869).

In the case of metals of other groups, the salts are prepared either from the metal halide (MX) and the appropriate
thiol in the presence of a hydrogen halide acceptor such as
triethylamine,

$$MX + PhSH \xrightarrow[Et_3N]{} PhSM + Et_3NH^{\oplus}X^{\ominus}$$

or by interaction of the metal halide and a lead thiolate, as shown

$$2MX + Pb(SPh)_2 \longrightarrow 2PhSM + PbX_2$$

Nucleophilic reactivity has been quantified by N_+ values and in the reactions of the various nucleophilies listed with p-nitromalachite green the following relative N_+ values (solvent in brackets) were found for MeOH (MeOH) 0.5; $MeO^{\ominus}$ (MeOH), 7.5; $N_3^{\ominus}$ (MeOH), 8.5; $PhS^{\ominus}$ (MeOH), 10.7; $PhS^{\ominus}$ (DMSO), 13.1; $CN^{\ominus}$ (DMSO), 8.6, where $N_+ = \log |K_n/K_{H_2O}|$ in which K_n = the rate constant for reaction of a cation with a specific nucleophilic system and K_{H_2O} = the rate constant for reaction of the same cation with water in water (C.D. Ritchie, Acc. Chem. Res., 1972, 5, 348; C.D. Ritchie and P.O.I. Virtanen, J. Amer. chem. Soc., 1972, 94, 4966).

Steric effects have been found of importance as in the reaction of RC_6H_4SH (R=H, 2-t-Bu, 4-t-Bu) additively with N-ethylmaleimide or in the displacement of 2,4-dinitrothiophenoxide from ethyl 2,4-dinitrophenyldisulphide (D.S. Garwood and D.C. Garwood, Tetrahedron Letters, 1970, 4959; J. org. Chem., 1972, 37, 3804).

(a) *Alicyclic and acyclic examples*

Facile displacement of the methane sulphonate group by thiophenoxide occurs in the bis-methanesulphonates of 3-arylthiopropane-1,2-diols, 2-arylthiopropane-1,3-diols and 1-arylthiopropane-2-ols (M.S. Khan and L.N. Owen, J. chem. Soc., Perkin I Trans., 1972, 2060; 1972, 2067).

In nucleophilic displacement at saturated carbon in p-substituted benzyl chlorides with thiophenoxide and analogous oxygen nucleophiles a chlorine kinetic isotope effect (K_{35}/K_{37}) has been observed, the reactions proceeding by a concerted transition state (E.P. Grimsrud and J.W. Taylor, J. Amer. chem. Soc., 1970, 92, 739).

$$R^1S^{\ominus} + ArCH_2Cl \longrightarrow \left[R^1S...ArCH...Cl\right] \longrightarrow R^1CH_2Ar + Cl^{\ominus}$$

With change of the p-substituent from electron donating to

withdrawing, the relative magnitude of bond breaking and making alters in the transition state and with thiophenoxide and
with methoxide, K_{35}/K_{37} increases following the series $p-NO_2<$
H<p-OMe indicating greater bond breakage accompanying increasing electron withdrawal by the p-substituent. For both oxygen and sulphur nucleophile the isotope effect decreases with
increase in basicity $^{\Theta}OMe>^{\Theta}OPh$ and $^{\Theta}SBu^n>^{\Theta}SPh$, indicating less
bond cleavage in the transition state with the stronger base.
The reaction is slower with oxygen nucleophiles (possibly due
to solvation) and the isotope effect is smaller, suggesting
that bond breaking is smaller and the oxygen nucleophile is
stronger.

The reactions of 1,1-diaryl-2,2,2-trichloro and 1,1-
diaryl-2,2-dichloroethane with thiophenoxide have been studied and the Hammett equation applied, two types of reaction,
one elimination and the other S_N2 substitution being observed.
(D.J. McLennan and R.J. Wong, J. chem. Soc., Perkin II, 1972,
279).

$(p-XC_6H_4)_2CHCHCl_2 + 2PhS^{\Theta} \longrightarrow (pXC_6H_4)_2C=CHSPh + 2Cl^{\Theta} + PhSH.$

Heterocyclic derivatives can serve as nucleophiles as in the
reaction of thiophenoxide with the hydrochloride of 3-chloro-
methylpyridazine (K. Yu Novitskii et $al.$, Chem. Abs., 1970,
$\underline{36}$, 399). In 2,3,3-trichloro-N-acetylpiperidine only the
2-chlorine appears to be displaced with various nucleophiles
(H. Boelme and H. Dehmel, Arch. Pharm., 1971, $\underline{304}$, 403).

However in the reaction of 3-chloro-2-hydroxytetrahydropyran,
the 3-chlorine atom and the 2-hydroxy group are replaced
giving $trans$-3-hydroxy-2-phenylthiotetrahydropyran (G.Descotes
and D. Sinou, Bull. soc. Chim. Fr., 1971, 4116).

In the reaction of 1-(β-(phenylsulphonyl)ethyl)-piperidine
hydrochloride (or methiodide) with aromatic thiols in aqueous
dioxan,C-N bond cleavage apparently occurs by way of an addit-
ion/elimination mechanism(R. Andusano et $al.$, Chem. Abs., 1970

<u>73</u>, 25259). (Scheme 11).

(Scheme 11)

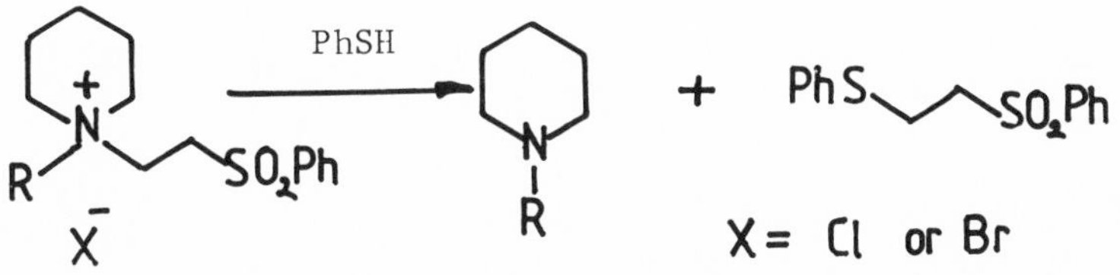

$$X = Cl \text{ or } Br$$

Thiophenoxide with 3-chlorothietane gives a mixture containing phenyl 3-thietanylsulphide and the independently synthesised disulphide shown probably by way of the cations indicated (B. Arbuzov and O.N. Nuretidinova, Izv. Akad. Nauk SSSR, Ser. Khim., 1971, 2594).

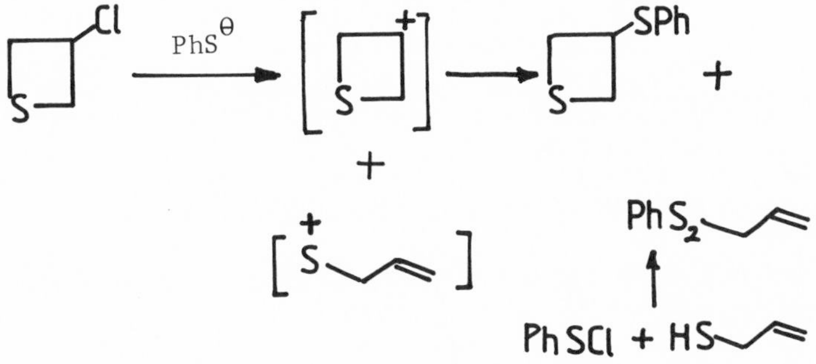

Many reactions of thiophenoxide with halogenated alkynes and alkenes have been described.

In the reaction shown, rate constants for which were measured, attempts to trap the ion $pZC_6H_4 C \equiv C^{\ominus}$ were unsuccessful (P. Beltrame $et.$ $al.$, Gazz. Chim. Ital., 1972, $\underline{102}$, 164).

$$p\text{-}ZC_6H_4\text{-}C \equiv C\text{-}X + p\text{-}MeC_6H_4SNa \xrightarrow{DMF} p\text{-}ZC_6H_4\overset{\ominus}{C}=C(X)SC_6H_4Me \ \overset{+}{Na}.$$

The reaction of $EtS^{\ominus}$ is accompanied by an attack on carbon also (M.C. Verploegh $et.$ $al.$, Rec. Trav. Chim., 1971, $\underline{90}$, 765). The mechanism is not certain but the preceding authors favour an addition/elimination mechanism with formation $p\text{-}ZC_6H_4C \equiv C^{\ominus}$ $(X)SC_6H_4Me$ and fast elimination of $X^{\ominus}$.

The reaction of thiophenoxide with propargyl halides results in a complex mixture in which substitution , addition and reducing action is observed (M. Verny, Bull. Soc. Chim. Fr.,

284

1970, 1942).

$$PhS^{\ominus} + HC{\equiv}CCMeXCO_2Et \longrightarrow HC{\equiv}CMeC(SPh)CO_2Et$$
$$H_2C{=}C{=}CMeCO_2Et$$
$$H_2C{=}C(SPh)CHMeCO_2Et$$
$$MeC(SPh){=}CMeCO_2Et$$

The mixed dihalo compounds shown react at the saturated carbon, although iodo compounds in the presence of potassium hydroxide tend to react at the unsaturated centre to give reduction products (F. Kai and S. Seki, Chem. Abs., 1970, <u>72</u>, 31696).

$$XC{\equiv}C{-}CH_2X^1 + ArS^{\ominus} \longrightarrow XC{\equiv}CCH_2SAr$$

The substitution reactions of unsaturated compounds with thiols have been reviewed (G. Modena, Acc. chem. Res., 1971, <u>4</u>, 73; Z. Rapoport, Advan. phys. org. Chem., 1969, <u>7</u>, 1).

1-Chlorocyclohex-1-ene with sodium thiophenoxide, reacts as shown (P. Caubère and J.-J Brunet, Bull. Soc. Chim. Fr., 1970, 2418).

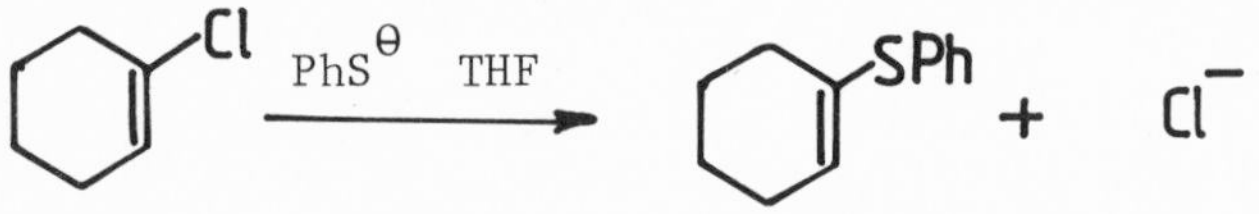

Generally for substituted alkenes the configuration is retained as shown in the reaction of thiophenol (J. Biougne and F. Theron, Compt. rend., 1971, <u>272C</u>, 858; G. Capozzi, G. Melloni and G. Modena, J. chem. Soc.,(C), 1970, 2625).

$$R^1(R^2S)C{=}CR^1(OSO_2R^2){+}PhSH \longrightarrow R^1(R^2S)C{=}CR^1(SPh){+}R^2SO_3H$$

With a thiolate however, mixtures of isomers are formed, and a ketonic product results, (Scheme 12)

In other cases the *trans* alkene gives the *trans* product but the *cis* isomer gives mixtures (predominantly *cis*), (G.Marchese and F. Naso, Chem. and Ind., (Milan), 1971, <u>53</u>, 760).

$$PhSO_2CH{=}CHF + PhS^{\ominus} \longrightarrow PhSO_2CH{=}CHSPh + F^{\ominus}$$

With mixed halogeno alkenes, replacement of fluorine occurs preferentially.

(Scheme 12)

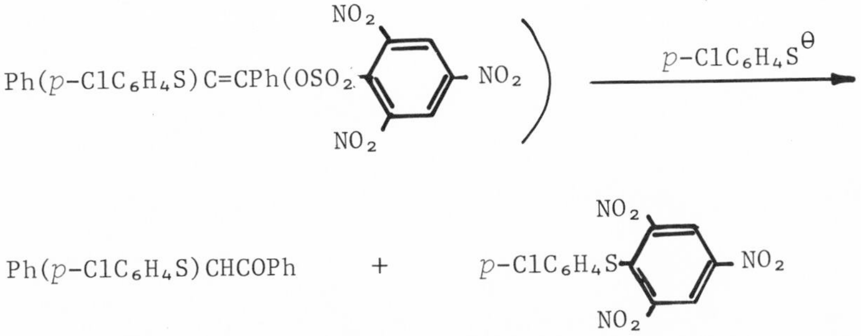

$$Ph(p\text{-}ClC_6H_4S)C=CPh(OSO_2\text{-}\underset{NO_2}{\underset{NO_2}{\overset{NO_2}{\bigcirc}}}\text{-}NO_2) \xrightarrow{\quad p\text{-}ClC_6H_4S^{\ominus}\quad}$$

$$Ph(p\text{-}ClC_6H_4S)CHCOPh \quad + \quad p\text{-}ClC_6H_4S\text{-}\underset{NO_2}{\underset{NO_2}{\overset{NO_2}{\bigcirc}}}\text{-}NO_2$$

$$CF_2 = CFCl + PhS^{\ominus} \longrightarrow PhSCF=CFCl + F^{\ominus}$$

The butane thiolate ion can react similarly but like $EtS^{\ominus}$ mentioned previously will add across the double bond as well as substitute a chlorine atom. 3-Chloro and 3-bromometh- acrylonitrile (*cis* and *trans*) undergo nucleophilic substitution at C3 with more than 95% retention of configuration. (D.V. Gardner and D.E. McGreer, Canad. J. Chem., 1970, <u>48</u>, 2104),

$$Cl_2C=C(Cl)CN + PhS^{\ominus} \longrightarrow (PhS)_2C=C(Cl)CN$$
$$+$$
$$(PhS)_2C=C(\overset{+}{S}Ph)CN(trace)$$

2,3,3-Trichloroacrylonitriles of the type shown react pre- dominantly at the 3- position and this has been attributed, amongst other factors, to the stabilisation of the intermed- iate ion (R.L. Soulen *et. al.*, J. org. Chem., 1971, <u>36</u>, 3386).

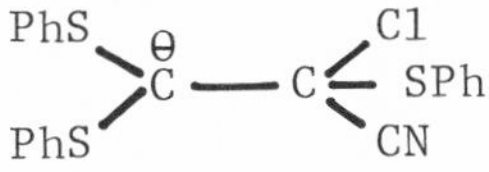

$$\underset{PhS}{\overset{PhS}{>}}\overset{\ominus}{C}\text{---}C\overset{Cl}{\underset{CN}{<}}SPh$$

Thiophenoxide tends to result in conventional substitution reactions in contrast to ethane thiolate which can in the sa- me compound cause reduction as illustrated (M. Verny, Bull. Soc. Chem. Fr., 1970, 1946).

$$Me(SEt)C=CBrCO_2Et \xrightarrow[PhS^\ominus]{} Me(SEt)C=C(SPh)CO_2Et$$

$$EtS^\ominus / EtSH$$

$$Me(SEt)C=CHCO_2Et + (EtS)_2 + Br^\ominus.$$

Unlike propargyl halides, certain allylic halides with chlorine at the unsaturated terminal position, such as the hexachlorofulvene shown react with p-toluenethiolate in the presence of triethylamine at the vinylic position (E.T. McBee, *et al.*, J. org. Chem., 1972, <u>37</u>, 1100).

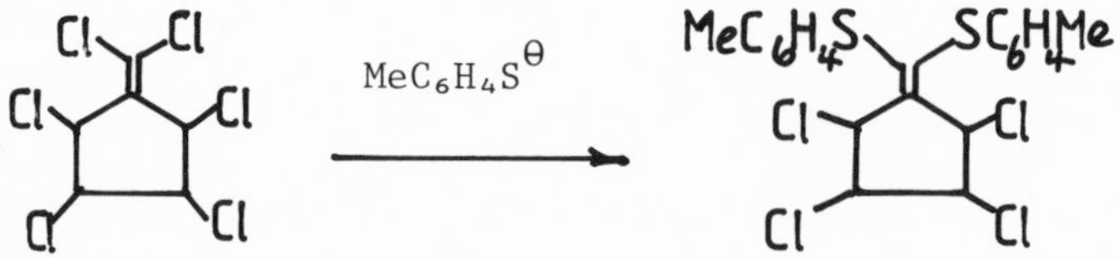

(b) *Aromatic and heterocyclic examples:*

Aromatic nucleophilic substitution with the thiolate anion has been mentioned in the preparative section with particular reference to thiophenoxide. The subject of sulphur reagents in this form of substitution reaction has been reviewed (G. Bartoli *et. al.*, Int. J. sulphur Chem., C, 1971, <u>6</u>, 77). General reviews have appeared: J.R. Beck, Tetrahedron, 1978, <u>34</u>, 2057 and J. Miller, 'Aromatic Nucleophilic Substitution', Elsevier, Amsterdam and New York, 1968. Surprisingly, although a great deal of both preparative and kinetic work has been carried out with oxygen nucleophiles, thiolate reactions have received much less attention from the quantitative aspect. In the following series with 1-X-2,4-dinitrobenzene, where X is the leaving group, it seems probable that the relative rate of reaction with thiolate would be $F>NO_2>OSO_2C_6H_4Me(p)>$ $SOPh\backsim Br\backsim Cl>SO_2Ph\backsim OC_6H_4NO_2(p)>I$. The reaction with nitro compounds is considered to involve an addition/elimination mechanism with the formation of the Meisenheimer complex as the rate-determining step. Such complexes have been reviewed (M.R. Crampton, Adv. phys. org. Chem., 1969, <u>7</u>, 211; M.J. Strauss, Chem. Rev., 1970, <u>70</u>, 667) and detailed aspects relating to the thiophenoxide and the ethanethiolate ions discussed (M.R. Crampton, J. chem. Soc. (B), 1971, 2112;

M.R. Crampton and M. El Ghariani, *ibid.* 1971, 1043).

Solvents effects are also important. A kinetic investigation of the displacement of halide ion from *o*- and *p*-bromo and *o*- and *p*-fluoronitrobenzene by thiophenoxide has shown that a crown ether considerably accelerates reaction in *t*-butanol but not in methanol. Nucleophile ion-pairing is considered to occur only in the former solvent (G. Guanti *et. al.*, J. chem. Soc., Perkin II, 1975, 389). Anion interactions with other species are also of some significance. Thiophenoxide and phenoxide ions interact strongly with micelles of cetyl trimethylammonium bromide which also markedly catalyses reaction of these ions with 1-fluoro-2,4-dinitrobenzene (C.A. Bunton *et. al.*, Tetrahedron, 1975, 31, 1139).

In the case of polyhalogenated compounds, the effects of orientation and reactivity have been studied in detail (J. Burdon, Tetrahedron, 1965, 21, 3373) and a detailed account of recent work has been given (M.E. Peach, in the 'Chemistry of the Thiol Group', ed. S. Patai, Interscience, New York, 1974, ch. 16, p. 737). Most activating groups (e.g. NO_2) result in *p*-substitution but nevertheless some *o*-substitution can occur and the presence of deactivating groups (e.g. NH_2, $O^\ominus$, $S^\ominus$) causes some *m*-substitution (J. Miller, *loc. cit.* The stability of Wheland type intermediates has been reviewed (W. Pritzkow, Z. Chem., 1970, 10, 330). The dielectric constant of the solvent is crucial in determining the relative amounts of *o*- and *p*-substitution. The discrete formation of the nucleophile as an anion appears to be the operative factor.

Extensive work on the interaction of hexafluorobenzene with thiolates has been carried out, With thiophenoxide (K.R. Langille and M.E. Peach, J. fluorine Chem., 1971, 1, 407) in ethylene glycol and/or pyridine the products were C_6F_5SPh, $C_6F_4(SPh)_2$ and $C_6F_2(SPh)_4$, the latter being formed by way of the trifluoro compound. In the reaction of pentafluoropyridine with thiophenoxide, substitution occurred *p*- to the nitrogen (R.N. Hazeldine *et. al.*, J. chem. Soc. (C), 1969, 1660).

The reaction of mixed hexahalobenzenes with copper thiolates can result in normal substitution or in certain cases

reduction has been observed dependent on the solvent
(J.C. Tatlow *et al.*, J. chem. Soc., Perkin I, 1972, 763). Dehalogenation occurs in the cases of 1-halo-2-naphthol, 2-halo-
1-naphthol and 1-halo-4-naphthol with thiophenoxide (V. Calo,
et al., Gazz. Chim. Ital., 1971, 101, 685).

Interaction of 3,4,5,6-tetrafluorophthalonitrile with
thiophenoxide in water as solvent resulted in the tetrasubsti-
tuted compound whereas in methanol the disubstituted compound
predominated. Replacement of the 4,5 fluorine atoms may be
attributable to a more stable *p*- than *o*-intermediate at the
first substitution stage (J.M. Birchall, R.N. Hazeldine and
J.O. Morley, J. chem. Soc., (C), 1970, 456).

With a variety of thiolates (including thiophenoxide)
fluorobenzene or bromobenzene in the presence of sodamide
afford the sulphides most efficiently in a solvent comprising
hexamethylphosphoric triamide/tetrahydrofuran (1:5)
(P. Caubère and M.-F. Hochu, Bull. Soc. Chim. Fr., 1969, 2854).
Pyrrolidine has also been found valuable as a solvent for
certain dihaloaromatic compounds (Monsanto Co., Chem. Abs.,
1970, 73, 109489).

The greater reactivity of thiophenol compared with its
anion towards bromopyridine is considered to be due to protonation of the bromopyridine (R.A. Abramovitch, F. Helmer and
M. Livetts, J. org. Chem., 1969, 34, 1730).

Although the displacement of halide by thiolate has been
most widely studied probably because of the ready availability
of such compounds, other leaving groups have been examined.
In the compounds p-$XC_6H_4SO_2CF_3$, the rate constants for the replacement of X by thiophenoxide in methanol showed the following order, X = SO_2CF_3>NO_2>F>Cl (S.M. Shein and K.V. Solodova,
Zh. org. Khim., 1970, 6, 1465).

The replacement of nitro in a furano compound by thiolate
has been observed in several instances where the corresponding
halogen derivative is stable (F. Lieb and K. Eiter, Ann. Chem.
, 1972, 761, 130).

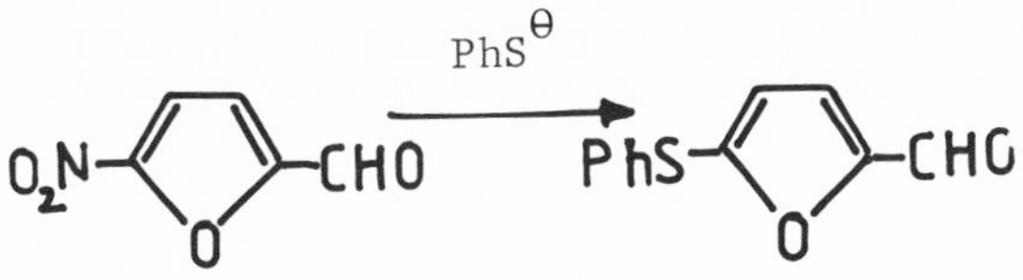

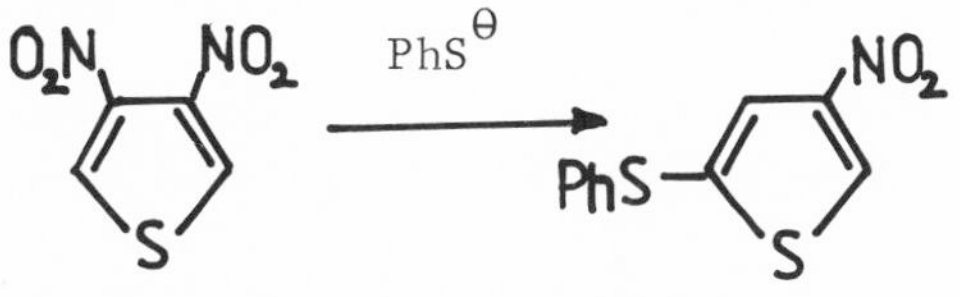

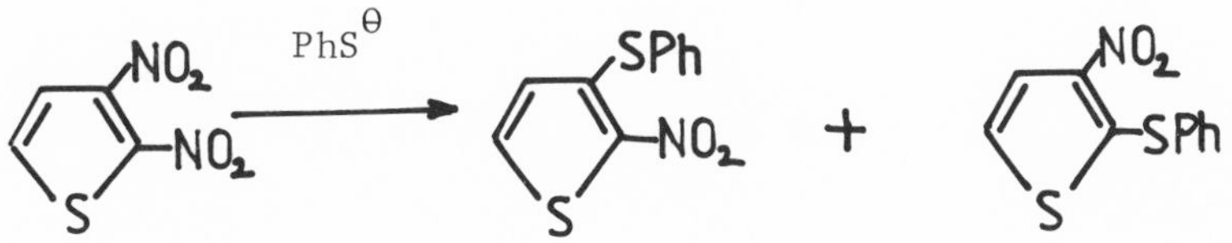

3,4-Dinitrothiophene undergoes replacement of one nitro group with simultaneous rearrangement (C. Dell'Erba, D. Spinelli and G. Leandri, Gazz. Chim. Ital., 1969, 99, 535) while in 2,3-dinitrothiophene either one but not both of the nitro groups can be replaced (C. Dell'Erba and G. Guanti,*ibid.*, 1970, 100, 223).   These results contrast with the reaction of 2,3-dibromo-5-nitrothiophene, in which only the 2-bromo is displaced with thiophenoxide (D. Spinelli, G. Consiglio and C.A. Corrao, Tetrahedron Letters, 1972, 4021).

(R=H or Me)

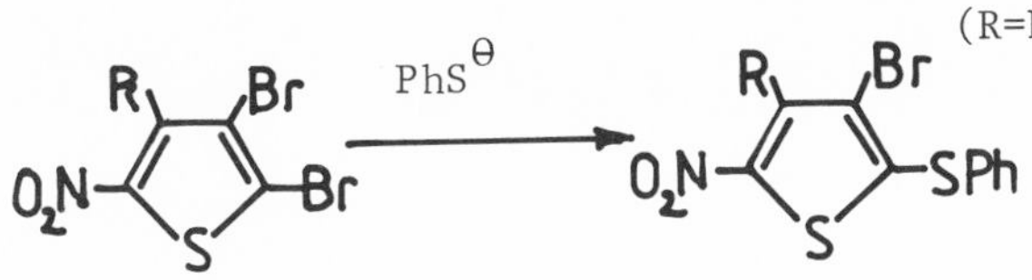

(c) *Dealkylation and deacylation reactions*

Demethylation with salts of ethanethiol or ethanedithiol in aprotic solvents is well known.   Sodium thiophenoxide can effect a number of dealkylations on a preparative basis.   Selective removal of a methyl group from a quaternary salt can be achieved (M. Mori, Chem. Abs., 1971, 75, 140376).

$$(Et)_3\overset{\oplus}{N}MeCl \xrightarrow{\ PhS\overset{\ominus}{\ }\ } Et_3N + PhSMe$$

A similar reaction with a diquaternary salt is shown below
(S. Hayashi *et. al.*, Chem. pharm. Bull., Tokyo, 1972, 20, 15).

$$Et_2\overset{\oplus}{Me}N(CH_2)_2\overset{\oplus}{N}MeEt_2 \quad 2I^{\ominus} \xrightarrow{\ PhSH\ } Et_2N(CH_2)_2NEt_2 + MeSPh$$

Laudanisine methochloride is selectively *N*-demethylated by
thiophenoxide in methyl ethyl ketone (M. Shamma, N.C. Deno and
J.F. Reman, Tetrahedron Letters, 1966, 1375).

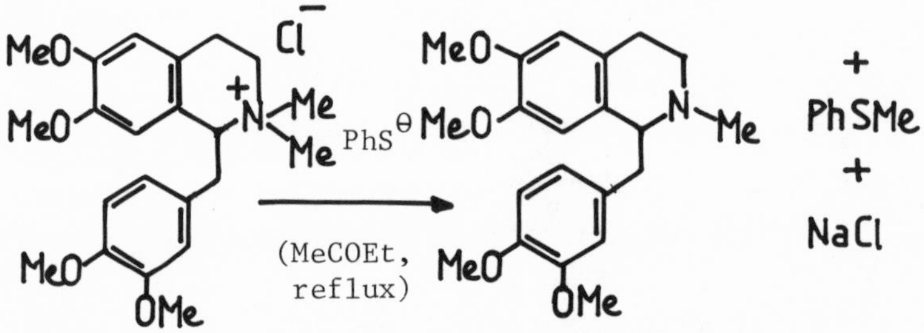

The alkoxy tri(dimethylamine)phosphonium chloride shown is, by
contrast, debenzylated with thiophenoxide (B. Castro and
C. Selve, Bull. Soc. Chim. Fr., 1971, 2296).

$$PhCH_2O\overset{\oplus}{P}(NMe_2)_3Cl^{\ominus} \xrightarrow[Et_3N,\ 60°C]{\ PhSH\ } OP(NMe_2)_3 + PhCH_2SPh + Et_3\overset{\oplus}{N}HCl^{\ominus}$$

In a similar way methylene ethers, methyl ethers and methyl
esters can be demethylated with thioethoxide (G.I. Feutrill
and R.N. Mirrington, Austral. J. Chem., 1972, 25, 1719, 1731),

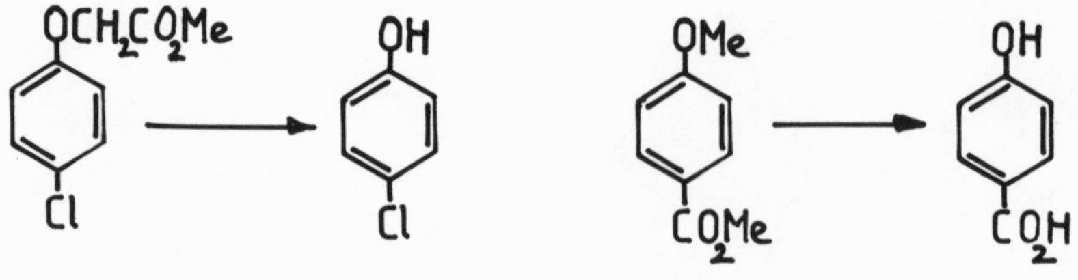

Thiophenoxides, surprisingly, can cause aryl oxygen bond

fission in competition with attack at the carbonyl group in the reaction with some nitrophenyl acetates (C. Dell'Erba *et. al.*, Chem. Comm., 1975, 823).

Dealkylation of the 2-alkoxy-1-methylbenzimidazoles can be achieved with thiophenol (P. Dembech, *et. al.*, J. chem. Soc. (B), 1971, 2299).

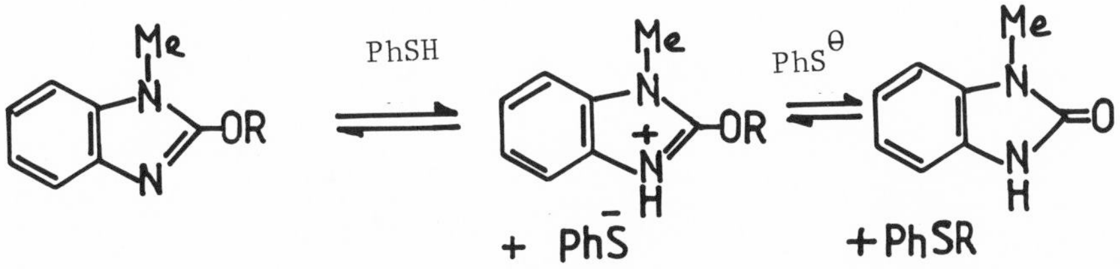

This is one aspect of the general reaction Het-OMe+ArSH⟶ Het-OH+ArSMe. Thiophenoxide unlike butanethiolate does not appear to behave as a reducing agent in addition to its de-alkylation function (M.F. Shostakovskii *et.al.*, Ivr, Akad. Nauk, SSSR, Ser. Khim., 1972, 90; and is therefore a mild reagent.

A selective deacylation function is shown by thiophenol in its reaction with acetic formic anhydride in a synthesis of arylthioformates whereby the formyl group reacts preferent-ially as shown (P.C. Bax and W. Stevens, Rec. Trav. Chim., 1970, 89, 265).

HCOOCOMe $\xrightarrow{\text{PhSH}}$ HCOSPh (90%) + MeCOSPh (7%)

(d) *Addition Processes*

(i) *Alkenes and alkynes*

A great variety of addition reactions to unsaturated cen-tres occurs with thiophenol and thiophenoxide in which sulphur is incorporated into the molecule.

Most addition reactions are believed to proceed by a rad-ical mechanism and these reactions have been reviewed (K. Griesbaum, Angew. Chem. internat., Edn., 1970, 9, 273; P.I. Abell, in 'Free Radicals', ed. J.C. Kochi, Wiley-Inter-science, 1973, vol. II, p. 63). The addition of thiols to alkenes, alkynes, carbonyl compounds, conjugated systems, and alkylene oxides are considered in this section.

The mechanism for the addition of thiols to alkenes, in the absence of bases, but in the presence of traces of peroxides is free radical in type giving sulphides as anti-Markovnikov adducts. By the use of very carefully purified reagents in the presence of acid, Markovnikov addition takes place (E.E. Reid, 'Organic Chemistry of Divalent Sulphur', Chem. Publishing Co. Inc., New York, 1969, vol. 2, p. 13), as shown in the following example,

$$Me_2C=CHMe \xrightarrow{RSH} Me_2C(SR)CH_2Me$$

The anti-Markovnikov addition is shown in the addition of substituted thiophenols to derivatives of phenyl vinyl sulphone, (P. De Maria and A. Fini, J. chem. Soc., (B), 1971, 2335; P. De Maria and M. Falzone, Chim. Ind., (Milan), 1972, 54, 791).

$$XC_6H_4SH + YC_6H_4SO_2CH = CH_2 \longrightarrow YC_6H_4SO_2CH_2CH_2SC_6H_4X$$

p-Isopropenylphenol with aliphatic activated thiols in chloroform solution containing p-toluenesulphonic acid gave a normal Markovnikov product while with thiophenol and p-chlorothiophenol, Friedel-Crafts type alkylation occurred (F. Wolf and H. Finke, Z. Chem., 1972, 12, 180).

The progressive addition of thiols to alkynes occurs, leading by way of the vinylic sulphide to a saturated thioether. In the addition of p-toluenethiol derivatives in methanolic solution to negatively substituted alkynes, $HC\equiv CY$ (Y = -CN, $-SO_2C_6H_4Me-p$, $-C_6H_4NO_2-p$, $-CO_2Me$, $-CONH_2$, $-COMe$) the degree of *trans* selectivity depends on the nature of Y and decreases when Y can delocalise the incipient negative charge (W.E. Truce and G.J.W. Tichenor, J. org. Chem., 1972, 37, 2391). (Scheme 13)

With the propyne derivatives shown Markovnikov addition of thiophenol takes place. The initial product is not isolated

(Scheme 13)

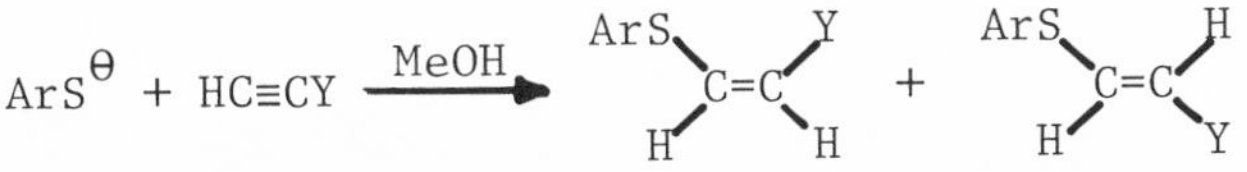

$$ArS^{\ominus} + HC\equiv CY \xrightarrow{\text{MeOH}} \begin{matrix} ArS \\ H \end{matrix} C=C \begin{matrix} Y \\ H \end{matrix} + \begin{matrix} ArS \\ H \end{matrix} C=C \begin{matrix} H \\ Y \end{matrix}$$

but isomerisation and dealkylation occurs.

$$PhS^{\ominus} + HC \equiv CCH_2\overset{\oplus}{S}Me_2 \longrightarrow CH_2 = (PhS)CCH_2\overset{\oplus}{S}Me_2$$

$$Me(PhS)C = CH\overset{\oplus}{S}Me_2 \longrightarrow Me(PhS)C=CHSMe$$

Subsequent to addition, with appropriate systems, cyclisation of adducts can occur as illustrated for the synthesis of a thiochromone carboxylic acid (R. Hazard and J. King, Chem. Abs., 1973, 78, 43276).

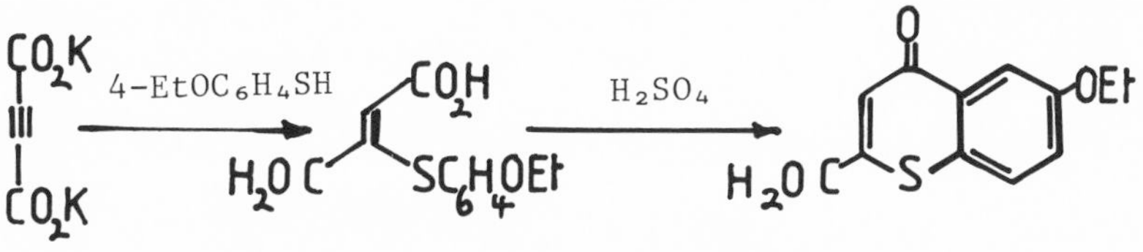

o-Aminobenzenethiol adds to dimethyl acetylenedicarboxylate followed by cyclisation and photochemical isomerisation of the product to give the benzothiazine shown (Y. Maki and M. Suzuki, Chem. Pharm. Bull., 1972, 20, 832).

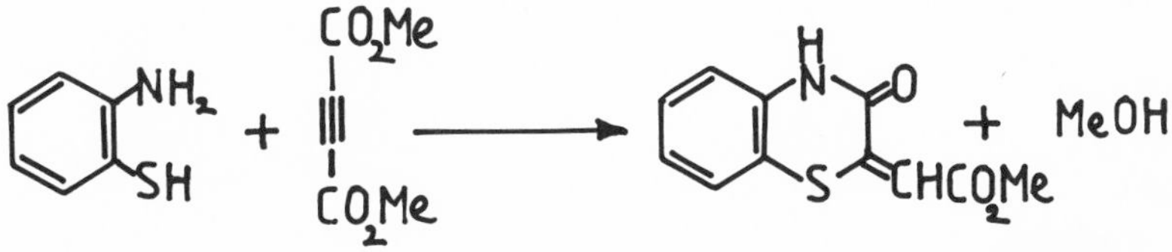

Azomethines such as benzylidene o-nitroaniline can undergo addition with p-toluenethiol (R. Marshall and D.M. Smith, J. chem. Soc. (C)., 1971, 3510).   In other compounds both C=N bond fission and reduction of the halo group in one of the fragments occurs. (M.W. Barber, S.C. Lauderdale and J.R. West, J. org. Chem., 1972, 37, 3555).

$$Ph_2C = C = NC_6H_4Br \ (p) \xrightarrow{PhSH} Ph_2C = C(SPh)_2 + PhNH_2HBr$$

In cyclic systems such as 3-benzyl-2-phenylthiazolinium bromide addition occurs to the C=N double bond as shown (A.D. Clark and P. Sykes, J. chem. Soc., (C), 1971, 103).

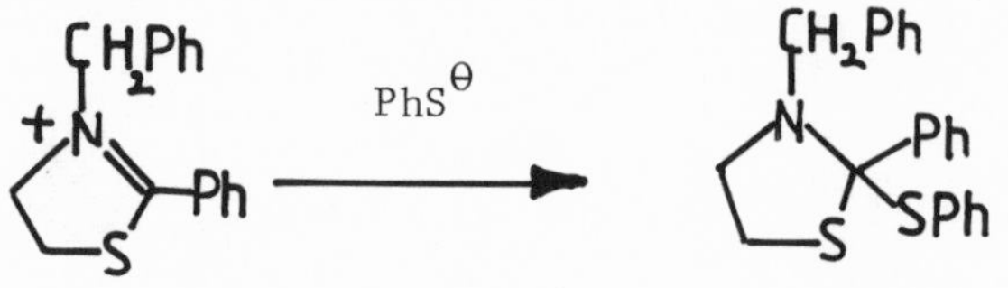

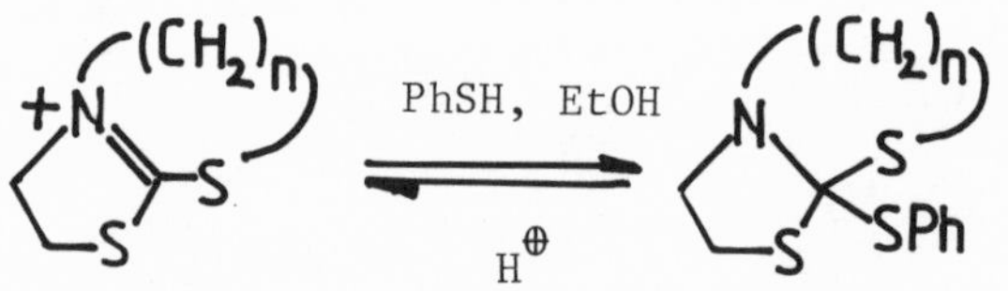

(ii) *Carbonyl compounds*

Studies have been made of hemithioacetal formation in the system thiophenol-acetaldehyde (R.E. Barnet and W. P. Jencks. J. Amer. chem. Soc., 1969, 91, 6758). The addition reaction can lead to an azathiobicyclic compound as shown (J. Szabo *et. al.*, Chem. Abs., 1972, 76, 153687).

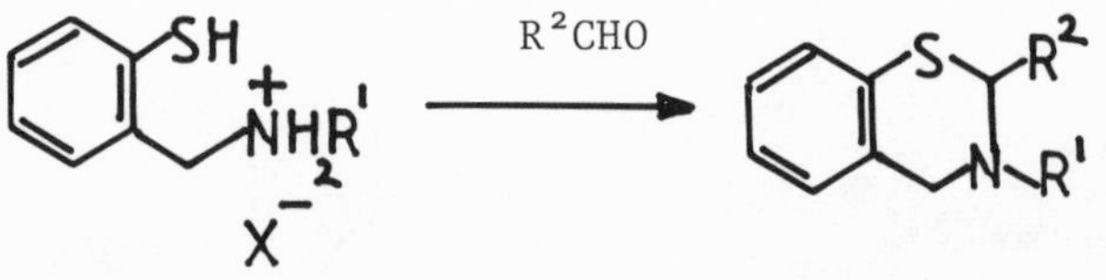

Formaldehyde, aromatic thiols and secondary amines can interact as illustrated (A.M. Kuliev, A.K. Kyazim-Zade and K.Z. Guseinov, Dokl. Akad. Nauk. Azerb SSSR, 1972, 27, 20).

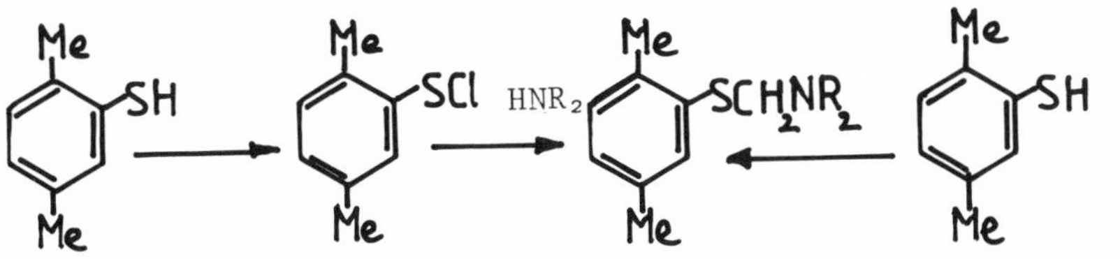

Unusual cyclic dithia-compounds can also be derived by way of, presumably, hemidithioacetal formation (H. Tokunaga, T. Kawashima and N. Inamoto, Bull. chem. Soc., Japan, 1972, 45, 2220).

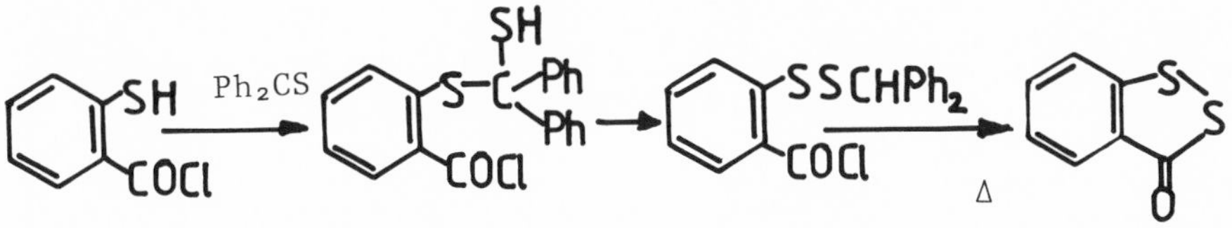

(iii) *Conjugated systems*

In such systems both 1,2 and 1,4 additions have been observed. Thiols add principally to the C=C bond in the C=C-C=O system of quinones and certain α,β-unsaturated lactones (W.S. Powell and R.A. Heacock, Experientia, 1972, 28, 124; S.M. Kupchan *et. al.*, J. org. Chem., 1970, 35, 3539), the additions occurring in neutral or alkaline solution by way of the thiolate anion. However in other compounds, 1,4-addition occurs as in the example shown (J.K. Sugden, J.E. Hogan and N.J. Van Abbe, J. chem. Soc., (C), 1971, 3875).

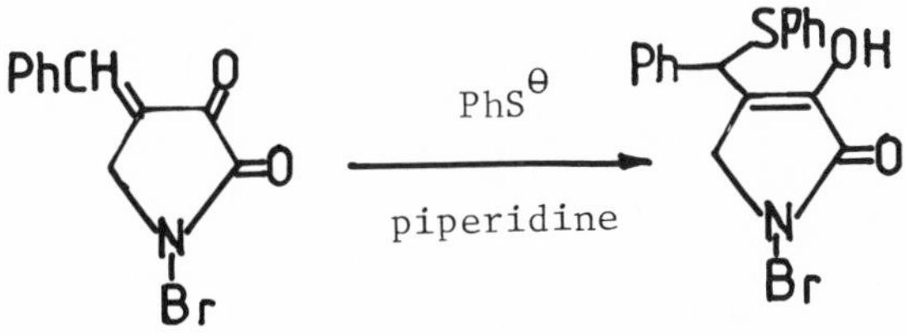

The addition of thiophenols to N-ethylmaleimide in ethanolic solution (95%) at pH 5-7 has been examined. The mechanism shown was considered to be operative (D. Semenow-Garwood J. org. Chem., 1972, 37, 3797) as the involvement in the addition process of ArSH could not be detected.

$$ArSH \xrightarrow{OH^\ominus} ArS^\ominus + H_2O$$

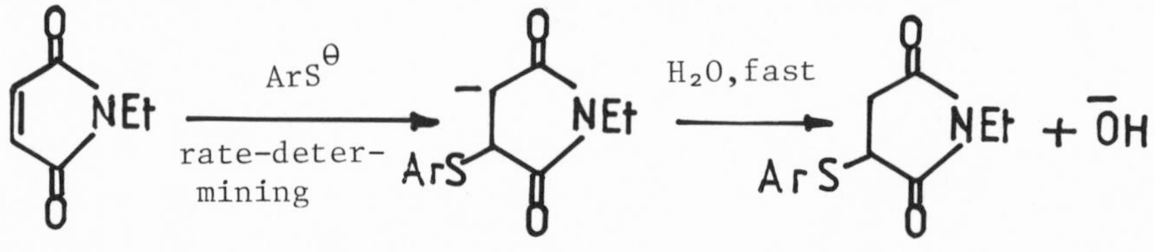

Under the conditions described however a more realistic interpretation could be the existence of the equilibrium,

$$ArSH + H_2O \rightleftharpoons H_3O^\oplus + ArS^\ominus$$

Steric effects are observed with o-substituted benzenethiols, leading to inhibition of solvation of the thiolate with consequent increase of nucleophilicity resulting in a rate acceleration together with a rate retardation on account of interference between the thiolate ion and the double bond in the transition state. o-t-Butyl thiophenol is more reactive than other derivatives due to the steric acceleration.

In the case of the C=C-C=N system addition to the C=C bond occurs as illustrated, ethanol favouring the formation of the axial sulphides while with tetrahydrofuran some equatorial sulphide is also obtained. (R.A. Abramovitch et. al., J. org. Chem., 1972, 37, 3577).

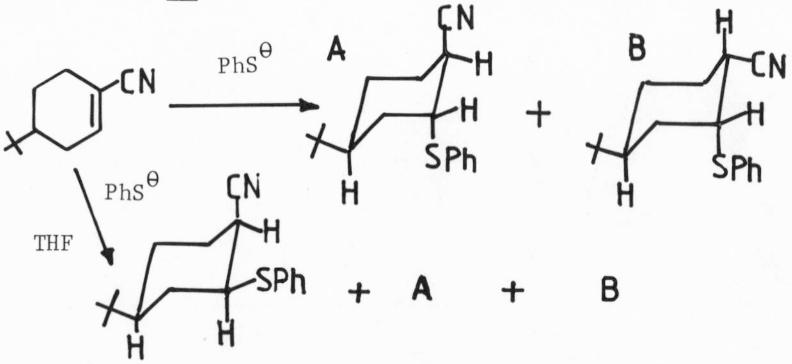

With C=C-N=C however the type pf product depends on the structure of the reactant and the thiol (K. Bürger, G. George and J. Fehn, Ann. Chem., 1972, _757_, 1).

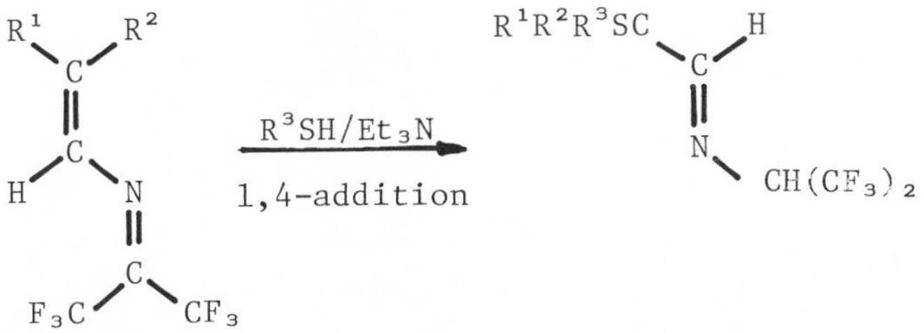

R^1	R^2	R^3
Me	Me	Ph
$CH(CF_3)_2$	i-Pr	H	Ph
	i-Pr	H	t-Bu

Addition of thiophenol resulted, when R^1 = Ph, R^2 = H, in the formation of the enamine shown while with ethanethiol the mode of 1,4-addition was reversed.

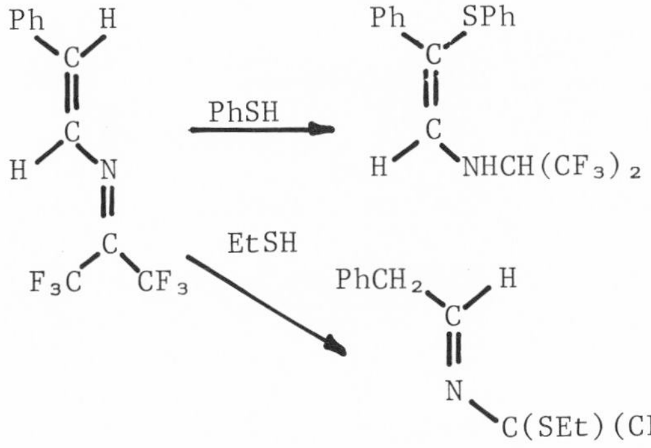

(or the geometrical isomer)

In the reaction of thiophenol with the cyclic system shown containing C=C-C=N, reversible reductive ring opening takes place while aliphatic thiols do not react (P. Sykes and H. Ullah, J. chem. Soc., Perkin I, 1972, 2305).

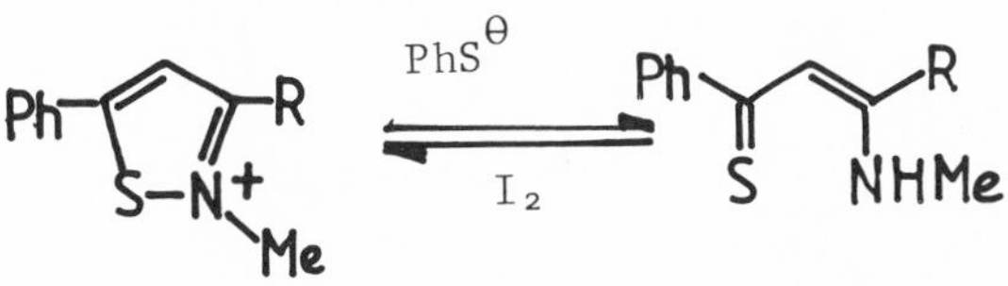

(R=H,Me)

The quinone, 4,7-benzimidazoledione, reacts with aromatic thiols at the C=C double bond to give the hydroquinone and the

C=C-C=N system takes no part in the reaction. The hydroquinone structure was not formed when certain aliphatic thiols
were used but instead the quinone resulted (L.C. March and
M.M. Joullié, J. heterocycl. Chem., 1970, 7, 249).

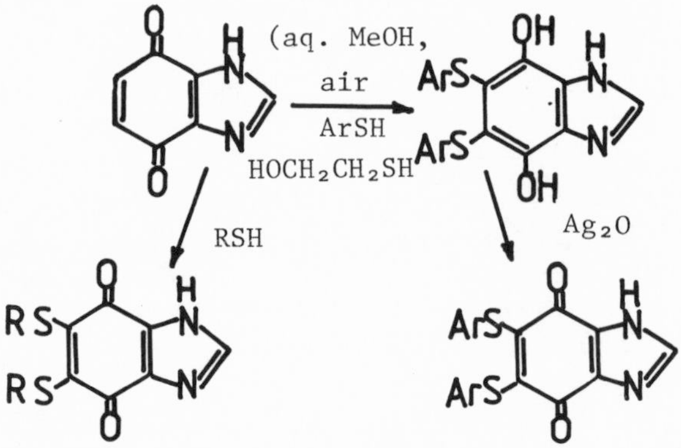

(iv) *Alkylene oxides*

   Alkylene oxides, R being a donating group form, under basic
or neutral conditions with thiols, by ring cleavage yield
mainly the product produced by attack at the least substituted carbon (A.M. Kuliev, *et al.*, Dokl. Akad. Nauk. Azerb.
SSSR, 1971, 27, 24).

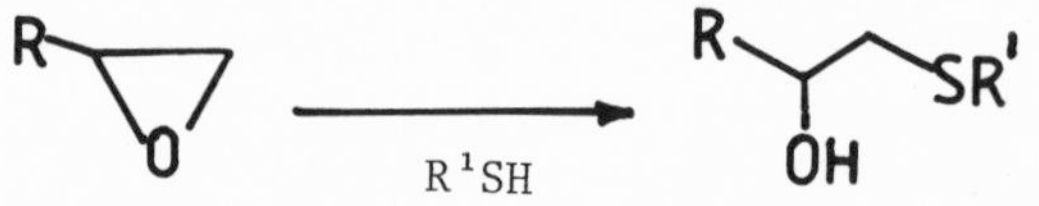

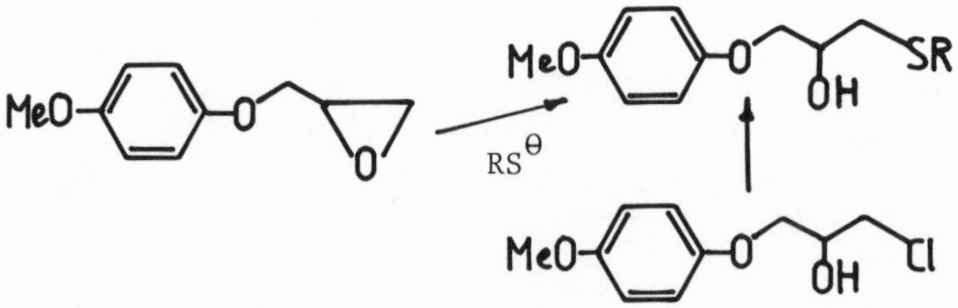

   Both the C-O and C-C ring bond can however be cleaved dependent on the nature of X and Y (S. Ukai *et. al.*, Chem. Abs.,
1972, 77, 34207).

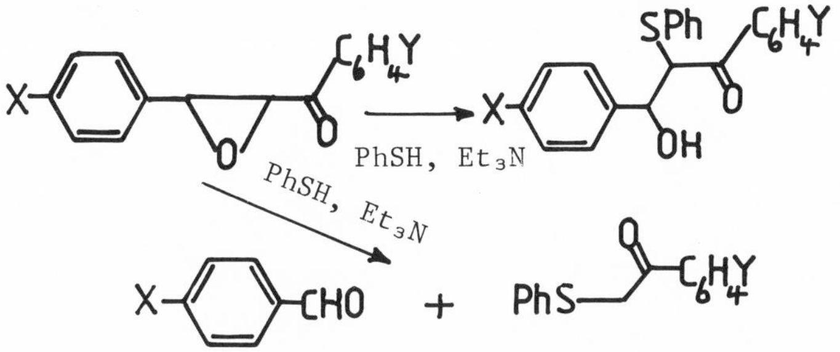

X—⟨C₆H₄⟩—CH(epoxide)CH—C₆H₄Y → PhSH, Et₃N → X—⟨C₆H₄⟩—CH(SPh)...

PhSH, Et₃N

X—⟨C₆H₄⟩—CHO + PhS—CO—C₆H₄Y

In compounds containing aliphatic chlorine and an epoxide ring
the former reacts preferentially with thiolate (M.G. Voronkov
and Z.I. Mikhailov, Zh. Obshch. Khim., 1972, 42, 615) and the
epoxide ring is opened by thiol in the presence of thiolates
in preference to addition to a double bond (G.I. Zaitseva and
V.M. Albitskaya, Zh. org. Khim., 1969, 5, 612).

(6) *Synthetic uses of aromatic thiols*

Important advances have been made in the use of thiols, as
reagents by their incorporation into the molecule in order to
effect the desired synthetic process followed by their removal
having fulfilled their useful function.   The use of thiols,
as a blocking group, in sulphur extrusion reactions and formation of sulphur ylids is now basic to organic synthesis.
Benzyl mercaptan has an advantage over aliphatic thiols in
giving crystallisable products.   Thiophenols (and also selen-
ophenols) have only been extensively exploited in the last
half decade.   Organic syntheses based on the use of the
functional SPh and Ph₃PO groups have been reviewed (S. Warren,
Acc. chem. Res., 1978, 11, 401).   A wide variety of different
classes of organic compound can be synthesised as will be seen
by the following examples in which both thiophenols and selenophenol are referred to.

(a) *Unsaturated hydrocarbons*

A regiospecific diene synthesis has been devised based on
the use of 2,4-dinitrophenylsulphenyl chloride (H.J. Reich,
I.L. Reich and S. Wollowitz, J. Amer. chem. Soc., 1978, 100,
5981).

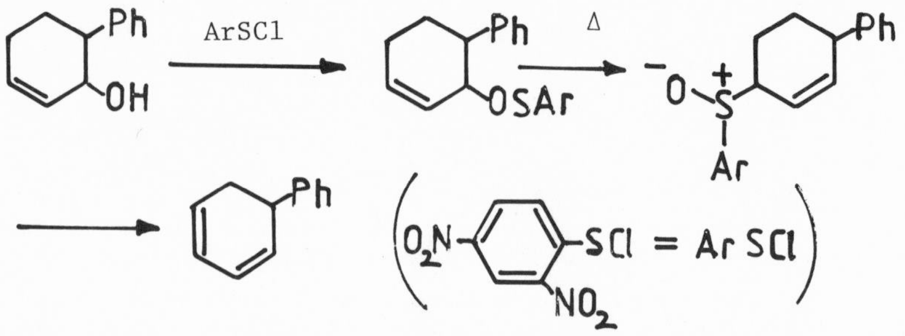

An electrochemical synthesis of cyclopropanes from α,β-unsaturated carbonyl compounds in good yield (69%) is based on using the thiophenyl group (T. Shono *et. al.*, Tetrahedron Letters, 1978, 1205).

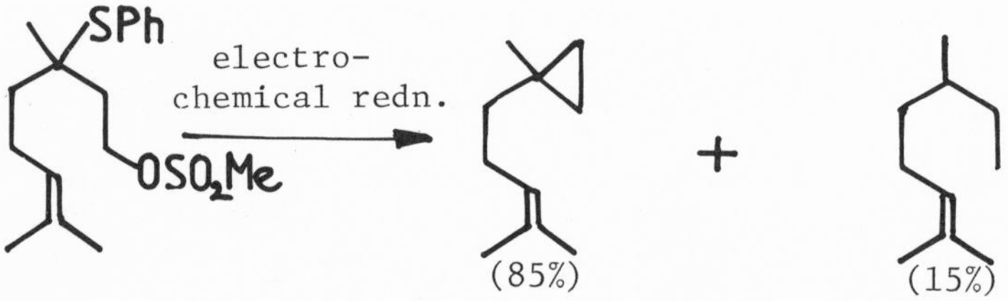

An example of the use of thiophenol to generate an α-sulphenyl carbanion, from farnesyl phenyl sulphide, for the synthesis of squalene is shown (J.E. Biellmann and J.B. Ducep, Tetrahedron Letters, 1969, 3707). (Scheme 14).

12-Phenylthiosqualene is obtained in a synthesis of squalene (W.D. Ollis *et. al.*, Chem. Comm., 1969, 99) which proceeds by way of the Stevens rearrangement, in which instead of the methylsulphonium salt, which cannot be formed from trimethyloxonium tetrafluoro borate, the phenylsulphonium salt is used. (Scheme 15).

The application of the Stevens rearrangement to other isoprenoid compounds has been referred to (R.K. Olsen and J.O. Currie, in 'The Chemistry of the Thiol Group', ed. S. Patai, Interscience, 1974, ch. 12, p. 563). The rearrange-

Scheme 14

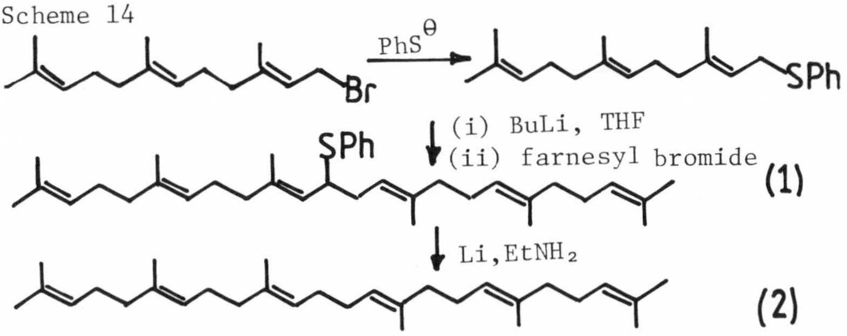

$$\xrightarrow{\text{PhS}^{\ominus}}$$

(i) BuLi, THF
(ii) farnesyl bromide **(1)**

Li, EtNH₂

(2)

Scheme 15

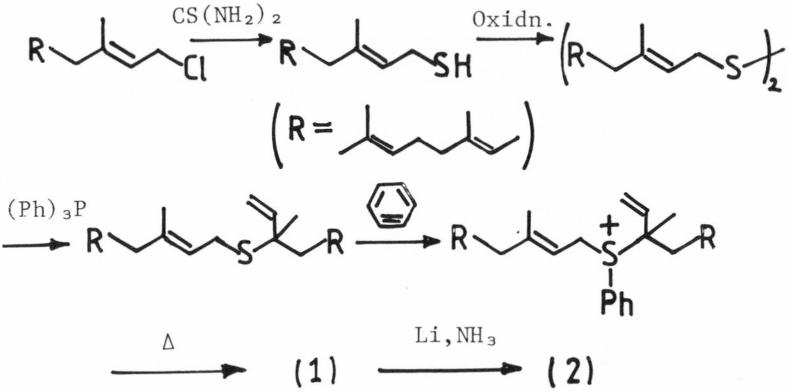

$$R \xrightarrow{\text{CS(NH}_2\text{)}_2} R \xrightarrow{\text{Oxidn.}} \left(R\text{—S}\right)_2$$

$$\left(R = \underset{}{\text{/\\/\\/}} \right)$$

(Ph)₃P

Δ ⟶ **(1)**   Li, NH₃ ⟶ **(2)**

ment of allylic sulphonium salts usually proceeds in excellent
yield although the thiophenyl compound $CH_2 = CHCH_2 \overset{\oplus}{S} (Ph)_2$
gives a number of products (B.M. Trost and R. La Rochelle,
Tetrahedron Letters, 1968, 3327) and this may explain why
methyl sulphonium salts have proved more attractive.   The
Stevens rearrangement of allylic sulphonium salts typified by
the example below and the extensive work by Baldwin and co-
workers and others, has been described (R.K. Olsen and
J.O. Currie, *loc. cit.*)

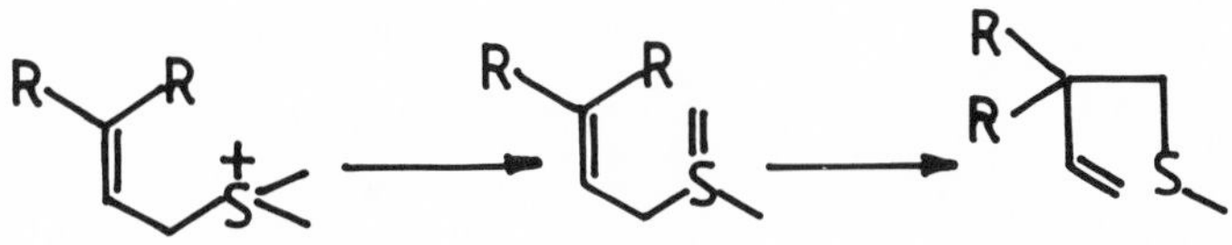

(b)   *Carbonyl compounds and derivatives*

Much work has been published on the synthesis of aldehydes, ketones, carboxylic acids, esters and their homologues
and some typical examples are illustrated, many of which proceed through formation of an α-sulphenyl carbanion.

A bis-(phenylthio) carbanion offers some advantage over
a dithian in the synthesis of ketones of the type, $R^1COCHR^2R^3$
by the route shown (O.W. Lever, Tetrahedron, 1976, 32, 1943).
The synthesis of carbonyl compounds from sulphur-containing
intermediates is reviewed.

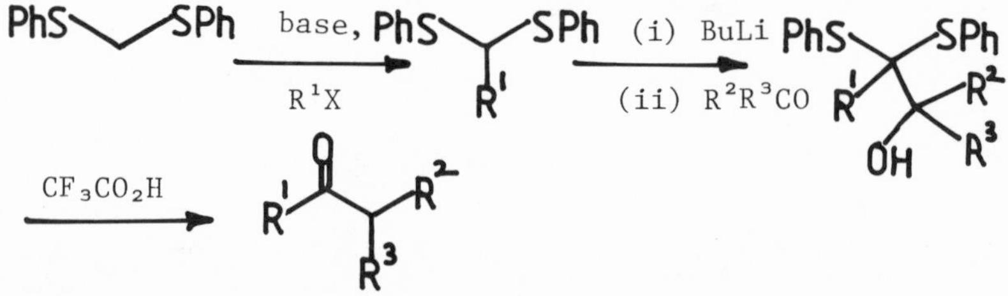

An olefin can be converted into an aldehyde by a method which
provides a useful alternative to the hydroboration-oxidation
route (P. Bakuzis *et al.*, J. org. Chem., 1976, 41, 2769).

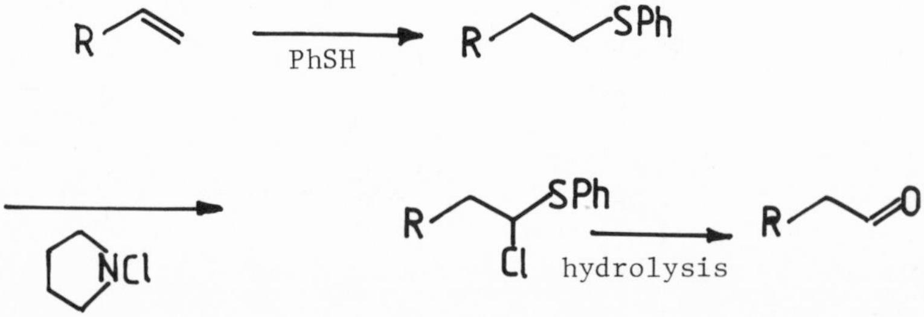

A variation of the Wittig reaction for homologation of
an aldehyde has made use of thiophenoxymethylenephosphoran in
the synthesis of 11-oxaprostaglandins (I. Vlattas and A.O. Lee,
Tetrahedron Letters, 1974, 445).

The o-formylation of thiophenols has been effected in a
similar way to that for phenols, discussed in Chapter 4 on
'Phenols' (P.G. Gassman and D.R. Amick, Tetrahedron Letters,

1974, 889; 1974, 3463).

Thiophenol has been used in the preparation of secondary and tertiary alkyl ketones (J.H. Posner and C.E. Whitten, Org. Synth., 1976, 55, 122; G.H. Posner, D.J. Brunelle and L. Sinoway, Synthesis, 1974, 662).

$$PhSLi + CuI \xrightarrow[(25°)]{THF} PhSCu + Li$$

$$PhSCu + (CH_3)_3CLi \xrightarrow[THF]{-60° \text{ to } -65°} PhS\left[(CH_3)_3C\right]CuLi$$

$$\xrightarrow[-60° \text{ to } -65°]{PhCOCl} PhCOC(CH_3)_3 \quad (84-87\%)$$

2,2-Dialkylcyclohexanones can be readily obtained by the method shown (R.M. Coates, H.D. Pigott and J. Ollinger, Tetrahedron Letters, 1974, 3955).

Alkylation of a cyclo-2-enone at the 3-position can be effected by the device of converting it to an anion which functions as the effective equivalent of a acyl vinyl anion, (T. Cohen, D.A. Bennett and A.J. Mura, J. org. Chem., 1976, 41, 2507).

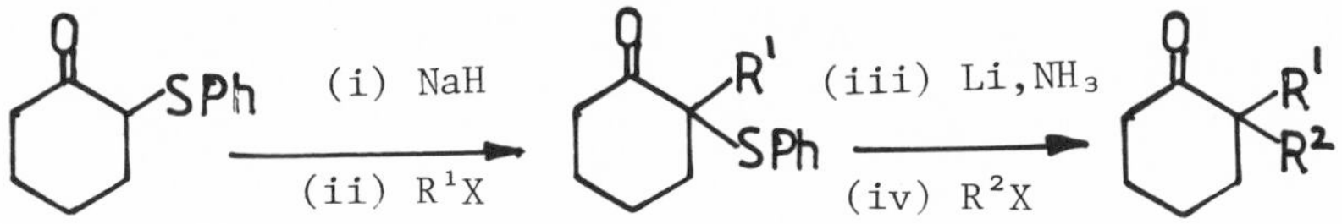

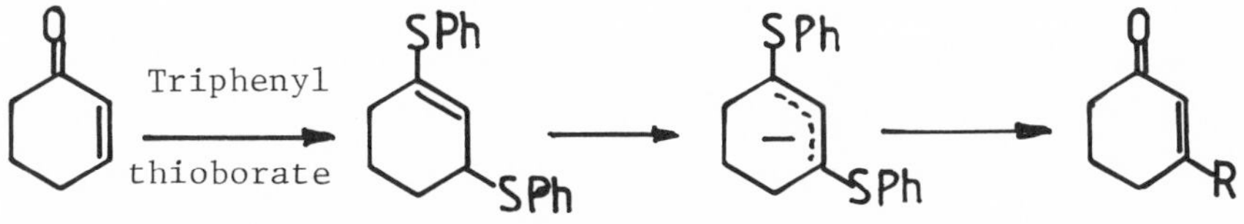

2-Alkylation of a similar system is readily effected by way of the following procedure for the conversion of 4,4-di-methylcyclohex-2-enone into 2,4,4-trimethylcyclohex-2-enone in 60% overall yield (J.-C. Richer and D. Perelman, Canad. J. Chem., 1966, 44, 2003).

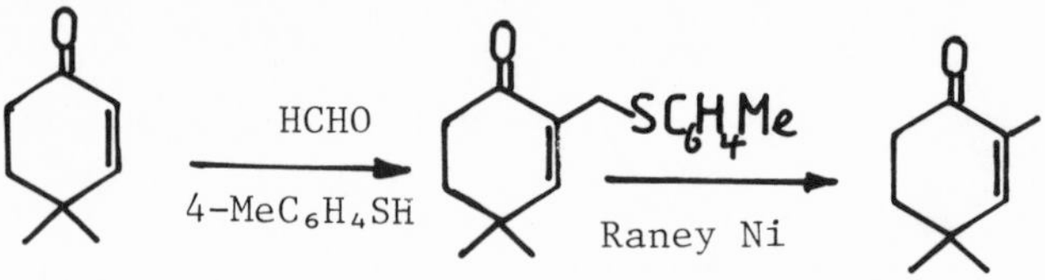

Certain bicyclic cylopentanones can be prepared by way of 1-lithiocyclopropyl phenylsulphide as shown (B.M. Trost and D.E. Keeley, J. Amer. chem. Soc., 1976, 98, 248; c.f. R.D. Miller, Chem. Comm., 1976, 277).

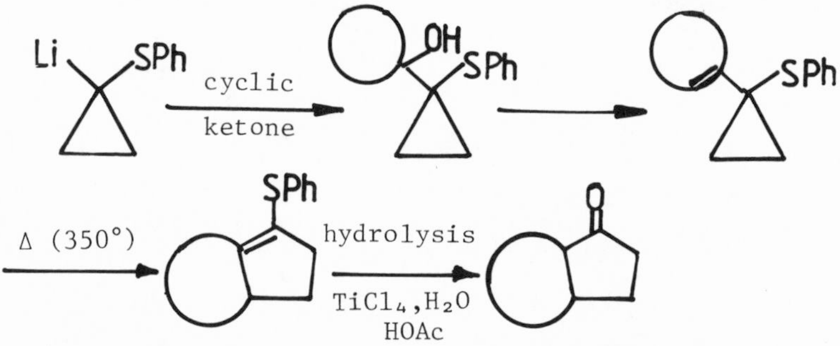

Diphenyldisulphide (from thiophenol) has been used in the following transformation of a 1-keto to a 2-keto compound (B.M. Trost, K. Hiroi and S. Kurozumi, J. Amer. chem. Soc., 1975, 97, 438). (Scheme 16).

Versatile syntheses with p-toluenesulphenyl derivatives (from MeC_6H_4SCl) of enone, dienones and dienals have been developed using sulphur-based methodology (R.C. Cookson and R. Yopalan, Chem. Comm., 1978, 608; R.C. Cookson and P.J. Parsons, Chem. Comm., 1978, 821,822).

γ-Keto esters can be conveniently derived from $\alpha\beta$-unsaturated ketones by the use of tris (triphenylthio)methane as follows (A.-R.B. Manas and R.A.J. Smith, Chem. Comm., 1975, 216).

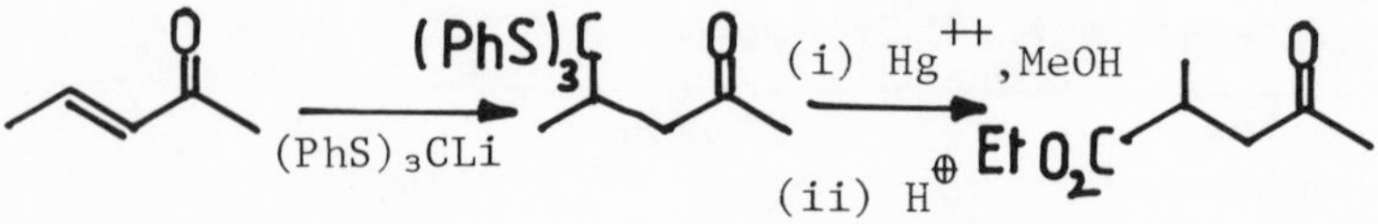

(Scheme 16).

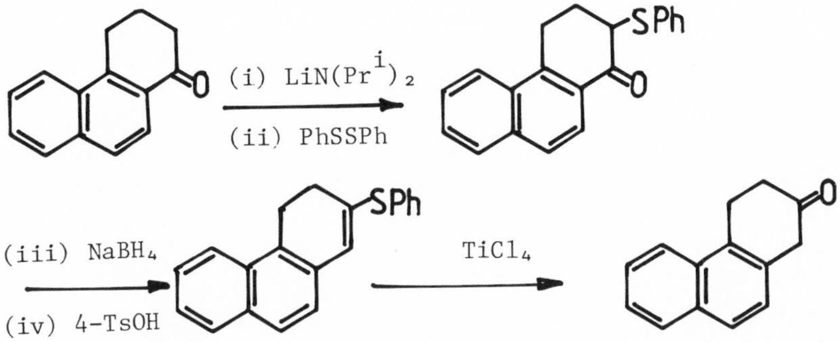

Cyclic γ-keto esters are available from β-keto esters by the annelation sequence shown (J.P. Marino and R.C. Landick, Tetrahedron Letters, 1975, 4531).

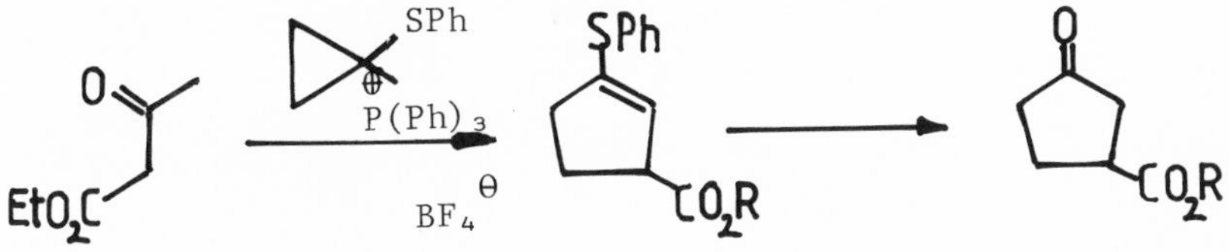

1-Lithiocyclopropyl phenylsulphide and diphenyl sulphonium cyclopropylide can be used in spiroannelation procedures (B.M. Trost and D.E. Keeley, J. Amer. chem. Soc.,1974, 96, 1252).
    The readily available thiophenoxyacetyl chloride has been used as a reagent in a 'macrolide' ester synthesis (T. Takahashi *et al.*, J. Amer. chem. Soc., 1978, 100, 7424). (Scheme 17).

The sulphoxide of methyl thiophenoxyacetate $PhSOCH_2CO_2Me$ has been used for a synthesis of 5-phenyl substituted resor-cinols, (see chapter 4 on 'Phenols').

(c) *Alcohols and amines*

(Scheme 17).

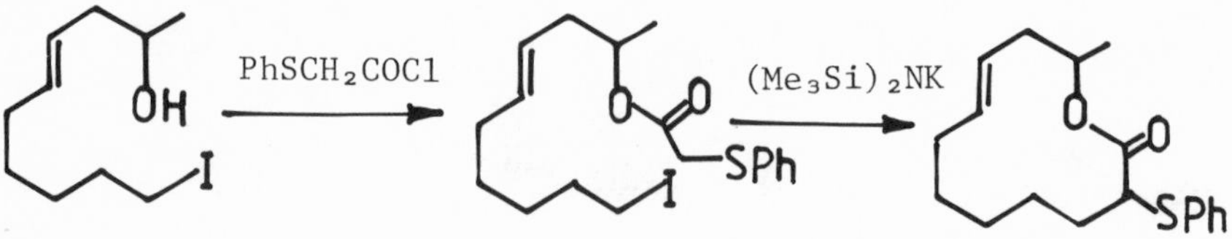

Sulphenylation of cyclopentadiene monoepoxide has been effected in the following way and provides a general synthesis of allylic alcohols and tri-substituted olefins. (D.A. Evans *et. al.*, Tetrahedron Letters, 1973, 1385, 1389; D.A. Evans and G.C. Andrews, Acc. chem. Res., 1974, 7, 147).

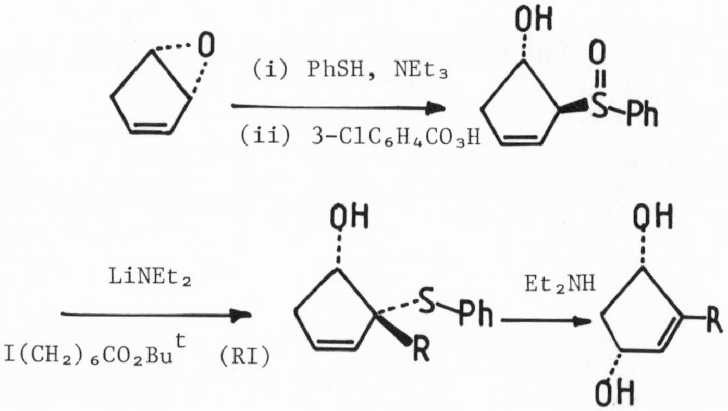

Bridged unsaturated piperidine derivatives can be prepared from cyclic dienes by transformation of the initially formed 1,2-addition product which itself can be converted to the lactam shown with thiophenol and pyridine (J.R. Malpas and N.J. Tweddle, J. chem. Soc., Parkin I, 1977, 874). (Scheme 18).

Phenyl sulphenyl chloride obtained as described earlier has been used for lactonisation (K.C. Nicolaou and Z. Lysenko, Chem. Comm., 1977, 293). (Scheme 19).

(Scheme 18)

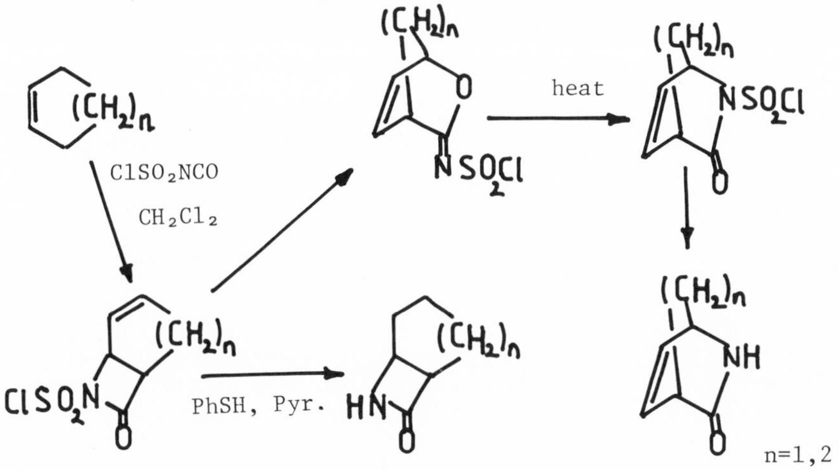

(Scheme 19)

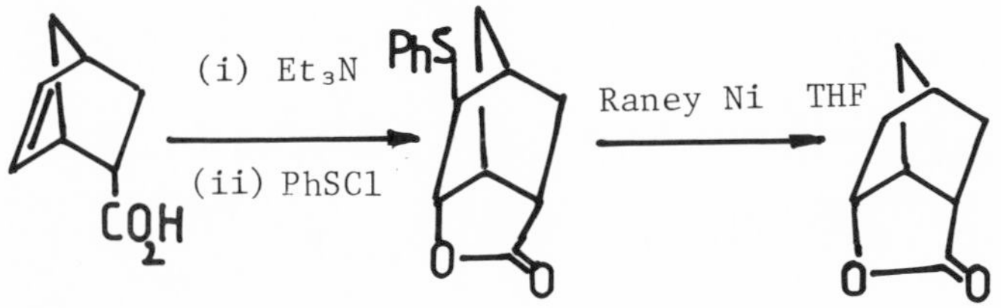

(d)   *Use of selenophenol derivatives*

Reference has been made to the use of selenophenol as a means of converting RCH_2Br to RCH_2CHO. This reagent has a number of uses which for one reason or another has led to its adoption often in the form of a simple derivative as can be seen from the following examples.   There is little doubt that the synthetic importance of these compounds will increase. Phenylselenyl chloride has been used to convert γ,δ-unsaturated acids to bicyclic lactones (D.L.J. Clive and G. Chittattu, Chem. Comm., 1977, 484). (Scheme 20)

The alkyl bromide (RCH_2Br) may be converted into the homogous aldehyde by way of diphenylselenide (prepared from selenophenol) by the method illustrated (K. Sachdev and H.S. Sachdev, Tetrahedron Letters, 1976, 4223).

(Scheme 20)

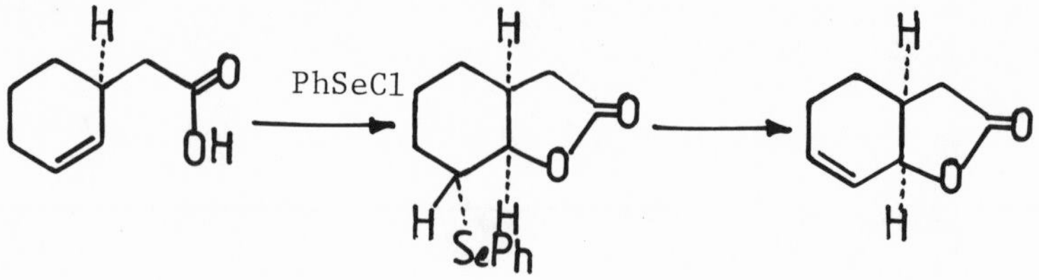

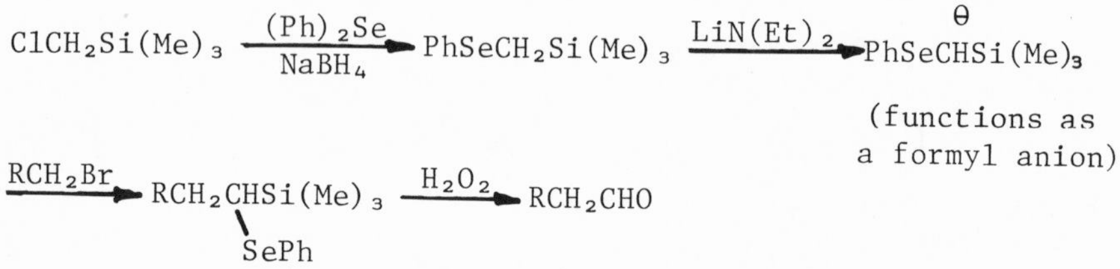

$$\text{ClCH}_2\text{Si(Me)}_3 \xrightarrow[\text{NaBH}_4]{\text{(Ph)}_2\text{Se}} \text{PhSeCH}_2\text{Si(Me)}_3 \xrightarrow{\text{LiN(Et)}_2} \overset{\ominus}{\text{PhSeCHSi(Me)}_3}$$

(functions as
a formyl anion)

$$\xrightarrow{\text{RCH}_2\text{Br}} \underset{\underset{\text{SePh}}{|}}{\text{RCH}_2\text{CHSi(Me)}_3} \xrightarrow{\text{H}_2\text{O}_2} \text{RCH}_2\text{CHO}$$

Phenylselenylcyanide may be used in the tranformation of aldehydes into unsaturated nitriles (P.A. Grieco and Y. Yokoyama, J. Amer. chem. Soc., 1977, 99, 5210).

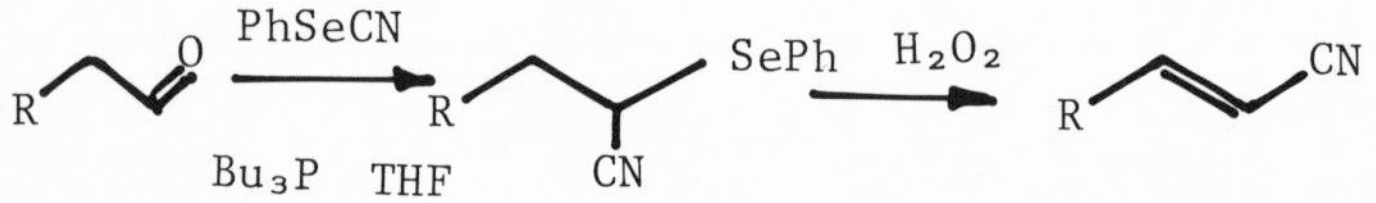

New routes to α,β-unsaturated carbonyl compounds have been based on the use of phenylselenyl chloride (K.B.Sharples, R.F. Laver and A.Y. Teranishi, J. Amer. chem. Soc., 1973, 95, 6137).

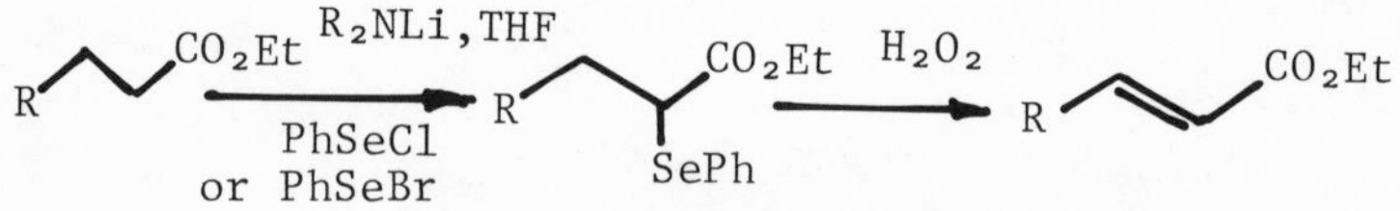

The conversion of carbonyl compounds to α,β-unsaturated analogues by selenation has been explored in detail (H.J. Reich, J.M. Renga and I.L. Reich, J. Amer. chem. Soc.,1975, 97, 5434).

A chiral synthesis of (+)-15 -(S)-Prostaglandin-A₂ from carbo-
hydrates has been effected with the aid of phenylselenyl
chloride (G. Stork and S. Raucher, J. Amer. chem. Soc., 1976,
98, 1583). (Scheme 21).
The role of organoselenium chemistry in synthesis has been ad-
mirably reviewed (D.L.J. Clive, Tetrahedron, 1978, 34, 1049).
Phenylselenation has been used in lipid syntheses to effect
conjugation (J.E. Baldwin, N.V. Reed and E.J. Thomas,
Tetrahedron 1981, 37, 263.)

Scheme 21

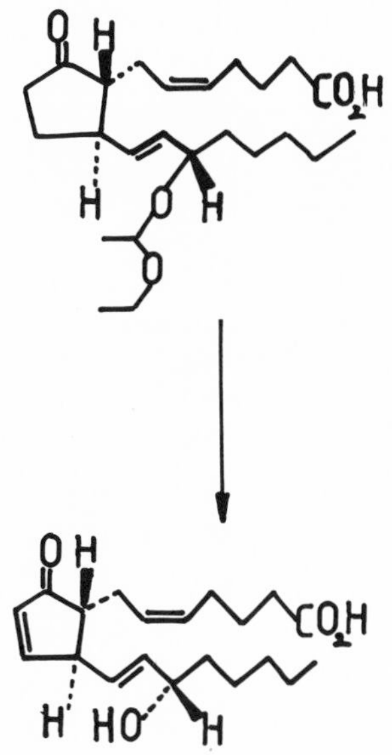

Reagents (1) LiNPr$^i_2$ (2) PhSeCl (3) NaIO₄ (4) removal of
protective group.

SULPHIDES

(1) *Saturated alkyl, aryl and diaryl compounds*

(a) *Preparation*

Alkylaryl sulphides can be prepared by a number of routes such as: alkylation of thiophenols, of aryl disulphides, reduction of appropriate sulphoxides and sulphones and addition to certain thiocarbonyl compounds.

(i) *By alkylation*

Neopentyl phenylsulphide has been prepared by the interaction of neopentylbromide and sodium thiophenoxide in the presence of cetyl tributylphosphonium bromide (D. Landini and T. Rolla, Org. Synth., 1978, $\underline{58}$, 143).

$$(CH_3)_3CCH_2Br + PhSNa \xrightarrow[C_{16}H_{33}(Bu)_3\overset{\oplus}{P} \overset{\ominus}{Br}]{} C_6H_5SCH_2C(CH_3)_3 + NaBr$$

Halogenobenzenes have been used, as mentioned in the section on nucleophilic reactions of thiophenols, but this procedure may not be the method of choice. Interaction of ethanethiol with bromobenzene in the presence of sodamide and hexamethylphosphoric triamide gives ethyl phenyl sulphide (P. Caubère, Bull. soc. Chem. France, 1967, 3446, 3451). Selective replacement of a fluorine atom can be achieved (M.E. Peach and A.M. Smith, J. fluorine Chem., 1974, $\underline{4}$, 341,399).

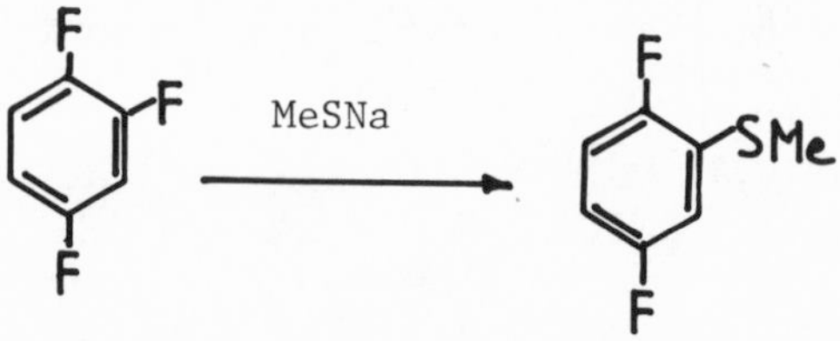

Iodobenzene reacts with copper (II) trifluoromethylthiolate to give phenyl trifluoromethyl sulphide (L.M. Yagupolskii, N.V. Kondratenko and V.P. Sambur, Synthesis, 1975, 721). (Scheme 1)

Thiocyanates have been used in place of thiophenols as indicated (Ethyl corporation, EP 4606; Chem. Abs., 1980, $\underline{92}$,

(Scheme 1)

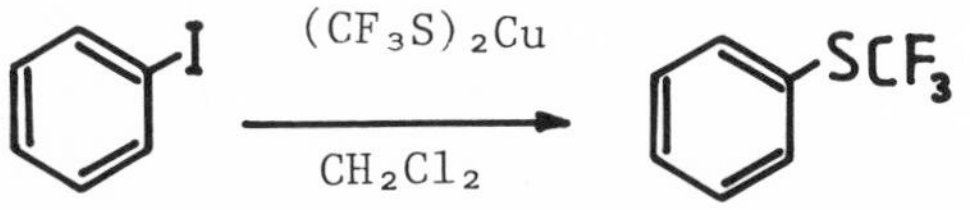

58420) in the presence of a phase transfer agent.

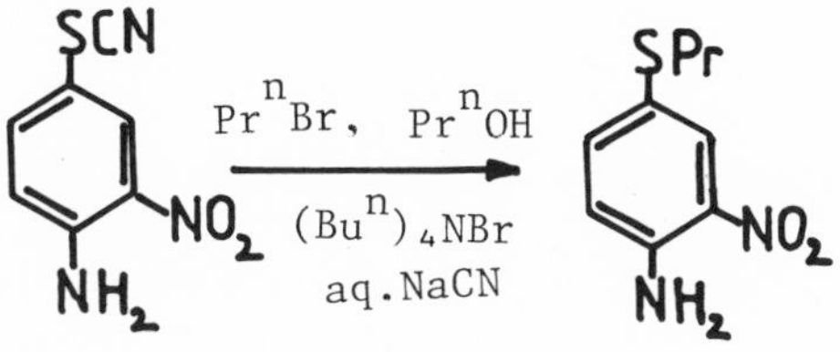

Diazonium salts react with silver trifluoromethylthiolate to yield phenyl trifluoromethyl sulphides, although there have been explosions in such reactions (N.V. Kondratenko and V.P. Sambur, Ukrain, khim. Zhur., 1975, 41, 516).

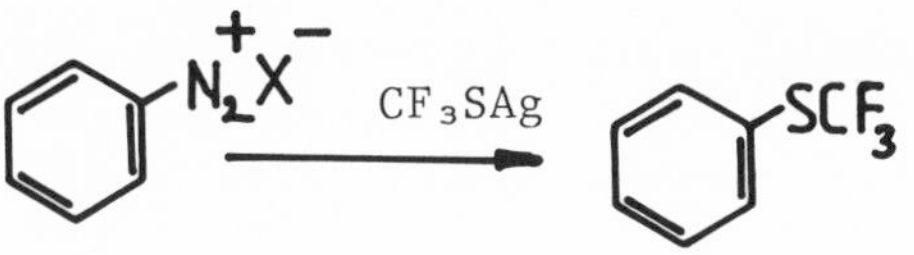

Benzenoid hydrocarbons on alkylation with dimethylsulphide sulphone in the presence of aluminium chloride yield the corresponding arylthiomethyl ether in good yield (J.K. Bosscher, E.W.A. Kraak and H. Kloosterziel, Chem. Comm., 1971, 1365).

$$ArH \xrightarrow[AlCl_3]{MeSSO_2Me} ArSMe$$

Vinylic sulphides can be prepared by the reaction of an appropriate sulphenyl halide with triphenylvinyltin (J.L. Wardell and A. Siddique, J. organometal. Chem., 1974, 78, 395).

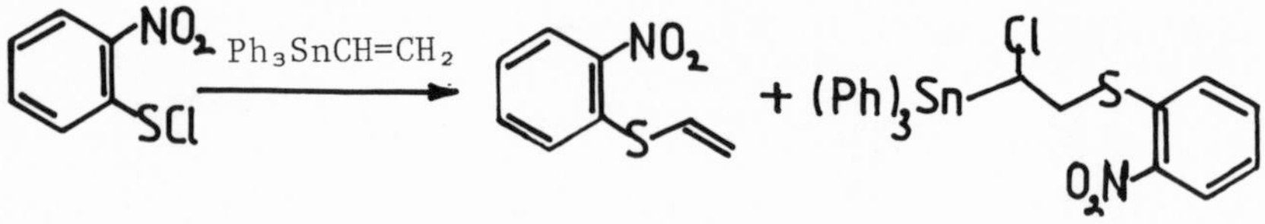

Addition reactions of alkenes with thiophenols provide another route to sulphides as discussed in the section on reactions of thiophenols.

(ii)   *From disulphides and thiolcarbonates*

Diphenyl disulphide with trialkylborane affords the corresponding alkyl phenyl sulphide by way of a free-radical chain reaction (H.C. Brown and M. Midland, J. Amer. chem. Soc., 1971, 93, 3291).

$$PhSSPh + BR_3 \longrightarrow PhSR + PhSBR_2$$

Methyl phenyl thiocarbonate (or xanthate) is converted into phenyl methyl sulphide by treatment with trialkylphosphines, tetraethylammonium fluoride or a number of other catalysts (F.N. Jones, J. org. Chem., 1968, 33, 4290).

$$PhSCOMe \xrightarrow[\substack{\oplus \quad \ominus \\ NEt_4 \ F}]{R_3P \ \text{or}} PhSMe$$

Interaction of diphenyl disulphide with triethyloxonium tetrafluoroborate gives a number of products including ethyl phenyl sulphide (B. Miller and C.-H. Han, Chem. Comm., 1970, 623; J. org. Chem., 1971, 36, 1513). The products are closely similar to those found in semidine rearrangements.

$$PhSSPh \xrightarrow[Et_3OBF_4]{\oplus \quad \ominus} PhSEt + p\text{-}PhSC_6H_4SEt + \text{other products}$$
$$(51\%) \qquad (30\%)$$

Diphenyl disulphide reacts with 2,6-di-t-butylphenol to give 4-hydroxy-3,5-di-t-butylphenyl phenyl sulphide by sulphenylation(T. Fujisawa and T. Kojima, J. Org. Chem.,1973, 38,687)
Diaryl disulphides and sodium cyanide in alcoholic solution afford the corresponding aryl alkyl sulphide (K. Tanaka,

J. Hayami and A. Kaji, Bull. chem. Soc., Japan, 1972, 45, 536).

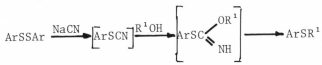

$$ArSSAr \xrightarrow{NaCN} [ArSCN] \xrightarrow{R^1OH} \left[ArSC \underset{NH}{\overset{OR^1}{\Big\langle}} \right] \longrightarrow ArSR^1$$

Phenylsulphenyl chloride reacts with benzylic cyanides to give α-thiophenyl ethers as shown (S.J. Selikson and D.S. Watt, Tetrahedron Letters, 1974, 3029).

$$ArCH(R)CN \xrightarrow{PhSCl,base} ArC(SPh)(R)CN$$

$$\text{Me}_2\text{Si(Bu}^t\text{)Cl} \qquad \text{PhSCl}$$

$$ArC(R)=C=NSi(Bu^t)Me_2$$

(iii) *By reduction*

Reduction of sulphones and of sulphoxides is not easily effected although it does provide a distinct route to aryl alkyl sulphides (J.N. Gardner, *et al.*, Canad. J. Chem., 1973, 51, 1419).

$$ArH \xrightarrow{ClSO_3H} ArSO_2Cl \xrightarrow{Na_2SO_3} ArSO_2Na \xrightarrow{Me_2SO_4} ArSO_2Me \xrightarrow{Bu_2AlH}$$

ArSMe

A number of methods have been examined for the reduction of sulphoxides to sulphides some of which are applicable to both aryl and alkyl sulphoxides.
Phosphorus trichloride/trichlorosilane (T.H. Chan, A. Melnyk and D.N. Harpp, Tetrahedron Letters, 1969, 201),
Dichlorosilane and triphenylphosphine in carbon tetrachloride, halogen in acidic media (T. Aida, N. Furukawa and S. Oae, *ibid.*,1973, 3853),
Phosphorus pentasulphide (R.G. Micetich, C.G. Chin and R.B. Morin, *ibid.*, 1976, 975; T.W.J. Still, S.K. Hasan and K. Turnbull, Synthesis, 1977, 468) have all been used. Other reductants for the conversion of sulphoxides into sulphides are sodium borohydride in the presence of cobalt (II) chloride (D.W. Chasar, J. org. Chem., 1971, 36, 613),with 2-chloro-1,3,2-benzodioxophosphole (D.W. Chasar and D.M. Pratt, Synthesis, 1976, 262).

314

Dichloroborane (H.C. Brown and N. Ravindram, Synthesis, 1973, 42),
Hydrogen with palladium/carbon or with iron pentacarbonyl
(H. Alpen and E.C.H. Keung, Tetrahedron Letters, 1970, 53),
Trifluoroacetic anhydride in the presence of iodide ion
(J. Drabowicz and S. Oae, Synthesis, 1977, 404),
Phosphites (M. Dreux, Y. Leroux and P. Savignac, *ibid.*, 1974, 506),
Titanium (III) chloride (T.-L. Ho and C.M. Wong, Synth. Comm., 1973, <u>3</u>, 37) and
Trimethyliodosilane, trimethylbromosilane and trimethylphenyl
silane (G.A. Olah, *et al.*, Synthesis, 1977, 583) are useful.
Methods for the reduction of sulphoxides to sulphides have
been discussed and various complx ions of molybdenum (II) and
(III) found useful (R.G. Nuzzo, H.J. Simon and J. San Filippo,
J. org. Chem., 1977, <u>42</u>, 568).

(iv) *From thiocarbonyl compounds*

Thiones react with phenyllithium in two ways leading to
tertiary thiols and the isomeric sulphides (A. Ohno, *et al.*,
Chem. Letters, 1975, 983; A. Ohno, *et al.*, Bull. chem. Soc.,
Japan, 1975, <u>48</u>, 3718).

Addition of Grignard reagents to thiones results in mixtures in which sulphide is present (M. Dagonneau and J. Vialle,
Tetrahedron Letters, 1973, 3017; and with thione keto compounds, sulphide formation occurs, (J. Vialle, *et al.*, *ibid.*,
1976, 4275).
Diaryl sulphides are available by slightly different
routes. Alkylation methods have been discussed under the section on nucleophilic reactions of thiophenols.
Diphenyl sulphide can be obtained from thiophenol by heating it at 530-60° in an inert atmosphere with a contact time
of 35 sec. (Av Sibe Irkut org., S.U. 431 769). Di-(2,4-di-
nitrophenyl)disulphide upon heating in various alcohols forms
di-(2,4-dinitrophenyl) sulphide (B.I. Stepanov, V. Ya. Rodionov
and T.A. Chibisova, Zhur. org. Khim., 1974, <u>10</u>, 79). This
and other observations are relevant to the possible thiosul-

phoxide structure of diaryl disulphides.

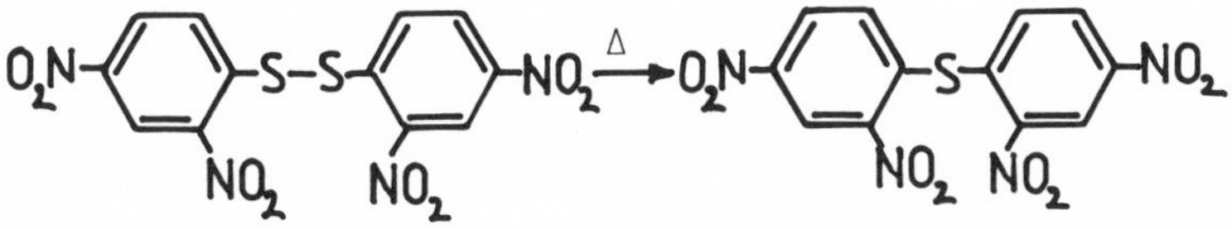

3-Methylsalicyclic acid reacts with 4-chlorophenylsulphenyl chloride as shown (J.K. Landquist and D.R. Summers, VII International Symp. on org. sulphur Chem.. Hamburg, 1976, Abstracts p. 176).

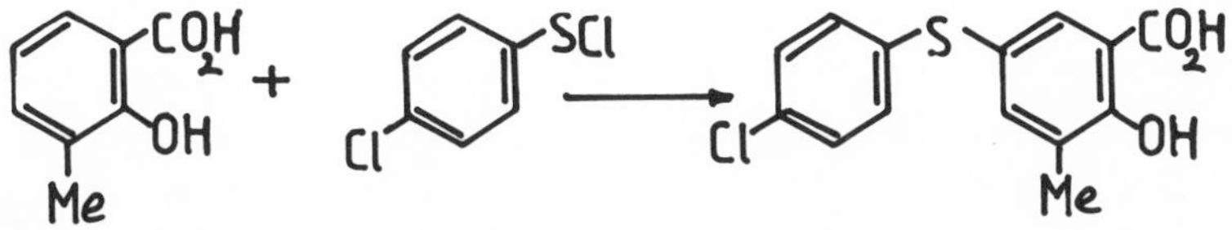

Interaction of diphenyl sulphoxide with acetyl chloride, which reacts as a reducing agent, results in formation of diphenyl sulphide (T. Numata and S. Oae, Chem. and Ind., (London), 1973, 277).

$$\mathrm{PhSOPh} \xrightarrow{\mathrm{AcCl}} \left[\mathrm{Ph{-}\overset{\oplus}{S}(OAc){-}Ph}\ \ \mathrm{Cl}^{\ominus} \right] \longrightarrow \left[\mathrm{Ph{-}\overset{\oplus}{S}(Cl){-}Ph}\ \ \mathrm{OAc}^{\ominus} \right] \xrightarrow{\mathrm{AcCl}} \mathrm{Ph_2S + Cl_2 + Ac_2O}$$

(b) *Reactions*

(i) *Substitution*

α-Halogeno sulphides can be prepared as indicated earlier in the preparative section of thiophenols. Additionally other reagents such as sulphur dichloride, and 3-(dichloroiodo) pyridine with aliphatic sulphides, (E. Vilsmaier and W. Sprugel, Ann., 1971, 749, 62), N-chlorosuccinimide with aliphatic sulphides (E. Vilsmaier and W. Sprugel, Ann., 1971, 747, 151) and sulphuryl chloride with dibenzyl sulphide (R.H. Mitchell, Tetrahedron Letters, 1973, 4395) have been used. Chloromethyl phenyl sulphide has important uses which are discussed in a subsquent section. Thioanisole (methyl phenyl sulphide) undergoes complexation with Lewis acids and the type and extent of this can affect the products formed in acetylation and chloromethylation reactions and the rates of such reactions (S.H. Pines, J. org. Chem., 1976, 41, 884).

(ii) *Oxidation*

With sulphides energetic oxidation results in the formation of sulphones but specific procedures result in the formation of sulphoxides. Hydrogen peroxide with benzyl methyl sulphide gives the sulphoxide or the sulphone dependent on whether acetone or acetic acid is used as solvent. Tungsten or vanadium compounds with hydrogen peroxide result in improved yields.

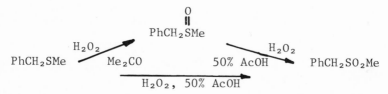

Alternative reagents for the oxidation of sulphides to sulphoxides are peroxy acids (*m*-chloroperbenzoic acid, peroxydodecanoic acid), sodium metaperiodate, *t*-butylhypochlorite (T. Durst in 'Sulphoxides', Organic Sulphur Compounds, Vol. 3, Comprehensive Organic Chemistry, Pergamon, 1979). *N*-Bromosuccinimide, *N*-chlorobenzothiazole have been used. Phenyl isopropyl sulphide at 25°C with periodate in 0.05 M borate containing bovine serum albumin gives *R*-phenyl isopropyl sulphoxide (70% yield) in 81% optical purity (T. Sugimoto *et. al.*, Chem. Comm., 1979, 402). For diaryl sulphides, ceric ammonium nitrate is an effective oxidant.

Thioanisole and *t*-butyl peroxide at 140° give 1,2-di-(phenylthio) ethane whereas by photochemical reaction at 40° diphenyl disulphide is formed (H.B. Henbest, J.A.W. Reid and C.J.M. Stirling, J. chem. Soc., 1964, 1220).

Photo-oxidation of benzyl methyl sulphide resulting in the formation of benzaldehyde has been studied (E.J. Corey and C. Ouannes, Tetrahedron Letters, 1976, 4263).

(iii) *Rearrangement*

In the Claisen rearrangement of allyl phenyl sulphide
the intermediacy of 2-allylthiophenol(proved by the formation
of its methylether with methyl iodide) in the reaction
has been established.   It has been isolated and converted to
the same mixture which results directly from the sulphide
(H. Kwart and J.L. Schwartz, J. org. Chem., 1974, __39__, 1575).

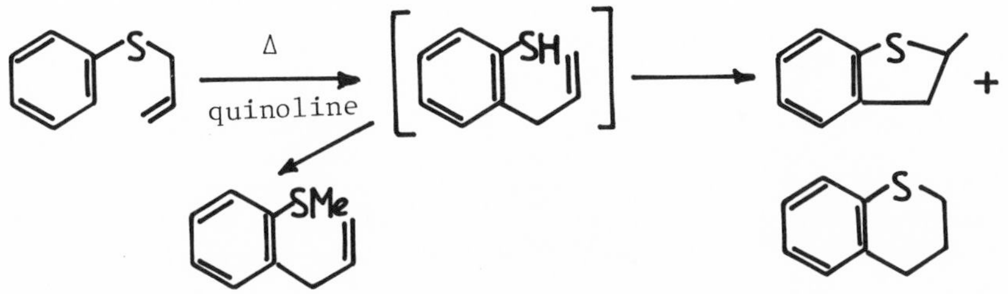

In the rearrangement of 2,6-dimethylphenyl allyl sulphide no
p-Claisen product is found but a cleavage product and four
cyclic products resulting from formation of a thio-2,4-dienone
followed by 1,3 and 1,4-methyl group migrations (H. Kwart and
M.H. Cohen, Chem. Comm., 1968, 1296).

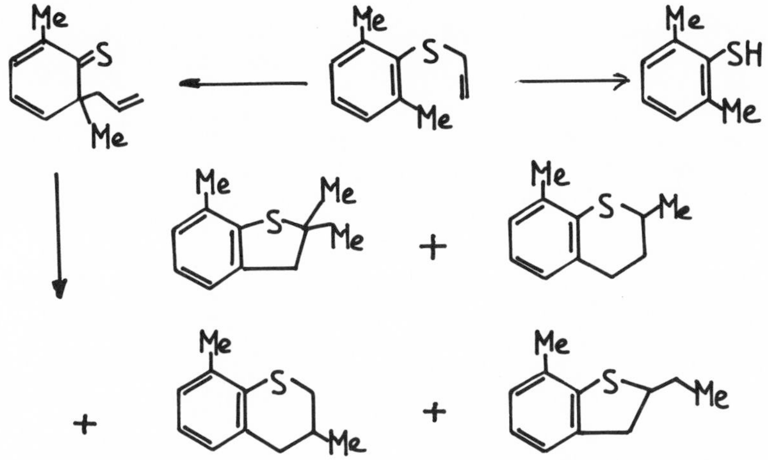

Phenyl propargyl sulphide gives two monocyclic products
as a result of Claisen rearrangements which then undergo isomerisation as shown (H. Kwart and T.J. George, Chem. Comm.,

1970, 433).   In the cyclisation of 2-propynylthiophenol only
a five-membered ring product apparently results.

The Pummerer rearrangement of an allyl α-phenylsulphinyl
ester has been used synthetically and the product can be
transformed into β,γ-unsaturated aldehyde (B. Lythgoe,
J.R. Milner and J. Tideswell, Tetrahedron Letters, 1975, 2593).

Synthetic uses of allylic sulphides are discussed in a
later section.
Benzylic sulphides are formed in certain rearrangements
of sulphonium salts as shown in the preparation of 2-methyl-
aniline (H.H. Wasserman and I. Saito, J. Amer. chem. Soc.,
1975, 97, 905).

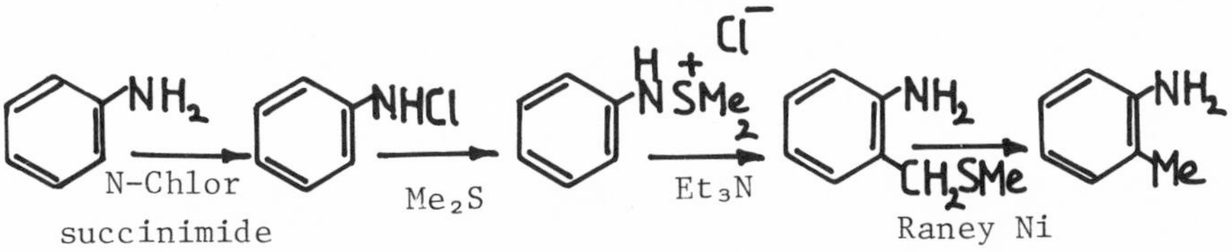

Dibenzyl sulphide undergoes the Stevens rearrangement probably by way of a two-radical reaction step (G.C. Barrett, Sulphides' in'Organic Sulphur Compounds', Comprehensive Organic Chemistry, Barton and Ollis, vol. 3, p. 33, Pergamon, 1979) as shown,

$$PhCH_2SCH_2Ph \xrightarrow[-78°C]{BuLi \atop TMEDA} [PhCHSCH_2Ph] \xrightarrow{warm} PhCHCH_2Ph \xrightarrow{MeI} PhCHCH_2Ph$$

and the Sommelet rearrangement by a concerted [2,3]-sigmatropic process as shown

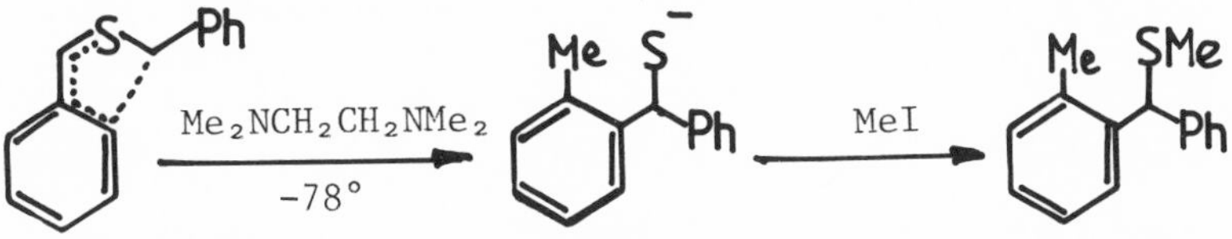

β-Hydroxyalkyl sulphides after conversion into the corresponding sulphoxides undergo the Pummerer rearrangement upon treatment with hot acetic anhydride (S. Iriuchijima, K. Maniwa and G.T. Tsuchihashi, J. Amer. chem. Soc., 1974, 96, 4280).

$$RCH(OH)CH_2SR^1 \longrightarrow RCH(OH)CH_2SOR^1 \xrightarrow{Ac_2O} RCH(OAc)CH(OAc)SR^1$$
(1:1-diastereoisomeric mixture)

(c)   *Synthetic uses of thioethers*

Like thiols, sulphides (thioethers) have a useful role in organic synthesis. By comparison their oxygen analogues have no similar function reflected in the vast difference in activity of the SMe group compared with the OMe group towards bases such as alkyllithiums. Additionally as a result of the polymerisability of the sulphur atom (F. Bernardi, *et al.*, J. Amer. chem. Soc., 1975, 97, 2209) an α-carbanion can be stabilised and together with, in the final stage after react-

ion, the ready cleavage of the weak C-S bond, an ideal system
is available for exploitation in organic synthetic transformations.   Some of the many examples of this facility are considered in the following sub-section

(i) *Phenyl methyl sulphide (and its derivatives)*

     By metallation of thioanisole (methyl phenyl sulphide)
with n-butyllithium in the presence of diazabicyclo[2,2,2]-
octane (DABCO) a high yield of phenylthiomethyllithium is obtained.   In the absence of the diamine a low yield results
(E.J. Corey and D. Seebach, J. org. Chem., 1966, 31, 4097) although at -78°C hexamethylphosphoric triamide also results in
a good yield of the α-carbanion (T.M. Dolak and T.A. Bryson,
Tetrahedron Letters, 1977, 1961). Interaction of the α-carbanion shown with ketones leads to the tertiary alcohol and by
further base treatment an epoxide (J.R. Shanklin, *et al.*,
J. Amer. chem. Soc., 1973, 95, 3429).   Upon successive treatment of the alcohol with methyllithium and *o*-phenylenephos-
phorochloridate an olefin is obtained in 70-90% yield
(I. Kuwajima, S. Sato and Y. Kurata, Tetrahedron Letters,
1972, 737).

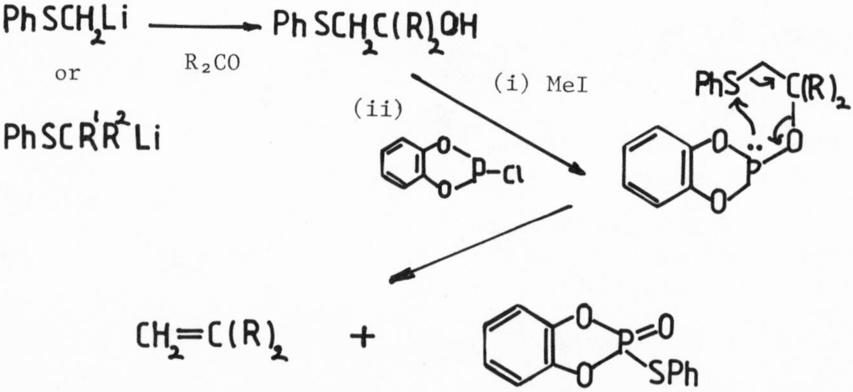

The sulphoxide $PhS(=O)CH_2Li$ can be used in place of the sulphide (I. Kuwajima and M. Uchida, Tetrahedron Letters, 1972,
649).

     A scheme has been devised in which titanium (IV)chloride
followed by lithium aluminium hydride-triethylamine or acetylation followed by lithium-ammonia afford the same product or

an oxiran by the route shown (Y. Watanabe, M. Shiono and
T. Makaiyama, Chem. Letters, 1975, 871).

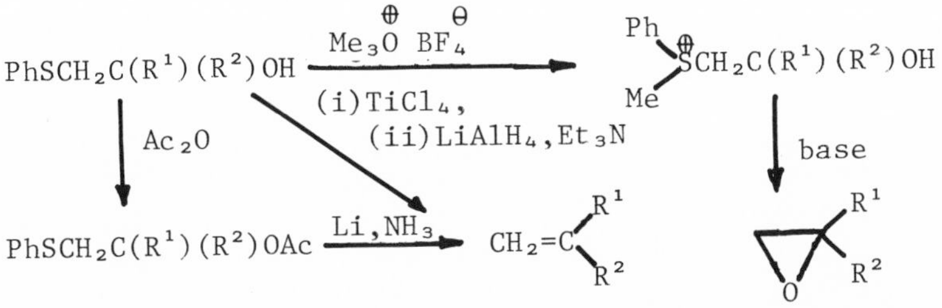

$$PhSCH_2C(R^1)(R^2)OH \xrightarrow[\text{(i)}TiCl_4, \text{(ii)}LiAlH_4,Et_3N]{Me_3O^{\oplus}\ BF_4^{\ominus}} \overset{Ph}{\underset{Me}{\nearrow}}\overset{\oplus}{S}CH_2C(R^1)(R^2)OH$$

$$\Big\downarrow Ac_2O$$

$$PhSCH_2C(R^1)(R^2)OAc \xrightarrow{Li,NH_3} CH_2=C\overset{R^1}{\underset{R^2}{\diagdown}}$$

base → oxiran with R^1, R^2

A 2-substituted prop-1-ene can be derived from an ester
as shown (R.L. Sowerby and R.M. Coates, J.Amer. chem. Soc.,
1972, 94, 4758).

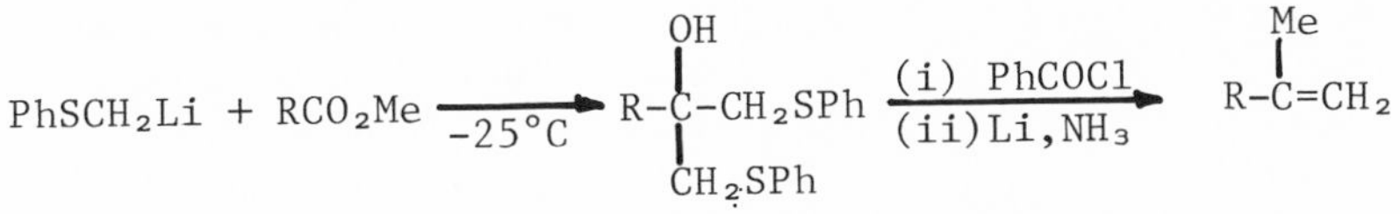

$$PhSCH_2Li + RCO_2Me \xrightarrow{-25°C} \underset{CH_2SPh}{\overset{OH}{R-\underset{|}{\overset{|}{C}}-CH_2SPh}} \xrightarrow[\text{(ii)}Li,NH_3]{\text{(i) }PhCOCl} \underset{}{\overset{Me}{R-\overset{|}{C}=CH_2}}$$

Homologation of an alkyl iodide can be realised
(E.J. Corey and M. Jautelat, Tetrahedron Letters, 1968, 5787).

$$PhSCH_2Li + RCH_2I \longrightarrow PhSCH_2CH_2R \xrightarrow{MeI}{NaI} ICH_2CH_2R + C_6H_5SCH_3$$

Cyclic β,γ-unsaturated aldehydes can be derived in the
following way (M.E. Garst and T.A. Spencer, J. Amer. chem.
Soc., 1973, 95, 250).

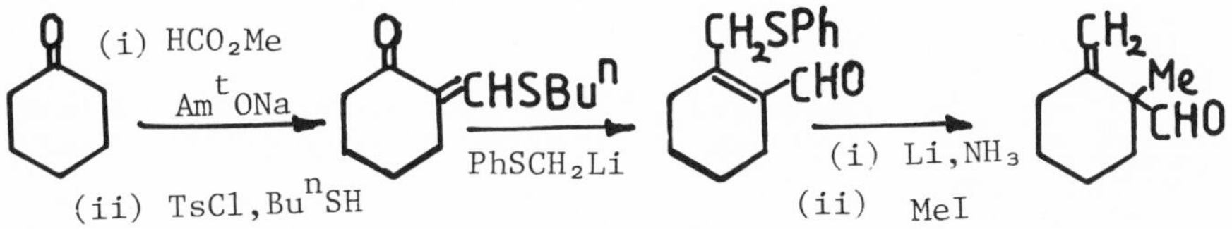

The carbanion derived from an aryl alkyl sulphide behaves as a nucleophilic source of alkyl groups and in a similar way chloromethyl phenyl sulphide can be used (see below).

(ii)   *Chloromethyl phenyl sulphide and derivatives*

In the Woodward-Eschenmoser synthesis of vitamin B_{12}, methylation in the corrin ring is carried out with chloromethyl phenyl sulphide and the reaction shown with thiophenol followed by reduction (A. Eschenmoser, Quart. Rev., 1970, 24, 366)

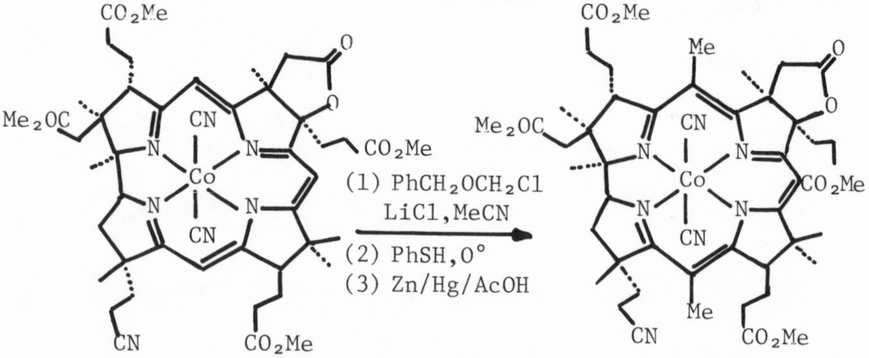

Oxirans are formed from chloromethyl phenyl sulphide as well as from the lithium derivative of thioanisole as illustrated in the following example (L. Barnardi, R. de Castiglione and U. Scarponi, Chem. Comm., 1975, 3120).

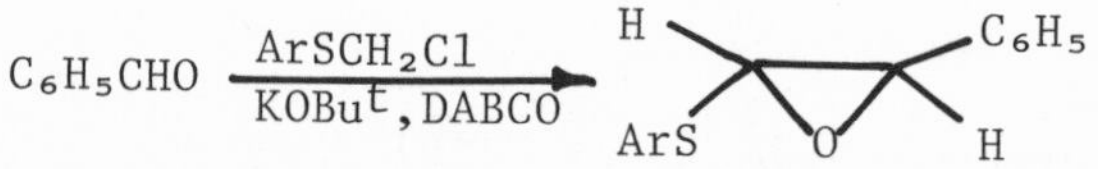

Benzyl bromomethyl sulphide can be used to effect the methylenation of phenylacetic acid (H.J. Reich and J.M. Renga, Chem. Comm., 1974, 135).

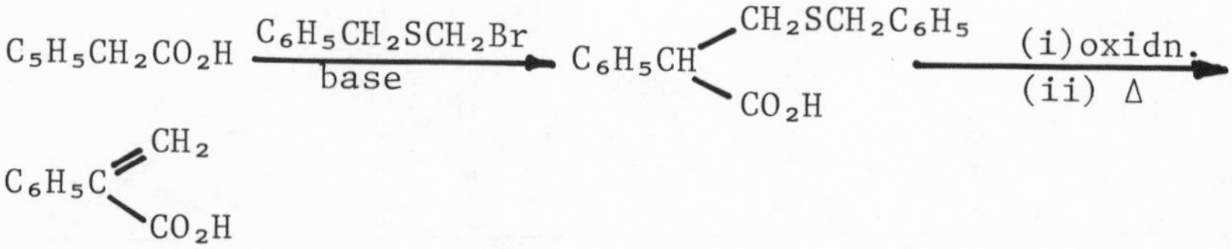

Apart from its use in methylation, chloromethyl phenyl sulphide has been used in the formation of cyclopropyl sulphides and in the extrusion reactions leading to unsaturated hydrocarbons as seen in the following examples.

Interaction with an olefin under phase-transfer catalysis yields the cyclopropylsulphide shown (G. Boche and D.R. Schneider, Tetrahedron Letters, 1975, 4247).

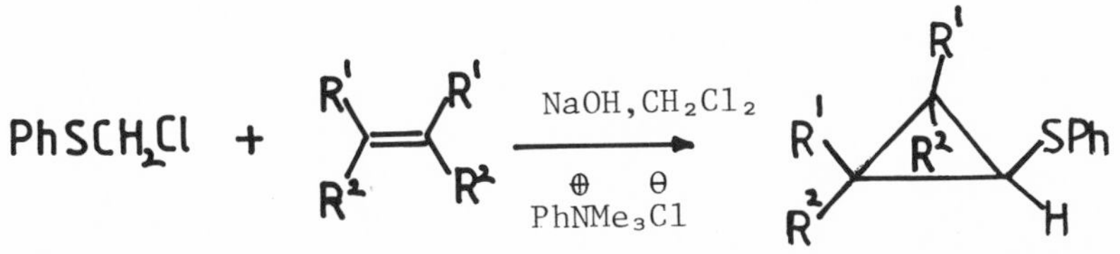

Chloromethyl phenyl sulphide has been used in a study of γ- versus α- alkylation of silyldienolethers as shown (I. Fleming and T.V. Lee, Tetrahedron Letters, 1981, 705) with various Lewis acids such as zinc bromide and titanium tetrachloride.

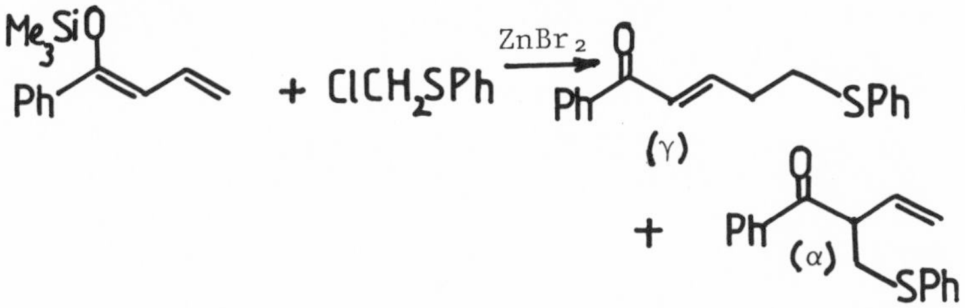

α-Chlorodibenzyl sulphide with potassium t-butoxide and triphenylphosphine affords *E*-stilbene by way of the intermediate shown (R.H. Mitchell, Tetrahedron Letters, 1973, 4395).

$$C_6H_5CH_2SCHClC_6H_5 \xrightarrow[Ph_3P]{KOBr^t,} C_6H_5\text{-}\underset{\diagdown S \diagup}{CH\text{-}CH}\text{-}C_6H_5 \longrightarrow \overset{E}{C_6H_5CH{=}CH_6H_5}$$

The α,α-dichloro analogue gives tolan (R.A. Mitchell, Chem. Comm., 1973, 955).

$$C_6H_5CH_2SC(Cl)_2C_6H_5 \longrightarrow C_6H_5C{\equiv}CC_6H_5$$

(2) *Thioacetals and related compounds*

The presence of two sulphur atoms attached to the same

carbon atom imparts increased reactivity to the shared carbon
atom and carbanion formation readily occurs. The methylene
group or substituted methylene group behaves as a masked formyl group or nucleophile because effectively the transformation
RX → RCHO can be achieved. Thioacetals may be regarded as
substances in which the polarisation of the carbonyl group is
reversed (Umpolung) leading to a situation with great synthetic potential. This has been reviewed at length (B. Gröbel
and D. Seebach, Synthesis, 1977, 357; D. Seebach and H. Kolb,
Chem. and Ind., (London), 1974, 687). This facility has
come about mainly through the application of cyclic dithio
compounds such as 1,3-dithians as the reagents of choice rather
than through aryl compounds, thiolanes. Thiophenol is very
considerably cheaper (tenfold) than propane-1,3-dithiol, or
ethane-1,2-dithiol and di(phenylthio) formal, $C_6H_5SCH_2SC_6H_5$,
would seem to have potentially many of the applications of
(cyclic) dithians one of which is seen in the following ketone
synthesis, effecting a conversion of RX to $RCOR^1$.

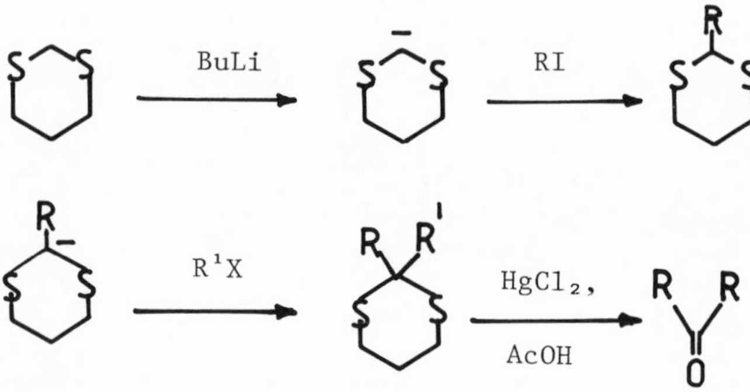

(a) *Preparation*

    Essentially many of the applications of aromatic thio-
acetals involve a preliminary formation of the intermediate
by an addition process. Di(thiophenyl) formal is readily
obtained by the insertion interaction of diphenyl disulphide
with carbene (L. Field and C.H. Banks, J. org. Chem., 1975,
40, 2774).

$$PhSSPh + CH_2: \longrightarrow PhSCH_2SPh$$

Lactones can interact (sulphenylate) with an appropriate reagent in the manner shown (M. Watanabe, K. Shirai and
T. Kunamoto, Chem. Letters, 1975, 855).

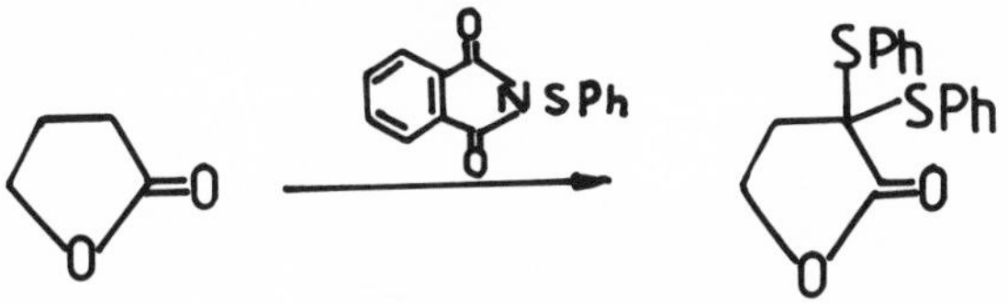

Thioesters such as phenyl dithiobenzoate interact with phenyl lithium to give the same product as obtained from an aromatic aldehyde and thiophenol, together with a vinylic compound (P. Beak and J.W. Worley, J. Amer. chem. Soc., 1972, 94, 597).

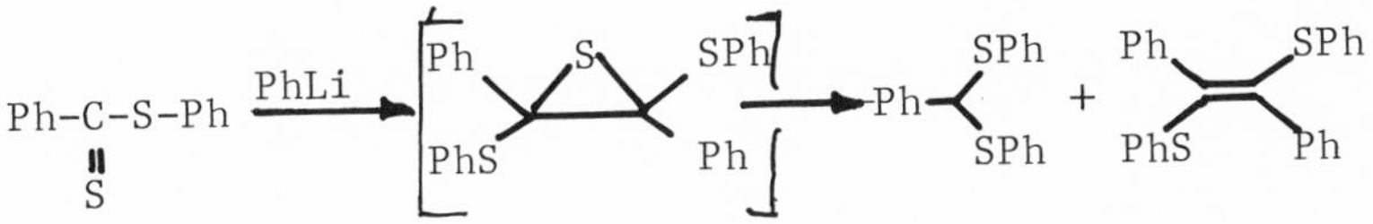

Thiols and carbonyl compounds, with excess of the former, give thioacetals in the presence of an acidic catalyst and with an inert atmosphere to avoid oxidation (T.R. Oakes and G.W. Stacy, J. Amer. chem. Soc., 1972, 94, 1594; Y. Ogato and A. Kawasaki, J. chem. Soc., Perkin II, 1975, 134).

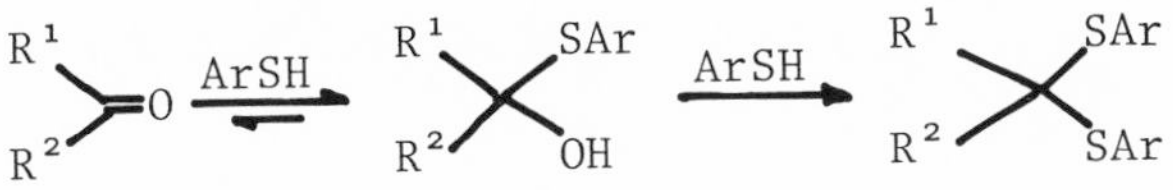

Ketones and thiophenols in the presence of trimethylsilyl chloride give silylated hemithioacetals as shown (T.H. Chan and B.S. Org., Tetrahedron Letters, 1976, 319).

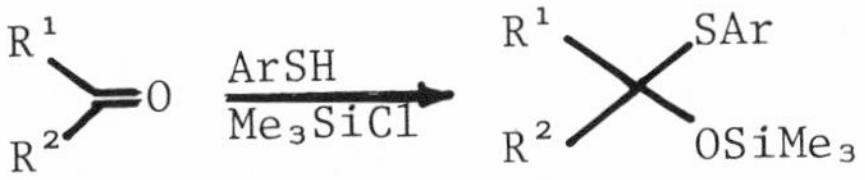

Ketene interacts with tri(thiophenyl)methane to give the thioacetal shown (H. Eck and H. Prigge, Ann., 1972, 755, 177).

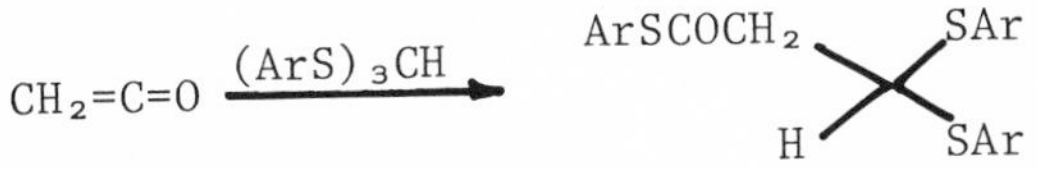

Acetals, and their N-analogues undergo conversion into

thioacetals as illustrated in the following example (Y. Le Floc'h, A. Brault and M. Kerfanto, Compt. rend., 1969, 268C, 1718).

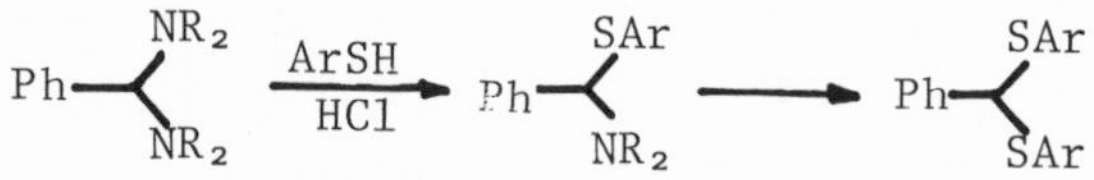

α-Halogenated arylsulphoxides react in the presence of molybdenum hexacarbonyl as illustrated (H. Alper and G. Wall, Chem. Comm., 1976, 273) with dimethoxyethane as solvent.

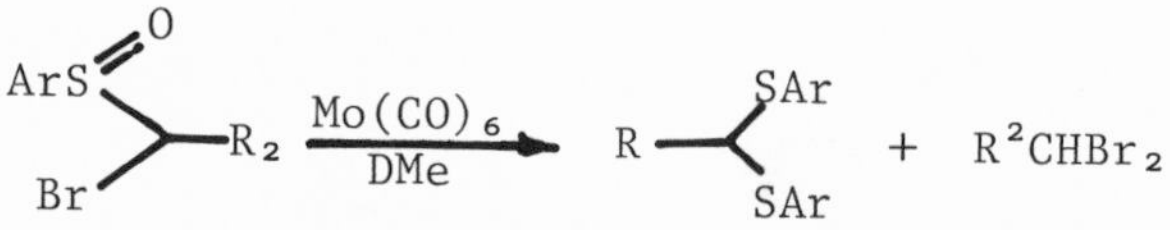

(b)   *Synthetics uses of thioacetals*

Unlike their oxygen analogues, the stability of thioacetals towards acid and their ease of carbanion formation permit a considerable number of synthetic uses.   By alkylation a sulphonium salt is formed from which an ylid can be derived giving the following preparation of a cyclopropyl sulphide (Y. Hayasi, M. Takaku and H. Nozaki, Tetrahedron Letters, 1969, 3179).

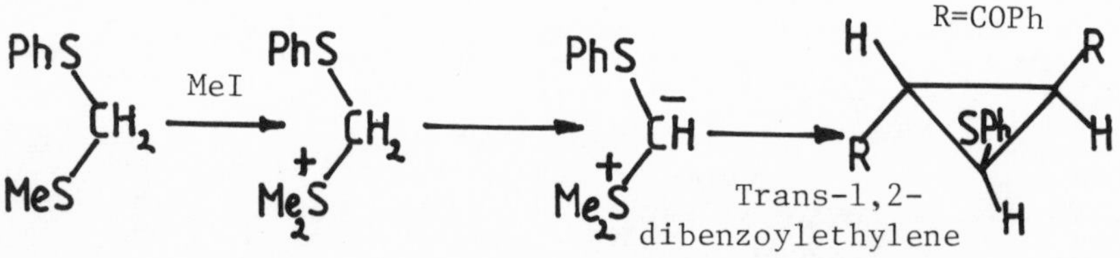

S-alkylation can sometimes take place as well as C-alkylation as in the attempted synthesis of cyclic ketones from α,ω-dibromoalkanes (K. Ogura, M. Yamashita, M. Suzuki and G. Tsuchihashi, Tetrahedron Letters, 1974, 3653), although the use of methyl methyl thiomethylsulphoxide overcame this and led with 1,3-dibromopropane to a synthesis of cyclobutane. Ways around the difficulty have been found by using mixed halogeno α,ω-dihalides (D. Seebach and A.K. Beck, Org. Synth., 1971, 5176.

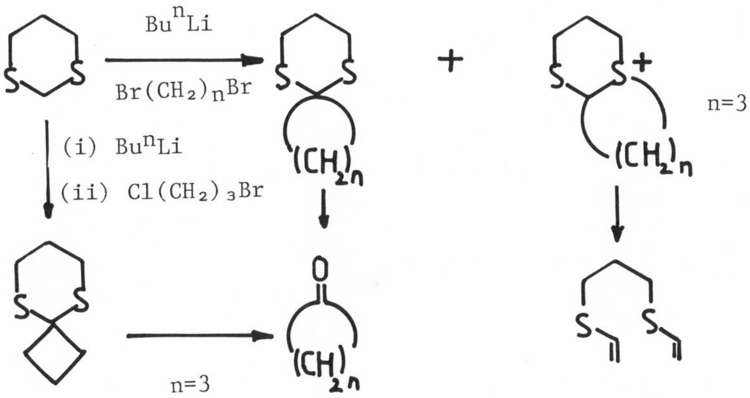

Cation formation at carbon by way of tritylfluoroborate can also be utilised to derive a carbene which forms an alkene as shown (D. Seebach, Angew Chem. Internat. Edn., 1969, 6, 443; Synthesis, 1969, 13).

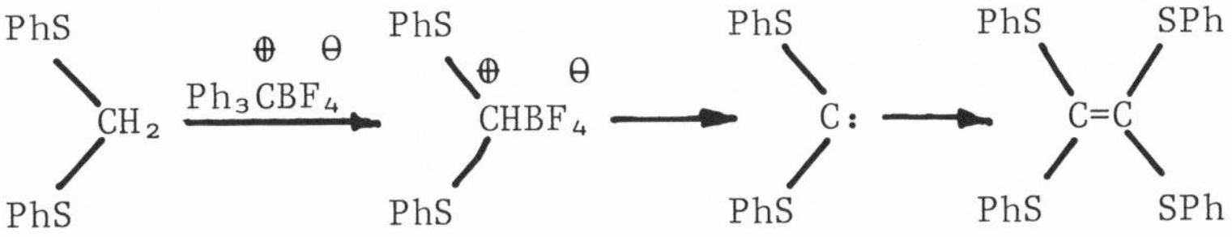

As an alternative to formation of a carbanion,chlorination of 1,3-dithians with N-chlorosuccinimide at the methylene group, giving effectively a formyl cation equivalent, followed by reaction with a vinylic amine $R^1R^2C=CR^3N(R^4)_2$ and acidic treatment gives an α-(1,3-dithian-2-yl) aldehyde or ketone (E.C. Taylor and J.L. La Mattina, Tetrahedron Letters, 1977, 2077).(Scheme 2).

Rearrangement as shown in the following example can lead to the synthesis of *sec* or *t*-alcohols from dithiolanes (R. Hughes *et. al.*, J. chem. Soc., Perkin 1, 1977, 1172). (Scheme 3).

(Scheme 2)

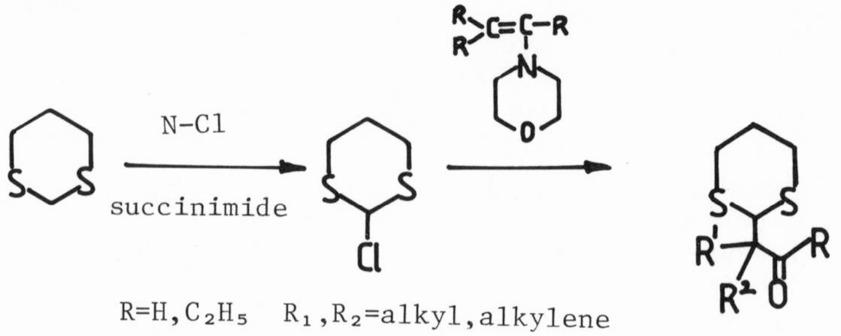

R=H,C₂H₅ R₁,R₂=alkyl,alkylene

(Scheme 3)

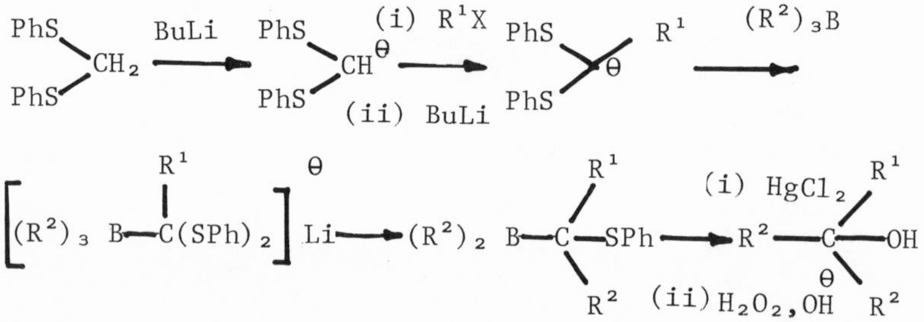

Nucleophilic substitution reactions enabling RX to be converted to RCHO or RCOR^1 have their counterpart in the nucleophilic addition shown, by which RCHO can be transformed to RCOCH₂SPh (I. Kuwajima and Y. Kurata, Chem. Letters, 1972, 291) probably by way of the intermediate RCH(OLi)C(SPh)₂Li.

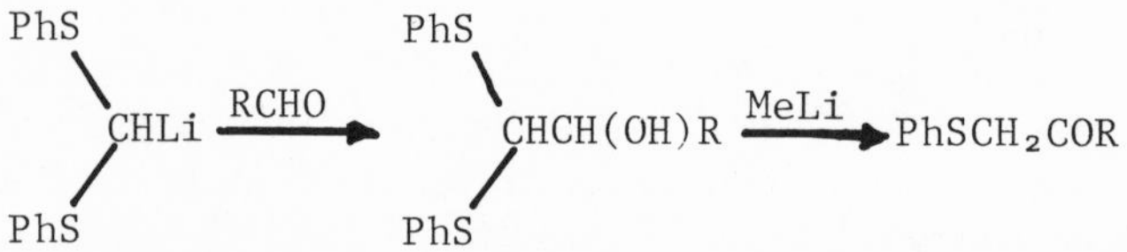

It seems likely that many of the syntheses effected by dithians could also be carried out with dithiolanes, disadvantages being their higher molecular weight and problems of cleavage. Many of the synthetic sequences possible with 1,3-dithians have been summarised (G.C. Barrett, 'Polyfunctional sulphides' in vol. 3, Organic Sulphur Compounds; Comprehensive Organic Chemistry, Barton and Ollis, Pergamon, 1979).

Frequently, efficient removal of the thioacetal portion

after completion of the synthetic transformation is necessary
and a variety of methods exist for this. Unless C-S bond
cleavage can be readily brought about after these stages the
overall value of these intermediates is lost. Methods which
are suitable for formation of a ketone, e.g. aqueous mercuric
chloride, are ineffective for the conversion of monoalkyl derivatives into aldehydes and mercuric oxide with boron trifluoride etherate in aqueous tetrahydrofuran affords yields
in the range 60-90% (J.B. Jones and R. Grayshan, Chem. Comm.,
1970, 741). The use of titanium IV chloride/water (with or
without mercuric oxide) is a newer approach (A.J. Mura *et al.*,
Tetrahedron Letters, 1975, 4435).

Other procedures have been instanced (T. Makaiyama *et
al.*, Bull. chem. Soc., Japan, 1972, 45, 3723).

When 1,3-dithiolans were used in place of the 1,3-di-
thian in a study of the synthesis of dihydrojasmone shown below, the routes were comparable but hydrolysis was incomplete
with cold concentrated sulphuric in the former case.

Occasionally removal of PhS from a vinylic sulphide
structure resulting from a synthesis has proved intractable
and in one case 20% hydrochloric acid in dioxan is needed
since the mercuric ion-catalysed procedure is ineffective
(J.P. Marino and R.C. Landick, Tetrahedron Letters, 1975,
4531).

Silver oxide with methanol (D. Gravel, C. Vaziri and
S. Rahal, Chem. Comm., 1972, 1323) is an alternative cleavage
agent but requires a long reaction period.

Oxidative hydrolysis with N-halosuccinimide agent can
be much more rapidly effected and may be of value where acidic
conditions are prohibitive (E.J. Corey and B.W. Erickson,
J. org. Chem., 1971, 36, 3553). Ceric ammonium nitrate has
been employed (T.-L. Ho, H.C. Ho and C.M. Wong, Chem. Comm.,

1972, 791), and chloramine T (W.F.J. Hurdeman, H. Wynberg and
D.W. Emerson, Tetrahedron Letters, 1971, 3449). Oxidation of
1,3-dithiolans with monoperphthalic acid gives the corresponding sulphones which can be hydrolysed under alkaline conditions (S.J. Daum and R.L. Clarke, *ibid.*, 1967, 165). The use
of iodine-dimethylsulphoxide at 100° results in 80% yields of
final products in the course of one hour (J.B. Chattopadhyaya
and A.U. Rama Rao, Tetrahedron Letters, 1973, 3735). Smooth
C-S cleavage by anodic oxidation has been described giving in
dry solvent an α-disulphide or in wet solvent a thiosulphonate
(J.-G. Gouray, G. Jeminet and J. Simonet, Chem. Comm., 1974,
634).
Mild alkaline cleavage can be employed on the mono or di-sul-
phonium salts of 1,3-dithiolans formed by the use of the
Meerwein reagent. (T. Oishi, H. Takechi, K. Kamemoto and Y.
Ban, Tetrahedron Letters, 1974, 11) to afford as shown good
yields of aldehydes or ketones (M. Fétizon and M. Juron, Chem.
Comm., 1972, 382).

In all the preceding applications the objective has been
recovery of a carbonyl compound by essentially a hydrolytic
procedure. By reductive cleavage the synthesis of hydrocarbons and functionalised hydrocarbons is possible and a wide
variety of reagents is available. Many of those, such as
lithium/ethylamine at low temperatures, hydrazine or Raney nickel, are well known and their use has become standard. Lithium aluminium hydride with copper (II) chloride and zinc chloride has been shown to be more effective (T. Mukaiyama, Internat. J. sulphur Chem., 1972, 7, 173). Calcium in liquid ammonia yields alkylthioalkane thiols from 1,3-dithiolans and 1,3-
dithians by cleavage of only one C-S bond. (B.C. Newman and
E.L. Eliel, J. org. Chem., 1970, 35, 3641). Nickel boride has
been used as a desulphurising agent (R.B. Boar, *et al.*, J.
chem. Soc., Perkin I, 1973, 634).

(c) *Synthetic uses of tris(arylthio)methanes*

By the further substitution of an arylthio group into
dithiolanes, tris(arylthio)methanes are obtained. The react-

ion of phenylsulphenyl chloride with the carbanion of a di-
thiolane gives one route to the required starting material as
shown.

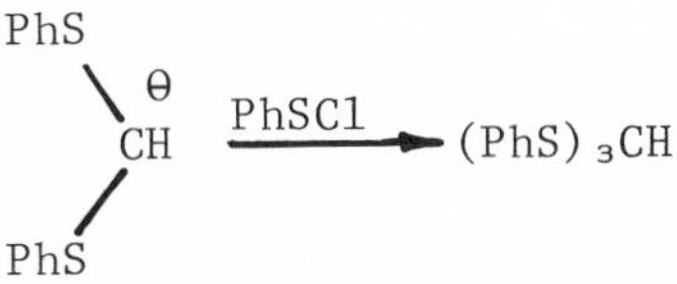

PhS
  \\ $\ominus$
   CH $\xrightarrow{\text{PhSCl}}$ (PhS)$_3$CH
  /
PhS

Other routes are the interaction of methyl or ethyl
orthoformate with a thiol in the presence of boron trifluoride
etherate or from a phenyl thioformate (D.H. Holsboer and
A.P.M. van der Veek, Rec. trav. chim., 1972, 91, 349).

HC(OMe)$_3$ $\xrightarrow[\text{BF}_3,\text{Et}_2\text{O}]{\text{PhSH}}$ HC(SPh)$_3$ $\xleftarrow[\text{HgCl}_2]{\text{ArSCHO}}$ Cl$_2$CHSAr $\xleftarrow{\text{PCl}_5}$ OHCSPh

Tris(arylthio)methanes are oxidised by iodine to hexa-
(thioaryl)ethanes.   These in turn decompose thermally to give
tri(thiophenyl)methyl radicals which undergo conversion into
the carbene and thence by Wolff rearrangement into phenyl dithiobenzoate as shown (D. Seebach and A.K. Beck, Ber., 1969,
105, 3892).

(PhS)$_3$CC(SPh)$_3$ $\xrightarrow{100°}$ (PhS)$_3$C· $\longrightarrow$ PhSC̈SPh $\longrightarrow$ PhCSPh
$\qquad\qquad\qquad\qquad\qquad\qquad\qquad\qquad\qquad\qquad\qquad\quad$ ‖
$\qquad\qquad\qquad\qquad\qquad\qquad\qquad\qquad\qquad\qquad\qquad\quad$ S

The product from the interaction of tris(arylthio)meth-
ane with butyllithium (D. Seebach, Angew. Chem. Internat. Ed.,
1967, 6, 442) may be considered as a methyl carbanion equivalent.   Michael addition and carbonyl addition reactions respectively are exemplified in the reactions shown below
(A.R.B. Manas and R.A.J. Smith, Chem. Comm., 1975, 216).

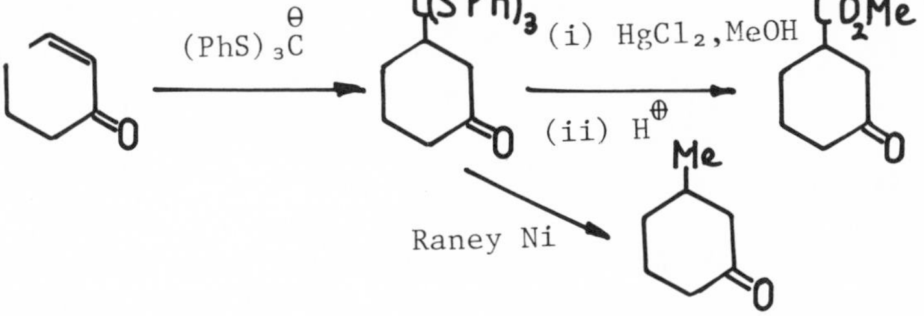

(3) *Unsaturated sulphides*

A number of unsaturated sulphides such as 1,1-bis(aryl-thio)alkenes, sometimes called keten thioacetals, and other vinylic and acetylenic sulphides have an important place in organic synthesis. Although some of these arise from the re-actions of materials already mentioned in prior sections, their special applications are emphasised in outline in this present section.

The use of ketene thioacetals is almost entirely con-fined to the alkylthio compounds (3) to the 1,3-dithians (1), (2) shown and a few aryl members (4).

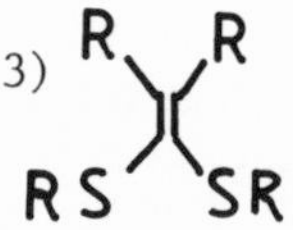

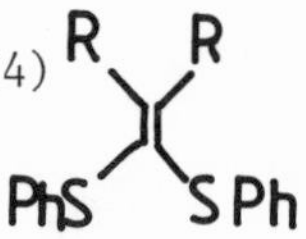

The bis(thiophenyl) compound shown has been obtained from the reaction of 1,1,1-tri(thiophenyl)ethane with tri-fluoroacetic anhydride (M. Hojo, R. Masuda and Y. Kamitori, Tetrahedron Letters, 1976, 1009).

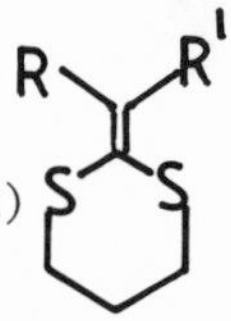

$$(ArS)_3CCH_3 \xrightarrow{(CF_3CO)_2O} (ArS)_2C=CHCOCF_3$$

(a) *Preparation*

Vinylic sulphides can be obtained from a number of diff-erent routes. The interaction of ketones with the thioanisole derivative shown provides an efficient synthesis (D.J. Peters-on, J. org. Chem., 1968, 33, 780; F.A. Carey and A.S. Court, *ibid.*, 1972, 37, 939; T.M. Dolak and T.A. Bryson, Tetrahedron Letters, 1977, 1961).

$$R_2C=O + PhSCH_2LiSiMe_3 \longrightarrow R_2C=CHSPh$$

The ylid illustrated reacts with aldehydes to afford similar compounds

$$RCHO + Ph_3P=CHSPh \longrightarrow RCH=CHSPh + Ph_3PO$$

Alkynes undergo addition of thiophenols (and of aliphatic disulphides). The acetylenic ketone shown gives predominantly
a *cis*-product but addition was not stereospecific (E.L.
Prilezhaeva, I.L. Mikhelsavili and V.S. Bogdanov, Zhur. org.
Khim., 1972, $\underline{8}$, 1505).

$$HC\equiv CCOR + PhSH \xrightarrow{Z/E} PhSCH=CHCOR$$

Phenylsulphenyl chloride adds to the α,β-acetylenic ester indicated to give mainly a *trans*-product (M. Verny and R.
Vessiere, Bull. Soc. Chim., France, 1970, 746).

$$MeC\equiv CCO_2Et \xrightarrow{PhSCl} ClC(Me)=C(SAr)CO_2Et$$
$$E/Z$$

The copper (II) derivative shown interacts with arylsulphenyl
chlorides to give an acetylenic sulphide (C.E. Castro, *et. al.*,
J. Amer. chem. Soc., 1969, $\underline{91}$, 6464).

$$(RC\equiv C)_2Cu + ArSCl \xrightarrow[\text{acetonitrile}]{\text{pyridine or}} RC\equiv CSAr$$

Simultaneous elimination and substitution can be used to convert a vinylic into an acetylenic sulphide as shown
(S.Y. Delavarenne and H.G. Viehe, Tetrahedron Letters, 1969,
4761).

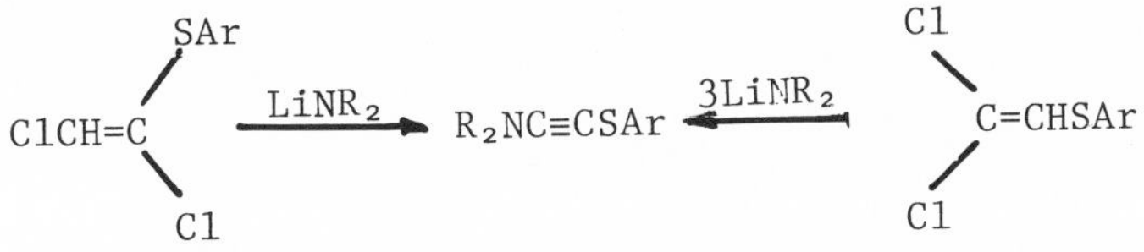

Tolan with sulphur dichloride gives 2-phenylbenzothiophene
while alkyl derivatives of phenylacetylene give products whose
configuration depends on the solvent, as shown (T.J. Barton and
R.G. Zika, J. org. Chem., 1970, $\underline{35}$, 1729). (Scheme 4).

Organosulphur compounds provide a third source of vinylic sulphides. Thiobenzophenone with the allenic compound shown affords a four membered cyclic vinylic sulphide and a dienoid
compound (H. Gotthardt, Tetrahedron Letters, 1971, 2343, 2345).
(Scheme 5).

(Scheme 4)

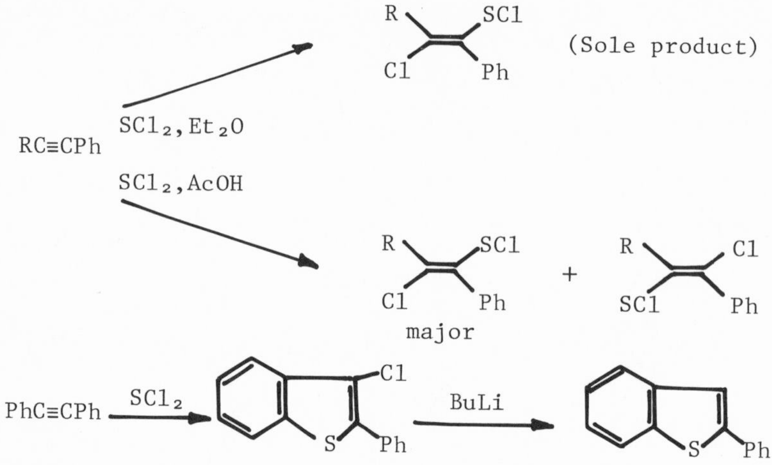

RC≡CPh

SCl₂, Et₂O

SCl₂, AcOH

(Sole product)

major

PhC≡CPh →SCl₂→ →BuLi→

(Scheme 5).

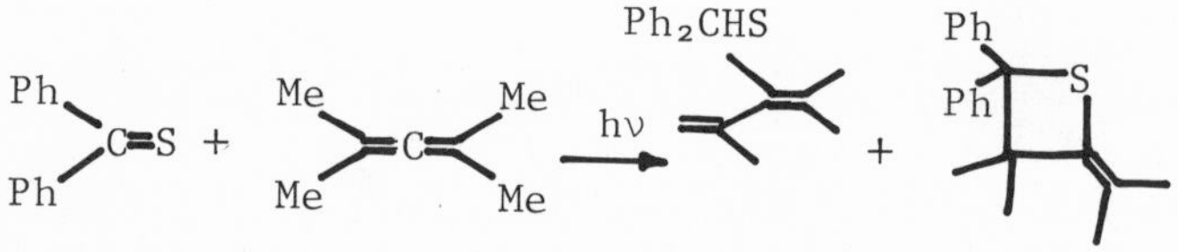

Thioketens interact with phenyllithium, and carbonation of the product yields the carboxyvinylic sulphide. (E. Schauman and W. Walter, Ber., 1974, 107, 3562).

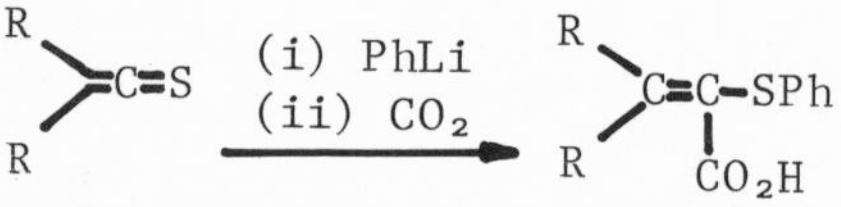

The reaction of phenyl methylthioacetylene with lithium aluminium hydride gives the E-vinylic sulphide whereas copper (I) hydride gives the Z-adduct. (R. Vermeer, *at. al.*, Rec. trav. Chim., 1976, 95, 25). (Scheme 6).
Phenylthiodiethylaminoacetylene in the presence of boron fluoride dimerises to a cyclobutadiene derivative (R. Gompper, S. Mensch and G. Leybold, Angew. Chem., 1975, 87, 711).

(Scheme 6).

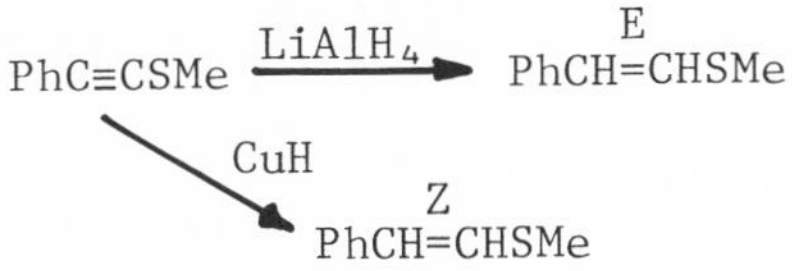

PhC≡CSMe $\xrightarrow{\text{LiAlH}_4}$ $\overset{\text{E}}{\text{PhCH=CHSMe}}$

$\searrow$ CuH

$\overset{\text{Z}}{\text{PhCH=CHSMe}}$

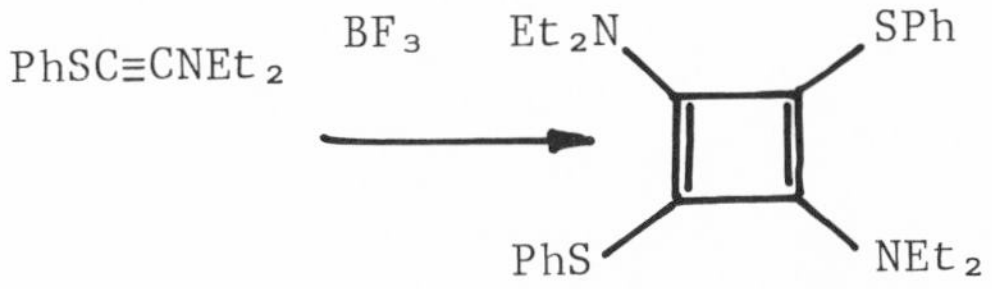

PhSC≡CNEt$_2$ $\xrightarrow{\text{BF}_3}$

(b) *Synthetic uses of unsaturated sulphides*

The main use of 1,1-bis(alkylthio)alkenes lies in their acidic hydrolysis to give aldehydes, or ketones according to the method employed, as illustrated (E.J. Corey and R.A.K. Chen, Tetrahedron Letters, 1973, 3817; E.J. Corey and A.P. Kozikowski, *ibid.*, 9253).

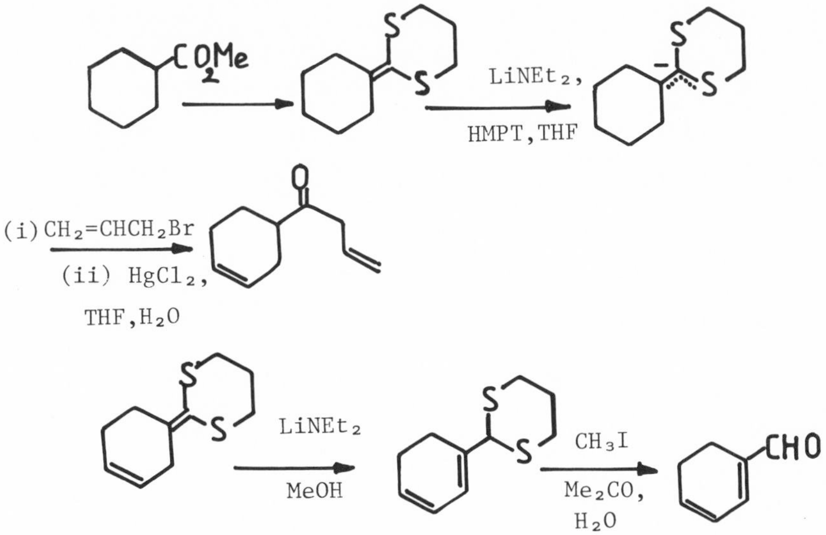

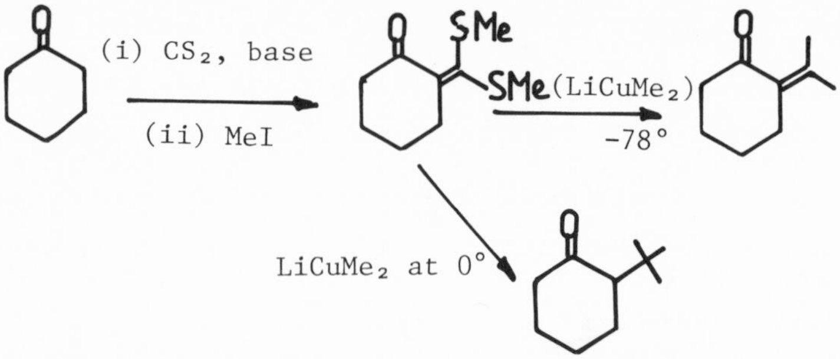

Employment of PhSCH₂SPh in the above reaction and in the example shown below would seem to be possible but it could be anticipated that C-S bond cleavage would be less straightforward (F.A. Carey and A.S. Court, J. org. Chem., 1972, 37, 1926, 4474).

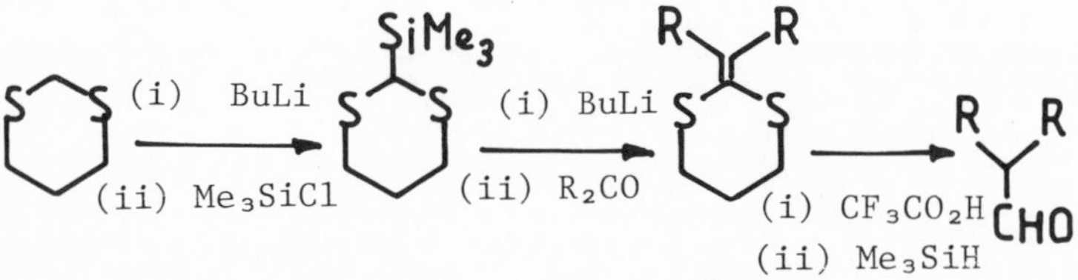

Vinylic sulphides such as ArSCH=CHR are intermediates in the homologation of aldehydes (T. Yamamoto, M. Kakimoto and M. Okawara, Tetrahedron Letters, 1977, 1659; A.J. Mura, D.A. Bennett and T. Cohen, *ibid.*, 1975, 4433).

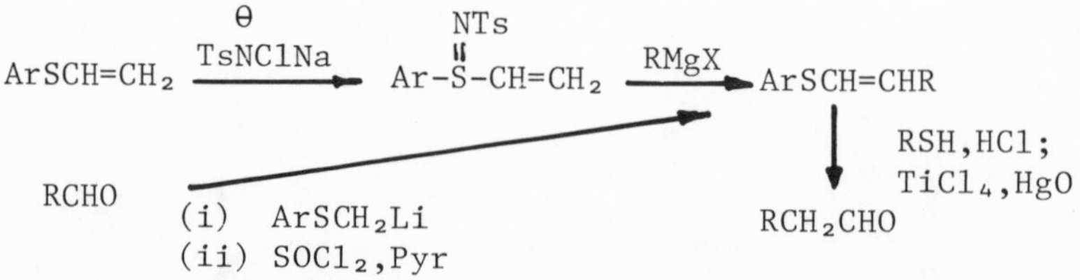

The intermediate can be derived by a somewhat simpler sequence from RCHO as shown (B.M. Trost and D.E. Keeley, J. Amer. chem. Soc., 1976, 98, 248).

The use of the corresponding substituted (R^1) analogue shown gives a route to ketones (A.J. Mura *et. al.*, Tetrahedron Letters, 1975, 4437), dimethyllithiocuprate affords α,β-unsaturated esters (G.H. Posner and D.J. Brunelle, Chem. Comm., 1973, 907) and Raney nickel gives a route to alkenes

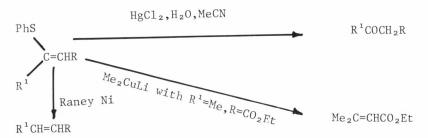

$$HgCl_2, H_2O, MeCN$$

$$PhS$$

$$C=CHR$$

$$R^1$$

$$R^1COCH_2R$$

Raney Ni

$$Me_2CuLi \; with \; R^1=Me, R=CO_2Et$$

$$Me_2C=CHCO_2Et$$

$$R^1CH=CHR$$

Allylic sulphides represent another group of compounds which can be utilised in organic synthesis. Certain examples involving the group >C=C-CH$_2$SPh have been referred to in the section describing the use of PhSCH$_2$Li and others are considered below. Some of the best aliphatic examples of the use of the allylic sulphide group are outside the scope of this review on aryl compounds.

In the synthesis of germaclane type sesquiterpenes, the key stage involves opening an epoxide ring by a lithio derivative of an allylic sulphide (M. Kodama, J. Matsuki and S. Ito, Tetrahedron Letters, 1976, 1121).

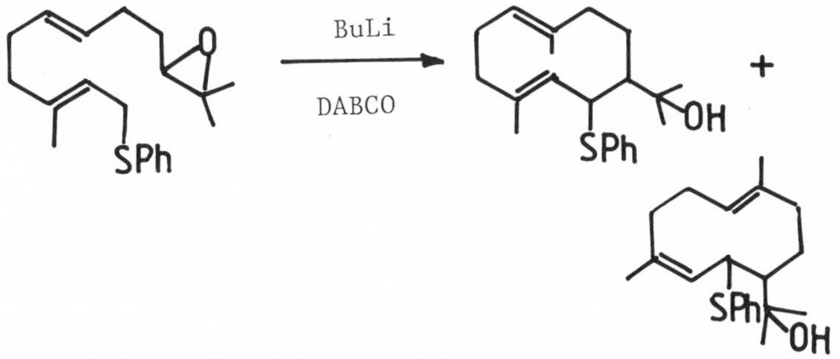

BuLi

DABCO

+

Furanoterpenes can be derived by the route shown (M. Kondo and M. Matsumoto, Tetrahedron Letters, 1976, 391). (Scheme 7).

Natural product synthesis has been considerably stimulated by the advent of these sulphur-containing reagents as shown by the following intramolecular rearrangements (K. Kondo

and I. Ogima, Chem. Comm., 1972, 62). (Scheme 8).
Scheme 7

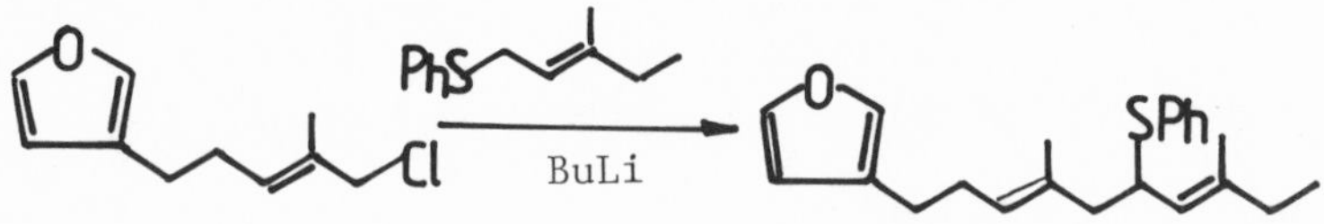

Scheme 8

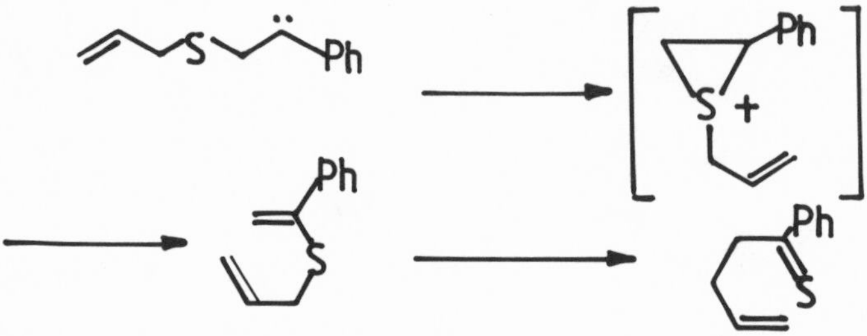

The malonate carbene shown interacts with a substituted allylic sulphide followed by a rearrangement (P.A. Grieco,
D. Boxler and K. Hiroi, J. org. Chem., 1973, 38, 2572).

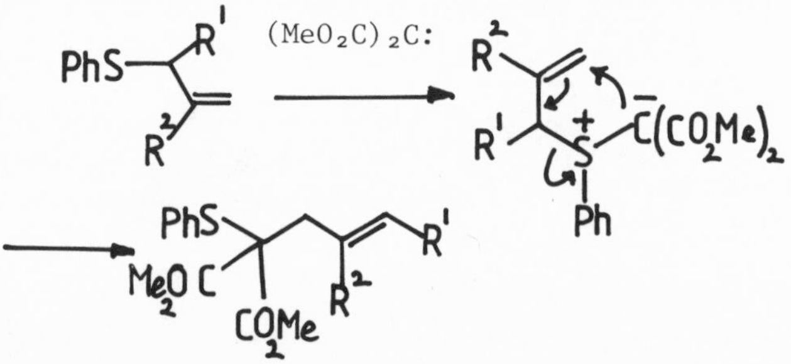

(4)   *Benzylic sulphides*

Related in reactivity to the allylic sulphides, benzylic
sulphides have been widely employed in the synthesis of cyclophanes notably by V. Boekelheide and coworkers. *o*-Xylylene dibromide with sodium sulphide affords the disulphide shown
(R.A. Mitchell and V. Boekelheide, J. Amer. chem. Soc., 1970,
92, 3510).

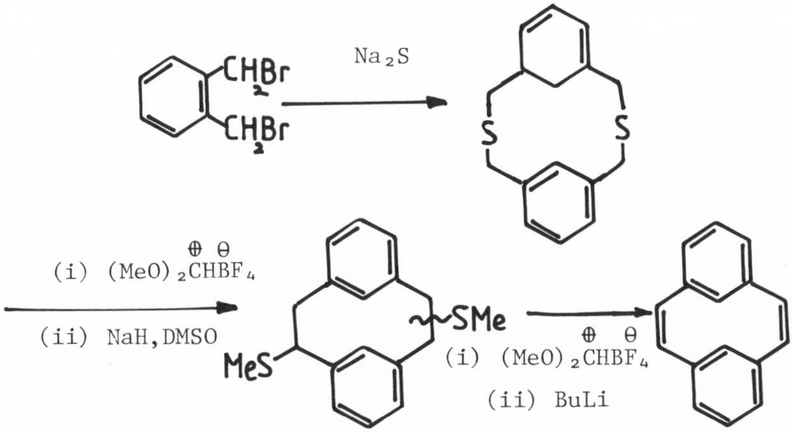

Upon methylation, followed by Stevens rearrangement (the method of choice) the intermediate shown gives (2,2-metaparacyclophane-1,9-diene in good yield (V. Boekelheide and P.H. Anderson, Tetrahedron Letters, 1970, 1207). Sulphones have also been used to produce similar cyclophanes (F. Vögtte, Angew. Chem., Internat. Edn., 1969, 8, 274).

Extrusion reactions with substituted benzyl and dibenzyl sulphides have been employed for the synthesis of the hindered olefins (D.H.R. Barton and B.J. Willis, Chem. Comm., 1970, 1225).

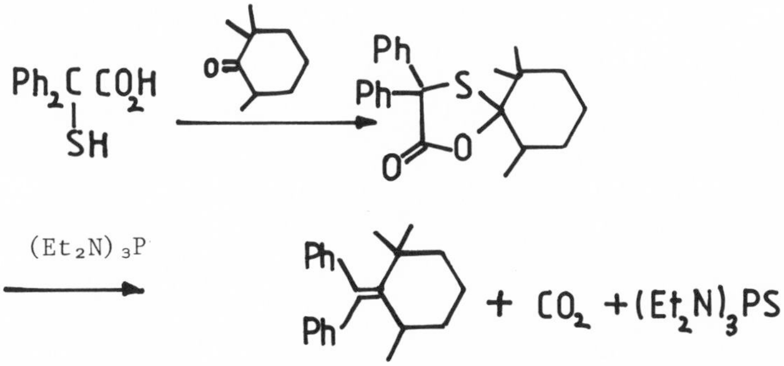

Chapter 6

MONONUCLEAR HYDROCARBONS CARRYING NUCLEAR
SUBSTITUENTS CONTAINING SULPHUR: SULPHONIC ACIDS,
SULPHINIC ACIDS,   AND SULPHENYL COMPOUNDS OF THE
BENZENE SERIES

G.C.BARRETT

1 *Nuclear Sulphonic Acids of Benzene and its Homologues;*
  Arene Sulphonic Acids

(a)   Preparation

Direct or indirect introduction of the SO₃H group into the benzene
nucleus, brought about by sulphonation or by the oxidation of a sulphur
functional group, respectively, continue to be used for the preparation
of sulphonic acids of the benzene series.   Insertion of SO_3 into metal –
carbon bonds (W.Kitching and C.W.Fong, Organometal.chem.Reviews,
1970, A5, 281; A.Wojcicki, Acc.chem.Res., 1971, 4, 344), though
explored mostly only for the preparation of alkanesulphonic acids, has
yielded $PhSO_3Re(CO)_5$ and the p-tolyl analogue (E.Lindner and R.Grim-
mer, Ber., 1971, 104, 544) from which the arenesulphonic acids may be
obtained.

Direct Sulphonation

   While H_2SO_4, $H_2S_2O_7$, or SO_3 are commonly used for sulphonation
(for mechanistic studies, see H.Cerfontain, A.Koeberg-Telder, and E.
van Kuipers, J.chem.Soc., Perkin II, 1972, 2091; P.K.Maarsen and
Cerfontain, ibid., 1977, 1008), excellent yields have been claimed

for the use of an alternative reagent, an N-alkylsulphamic acid at 150-195°, in the preparation of substituted benzenesulphonic acids (F.L.Scott, J.A.Barry, and W.J.Spillane, J.chem.Soc., Perkin I, 1972, 2663):

$$R_2N.SO_3H + ArH \longrightarrow ArSO_3^- H\overset{+}{N}R_2$$

(ii) Indirect Sulphonation

Oxidation of disulphides to arenesulphonic acids can be brought about through the use of Cl_2-H_2O, or 30% aqueous H_2O_2, or 90% H_2O_2-trifluoroacetic anhydride-CH_2Cl_2, the last-mentioned system being superior by virtue of its simplicity and the quality of the product (W.H.Dennis, D.H.Rosenblatt, and J.R.Simcox, Synthesis, 1974, 295).

Copper arenesulphinates, formed from the organocuprate and SO_2 at -40°, can be oxidized to the corresponding sulphonyl bromides with bromine and thence to arenesulphonic acids by hydrolysis (G.Cahiez, D.Bernard, J.F.Normant, and J.Villieras, J.organometal.Chem., 1976, 121, 123). Arenesulphinic acids themselves can be oxidized to arenesulphonic acids using benzeneseleninic acid (R.A.Gancarz and J.L.Kice, Tetrahedron Letters, 1980, 1697).

$$ArSO_2H + PhSeO_2H \longrightarrow ArSO_3^- PhSeOH_2^+ + ArSO_2SePh$$

Further examples of nucleophilic substitution approaches to arene-sulphonic acids through the use of sodium sulphite as reagent include reactions leading to sodium hexabenzenesulphonate (N.S.Dokunikhin and G.A.Mezentseva, Chem.Abs., 1974, 81, 63282; idem, Zhur.org. Khim., 1976, 12, 621).

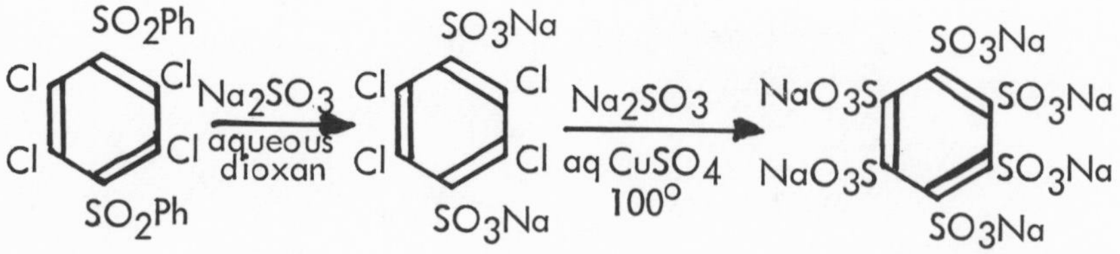

(b) General Properties and Reactions

The first example of the direct reduction of a sulphonic acid to the corresponding thiol has been reported (T.Numata, H.Awano, and S.Oae, Tetrahedron Letters, 1980, 1235).

$$ArSO_3H + (CF_3CO)_2O + Bu_4N^+I^- \xrightarrow[CH_2Cl_2]{\text{room temp}} ArSH + ArSCOCF_3$$

Displacement of the sulphonate grouping from an arenesulphonic acid has been illustrated frequently over the years; a novel photolytic procedure leading to reductive desulphonylation products (O.P.Studzinskii, N.I. Rtishchev, and A.V.Eltsov, Zhur.org.Khim., 1977, 13, 160; N.Suzuki, K.Ito, A.Inoue, and Y.Izawa, Chem. and Ind., 1977, 399)

$$ArSO_3^-Na^+ \xrightarrow[DMSO]{h\nu} (\pi - \pi^*) \longrightarrow ArH$$

or to the products of replacement by chloro- or amino-groups (Eltsov, Studzinskii, and A.V.Devekki, Zhur.org.Khim., 1973, 9, 740) has been illustrated with representative arenesulphonic acids.

The sulphonate anion $-SO_2O^-$ is a deactivating substituent, even though it carries a negative charge, as demonstrated in the rate and course of the nitration of arenesulphonic acids (R.B.Moodie, K.Schofield, and T.Yoshida, J.chem.Soc., Perkin II, 1975, 788). [1]H- and ^{13}C-n.m.r. studies (Koeberg-Telder and Cerfontain, ibid., 1975, 226) and u.v. spectrometric studies (Cerfontain and B.W.Schnitger, Rec. trav.Chim., 1972, 91, 199) provide information on pK values of arenesulphonic acids in sulphuric acid solutions.

The relative nucleophilicities of arenesulphinate anions have been determined through a kinetic study of the following reaction (D.N.Kerill and A.Wang, Chem.Comm., 1976, 618), indicating that substituents in the aryl grouping transmit the effects of electron release or withdrawal through the sulphur atom of the sulphonate anion.

$$ArSO_3^- Bu_4N^+ + CF_3SO_3CH_3 \longrightarrow ArSO_3CH_3 + CF_3SO_3^- Bu_4N^+$$

(c) Sulphonic anhydrides $ArSO_2.O.SO_2$ and $ArSO_2.O.COR$

Most of the published work deals with aliphatic sulphonic anhydrides, but includes results which apply to these classes of compound generally. Hydrolysis mechanisms (R.M.Laird and M.J.Spence, J.chem.Soc.B,

1971, 454, 1434) and comparative studies of aminolysis and methanolysis (N.T.Maleeva, V.A.Savelova, L.M.Litvinenko, and L.G.Kuryakova, Zhur.org.Khim., 1975, 11, 1015) have been reported for sulphonic anhydrides, while for mixed sulphonic - carboxylic anhydrides there have been studies of disproportionation, ketene extrusion leading to the formation of sulphonate esters, and the analogous formation of methoxymethyl sulphonates with $CH_2(OCH_3)_2$ (M.H.Karger and Y.Mazur, J.org. Chem., 1971, 36, 528,532,540).

Unsymmetrical arenesulphonic anhydrides $ArSO_2.O.SO_2CF_3$ are the relatively unstable products of the reaction:

$$ArSO_2Br + CF_3SO_3^- Ag^+ \longrightarrow ArSO_2.O.SO_2CF_3 + AgBr$$

They are powerful Friedel-Crafts sulphonylation reagents (K.Huthmacher, G.Koenig, and F.Effenberger, Ber., 1975, 108, 2947; Effenberger and Huthmacher, ibid., 1976, 109, 2315).

$$ArSO_2.O.SO_2CF_3 + Ar'H \longrightarrow ArSO_2Ar' + CF_3SO_3H$$

(d) Sulphonyl peroxides $ArSO_2.O.OSO_2R$ and $ArSO_2.O.O.COR$

Products of radical decomposition of benzoyl toluene-p-sulphonyl peroxide involving attack on solvent include chlorobenzene and hexachloroethane when conducted in carbon tetrachloride (R.Hisada, H.Minato, and M.Kobayashi, Bull.chem.Soc.Japan, 1971, 44, 2541). Arenesulphonoxylation products are formed from symmetrical sulphonyl peroxides with alkenes (J.Bolte, A.Kergomard, and S.Vincent, Bull.Soc. chim.France, 1972, 301) and with aromatic substrates (R.L.Dannley, R.V.Hoffman, P.K.Torstrom, R.L.Waller, and R.B.Srivastava, J.org. Chem., 1974,39, 2543; Dannley and Tornstrom, *ibid*.1975, 40, 2278; Dannley and Hoffman, *ibid*. p.2426).

$$(ArSO_2.O)_2 + PhCOCH_3 \longrightarrow ArSO_2.O\text{-}C_6H_4\text{-}COCH_3$$
$$+ PhCOCH_2O.SO_2Ar$$

(e) Sulphonate Esters

A novel variation of a standard method of synthesis of sulphonate esters involves transesterification of arenesulphonic acids with an acyl phosphonate in benzene (P.Golborn, Synthetic Comm., 1973, $\underline{3}$, 273).

$$ArSO_3H + R^1CH_2CO.P(O)(OR^2)_2 \rightarrow ArSO_3R^2$$

Arenesulphonoxylation of arenes has been discussed in the preceding section. Electrophilic substitution of activated benzenes with methyl fluorosulphonate ('magic methyl') at 100° leads to methyl arenesulphonate esters (T.Kametani, K.Takahashi, and K.Ogasawara, Synthesis, 1972, 473).

Arenesulphonylation would be a more widely-used device for the protection of alcohols if improved methods for the cleavage of sulphonate esters could be found. Cleavage using solutions of alkali metals in hexamethylphosphoric triamide (T.Cuvigny and M.Larcheveque, J.organometal.Chem., 1974, $\underline{64}$, 315), electrochemical reduction (P.H. Boyle and J.H.Coy, J.org.Chem., 1973, $\underline{38}$, 826), or photolysis in ether at 254nm (A.Abad, D.Mellier, J.P.Pete, and C.Portella, Tetrahedron Letters, 1971, 4555, 4559) have been generally disappointing because of low yields or extended reaction times.

The extremely high alkylation reactivity of alkyl trifluoromethanesulphonate esters ('triflates') and fluorosulphonates contrasts with the mild behaviour of a deactivated benzenesulphonate ester, illustrated in the use of methyl p-nitrobenzenesulphonate for the specific methylation of the imidazole grouping in the side-chain of the histidine residue at position 57 in chymotrypsin (Y.Nakagawa and M.L.Bender, J.Amer.chem.Soc., 1969, 91, 1566). This reaction course competes with the more familiar nucleophilic S-O cleavage, for which the flow of papers dealing with structure – rate relationships is as voluminous as ever (e.g. G.C.Barrett, in 'Organic Compounds of Sulphur, Selenium, and Tellurium', a Specialist Periodical Report of The Chemical Society, Vols. 1 – 5, 1970-1979). Hammett and Taft substituent constants have been determined for toluenep-sulphonate, methanesulphonate, and triflate groups (P.J.Stang and A.G.Anderson, J.org.Chem., 1976, $\underline{41}$, 781).

A solid-state rearrangement of p-dimethylaminobenzenesulphonate esters is made possible by an appropriate arrangement of molecules in the crystal structure (C.N.Sukenik et al., J.Amer.chem.Soc., 1977, $\underline{99}$, 851).

$$p\text{-}(CH_3)_2N.C_6H_4.SO_3CH_3 \longrightarrow p\text{-}(CH_3)_3\overset{+}{N}.C_6H_4.SO_3^{-}$$

(f)   Sulphonyl halides

Preparative procedures for arenesulphonyl chlorides are well-established
and these compounds provide a relay for the synthesis of corresponding
fluorides, involving ion-exchange aided by a cation-exchange resin
(C.L.Borders, D.L.McDonell, and J.L.Chambers, J.org.Chem., 1972,
37, 3549) or by an 18-crown-6-ether (T.A.Bianchi and L.A.Cate, ibid.,
1977, 42, 2031), or the use of xenon difluoride (S.A.Volkova, Z.M.
Sinyntina, and L.N.Nikolenko, Zhur.obshch.Khim., 1974, 44, 2592).
Sulphonyl fluorides have been prepared by electrofluorinolysis of sulphones (H.Buerger, F.Heyder, G.Pawelke, and H.Niederprum, J.Fluorine
chem., 1979, 13, 251). Sulphonyl halides are intermediates in certain
methods, described earlier in this Chapter, for the conversion of divalent
sulphur compounds into sulphonic acids.

The reaction of a sulphuryl halide with an organolithium compound
has been illustrated so far only for the synthesis of alkanesulphonyl
chlorides (H.Quast and F.Kees, Synthesis, 1974, 489).

$$RLi \ + \ SO_2Cl_2 \xrightarrow{\ -65^o \ to \ -20^o\ } RSO_2Cl$$

Photolysis of benzenesulphonyl chloride yields the products of both
C-S and S - Cl homolysis, whereas the corresponding fluoride suffers
only C-S cleavage (I.I.Kandror, ...G.Gasanov, and R.K.Friedlina,
Tetrahedron Letters, 1976, 1075; *idem,* Izvest.Akad.Nauk SSSR, Ser.
khim., 1977, 550).

$$RSO_2{}^{\cdot} \ + \ hal^{\cdot} \longleftarrow RSO_2hal \longrightarrow R^{\cdot} \ + \ {}^{\cdot}SO_2hal$$

The use of arenesulphonyl halides for the synthesis of sulphinyl compounds is covered elsewhere in this Chapter. Electrochemical reduction
of arenesulphonyl chlorides in acetonitrile — chloromethane gives a
mixture of methyl arenesulphonate and aryl methyl sulphide (J.G.
Gourcy, G.Jeminet, and J.Simonet, Compt.rend., 1973, 277C, 1079).
Further radical reactions which have been illustrated recently include
photolysis (Kandror *et al.,loc.cit.*) and thermolysis (A.Horowitz and
L.A.Rajbenbach, J.Amer.chem.Soc., 1975, 97, 10) leading to sulphinic acids when a suitable proton source is available, and thermal or
copper(II)-catalyzed addition of arenesulphonyl bromides to phenylacetylene (Y.Amiel, J.org.Chem., 1974, 39, 3867) leading respectively to
trans or nonstereospecific addition products. In the homolytic addition
category, reports of the photo-addition of toluene-p-sulphonyl iodide to

styrenes (C.M.M. da Silva Correa and W.A.Waters, J.chem.Soc., Perkin II, 1972, 1575) and copper(II) catalyzed addition of arenesulphonyl chlorides to alkenes (J.Sinnreich and M.Asscher, J.chem.Soc., Perkin I, 1972, 1543) describe mechanistic studies of useful synthetic routes to sulphones. The value of these routes is limited, however, by the fact that the process is energetically unfavourable (Sinnreich and Asscher, *loc.cit*.), and this is made apparent in low yields of products and the formation of mixtures of isomers arising from isomerized alkenes form ed by the reversal of the addition reaction.

$$ArSO_2{\cdot} \ + \ {>}C = C{<} \longrightarrow ArSO_2{-}\overset{|}{\underset{|}{C}}{-}\overset{|}{\underset{|}{C}}{\cdot}$$

(g)   Sulphonamides

Friedel-Crafts sulphonamidation of benzene derivatives has been reviewed (S.K.Gupta, Synthesis, 1977, 39).

$$R^1R^2NSO_2Cl \ + \ \underset{R^3}{\bigcirc} \ \xrightarrow{\ AlCl_3\ } \ \underset{R^3}{\bigcirc}{}^{SO_2NR^1R^2}$$

Oxidation of sulphenamides with m-chloroperbenzoic acid (R.S. Glass and R.J.Swedo, Synthesis, 1977, 798) or with active MnO_2 at 70° (I.Glande and A.Golloch, J.Fluorine chem., 1975, 5, 83) gives sulphonamides. The easy availability of sulphenamides opens up a viable general synthesis by this route, a useful alternative to standard preparative procedures based on am inolysis of sulphonyl halides or sulphonate esters. The form ation of a sulphonamide from an arenesulphinate and an azodicarbonyl compound (J.E.Herweh and R.M.Fantazier, J.org.Chem., 1976, 41, 116) cannot offer a widely-applicable route.

$$ArSO_2^- \ K^+ \ + \ \overset{NCO.R}{\underset{NCO.R}{\|}} \longrightarrow \overset{ArSO_2.N.CO.R}{\underset{K^+\ {}^-N.CO.R}{|}}$$

N-Alkylation of a sulphonamide is brought about directly when form aldehyde is used as the reagent; in neutral or alkaline solutions, toluene-p-sulphonamide gives the N,N-bis(toluene-p-sulphonyl)aminal (C.D.Egginton and A.J.Lambie, J.chem.Soc.(C), 1969, 1623), and not hydroxymethyl toluene-p-sulphonamide as claimed earlier.

$$\underline{p}\text{-}CH_3C_6H_4SO_2NH_2 \quad + \quad HCHO \longrightarrow (\underline{p}\text{-}CH_3C_6H_4SO_2NH)_2CH_2$$

The corresponding cyclic trimers, also polymers, form from these reactants in aqueous acid solutions (Egginton and Lambie, loc.cit.) and
corresponding reactions have been reported for trifluoroacetic acid as
reaction medium (O.O.Orazi and R.A.Corral, J.chem.Soc., Perkin I,
1975, 772). N-(2-Hydroxyalkyl)toluene-p-sulphonamides are formed
from the sulphonamide via its N-chloro-N-sodio derivative (Chloramine
T) by reaction with an alkene in the presence of osmium tetroxide
(K.B.Sharpless, A.O.Chong, and K.Oshima, J.org.Chem., 1976,
41, 177). N-(2-Bromoalkylsulphonamides are formed from corresponding N,N-dibromosulphonamides and alkenes (H.Terauchi and S.Takemura,
Chem.pharm.Bull.Japan, 1975, 23, 2410). Direct N-arylation of an
N-aryl arenesulphonamide through copper(II) catalyzed reaction with an
aryl bromide is probably a radical process (I.G.C.Coutts and M.Hamblin,
J.chem.Soc., Perkin I, 1975, 2445); the equivalent indirect reaction is
illustrated in the prior conversion of the sulphonamide into its N-sulph-
inyl derivative (through refluxing with thionyl chloride) and the use of
this derivative as an electrophiloid reagent towards various types of
carbanionic centres (T.Minami, Y.Tsumori, K.Yoshida, and T.Agawa,
J.org.Chem., 1974, 39, 3412).

$$TsNSO \;+\; PhCO.CH_2S(O)CH_3 \longrightarrow \begin{cases} (TsNH)_2CHCOPh \\ + \\ TsNHCH(SCH_3)COPh \end{cases}$$

   A further range of N-substituted sulphonamides is formed from N,N-
dichlorosulphonamides with reagents carrying sulphur or phosphorus
functional groups. With dimethyl sulphoxide, the sulphoxonium ylide
$RSO_2NS(O)(CH_3)_2$ is formed (A.Schönberg and E.Singer, Ber., 1969,
102, 2557), and the phosphonium ylide $RSO_2NPR'_3$ in the corresponding
reaction with a phosphine and copper powder. N-(p-Nitrobenzenesulphonyloxy)benzenesulphonamide reacts similarly with dimethyl sulphide or dimethyl sulphoxide, in the presence of triethylamine, but yields
are rather lower (M.Okahara and D.Swern, Tetrahedron Letters, 1969,
3301).

   The two modes of sulphonamide cleavage, involving C-N or S-N bonds respectively, are both represented in reactions of N-alkyl-N,N-bis(sulphonamide)s (V.A.Curtis, A.Raheja, J.E.Rejowski, R.W.Majewski, and R.J.Baumgarten, Tetrahedron Letters, 1975, 3107).

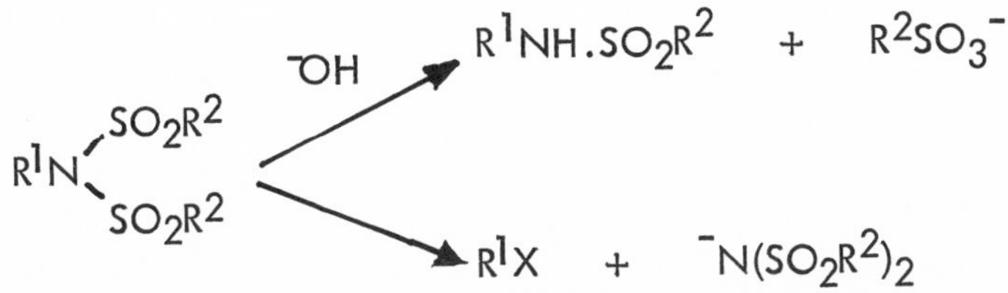

$$R^1N \underset{SO_2R^2}{\overset{SO_2R^2}{\diagdown}} \quad \overset{\text{-OH}}{\nearrow} \quad R^1NH.SO_2R^2 \ + \ R^2SO_3^-$$

$$\searrow \quad R^1X \ + \ ^-N(SO_2R^2)_2$$

The relatively unusual C - N cleavage process represented above is made possible by the good leaving group ability of the $-N(SO_2R^2)_2$ moiety. The more familiar S - N cleavage route is still difficult to bring about under mild conditions which are compatible with the use of sulphonylation for the protection of amino-groups in organic synthesis. Reductive cleavage of sulphonamides, to give amines, can be accomplished using alkali metals in hexamethylphosphoric triamide (T.Cuvigny and M. Larcheveque, J.organometal.Chem., 1974, 64, 315).

$$ArSO_2NR^1R^2 \xrightarrow[\text{(ii) } H_2O]{\text{(i) M/HMPT}} ArH + SO_2 + R^1R^2NH$$

The use of the naphthalenyl anion radical (from naphthalene and sodium in ether (W.D.Closson, S.Ji, and S.Schulenberg, J.Amer.chem.Soc., 1970, 92, 650) is as effective for the reductive cleavage of sulphonamides as the long-known electrochemical method (L.Horner, H.Hoenl, and E.Schmitt, Naturwissenschäften, 1976, 63, 577) or the use of bis(2-methoxyethoxy)aluminium hydride (E.H.Gold and E.Babad, J.org.Chem., 1972, 37, 2208).

Photochemical S - N cleavage of arenesulphonamides has been illustrated in the earlier literature, but the unusual desulphonylative photorearrangement of a vinyl arenesulphonamide (J.C.Arnold, J.Cossy, and J.P.Pete, Tetrahedron Letters, 1976, 3919) shown below, indicates that there is still a good deal to be learned about the influence of the structure of the sulphonamide on the course of the reaction.

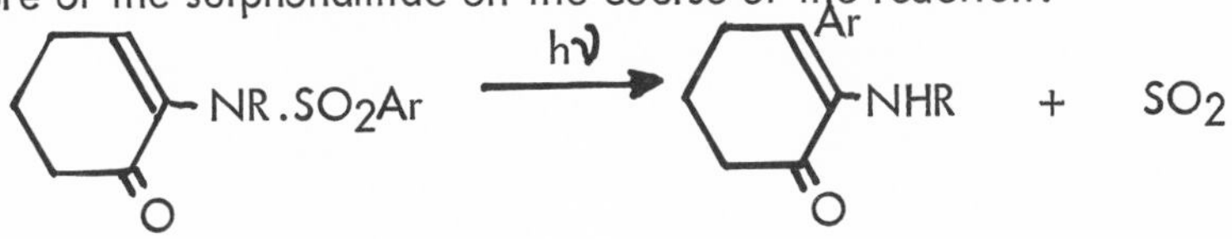

Rearrangement of N-(2-hydroxyalkyl)arenesulphonamides to N-(2-hydroxyalkyl)anilines also involves the desulphonylation of an arenesulphonamide, proceeding through two Smiles rearrangements (A.C.Knipe and J.Lound-Keast, Tetrahedron Letters, 1976, 2289).

$$ArSO_2NR^1CR^2R^3CHR^4OH \xrightarrow{^-OH} ArOCHR^4CR^2R^3NR^1SO_2^-$$

$$\xrightarrow[+H^+]{-SO_2} ArOCHR^4CR^2R^3NHR^1 \longrightarrow ArNR^1CR^2R^3CHR^4OH$$

Photolysis of N-arenesulphonylazoxy compounds leading to the arenesulphonic acid and products of decomposition of the accompanying diazonium cation also involves a rearrangement step (M.Kobayashi, K.Ochiai, and H.Minato, Chemistry Letters, 1976, 433).

$$Ar^1SO_2.N = N(O)Ar^2 \xrightarrow{h\nu} Ar^1SO_3^- \quad Ar^2N_2^+$$

Further S-N cleavage reactions of sulphonamides include the rearrangement of N,N-disubstituted benzenesulphonamides into o-aminophenyl aryl sulphones under the influence of an alkyl-lithium (D.Hellwinkel and M.Supp, Ber., 1976, 109, 3749).

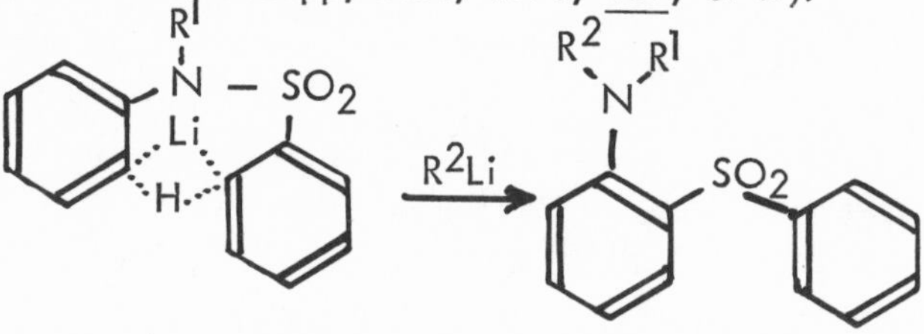

This is analogous to the Smiles rearrangement of di-aryl sulphones (W.E. Truce, B. van Gemert, and W.W.Brand, J.org.Chem., 1978, 43, 101) in which the process is initiated by o-metallation of the arene grouping attached to sulphur, although Hellwinkel and Supp consider metallation of the N-aryl group is also implicated.

The formation of N-methylindoles from 2-diazo-2'-(N-methyl-p-toluenesulphonamino)acetophenones involves a novel cleavage of an arenesulphonamide through electrophilic substitution (W.T.Flowers, G.Holt, C.P.Poulos, and K.Poulos, J.Chem.Soc., Perkin I, 1976, 1757).

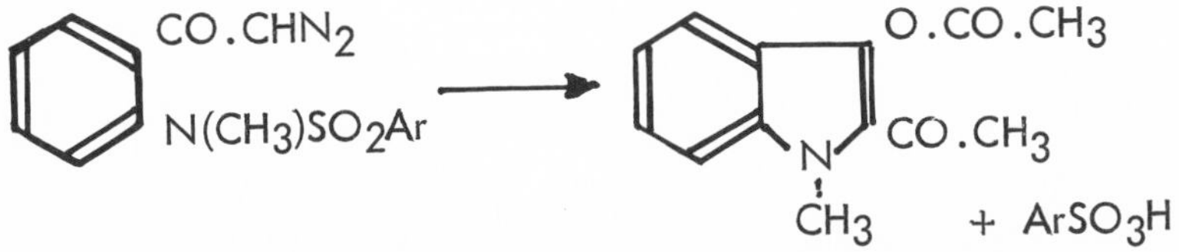

Cleavage of sulphonimides $PhCH_2N(SO_2Ar)_2$ by cyanide ion involves several steps (R.S.Glass and R.C.Hoy, Tetrahedron Letters, 1976, 1777).

$$PhCH_2N(SO_2Ar)_2 \longrightarrow PhCH{=}NSO_2Ar \xrightarrow{^-CN} PhCH(CN)\overset{-}{N}SO_2Ar$$

$$\longrightarrow PhC(CN){=}\overset{-}{N} \longrightarrow PhCN + {^-}CN + ArSO_2^-$$

An alternative cleavage pathway followed by analogous sulphonimides through reaction with iodide ion offers a valuable procedure for the conversion of an amine into the corresponding iodide (P.J. De Christopher, J.P.Adamek, G.D.Lyon, J.J.Galante, H.E.Haffner, R.J. Boggio, and R.J.Baumgarten, J.Amer.chem.Soc., 1969, 91, 2384).

$$RNH_2 \longrightarrow RN(SO_2Ar)_2 \xrightarrow[DMF]{I^-} RI$$

Hydrolysis of sulphonamides by dilute acid is normally a very slow reaction, but it can be accelerated substantially through participation by a suitably placed carboxy group (A.Wagenaar, A.J.Kirby, and J.B.F. N.Engberts, Tetrahedron Letters, 1976, 489); o-carboxybenzenesulphonamides undergo hydrolysis 10^5-10^6 times faster than p-carboxy-analogues.

The powerful electron-withdrawing effect of the arenesulphonyl group in arenesulphonamides and bis(arenesulphonyl)imides endows these compounds with high acidity, so much so that the latter compounds (pK$_a$ ≈ 1.45) are about as strongly acidic as phosphoric acid (F.A.Cotton and P.F.Stokely, J.Amer.chem.Soc., 1970, 92, 294). A consequence of this is the easy base-catalyzed N-alkylation reactions which are a continuing major theme is sulphonamide chemistry, e.g. a novel isoindole synthesis (J.Bornstein, D.A.McGowan, A.L.DiSalvo, J.E.Shields, and J.Kopecky, Chem.Comm., 1971, 1503).

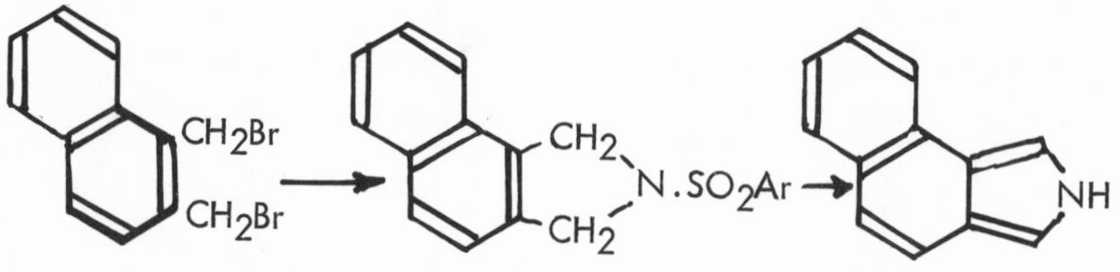

A novel method for the synthesis of pure samples of tertiary amines is also based upon this property (T.Oishi, K.Kamata, and Y.Bau, Chem. Comm., 1970, 777).

$$ArSO_2NR^1R^2 + (CH_3O)_2\overset{+}{C}R \ SbCl_6^- \longrightarrow ArSO_2\overset{+}{N}(CH_3)R^1R^2 \ SbCl_6^-$$

$$\xrightarrow{H_2O} \ CH_3\overset{+}{N}HR^1R^2 \quad ArSO_3^-$$

(h) Structural relatives of sulphonamides; sulphonylhetero-cumulenes

2,4,6-Tri-isopropylbenzenesulphonyl hydrazide is decomposed thermally into di-imide and the sulphinic acid, more readily than the benzene and toluene analogues. It is therefore a more effective reagent for the generation of di-imide in situ for the hydrogenation of alkenes (N.J. Cusack, C.B.Reese, and B.Roozpeikar, Chem.Comm., 1972, 1132).

An unusual nucleophilic aromatic substitution induced by an alkyl-lithium involves displacement of SO_2 from an intermediate arenesulph-inate ion formed by cleavage of the starting material, an arenesulphonyl hydrazone (K.B.Tomer and A.Weisz, Tetrahedron Letters, 1976, 231).

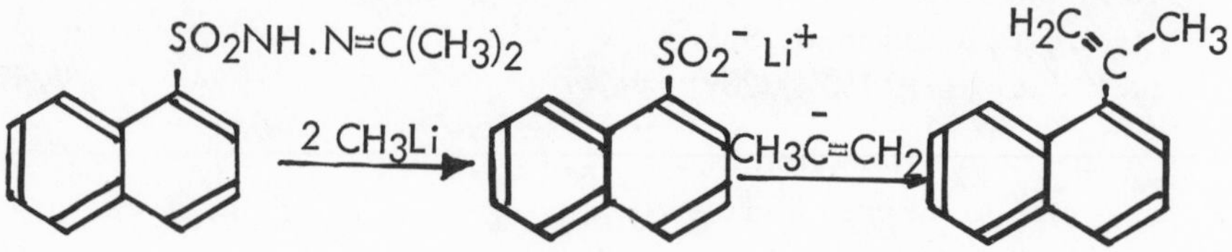

Arenesulphonyl azides, $ArSO_2N_3$, are precursors of arenesulphonyl nitrenes which, when generated in solutions of hydrocarbons or alcohols under reflux, yield C - H insertion products, i.e. sulphonamides (R.A. Abramovitch and R.G.Sutherland, Fortschr.chem.Forsch., 1970, 16, 1). The net result of the corresponding reaction with indoles, the formation of 3-arenesulphonylamino-derivatives, is not achieved with indolizines (A.S.Bailey, R.Scattergood, and W.A.Warr, J.chem.Soc. (C), 1971, 2479, 3769; A.S.Bailey, B.R.Brown, and M.C.Churn, ibid., p.1590). Alkenes can be converted into aldehydes and ketones through addition of an arenesulphonyl nitrene, followed by tautomerization of the resulting N-arenesulphonylimine into the isomeric enamine and its hydrolysis (R.A.Abramovitch, G.N.Knaus, M.Pavlin, and W.D.Holcomb, J.chem.Soc., Perkin I, 1974, 2169).

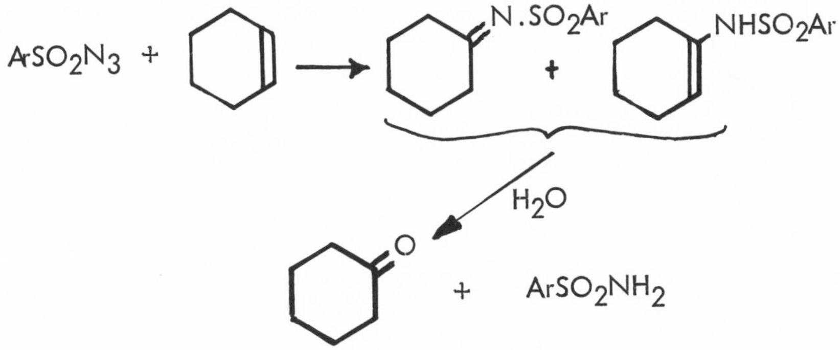

The reaction of arenesulphonylisocyanate with electron-rich alkenes gives 1,4-dipolar adducts.

$$ArSO_2NCO + CH_2{=}C(SCH_3)N(CH_3)_2 \longrightarrow ArSO_2\overset{-}{N}.CO.CH_2C({=}\overset{+}{S}CH_3)N(CH_3)_2$$

(R.Gompper and B.Wetzel, Tetrahedron Letters, 1971, 529)

$$ArSO_2NCO + R^1R^2C{=}CR^3NR^4_2 \rightarrow ArSO_2N{=}C(O^-)CR^1R^2\overset{+}{C}R^3NR^4_2$$

(E.Schaumann, S.Sieveking, and W.Walter, Tetrahedron Letters, 1974, 209)

β-Lactam formation through cycloaddition of toluene-p-sulphonyl isocyanate to alkenes (R.Lattrell, Ann., 1969, 722, 132) illustrates an alternative addition reaction of this class of compound.

N-Sulphinylsulphonamides $ArSO_2N{=}S{=}O$ are formed from the sulphonamides in refluxing thionyl chloride (F.Bentz and G.-E.Nischk, Angew.chem.internat.edit., 1970, 9, 66). These are useful intermediates for the preparation of N-acyl and N-arenesulphonyl-sulphonamides, through reaction with a carboxylic acid or a sulphonic acid, respectively (Bentz and Nischk, loc.cit.), and of sulphur di-imides $ArSO_2N{=}S{=}N.SO_2Ar$ on treatment with base (H.C.Bucholt, A.Senning, and P.Kelly, Acta chem.Scand., 1969, 23, 1279; H.H.Hoerhold and J.Beck, J.prakt.Chem., 1969, 311, 621), and sulphonyl isocyanates with phosgene (H.Ulrich, B.Tucker, and A.A.R.Sayigh, J.org.Chem., 1969, 34, 3200). These sulphur di-imides act as starting materials for the preparation of sulphenyl analogues $ArSO_2N{=}S{=}N.S.R$ and $ArSN{=}S{=}NSAr$ (Bucholt, Senning, and Kelly, loc.cit.).

N-Arenesulphonylsulphimides $ArSO_2N{=}S(CH_3)_2$ are obtained from N-halogeno-N-sodio-arenesulphonamides and dimethyl sulphide (F.E. Hardy, J.chem.Soc.(C), 1970, 2087).

(i) Sulphonyl cyanides

Preparations from arenesulphonyl chlorides are straightforward (M.S.A. Vrijland, Org.Synth., 1977, 57, 88).

$$PhSO_2Cl \xrightarrow[25^\circ]{Na_2SO_3/NaHCO_3/H_2O} PhSO_2Na$$

$$\xrightarrow[10-15^\circ]{ClCN/H_2O} PhSO_2CN \quad (92\%)$$

Sulphonyl cyanides are useful in synthesis, undergoing a range of cycloaddition reactions facilitated by the distorted electron density in the cyano-moiety caused by the strong electron withdrawal by the sulphonyl group (in contrast, aroyl cyanides do not show cycloaddition reactivity). Cycloaddition of sulphonyl cyanides to conjugated dienes under very mild conditions gives pyridines or pyridones after dehydrogenation and hydrolysis of the adducts (J.C.Jagt and A.M. van Leusen, Rec. trav.Chim., 1973, 92, 1343; see also R.G.Pews, E.B.Nyquist, and F.P.Corson, J.org.Chem., 1970, 35, 4096). Whereas the addition of chlorine or a sulphur chloride S_nCl_2 (n = 1, 2) to a sulphonyl cyanide gives adducts ($ArSO_2C(Cl){=}N)_2S$ with SCl_2, for example), which reflects the ability of the cyano-group in these compounds to react independently of the sulphonyl group (M.S.A.Vrijland, Tetrahedron Letters, 1974, 209), photoaddition at 254 nm of a sulphonyl cyanide to an alkene gives an aryl 2-cyano-alkyl sulphone (R.G.Pews and T.E.Evans, Chem.Comm., 1971, 1397).

$$ArSO_2CN + R^1R^2C=CR^3R^4 \longrightarrow ArSO_2CR^1R^2CR^3R^4CN$$

2. *Nuclear Sulphinic Acids of Benzene and its Homologues; Arene sulphinic Acids*

(a) Preparation

Insertion of sulphur dioxide into tetra-aryl stannanes leads to bis(arenesulphinyl) di-aryl stannanes $R_2Sn(O_2SAr)_2$ (U.Kunze, E.Lindner, and J.Koola, J.organometal.Chem., 1972, 40, 327). Several studies of metal - carbon bond insertion by sulphur dioxide have been described, indicating that the general arenesulphinic acid synthesis which seems to be available, based on this process, may be inefficient since not every metal - carbon bond undergoes insertion, and with many organometallic compounds, O-sulphinates (i.e. C-O bond formation) are formed. However, promising results with organocopper(II) compounds (G.Cahiez, J.F.Normant, and J.Villieras, J.organometal.Chem., 1976, 121, 123), Grignard reagents (H.W.Pinnick and M.A.Reynolds, J.org.Chem., 1979, 44, 160), and organoaluminium compounds (A.V. Kuchin, L.I.Akhmetov, V.P.Yurev, and G.A.Tolstikov, Zhur.obshch. khim., 1979, 49, 401) have been reported.

Conventional routes to arenesulphinic acids are illustrated in the cleavage of di-aryl sulphones by electrolysis (L.Horner and R.-J.Singer, Tetrahedron Letters, 1969, 1545) or by the more common nucleophilic reagents, n-butyl-lithium ((F.M.Stoyanovich, R.G.Karpenko, G.I. Gorushkina, and Y.L.Goldfarb, Tetrahedron, 1972, 28, 5017), or alkaline hydrolysis (O.F.Bennett, M.J.Bouchard, R.Malloy, P.Dervin, and G.Saluti, J.org.Chem., 1972, 37, 1356). Alkaline hydrolysis of thiolsulphonates is also serviceable (W.Walter and P.M.Hell, Ann., 1969, 727, 35). Reduction of arenesulphonyl chlorides with zinc or stannous chloride (K.K.Andersen, in 'Comprehensive Organic Chemistry' eds. Sir Derek H.R.Barton and W.D.Ollis, Vol.3, ed. D.N.Jones,

Pergamon Press, Oxford, 1979, p.317) is a long-established general preparative method. Electrochemical reduction of an arenesulphonamide (Horner and Singer, loc.cit.) is a high-yield route to both reaction products, the arenesulphinic acid and the amine (P.T.Cottrell and C.K. Mann, J.Amer.Chem.Soc., 1971, 93, 3579).

(b) *General Properties and reactions*

Disproportionation of benzenesulphinic acid into benzenethiol and benzenesulphonic acid (Z.Yoshida, H.Miyoshi, and K.Kawamoto, Chem.Abs., 1969, 71, 101 463) under nitrogen, or into the sulphonic acid and thiolsulphonate, together with diphenyl disulphide and diphenyl disulphone, in oxygenated solutions, illustrates the instability of free sulphinic acids. ^{18}O-Labelling studies (M.Kobayashi, H.Minato, and Y.Ogi, Bull.chem.Soc.Japan, 1972, 45, 1224) suggest that the sulphinylsulphone $Ar.S(O).SO_2Ar$, is an intermediate in the disproportionation.

The arenesulphinate anion shows a low order of nucleophilicity, but participates in Michael addition reactions (Y.Ogata, Y.Sawacki, and M.Isono, Tetrahedron, 1970, 26, 3045; H.W.Pinnick and M.A. Reynolds, J.org.Chem., 1979, 44, 160), Mannich reactions (T.Olijnsma, J.B.F.N.Engberts, and J.Strating, Rec.trav.chim., 1972, 91, 209),

$$R^1SO_2^- + R^2CHO + R^3CONHR^4 \longrightarrow R^1SO_2CHR^2NR^4COR^3$$

and reversible addition to nitroso-compounds (to give $ArSO_2N(OH)R$; A Darchen and C.Moinet, Chem.Comm., 1976, 820).

Catalysis of the cis - trans equilibration of alkenes by arenesulphinic acids is accounted for by reversible sulphone formation, but complex mixtures of organosulphur compounds are formed (T.W.Gibson and P.Strassburger, J.org.Chem., 1976, 41, 791). Alternative sites through which the arenesulphinate anion may effect nucleophilic substitution of halogen in a halogenomethyl ether are revealed in the formation of either the aryl methoxymethyl sulphone or the (unstable) methoxymethyl arenesulphinate, depending on the reaction conditions (K. Schank and H.G.Schmitt, Ber., 1974, 107, 3026).

$$ArSO_2^- + CH_3OCH_2hal \longrightarrow ArSO_2CH_2OCH_3 + ArS(O)OCH_2OCH_3$$

Pyrolysis of sodium toluene-p-sulphinate below 300° gives toluene and toluene-p-thiol, but at higher temperatures 4,4'-bitolyl, di-tolyl sulphide, and di-tolyl disulphide are also formed (P.Y.Johnson, E.Koza, and R.E.Kohrman, J.org.Chem., 1973, 38, 2967).

The conversion of an arenesulphinic acid into a sulphonyl halide by reaction with an $\alpha\alpha$-dihalogeno-ether probably involves the alkoxy-halogenomethyl sulphinate as intermediate (K.Schank and F.Schroeder, Phosphorus and Sulphur, 1976, 1, 307).

(c) Sulphinate esters

Conventional routes to these compounds, from the corresponding acid salts or halides, have been supplemented by a direct preparation from disulphides (I.B.Douglass, J.org.Chem., 1974, 39, 563).

$$R^1SSR^2 \xrightarrow[(CH_3CO)_2O]{Cl_2} R^1SOCl \xrightarrow{R^2OH} R^1S(O)OR^2$$

Chlorinolysis at -20° in an alcohol solvent yields the alkyl arenesulphinate, probably by way of the sulphinyl chloride; this compound is otherwise usually prepared by oxidation of a sulphenyl chloride, itself prepared by chlorinolysis of a disulphide in an inert solvent.

N-Alkyl-N'-toluene-p-sulphonylhydrazines $ArSO_2NHNH(CH_2)_2R$ yield the corresponding alkyl toluene-p-sulphinates $ArS(O)O(CH_2)_2R$ through treatment with mild oxidizing agents normally used for the liberation of di-imide from arenesulphonylhydrazines (O.Attanasi, L. Cagliotti, and F.Gasparrini, Chem.Comm., 1974, 138). Identification of minor reaction products suggests a radical reaction pathway.

N-Bromosuccinimide cleavage of aryl benzyl or t-butyl sulphoxides in hydroxylic media, proceeding through the bromosulphonium salt and S_N1 alcoholysis, leads to alkyl arenesulphinates (F.Jung and T.Durst, Chem.Comm., 1973, 4).

$$R^1S(O)Ar \xrightarrow{NBS}{R^2OH} R^1Br \quad + \quad ArS(O)OR^2$$

The formal reverse of this reaction, through the treatment of the sulphinate ester with a Grignard reagent to give the sulphoxide resulting from the displacement of the alkoxy group, has been developed into a partial resolution of an arenesulphinate ester; treatment of the racemic arenesulphinate with a deficiency of (S)-2-methyl- or 2-phenyl-1-butylmagnesium chloride leaves residual arenesulphinate ester enriched in the (S)-enantiomer; the degree of resolution can be determined through n.m.r. spectra of the partly-resolved arenesulphinate ester in a chiral solvent (W.H.Pirkle and M.S.Hoekstra, J.Amer.chem.Soc., 1976, 98, 1832).

Thermal rearrangement of a prop-2-ynyl arenesulphinate to an aryl allenyl sulphone illustrates a variation of the rearrangement of a sulphinate ester into a sulphone (G.Smith and C.J.M.Stirling, J.chem.Soc., (C), 1971, 1530).

$$ArS\overset{O}{\underset{O.CH_2C\equiv CH}{<}} \longrightarrow Ar.\overset{O}{\underset{O}{\overset{\|}{S}}}.CH=C=CH_2$$

Related studies of this rearrangement (S.Braverman and H.Mechoulam, Tetrahedron, 1974, 30, 3883) and its use in the synthesis of allenyl sulphones (M.Cinquini, F.Cozzi, and M.Pelosi, J.chem.Soc., Perkin I, 1979, 1430), have been described.

Pyrolysis of an alkyl arenesulphinate gives the alkene with stereochemistry to be expected for a syn-elimination involving a five-membered transition state (D.N.Jones and W.Higgins, J.chem.Soc.(C), 1970, 81); in a chiral alkyl arenesulphinate, the rate of elimination is different for each diastereoisomer (enantiomeric configurations at the sulphinyl chiral centre).

$$ArS\overset{O}{\underset{O.CHR^1CHR^2}{<}} \longrightarrow ArSO_2H + R^1CH=CHR^2$$

The rearrangemement of O-sulphinyl oximes into N-sulphonyl imines involves a radical pathway (R.F.Hudson and K.A.F.Record, Chem.Comm., 1976, 831).

$$Ar_2C=N.O.S(O)R \overset{35°}{\longrightarrow} Ar_2C=N.SO_2R$$

Ionic intermediates are implicated in the rearrangement of 2-furfuryl benzenesulphinates into aryl 2-furfuryl sulphones, since aryl 3-(2-methylfuryl) sulphones are also formed (S.Braverman and T.Globerman, Tetrahedron Letters, 1973, 3023; Braverman and Mechoulam, loc.cit.). These sulphinates undergo ethanolysis through a C-O fission mechanism (Braverman and Globerman, Tetrahedron, 1974, 30, 3873), whereas in the corresponding reaction with benzyl arenesulphinates, S-O fission is the exclusive pathway (Braverman and Y.Duar, Tetrahedron Letters, 1975, 343).

Photolysis of alkyl toluene-p-sulphinates involves S-O fission; the process is reversible (optically-active compounds are partly racemized) but leads mainly to products derived from alkoxy- and toluene-p-sulphinyl radicals (M.Kobayashi, H.Minato, Y.Miyaji, T.Yoshioka, K.Tanaka, and K.Honda, Bull.chem.Soc.Japan, 1972, 45, 2817).

(d)   Sulphinamides

The mild oxidation of sulphenamides with m-chloroperbenzoic acid (D. N.Harpp and T.G.Back, Tetrahedron Letters, 1972, 5313), with bromine — $(Bu^n_3Sn)_2O$ (Y.Ueno, T.Inoue, and M.Okawara, Tetrahedron Letters, 1977, 2413), with active MnO_2 (I.Glander and A.Golloch, J.Fluorine chem., 1975, 5, 83), or with N-chlorosuccinimide — potassium hydrogen carbonate — water (M.Haake, H.Gebbing, and H. Benack, Synthesis, 1979, 97) provide convenient routes to arenesulphinamides.   These are simple methods, but the more obvious route, the aminolysis of an arenesulphinyl chloride (M.R.Jones and D.J.Cram, J.Amer.chem.Soc., 1974, 96, 2183), is the standard procedure.   Use of (-)-menthyl N-(p-nitrobenzenesulphonyloxy)carbamate as substrate for the aminolysis route, and the separation of the resulting pair of diastereoisomers, provides both enantiomers of a chiral sulphinamide.   Inversion of configuration has been established for the conversion of an optically-active toluene-p-sulphinate ester into the corresponding chiral sulphinamide with an iminomagnesium halide (M.Cinquini and F.Cozzi, Chem.Comm., 1977, 502).

Dicyclohexylcarbodiimide or 2-chloro-1-methylpyridinium iodide serve as coupling agents for the synthesis of sulphinamides from sulphinic acids and amines (M.Furukawa and T.Okawara, Synthesis, 1976, 339).

Several reactions of sulphinamides have been studied from the point of view of their stereochemical course.   (S)-Arenesulphinamides undergo acid-catalyzed alcoholysis to give (S)-arenesulphinate esters, i.e. with inversion of configuration (M.Mikolajczyk, J.Drabowicz, and B.Bujniki, Chem.Comm., 1976, 568).   O-Methylation of a chiral toluene-p-sulphinamide using methyl trifluoromethanesulphonate ('methyl triflate') and alkaline hydrolysis of the resulting methoxysulphonium salt converts the starting material into its enantiomer (H.Minato, K.Yama-guchi, and M.Kobayashi, Chemistry Letters, 1975, 991).

Asymmetric induction accompanying the reduction by lithium aluminium hydride of the (S)-sulphinamide ArS(O)N=CRPh depicted above leads to (S)-amines PhCHRNH$_2$ and returns the methyl arenesulphinate of opposite configuration, when the reduction product is methanolysed (Cinquini and Cozzi, Chem.Comm., 1977, 723).

Reduction of sulphoximines with aluminium — mercury amalgam in 90% aqueous tetrahydrofuran (C.W.Schroeck and C.R.Johnson, J.Amer. chem.Soc., 1971, 93, 5305), or reaction of sulphoximines with toluene-p-sulphonyl chloride in pyridine (T.R.Williams, R.E.Booms, and D.J. Cram, J.Amer.chem.Soc., 1971, 93, 7338) yield the corresponding arenesulphinamides. In fact, these routes were used to provide the first optically-active sulphinamides, involving retention of configuration; these methods were established during studies leading to several previously-unknown hexaco-ordinate sulphur functional groups. N-Alkylation of sulphinamides can be brought about by treating the N-lithio derivative with an alkyl halide (E.Wenschuh and B.Fritze, J. prakt.Chem., 1970, 312, 129; see also Wenschuh and W.D.Riedmann, Z.chem., 1972, 12, 29) A secondary sulphinamide is converted into a sulphonimidoyl chloride by treatment with chlorine in ether at -78°, or with 1-chlorobenzotriazole (E.U.Jonsson, C.C.Bacon, and C.R.Johnson, J.Amer.chem.Soc., 1971, 93, 5306), alternatively chlorine in pyridine (Jonsson and Johnson, J.Amer.chem.Soc., 1971, 93, 5308).

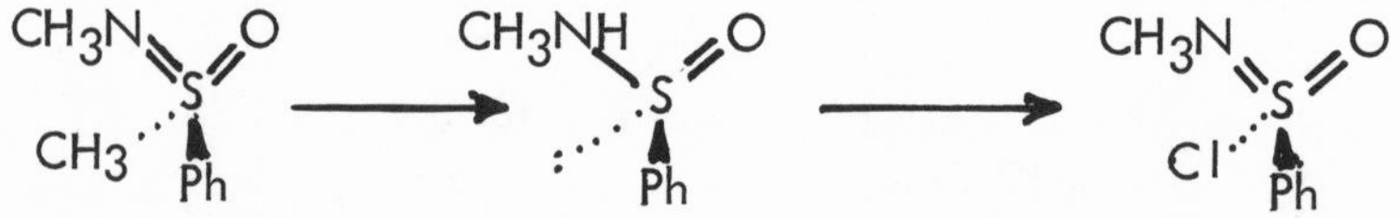

$$\underset{(+)-\underline{S}}{\underset{CH_3}{\overset{CH_3N}{\underset{Ph}{\diagdown S \diagup O}}}} \longrightarrow \underset{(+)-\underline{S}}{\underset{Ph}{\overset{CH_3NH}{\diagdown S \diagup O}}} \longrightarrow \underset{(-)-\underline{R}}{\underset{Cl}{\overset{CH_3N}{\underset{Ph}{\diagdown S \diagup O}}}}$$

Treatment of the sulphonimidoyl chloride with a secondary amine gives the sulphonamidine PhS(O)(NCH$_3$)NR$_2$ with inversion of configuration Jonsson and Johnson *loc.cit.*; see also F.Wudl, C.K.Brush, and T.B. K.Lee, Chem.Comm., 1972, 151) from which, by aluminium amalgam reduction, the enantiomer of the starting sulphinamide can be obtained (Jonsson and Johnson, *loc.cit.*).

The generation of a sulphinylnitrene or sulphinylnitrenium salt, in reactions of N-alkoxybenzenesulphinamides, is shown through the formation of the same products from benzenesulphinyl azide (T.J.Maricich, S.Madhusoodanan, and C.A.Kapfer, Tetrahedron Letters, 1977, 983). Thus, $PhSO_2N{=}S(CH_3)_2$ can be isolated from solutions of these benzenesulphinyl compounds in dimethylsulphoxide.

Photolysis of arenesulphinamides in alcohol solvents gives the corresponding arenesulphinate esters, while in aprotic solvents (benzene, acetonitrile), products typical of arenesulphinyl radicals (e.g. ArSO· $\longrightarrow$ ArSO$_2$SAr, etc) are formed (H.Tsuda, H.Minato, and M.Kobayashi, Chemistry Letters, 1976, 149).

(e) Sulphinyl azides

These compounds, easily prepared from an arenesulphinyl chloride with sodium azide, show the characteristic range of reactions of nitrenes, e.g. N-substitution leading to degradation products of the resulting sulphinamides (T.J.Maricich and V.L.Hofmann, J.Amer.chem.Soc., 1974, 96, 7770; see also Maricich, Madhusoodanan, and Kapfer, loc. cit.).

(f) Sulphinyl Sulphones

Several series of arenesulphinyl sulphones ArS(O)SO$_2$Ar have been prepared for mechanistic studies with two main themes: thermolysis into radicals (ArSO· + ArSO$_2$·) for comparisons of the ease of homolysis of series of disulphides, sulphinyl sulphones, and α-disulphones (J.L. Kice and N.A.Favstritsky, J.org.Chem., 1970, 35, 114), and nucleophilic cleavage reactions (J.L.Kice and L.F.Mullen, J.Amer.chem. Soc., 1976, 98, 4259) revealing the relative reactivities of sulphinyl sulphur and sulphonyl sulphur towards common nucleophiles. Sulphonyl sulphur is shown to be a 'harder' electrophilic centre than sulphinyl sulphur, and is very much like a carbonyl carbon atom in terms of its electrophilic reactivity (J.L.Kice and E.Legan, J.Amer.chem.Soc., 1973, 95, 3912).

3. Nuclear Sulphenic Acids of Benzene and its Homologues Arene Sulphenic Acids

(a) Preparation

The formation of sulphenate anions through the cleavage of diaryl disulphides in aqueous alkali (D.R.Hogg and A.Robertson, J.chem.Soc., Perkin I, 1979, 1125; D.R.Hogg, in 'Comprehensive Organic Chemistry', eds. Sir Derek H.R.Barton and W.D.Ollis, Vol.3, ed. D.N.Jones, Pergamon Press, Oxford, 1979, p.261),

$$ArSSAr + 2OH^- \longrightarrow ArSO^- + ArS^- + H_2O,$$

and through the cleavage of thiolsulphinates (E.Block and J.O'Connor, J.Amer.chem.Soc., 1974, 96, 3921, 3929),

$$ArS(O).SAr \xrightarrow{H^+} ArS(OH)SAr \xrightarrow{Nu^-} ArSNu + ArSOH,$$

provide straightforward routes to arenesulphenic acids, in principle, but in practice the media in which these species are formed rapidly promote further reactions. Therefore, no arenesulphenic acid has yet been isolated for physical and chemical study, but 2,2-dimethyl-propane sulphenic acid $(CH_3)_3CSOH$, formed through pyrolysis of t-butyl alkyl sulphoxides, is moderately stable in solvents which favour hydrogen-bonding (J.R.Shelton and K.E.Davis, Internat.J.sulfur chem., 1973, 8, 197, 205). Flash vacuum pyrolysis of an aryl t-butyl sulphoxide and collection of the products on a cold finger at $-196°$ provides aryl arenethiolsulphinates, the ultimate product in any attempted synthesis of an arenesulphenic acid (F.A.Davis, S.G.Yocklovich, and G.S. Baker, Tetrahedron Letters, 1978, 97). Arenesulphenic acids formed by the pyrolysis of arenesulphinimides $ArS(O)N=CHR$ have been trapped by alkenes (F.A.Davis, A.J.Friedman, and U.K.Nadir, J.Amer.Chem. Soc.,1978, 100, 2844). These results clearly show the initial formation of relatively high concentrations of an arenesulphenic acid in the pyrolysis reactions, since further self-condensation products which appear when arenesulphenic acids are generated in dilute solutions are not formed in the gas phase reactions.

$$ArS(O)SAr \longleftarrow ArSOH \xrightarrow{R^1R^2C=CR^3R^4} ArSCR^1R^2CHR^3R^4$$
$$\underset{O}{\overset{}{}}$$

(b) Sulphenyl halides

Cleavage of a disulphide with sulphuryl chloride and triphenyl phosphine to give an arenesulphenyl chloride (E.A.Parfenov and V.A.Fomin, Zhur.obshchei Khim., 1975, 45, 1129) is a variation of the long-established route based on the chlorinolysis of disulphides. Treatment of a thiol with N-chlorosuccinimide has been advocated as a simple alternative preparative method (P.B.Hopkins and P.L.Fuchs, J.org.Chem., 1978, 43, 1208).

$$ArSH + (CH_2CO)_2NCl \longrightarrow ArSCl + (CH_2CO)_2NH$$

Reactions of sulphenyl chlorides fall mainly into the categories of electrophilic addition to alkenes, leading to 2-chloroalkyl aryl sulphides; and nucleophilic substitution reactions leading to sulphenate esters and sulphenamides. The former category of reaction (for a review, see L.Rasteikiene, D.Greiciute, M.G.Linkova, and I.L.Knunyants, Uspekhii khim., 1977, 46, 1041) has been fully explored as far as its synthetic possibilities are concerned, and the emphasis has shifted towards mechanistic aspects, particularly the alternative formulations of the reaction intermediates formed in reactions with more complex alkenes.

$$ArSCl + R^1R^2C{=}CR^3R^4 \longrightarrow \begin{cases} R^1R^2C \overset{Ar}{\underset{}{\overset{\overset{+}{S}}{-}}} CR^3R^4 \quad Cl^- \\ \quad\quad or \\ ArS-CR^1R^2-\overset{+}{C}R^3R^4 \quad Cl^- \end{cases}$$

References to papers from the research groups mainly concerned (G.H. Schmid and T.T.Tidwell, J.org.Chem., 1978, 43, 460; N.S.Zefirov, N.K.Sadovaya, A.M.Magarramov, and I.V.Bodrikov, Zhur.org.Khim., 1977, 13, 245) provide entries to the substantial literature on this topic.

While aminolysis and alcoholysis of sulphenyl halides are covered in the following sections, since these reactions form the basis of standard methods of synthesis of other sulphenyl compounds, other nucleophilic substitution reactions are covered here. Comparisons of the reactivity of arenesulphenyl halides and their selenenyl analogues towards thiocarbonyl compounds or thiolsulphonate anions $ArSSO_3^-$ have been made (T.Austad, Acta chem.Scand.(A), 1977, 31, 93, 227). The use of sulphenyl halides for the synthesis of unsymmetrical disulphides is now well-established (B.I.Stepanov, V.Y.Rodionov, T.A.Chibisova, L.A.Yagodina, and A.D.Stankevich, Zhur.org.Khim., 1977, 13, 370).

$$Ar^1SCl + Ar^2SH \longrightarrow Ar^1SSAr^2 + HCl$$

(c) *Sulphenate esters*

The conventional synthesis employing an arenesulphenyl chloride and an alcohol has provided large numbers of arenesulphenate esters for mechanistic studies of hydrolysis and aminolysis reactions (D.R.Hogg and P.W.Vipond, J.chem.Soc.(C), 1970, 1 242; S.Braverman and D.Reisman, Tetrahedron, 1974, 30, 3891). New routes, the reaction of an N-alkylthiophthalimide with an alkoxide (D.H.R.Barton, G. Page, and D.A.Widdowson, Chem.Comm., 1970, 1466), not yet applied for the synthesis of arenesulphinate esters, and the rearrangement of a methoxymethyl aryl sulphoxide (T.J.Maricich and C.K.Harrington, J.Amer.chem.Soc., 1972, 94, 5115), are likely to develop into useful preparative methods.

$$\underset{\underset{O}{\|}}{Ph.S}.CH_2OCH_3 \longrightarrow Ph-S\cdots\cdots CH_2\cdots\cdots OCH_3$$

$$\longrightarrow PhS.O.CH_2OCH_3$$

Solvolysis of benzyl arenesulphenates involves S - O cleavage, in contrast with the C - O cleavage undergone by corresponding trichloro-methanesulphenates (Braverman and Reisman, *loc.cit.*).

Chlorinolysis of sulphenate esters in acetic acid proceeds via sulphonium ion intermediates (J.G.Traynham and A.W.Foster, J.Amer. chem.Soc., 1971, 93, 6216).

$$ArSOR \longrightarrow \underset{\underset{Cl}{|}}{Ar\overset{+}{S}OR} \ \ Cl^- \longrightarrow RCl + ArSCl + CH_3CO_2R$$

Cleavage of arenesulphinate esters by tri-alkyl phosphites yields alkoxy-thioaryloxyphosphoranes and their elaboration products (L.L. Chang and D.B.Denney, Chem.Comm., 1974, 84).

Allenyl arenesulphenates, formed from propargyl aryl sulphoxides by rearrangement (D.J.Abbott and C.J.M.Stirling, J.chem.Soc.(C), 1969, 818), have been shown to undergo *ortho*-Claisen rearrangement, since a benzodihydrothiophen is readily formed from representative examples (K.C.Majumdar and B.S.Thyagarajan, Chem.Comm., 1972, 83):

Pyrolysis of an aryl arenesulphenate gives a mixture of an o-hydroxy-phenyl aryl sulphide, together with di-aryl disulphides (D.R.Hogg, J.H. Smith, and P.W.Vipond, J.chem.Soc.(C), 1968, 2713). Heterolysis of the S – O bond leading to arylsulphenyl cations ArS⁺ accounts for this behaviour, and the same S$_N$1-type mechanism has been assigned to the decomposition of t-butyl 2-nitrobenzenesulphenate in anisole, which leads to t-butanol and 2-nitrophenyl methoxyphenyl sulphide via arene-sulphenyl cations and t-butoxide anions (Hogg and Vipond, J.chem. Soc.(C), 1970, 60).

(d) Sulphenamides

Aminolysis of sulphenyl halides continues to be used for the synthesis of sulphenamides, as the most convenient procedure, but other aminolysis reactions have been explored for the purpose; these involve disulphides in the presence of a silver or mercury(II) salt (F.A.Davis, A.Friedman, E.W.Kluger, E.B.Skibo, E.R.Fretz, A.P.Milicia, W.C.Le Masters, M.D.Bentley, J.A.Lacadie, and I.B.Douglass, J.org.Chem., 1977, 42, 967), aryl methyl sulphoxides with an aminoborane (R.H.Cragg, J.P.N.Husband, G.C.H.Jones, and A.F.Weston, J.Organometal. Chem., 1972, 44, C37), in certain cases through cleavage of a di-aryl sulphide with an N-chloroimide (M.Furukawa, Y.Fujino, Y.Kojima, M.Ono, and S.Hayashi, Chem. and pharm.Bull.Japan, 1972, 20, 2024). Transamination between a secondary amine and an N-arylthio-phthalimide proves to be a very satisfactory general synthesis (D.N. Harpp and T.G.Back, Tetrahedron Letters, 1971, 4953).

N-Arenesulphenylation of a primary sulphenamide with an arene-sulphenyl chloride gives tris(arenesulphenyl)amines, e.g. 'tribenzene-sulphenamide'(Ph S)$_3$N, which have a number of uses in synthesis (J.Almog, D.H.R.Barton, P.D.Magnus, and R.K.Norris, J.chem.Soc., Perkin I, 1974, 853), for example:-

$$R_3P + (PhS)_3N \longrightarrow R_3\overset{+}{P}SPh \ \overset{-}{N}(SPh)_2 \longrightarrow PhSN=PR_3$$

$$ArCH=NSPh \longleftarrow \overset{ArCHO}{}$$

and for conversion of enamines into 2-(phenylthio)alkanones. Specific o-amination of phenols can be effected via 2-(benzenesulphenyl)iminocyclohexadiene-1-ones

(D.H.R.Barton, I.A. Blair, P.D.Magnus, and R.K.Norris, J.chem. Soc., Perkin I, 1973, 1031, 1037).

Radical intermediates are implicated in the thermal rearrangement of arenesulphenanilides into 2- and 4-amino-diphenyl sulphides and disulphides and azobenzenes (F.A.Davis, E.R.Fretz, and C.J.Horner, J.org.Chem., 1973, 38, 690), and of tris(benzenesulphenyl)amine into diphenyl disulphide and nitrogen. Bis(benzenesulphenyl)amine gives a purple solution in benzene solution, after oxidation by lead oxide, ascribed to the formation of an unusually stable radical $(PhS)_2N \cdot$ (at least 7 days life in benzene or in n-hexane) (Y.Miura, N.Makita, and M.Kinoshita, Tetrahedron Letters, 1975, 127). This radical is also formed by pyrolysis of tris(benzenesulphenyl)amine at about 80° (Barton, Blair, Magnus, and Norris, loc.cit.). The relatively weak S – N bond undergoing homolysis in these reactions can incorporate a rotation barrier of up to 45 kJ mol^{-1} in N,N-disubstituted arenesulphenamides, and the source of the barrier has been a continuing topic of study (M.Raban and G. Yamamoto, J.Amer.chem.Soc., 1977, 99, 4160; D.Kost and E.Berman, Tetrahedron Letters, 1980, 1065).

(e) Thionitrites and thionitrates

Treatment of a thiol with dinitrogen tetroxide gives the thionitrite RSNO (Y.H.Kim, K.Shinhama, D.Fukushima, and S.Oae, Tetrahedron Letters, 1978, 1211). Further reaction resulting from oxidation by the nitrogen oxide leads to the sulphonic acid and the thiolsulphonate, and 'unsymmetrical' thiolsulphonates can be obtained from a thiol, a sulphinic acid, and dinitrogen tetroxide (S.Oae, D.Fukushima, and Y.H. Kim, Chem.Comm., 1977, 407), via the thionitrite and thionitrate.

Thionitrites and thionitrates show some promise as reagents for Ullmann reactions with anilines, which in conjunction with a copper(II) halide they convert into halogenobenzenes (Kim, Shinhama, and Oae, Tetrahedron Letters, 1978, 4519), and for benzoquinone-imine formation from p-aminophenols (Oae, Shinhama, and Kim, Chemistry Letters, 1979, 1077).

4. Nuclear Thiolsulphonic and Thiolsulphinic acids of Benzene and its Homologues; Arenethiolsulphonic and Arenethiolsulphinic Acids

Metal thiosulphonates $ArSO_2S^- \ M^+$ have been prepared by one of the standard methods (described in Rodd's Chemistry of Carbon Compounds, Second Edition, ed. S.Coffey, Vol. IIIA, p.486), and used in the synthesis of novel sulphur homologues of esters of these acids(H.C.Hansen and A.Senning, Chem.Comm., 1979, 1135). Further material collected here to supplement the coverage in the Second Edition only concerns the esters of the title acids.

(a) *Arenethiolsulphonate esters*

These compounds are formed by oxidation of disulphides with hydrogen peroxide in the presence of vanadium oxide catalyst (F.E.Hardy, P.R. H.Speakman, and P.Robson, J.chem.Soc.(C), 1969, 2334). With only one equivalent of oxidant, the thiolsulphonate is still the major oxidation product, with minor amounts of thiolsulphinate and unreacted starting material.

Alkylthiolation of a metal arenesulphinate using an N-alkylthio-succinimide (Y.Abe and J.Tsurugi, Chemistry Letters, 1972, 441) and alkylation of a metal thiosulphonate (B.G.Boldyrev, W.N.Obukhova, and W.N.Rochnyak, Zhur.priklad.Khim., 1972, 45, 888) are the commonly-used routes. Coupling of a thiol with a sulphinic acid using dinitrogen tetroxide leads to thiolsulphonates, as discussed in the preceding section of this Chapter.

Phenyl chlorothiolsulphonate PhS.SO2Cl provides access to a wide variety of thiolsulphonate esters through reaction with a Grignard reagent at -70° (J.Lazar and E.Vinkler, Acta chim.acad.Sci.Hung., 1974, 82, 87) but side-products such as SO2, disulphides, and hydrocarbons, are also formed.

Arenesulphenic acids generated in the presence of an arenesulphinic acid are trapped as arenethiolsulphonates (R.D.Allan, D.H.R.Barton, M.Girijavallabhan, and P.G.Sammes, J.chem.Soc., Perkin I, 1974, 1456).

Arenethiolsulphonates are valuable thioarylation reagents, giving 2-(arylthio)alkyl carbonyl compounds with α-keto-carbanions (G.C. Barrett, in 'Comprehensive Organic Chemistry', eds. Sir Derek H.R. Barton and W.D.Ollis, Vol.3, ed. D.N.Jones, Pergamon Press, Oxford, 1979, p.33) and sulphenamides, sulphenates, and disulphides with other nucleophiles. Cleavage by hypochlorite to give arenesulphonic acids has been subjected to detailed mechanistic scrutiny (J.L. Kice and A.R.Puls, J.Amer.chem.Soc., 1977, 99, 3455).

Desulphurization of arenethiolsulphonate esters with a tris(di-alkyl-amino)phosphine gives sulphones and sulphinate esters, through nucleophilic attack at sulphenyl sulphur (D.N.Harpp, J.G.Gleason, and D.K. Ash, J.org.Chem., 1971, 36, 322), whereas triphenylphosphine gives deoxygenation products, also obtained in an alternative process in which the thiolsulphonate ester is treated with reduced iron at 200° (T.Fujisawa, K.Sugimoto, and M.Ohta, Chemistry Letters, 1973,1241).

$$ArSO_2SR \longrightarrow ArSSR$$

Diaryl sulphides are formed by the pyrolysis of an aryl arenethiolsulphononate over copper (Fujisawa, Sugimoto, and Ohta, ibid., p.237).

$$PhSO_2SPh \longrightarrow PhSPh + SO_2$$

(b)   Arenethiolsulphinate esters

The standard synthesis, the reaction of a thiol with a sulphinyl chloride in the presence of a base, gives a small excess of one enantiomer when the base is an optically-active tertiary amine (M.Mikolajczyk and J.Drabowicz, Chem.Comm., 1976, 220). The formation of arenethiol-sulphinate esters as products of attempts to isolate arenesulphenic acids, and as minor products with thiolsulphonates through the oxidation of disulphides, has been mentioned in earlier sections of this Chapter. Mild oxidation of disulphides using m-chloroperbenzoic acid (D.N. Harpp and A.Granata, Synthesis, 1978, 782) or a 2-arylsulphonyl-3-aryl-oxaziridine (F.A.Davis, R.Jenkins, and S.G.Yocklovich, Tetrahedron Letters, 1978, 5171) gives serviceable yields of thiolsul-phinates, although it is difficult to avoid over-oxidation to the thiol-sulphonate in these routes (A.K.Bhattacharya and A.G.Hortmann, J. org.Chem., 1978, 43, 2728).

The lower stability of these esters, as compared to that of arene-thiolsulphonates, is implied in their disproportionation into thiolsul-phonates and disulphides (D.R.Hogg and J.Stewart, J.chem.Soc., Perkin II, 1974, 43, 436) via sulphenate anions (disulphides and sulphinic acids are formed from these initial S - S cleavage products). The racemization of a chiral thiolsulphinate is accounted for on the basis of an equilibrium involving the corresponding sulphenic acid (E. Block and J.O'Connor, J.Amer.chem.Soc., 1974, 96, 3921, 3929).

Protonation of an arenethiolsulphonate ester occurs at sulphenyl sulphur, and acid- and dialkyl sulphide-catalyzed decomposition in acetic acid follows several steps (T.-L.Ju, J.L.Kice, and C.G. Venier, J.org.Chem., 1979, $\underline{44}$, 610).

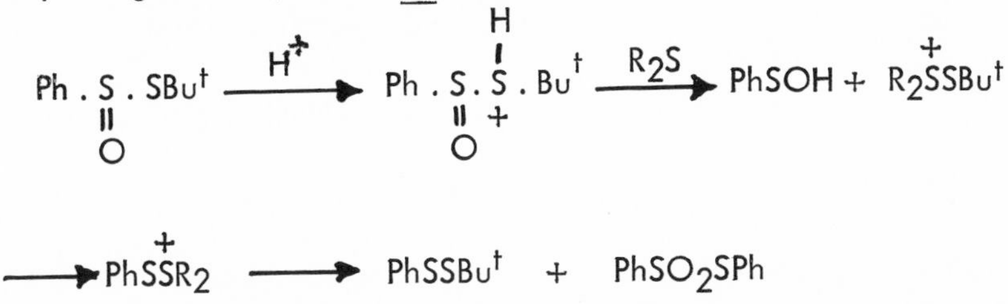

$$Ph \cdot \underset{\underset{O}{\|}}{S} \cdot SBu^t \xrightarrow{H^+} Ph \cdot \underset{\underset{O}{\|} +}{S} \cdot \overset{\overset{H}{|}}{S} \cdot Bu^t \xrightarrow{R_2S} PhSOH + R_2\overset{+}{S}SBu^t$$

$$\xrightarrow{} Ph\overset{+}{S}SR_2 \longrightarrow PhSSBu^t + PhSO_2SPh$$

Alkaline hydrolysis of arenethiolsulphonate esters involves nucleophilic attatck at the 'harder' sulphinyl sulphur atom (S.Oae, T.Takata, and Y.H.Kim, Tetrahedron Letters, 1977, 4219; see also J.L.Kice and C.-C.A.Liu, J.org.Chem., 1979, $\underline{44}$, 1918). Nucleophilic substitution with these compounds is accompanied by inversion of configuration, as with corresponding reactions with other sulphinyl compounds (Mikolajczyk and Drabowicz, *loc.cit.*).

Chapter 7

MONONUCLEAR HYDROCARBONS CARRYING NUCLEAR SUBSTITUENTS
CONTAINING SELENIUM OR TELLURIUM.*

P.G.HARRISON.

1. *Introduction*

Interest in the organic compounds of selenium and tellurium
has, over the last decade, been steadily increasing.
Tellurium, an element rather neglected in the past compared
to its lighter congeners, has experienced a systematic
investigation, and many classes of organotellurium compounds
are now well characterised. The analogous period of
development in organoselenium chemistry took place much
earlier, and more recent efforts have been directed, with
much success, towards the application of organoselenium
compounds as synthetic organic reagents.

2. *Selenium*

The organic chemistry of selenium has been discussed in
two books ('Organic Selenium Compounds', ed. D.L.Klayman and
W.H.Gunther, Wiley-Interscience, New York, 1973; 'Selenium',
ed. R.A.Zingaro and W.C.Cooper, Van Nostrand-Reinhold, New
York, 1974), and the literature upto March 1976 systemat-
ically reviewed ('Organic Compounds of Sulphur, Selenium and
Tellurium', Vols. 1-4, The Chemical Society, London).
Unfortunately, this latter series has been discontinued.

(a) *Arylselenium(II) Compounds*
The usual method of attaching an aryl group to selenium
involves the insertion of elemental selenium into the metal-
carbon bond of a Grignard or an organolithium reagent. A
modification of this procedure using two equivalents of

---

* Based on the literature published up to early 1979.

Grignard reagent and an aldehyde or ketone yields compounds
of the type ArSeCMeRAr (I.L.Lapkin, N.V.Bogoslovskii and N.
I.Zenkova, Zh. obshch. Khim., 1972,42,1972). Treatment of
arylselenium magnesium halides with chloroformic acid,
followed by aqueous sodium carbonate solution, affords aryl-
seleno esters such as p-MeOC$_6$H$_4$COSeC$_6$H$_4$Me-p (G.Heppke, J.
Mattens, K.Praefoke and H.Simon, Angew. Chem. internat. Edn.,
1977,16,318). An alternative method for the synthesis of
seleno esters involves the reaction of phenylselenocyanate
with carboxylic acids in the presence of tri-n-butylphosphine.

$$RCO_2H \ + \ ArSeCN \ \xrightarrow[CH_2Cl_2]{2Bu_3P} \ RCOSeAr$$

Yields of product are generally very high (≥78%), although
with p-chlorobenzoic acid the major product is diphenyl-
diselenide and the seleno ester is obtained in only 36% yield
(P.A.Grieco, Y.Yokoyama and E.Williams, J. org. Chem., 1978,
43,1283). Unsymmetrical phenylarylselenides may be prepared
by the reaction of sodium phenyl-selenide with aryl halides
in liquid ammonia if the reaction mixture is irradiated (A.B.
Pierini and R.A.Rossi, J. organometal. Chem., 1978,144,C12).
    Direct selenonation of aromatic molecules is of far less
synthetic utility than other established methods. Acetyl-
selenic acid (H$_2$SeO$_4$-acetic anhydride) gives an isomer
distribution with polyalkylbenzenes similar to that obtained
on sulphonation (C.Ris and H.Cerfontain, J. chem. Soc.,
Perkin 2, 1973,2129). Selenonation of 1,2-dimethoxybenzene
with aqueous SeO$_2$ gives a mixture of bis[3,4-dimethoxyphenol]-
selenide, m.p. 101-103°, and 2,3,7,8-tetramethoxy-5,10-
diselenanthrene, m.p. 172-173°, although the reaction may be
optimized to afford the latter product alone (J.Weiss, N.
Nitsche, F.Bohnke and G.Klar, Ann. 1973,1418).
    Several procedures have been devised for the introduction
of the phenylseleno group into organic molecules, principally
in order to capitalise on the readiness of selenoxides,
produced by further oxidation to undergo facile elimination
of benzeneselenic acid with the formation of carbon-carbon
double bonds (vide infra). The particular method adopted
varies with the organic substrate, but the following
examples are typical.

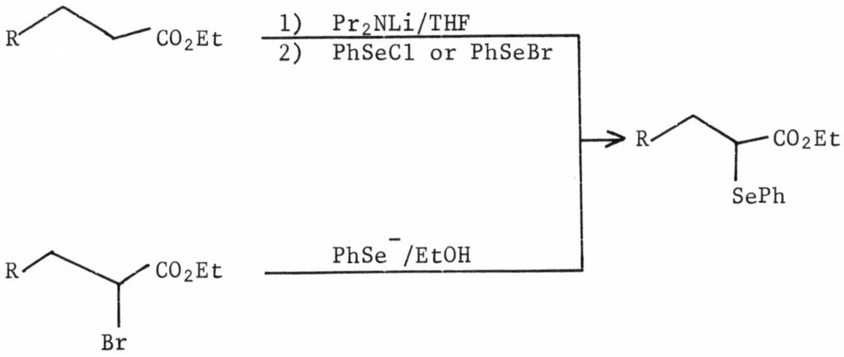

R⌒⌒CO₂Et → 1) Pr₂NLi/THF
              2) PhSeCl or PhSeBr

$$R\text{—}CH_2\text{—}CH(SePh)\text{—}CO_2Et$$

R⌒CH(Br)—CO₂Et → PhSe⁻/EtOH

(K.B.Sharpless, R.F.Lauer and A.Y.Teranishi, J. Amer. chem. Soc., 1973,95,6137).

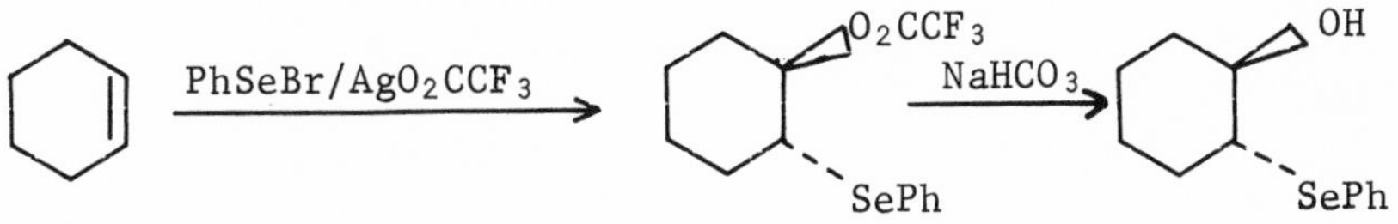

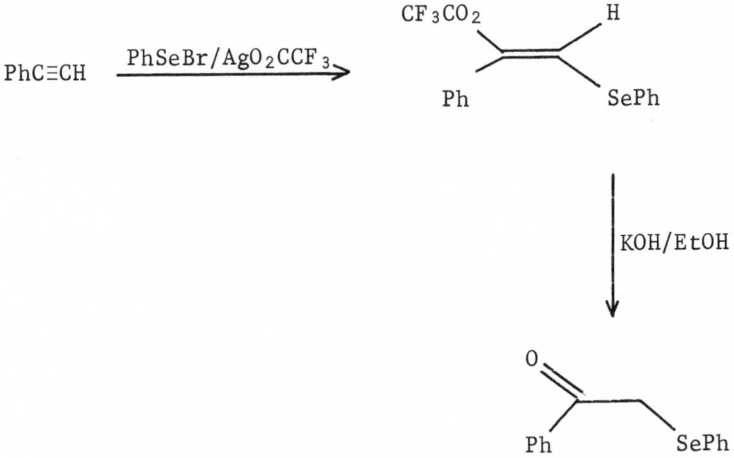

PhSeBr/AgO₂CCF₃ → NaHCO₃ →

PhC≡CH → PhSeBr/AgO₂CCF₃ →

↓ KOH/EtOH

(K.B.Sharpless and R.F.Lauer, J. org. Chem., 1974,39,429;

D.L.J.Clive, Chem. Comm., 1975,695; 1974,100; H.J.Reich, J. org. Chem., 1974,39,428.)

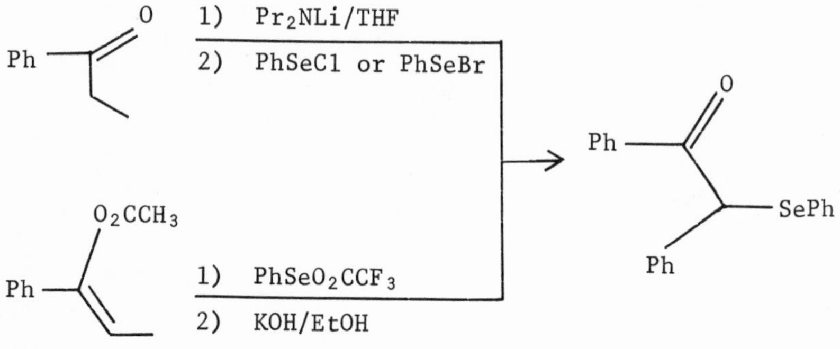

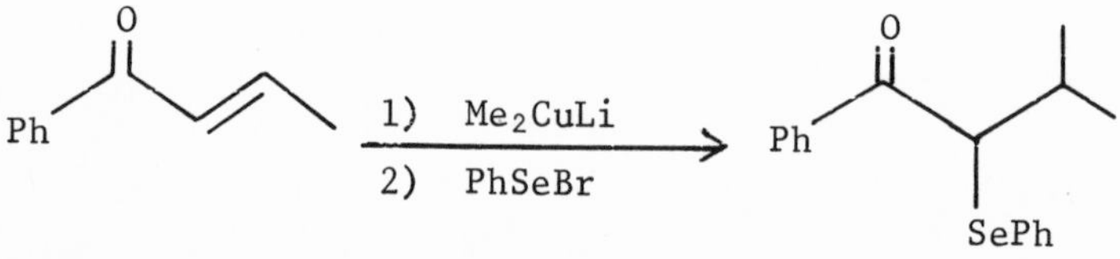

(H.J.Reich, J.M.Renga and I.L.Reich, J. Amer. chem. Soc., 1975,97,5434; J. org. Chem., 1974,39,2133.)

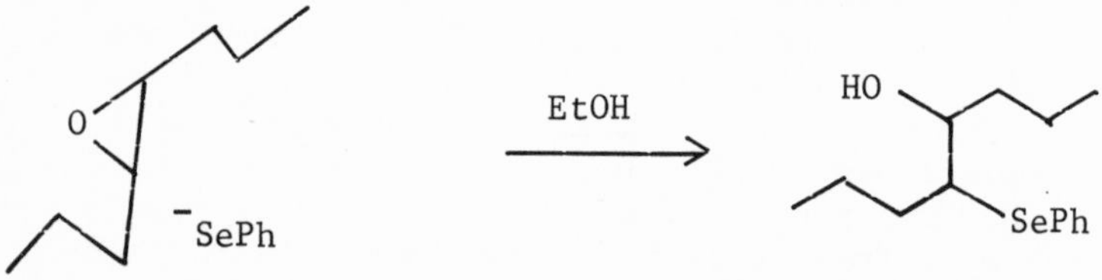

(K.B.Sharpless and R.F.Lauer, J. Amer. chem. Soc., 1973,95, 2697)

Like phenylselenenyl trifluoroacetate, phenylselenenyl bromide adds to monosubstituted alkenes to afford the products of either Markownikov or anti-Markownikov addition depending upon whether kinetically-controlled (THF, $-78°$) or thermodynamically-controlled (MeCN, $25°$) reaction conditions are employed (S.Raucher, J. org. Chem., 1977,42,2950). Dehydrohalogenation (using potassium tert-butoxide in THF)

of the resulting adducts provides a method for the regiospecific synthesis of 1- and 2-phenylselenoalkenes.

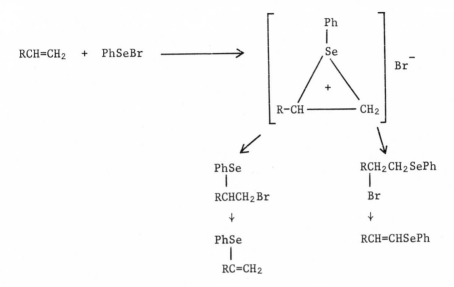

Typically, a mixture of both isomers is obtained. For
example, the reaction with propene under kinetic conditions
yields a 15:85 mixture of the 1- and 2-isomer compared to a
91:9 mixture under the conditions of thermodynamic control.
However, with some alkenes, e.g. 3,3-dimethylbut-1-ene, total
selectivity is obtained. The analogous reaction of ethylene
with p-tolylselenenyl chloride (a dark orange-red liquid
(b.p. $62°/0.07$mm) prepared by the reaction of di-p-tolyl-
diselenide with freshly-distilled sulphuryl chloride) does
not yield an episelenurane as first thought (D.G.Garrett and
G.H.Schmidt, Canad. J. Chem., 1974,52,1027), but rather
p-tolyl(chloroethyl)selenium(IV) dichloride (H.J.Reich and
J.E.Trend, Canad. J. Chem., 1975,53,1922).

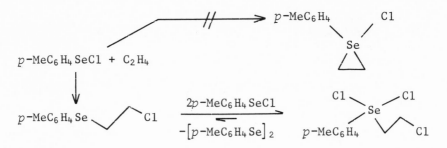

Raucher has also devised stereoselective synthesis of both the (E)- and the (Z)-phenylseleno alkenes starting from an alkyne (S.Raucher, M.R.Hansen and M.A.Colter, J. org. Chem., 1978,43,4885).

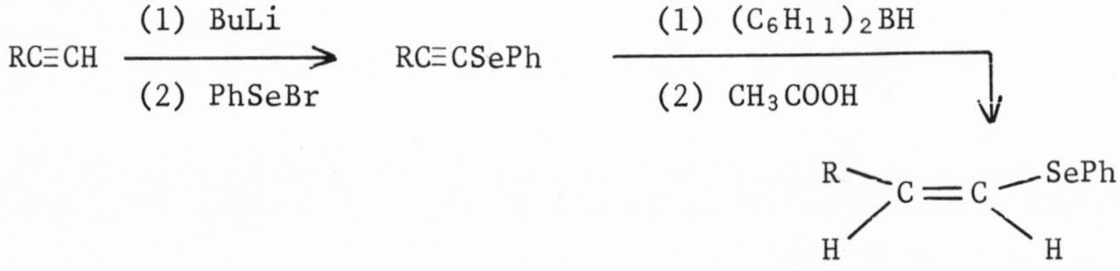

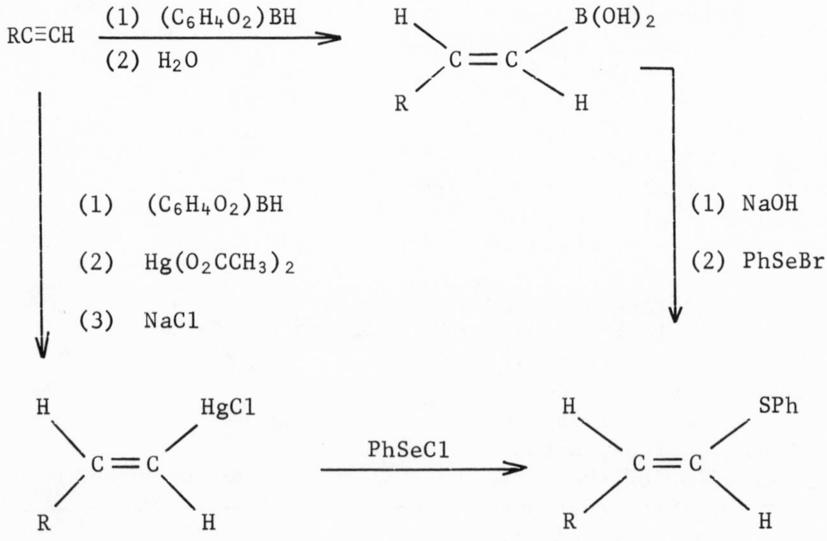

RC≡CH $\xrightarrow[\text{(2) PhSeBr}]{\text{(1) BuLi}}$ RC≡CSePh $\xrightarrow[\text{(2) } CH_3COOH]{\text{(1) } (C_6H_{11})_2BH}$

$$\begin{array}{c} R \\ \diagdown \\ H \diagup \end{array} C = C \begin{array}{c} SePh \\ \diagup \\ \diagdown H \end{array}$$

RC≡CH $\xrightarrow[\text{(2) } H_2O]{\text{(1) } (C_6H_4O_2)BH}$ $\begin{array}{c} H \\ \diagdown \\ R \diagup \end{array} C = C \begin{array}{c} B(OH)_2 \\ \diagup \\ \diagdown H \end{array}$

(1) $(C_6H_4O_2)BH$

(2) $Hg(O_2CCH_3)_2$

(3) NaCl

(1) NaOH

(2) PhSeBr

$$\begin{array}{c} H \\ \diagdown \\ R \diagup \end{array} C = C \begin{array}{c} HgCl \\ \diagup \\ \diagdown H \end{array}$$ $\xrightarrow{\text{PhSeCl}}$ $$\begin{array}{c} H \\ \diagdown \\ R \diagup \end{array} C = C \begin{array}{c} SPh \\ \diagup \\ \diagdown H \end{array}$$

The isomer purity obtained was >95% in each case.

Phenylvinylselenide, $PhSeCH=CH_2$, can be used as a $\left[^+CH=CH^-\right]$ synthon (S.Raucher and G.A.Koolpe, J. org. Chem., 1978,43, 4253), by reaction with an alkyllithium, trapping of the resulting α-lithioalkylohenylselenide with an electrophile, and subsequent oxidation/elimination of PhSeOH yielding a disubstituted alkene.

The choice of solvent in this reaction is very important in order to minimise selenium-carbon bond cleavage by the organolithium reagent, and the best results are obtained when the reaction is carried out in dimethyl ether or dimethoxyethane at 0°.

Arylseleniranium ions analogous to those postulated as reaction intermediates in the reaction of phenylselenenyl bromide with alkenes (vide supra) have been generated quantitatively by the interaction of the arylselenenyl hexafluorophosphate or hexafluoroantimonate (from aryl-selenyl-chloride and the appropriate silver salt in dichloromethane) with an alkene, or by the reaction of the silver salt with a suitable chlorine-containing selenium derivative (G.H.Schmid and D.G.Garratt, Tetrahedron Letters, 1975,1991).

The products were generally unstable with half-lives in solution of less than 48 hours at temperatures in excess of 20°, and therefore only identified by ^{1}H-n.m.r. spectroscopy.

Phenylallylselenides undergo $\left[1,3\right]$-allylic rearrangements with greater facility than do the analogous sulphur compounds. Thus, n-BuCH(SePh)CMe=CH_2 (obtained by substitution of the corresponding allyl chloride using sodium phenylselenide) rearranges completely in chloroform solution at 52° (half-life ca. 1.3 hours) to afford n-BuCH=CMeCH_2SePh (K.B.Sharpless and R.F.Lauer, J. org. Chem., 1972,37,3973).

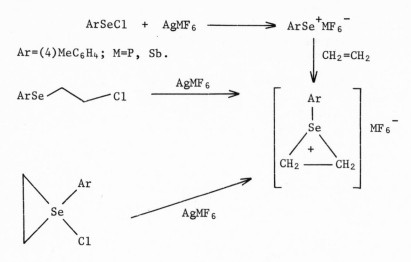

$$ArSeCl + AgMF_6 \longrightarrow ArSe^+MF_6^-$$

Ar=(4)MeC₆H₄; M=P, Sb.

Arylselenirenium ions may be obtained similarly.

Bis(phenylseleno)acetals are readily obtained by the
reaction of phenylselenol with ketones (J.W.Denis, W.Dumont
and A.Krief,Tetrahedron Letters, 1976,453). Mixed 0, Se-

$$RR'C=O \ + \ 2PhSeH \ \longrightarrow \ RR'C(SePh)_2$$

and S, Se-acetals are also accessible by the addition of
phenylselenol to a vinyl ether or sulphide using toluene-p-
sulphonic acid or boron trifluoride etherate as catalyst,
or by the reaction between potassium phenylsulphide and an
appropriate α-bromether or sulphide (A.Anciaux, A.Eman,
W.Dumont and D.Van Ende, Tetrahedron Letters, 1975,1613).

X = SR, OR

$$PhSe^-K^+ \ + \ BrCH_2X \ \longrightarrow \ PhSe-CH_2X$$

Treatment of selenoacetals with organolithium reagents
generally results in the cleavage of the selenium-carbon
bond with the formation of carbanions, which react with
a variety of electrophiles affording a route to α-seleno-
substituted compounds (A.Anciaux, A.Eman, W.Dumont and
A.Krief,Tetrahedron Letters, 1975,1617; D. Van Ende and
A.Krief, ibid, 1976,453; D. Van Ende and A.Krief, ibid, 1976,
457; W.Dumont and A.Krief, Angew. Chem., 1976,88,184).
The generation of phenylseleno-substituted carbanions
by deprotonation is not usually feasible using butyllithium
because of the ease with which selenium-carbon bond fission
takes place with this reagent, invariably resulting in a
mixture of cleavage and deprotonation products. The
cleavage reaction may, however, be completely suppressed
by the use of lithium dialkylamides (D.Seebach and
N.Peleties, Ber., 1972,105,511; D. Van Ende, A.Cravadov
and A.Krief, J. organometal. Chem., 1979, 177,1).

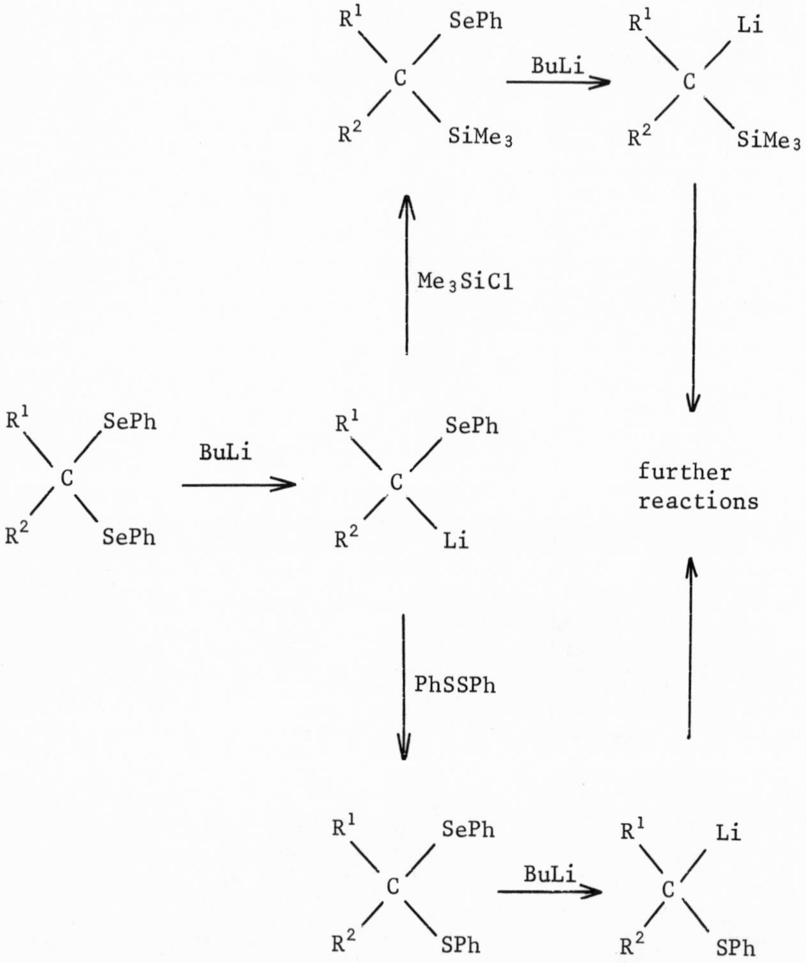

R1
SePh
C
R2
SiMe3
BuLi
R1
Li
C
R2
SiMe3
Me3SiCl
R1
SePh
C
R2
SePh
BuLi
R1
SePh
C
R2
Li
further
reactions
PhSSPh
R1
SePh
C
R2
SPh
BuLi
R1
Li
C
R2
SPh

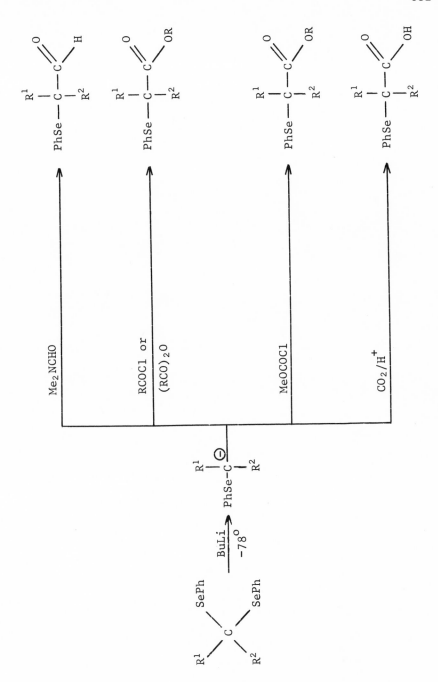

$$PhSeCH_3 \xrightarrow[\substack{TMEDA \\ \sim 40\%}]{BuLi/THF} PhSeCH_2Li \xleftarrow[\sim 90\%]{BuLi/THF} (PhSe)_2CH_2$$

$$(PhSe)_2CRH \xrightarrow{LiNR_2^!} (PhSe)_2CRLi$$

$$(MeO)_3CH + 3PhSeH \longrightarrow (PhSe)_3CH \xrightarrow{LiNR_2^!} (PhSe)_3CLi$$

$$(PhSe)_3CH \downarrow BuLi$$

$$(PhSe)_2CHLi + (PhSe)_3CLi + PhSeBu$$

Many phenylselena-metal derivatives have been synthesised, usually by substitution or transmetallation. The general methods employed are illustrated by the following typical examples:

$$Me_3SiCl + LiSePh \longrightarrow Me_3SiSePh$$

$$LiAlH_4 + 4PhSeH \longrightarrow LiAl(SePh)_4 \xrightarrow{Me_2HSiCl} Me_2SiH(SePh)$$

$$Me_3SiSePh + H_3GeF \longrightarrow H_3GeSePh$$

$$(H_3Ge)_2N_2C + 2PhSeH \longrightarrow 2H_3GeSePh$$

$$MCl_3 + Me_3SiSePh \longrightarrow M(SePh)_3$$

M = P, As, B.

(J.E.Drake and R.T.Hemmings, J. chem. Soc. Dalton., 1976,1730).

$$BX_3 \ + \ PhSeH \ \longrightarrow \ (PhSe)_3B$$

X = Cl, Br, I.

(M.Schmidt and H.D.Block, J. organometal. Chem., 1970,25,17; W.Siebert and A.Ospice, Ber., 1972,105, 464).

$$(R_3Sn)_2O \ + \ PhSeH \ \longrightarrow \ R_3SnSePh$$

$$R_2SnO \ + \ 2PhSeH \ \longrightarrow \ R_2Sn(SePh)_2$$

$$R_3SnSePh \ + \ AgNO_3 \ \longrightarrow \ AgSePh$$

$$R_3SnSePh \ + \ Cu(NO_3)_2.3H_2O \ \longrightarrow \ Cu(SePh)_2$$

$$R_3SnSePh \ + \ NiCl_2.6H_2O \ \longrightarrow \ Ni(SePh)_2.3H_2O$$

(E.C.MacMullin and M.E.Peach, J. organometal. Chem., 1973, 52,355).

$$PhSeM \ + \ C_7H_7Mo(CO)_2Br \ \longrightarrow \ C_7H_7Mo(SePh)_3Mo(CO)_3$$

M = H, Li or MgBr.

(D.Mohr, H.Wienand and M.L.Ziegler, J. organometal. Chem., 1977,134, 281; idem, Angew. Chem. internat. Edn., 1977,16, 261).

$$\left[C_5H_5Fe(CO)_2\right]_2 \ + \ Ph_2Se_2 \ \longrightarrow \ C_5H_5Fe(CO)_2SePh$$

(E.D.Schermer and W.H.Baddley, J. organometal. Chem., 1971, 27,83).

$$\underline{cis}-\left[PtMe_2L_2\right] \ \xrightarrow[-CH_4]{PhSeH} \ \underline{trans}-\left[PtMe(SePh)L_2\right]$$

$$\Big\downarrow PhSeH; \ -CH_4$$

L = PMePh$_2$, PMe$_2$Ph

$$\underline{trans}-\left[Pt(SePh)_2L_2\right]$$

$$AuMe(PMe_2Ph) \ + \ PhSeH \ \longrightarrow \ Au(SePh)(PMe_2Ph) \ + \ CH_4$$

$AuMe_3(PMe_2Ph)$ + PhSeH $\longrightarrow$ trans-$[AuMe_2(SePh)(PMe_2Ph)]$

+ CH_4

(R.J.Puddephatt and P.J.Thompson, J. organometal. Chem., 1976,117,395).

(b) The Selenoxide Elimination and Related Reactions

The extreme ease with which phenylalkylselenoxides undergo *syn*-elimination of benzeneselenininic acid has now been fully exploited to provide a convenient, versatile method for the introduction of an alkene function into organic molecules. The general procedure involves initial introduction of a phenylselenenyl group into the molecule, oxidation to the selenoxide (ozone, hydrogen peroxide, peracids or sodium metaperiodate), followed by facile or even spontaneous elimination of PhSeOH therefrom (K.B. Sharpless, R.F.Lauer and A.Y.Teranishi, J. Amer. chem. Soc., 1973,95,6137; K.B.Sharpless, M.W.Younf and R.F.Lauer, Tetrahedron Letters, 1973,1979; R.H.Mitchell, Chem. Comm., 1974,990; H.J.Reich, I.L.Reich and J.M.Renga, J. Amer. chem. Soc., 1973,95,5813; H.J.Reich and S.K.Shah, ibid., 1975,97, 3250; W.A.Kleschik and C.H.Heathcock, J. org. Chem., 1978, 43,1256; N.Ikora and B.Granerm, ibid., 1978,43,1607). A typical example of the conversion of phenylethylketone into phenylvinylketone:

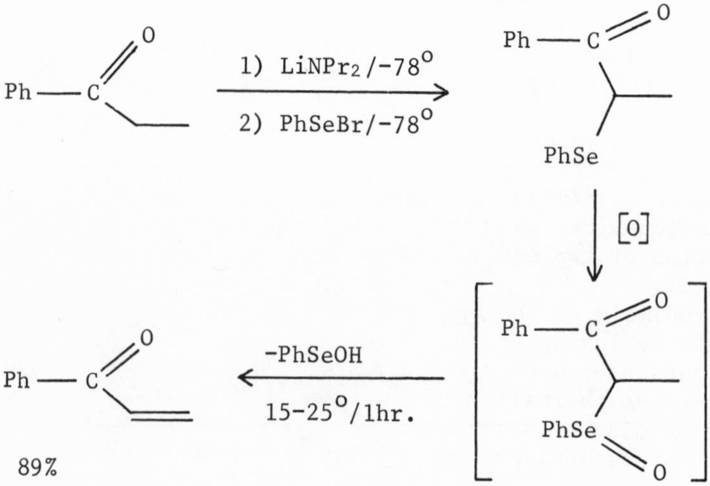

Although in the majority of cases the reactions proceed well to give high yields of olefinic products, several complications can occur. A frequent side-reaction is the subsequent addition of the elements of benzeneseleninic acid to the alkene to give β-hydroxyethylphenylselenides. This reaction may be completely suppressed by the presence of alkylamines during the elimination step. Protic reagents such as water, alcohols and carboxylic acids should in general be avoided since, besides facilitating "PhSeOH" addition to the alkene, they greatly slow down the rate of elimination as well as promoting solvolytic heterolysis of selenium-carbon bonds. Increased rates of *syn*-elimination may be obtained by using *o*-nitrophenylselenium derivatives, and *o*-nitroseleninic acid also has a reduced tendency for addition to alkenes. The choice of oxidising agent is also important. Hydrogen peroxide is less satisfactory since it is catalytically decomposed by selenoxides under some conditions, and when used as an oxidant care must be taken to ensure that excess is present until the elimination is complete otherwise "ArSeOH" readdition to alkene may occur. In addition, hydrogen peroxide and seleninic acids can also lead to epoxide formation, presumably via an intermediate areneperseleninic acid, when tetra-, tri- and even some disubstituted alkenes are being formed (P.A.Grieco, Y. Yokoyama, S.Gilman and M.Nishizawa, J. org. Chem., 1977,<u>42</u>, 2034). When peracids are used in the oxidation, the presence of amine during the elimination step is absolutely essential, because of the influence of carboxylic acids on the course of the reactions. Taking into account all of these possible problems, Reich <u>et al</u>. (J. org. Chem., 1978,<u>43</u>,1697) have proposed the following procedure for the optimum generation of alkene.

(i) Oxidation to selenoxide at as low a temperature as possible using *m*-chloroperbenzoic acid in dichloromethane or THF, or ozone in dichloromethane. Hydrogen peroxide may be used for selenoxides which are slow to eliminate.

(ii) Addition of two equivalents of amine (HNPr$_2$, NEt$_3$, HNEt$_2$, etc.).

(iii) Thermolysis of the basic selenoxide solution by addition to refluxing hexane or carbon tetrachloride.

The oxidation of α-silylselenides yields α-silylselenoxides which decompose thermally via both the <u>*syn*</u>-elimination reaction and also by a competing sila-Pummerer rearrangement to afford vinylselenides and carbonyl compounds (H.J.Reich and S.K.Shah, J. org. Chem., 1977,__42__,1773).

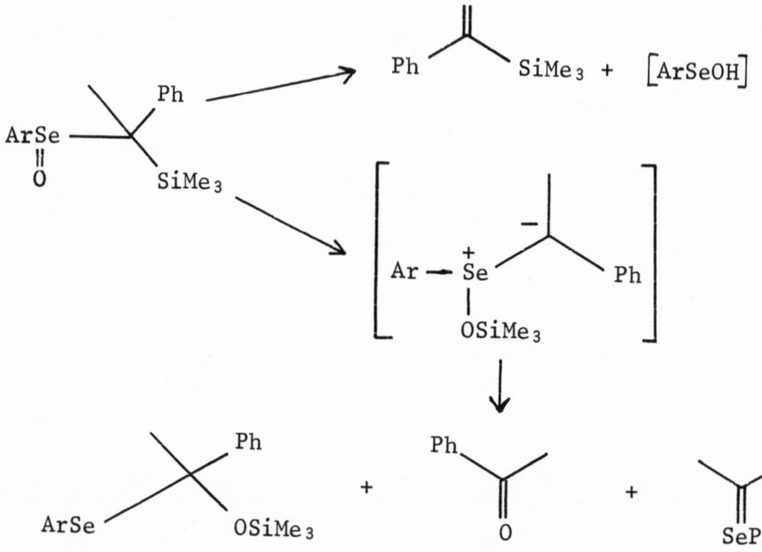

The ratio of products arising from the two competing reactions may be controlled to some extent by the reaction conditions, but a more pronounced change favouring *syn*-elimination occurs when the ArSe group is made more electronegative (e.g. $m\text{-}CF_3C_6H_4Se$ instead of C_6H_5Se).

A second alkene synthesis based on selenoxides has been described by Reich and Chow (Chem. Comm., 1975, 790). This method involves the generation of a β-hydroxyethylphenyl-selenide (by the reaction of an α-lithiated phenylalkyl-selenoxide and a carbonyl compound followed by reduction) which when treated with methylsulphonyl chloride and triethylamine undergoes reductive elimination.

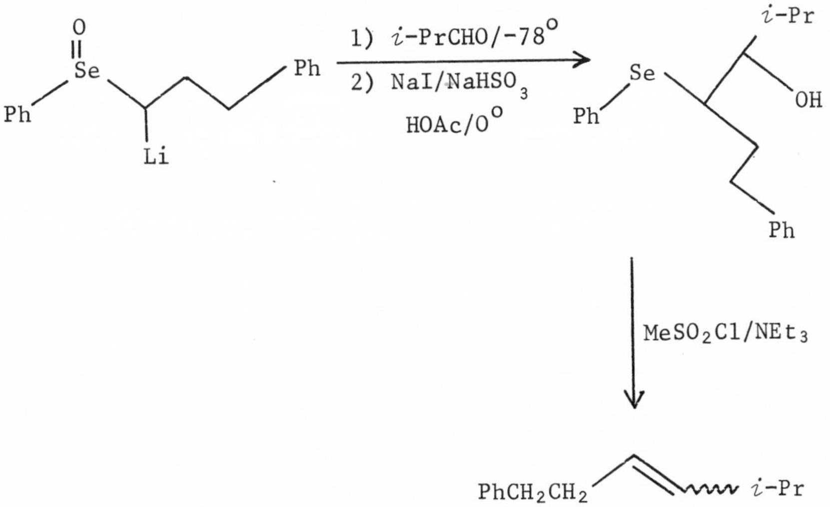

Substituted benzylphenylselenoxides rearrange and decompose on heating via elimination of phenylselenol affording the corresponding substituted benzaldehyde in high yields (I.D.Entwistle, R.A.Johnstone and J.H.Varley, Chem. Comm., 1976,61).

$$\underset{\overset{\|}{O}}{ArCH_2SePh} \xrightarrow[\text{2-3 mins}]{110\text{-}130^{\circ}} [ArCH_2OSePh] \xrightarrow{-PhSeH} ArCHO$$

(c) *Arylselenium(IV) Compounds*

Crivello and Lam (J. org. Chem., 1978,<u>43</u>,3055) have prepared triarylselenium salts via copper(II)-catalysed arylation of diphenylselenide by diaryliodonium salts.

$$Ar_2I^+AsF_6^- + Ph_2Se \xrightarrow[Cu(O_2CPh)_2]{120\text{-}125^{\circ}} ArPh_2Se^+AsF_6^- + ArI$$

Triphenylselenium(IV) hexafluoroarsenate, m.p. 185-187°;
p-tert-butylphenyldiphenylselenium(IV) hexafluoroarsenate,
m.p. 151-152°.

Both anisole and phenetole arylate selenium(IV) chloride
in the absence of solvent to yield the diarylselenium(IV)
dichloride (E.R.Clark and M.A.Al-Juraihi, J. organometal.
Chem., 1975,96,251):

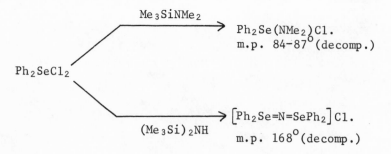

$$SeCl_4 + 2RO\text{—}\langle \rangle \longrightarrow \left[RO\text{—}\langle \rangle\text{—}\right]_2 SeCl_2 + 2HCl$$

R = Me or Et.

Transmetallation using aryllead compounds is also a useful
method for the preparation of arylselenium(IV) chlorides
since the diaryllead dichloride produced is insoluble and
easily separated.

$$SeCl_4 + Ph_4Pb \xrightarrow[\text{reflux 6h.}]{\text{toluene}} Ph_2PbCl_2\downarrow + Ph_2SeCl_2$$

Diphenylselenium dichloride, m.p. 177°; di-p-methoxyphenol-
selenium dichloride, m.p. 163°.

Diphenylselenium dichloride and dibromide are both
reduced by ammonia, methyl- or dimethylamines at -60° to
diphenylselenide. The reaction of the dichloride with
trimethylsilyldimethylamine or hexamethyldisilazane,
however, produces selenium-nitrogen bonded compounds (V.
Horn and R.Paetzold, Z. anorg. allg. Chem., 1974,404,213).

$$Ph_2SeCl_2 \nearrow^{Me_3SiNMe_2} Ph_2Se(NMe_2)Cl.$$
m.p. 84-87°(decomp.)

$$\searrow_{(Me_3Si)_2NH} \left[Ph_2Se=N=SePh_2\right]Cl.$$
m.p. 168°(decomp.)

Triphenylselenium chloride and diphenylselenium dibromide may be converted into a wide range of other derivatives by substitution using the appropriate silver salts.

$$Ph_3SeCl \ + \ AgX \ \longrightarrow \ Ph_3SeX$$

$$X = NO_3, \ O_2CCH_3, \ NCS.$$

$$Ph_2SeBr_2 \ + \ 2AgX \ \longrightarrow \ Ph_2SeX_2$$

$$X = NO_3, \ O_2CCH_3, \ O_2CCF_3, \ \tfrac{1}{2}SO_4, \ MeSO_3, \ NCO.$$

The attempted substitution of phenylselenium tribromide using silver nitrate or acetate, however, yields the products of dissociation of the intermediate tris-substituted compounds, even when the reactions are carried out at -60°.

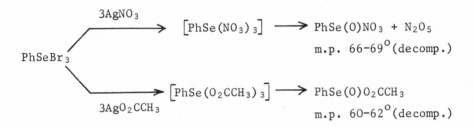

$$3AgNO_3$$
$$PhSeBr_3 \longrightarrow \left[PhSe(NO_3)_3\right] \longrightarrow PhSe(O)NO_3 + N_2O_5$$
$$\text{m.p. } 66\text{-}69^\circ \text{(decomp.)}$$
$$\longrightarrow \left[PhSe(O_2CCH_3)_3\right] \longrightarrow PhSe(O)O_2CCH_3$$
$$3AgO_2CCH_3 \qquad \text{m.p. } 60\text{-}62^\circ \text{(decomp.)}$$

Facile methoxy-halide ligand exchange occurs between aryl-selenium methoxides and halides leading to the formation of mixed products:

$$Ph_2SeX_2 \ + \ Ph_2Se(OMe)_2 \ \longrightarrow \ 2Ph_2Se(OMe)X$$

$$PhSe(OMe)_3 \ + \ 2PhSeX_3 \ \longrightarrow \ 3PhSe(OMe)X_2$$

$$2PhSe(OMe)_3 \ + \ PhSeX_3 \ \longrightarrow \ 3PhSe(OMe)_2X$$

$$X = Cl, \ Br.$$

Phenylselenium trimethoxide (from the reaction of phenyl-

selenium tribromide with sodium methoxide in methanol/ether
at $0°$) is an oil which is readily hydrolysed to PhSe(O)OH
(V.Horn and R.Paetzold, Z. anorg. allg. Chem., 1973,173,179,
186,398; P. Ash, J.S.Lee, D.D.Titus, K.B.Mertes and R.F.
Ziolo, J. organometal. Chem., 1978,135,91).
Triphenylselenium nitrate, m.p. 180-182°; triphenylselenium
acetate, m.p. 120-122°(decomp.); diphenylselenium
dinitrate, m.p. 153-154°(decomp.); diphenylselenium
methoxide chloride, m.p. 127°; phenylselenium dimethoxide
chloride, m.p. 129-131°.

Oxidation of diphenylselenide with ozone produces
diphenylseleninic anhydride, PhSe(O)OSe(O)Ph, which has
found some application as an oxidising agent, oxidising
primary amines to ketones, benzylamines to benzaldehydes,
and phenols to hydroquinones (D.H.R.Barton, P.D.Magnus and
M.N.Rosenfeld, Chem. Comm., 1975,301).

Diarylselenium dichlorides condense with a variety of
active methylene compounds in the presence of base to yield
stable, isolable selenonium ylides (N.N.Magdesieva, R.A.
Kyandzhetsian and A.A.Ibragimov, J. organometal. Chem., 1972,
42,399; N.N.Magdesieva and R.A.Kyandzhetsian, J. org.
Chem. (U.S.S.R.), 1971,7,2316; K.J.H.Wei, I.C.Paul, M.M.Y.
Chang and J.I.Musher, J. Amer. chem. Soc., 1974,96,4099).

$$Ph_2SeCl_2 \;+\; H_2CX_2 \;\xrightarrow[-2HCl]{KOH/MeOH}\; Ph_2\overset{+}{Se}-\overset{-}{C}X_2$$

$$H_2CX_2 = H_2C(CN)_2, \quad H_2C(COMe)_2, \quad$$

Dicyanomethylenediphenylselenurane, m.p. 135°(decomp.);
diacetylmethylenediphenylselenurane, m.p. 125-126.5°.

Similar diarylselenimides, ArXSe=NSO Ar' (X = R, Ar, Cl),
have been obtained by Derkach (Zhur. org. Khim., 1971,7,
1543; 1974,10,807,1873) by the reaction of trimethylsilyl-
arylselenides or diarylselenides with chloramine-T, by the
addition of iminoarylseleninyl chlorides to styrene in the
presence of copper(I) chloride, and by the reaction of
arylselenides with N-sodio-p-tosylsulphonamide in the
presnece of t-butyl hypochlorite (S.Tamagaki, S.Oae and

K.Sakaki, Tetrahedron Letters, 1975,649).
Methylphenylselenonium p-toluenesulphonimide white crystals,
m.p. 129-130°.

3. *Tellurium*

The chemistry of aryltellurium compounds, not surprisingly,
parallels that of the analogous selenium derivatives, but
the tellurium compounds, have no significant application in
organic synthesis. The area in general has been discussed
in a monograph ("The Organic Chemistry of Tellurium", K.J.
Irgolic, Gordon and Breach, New York, 1974) and is
periodically reviewed by the same author (J. organometal.
Chem., 1975,103,91; 1977,130,411; 1978,158,235; 1978,158,
267).

(a) *Aryltellurium(II) Compounds*

The reaction of an aryl-lithium or an aryl Grignard
reagent with elemental tellurium leads to the formation of
the corresponding metal aryltellurides (M.Sato and T.
Yoshida, J. organometal. Chem., 1975,87,217; D.Seebach and
A.K.Beck, Ber., 1975,108,314; K.J.Irgolic, P.J.Busse, R.A.
Grigsby and M.R.Smith, J. organometal. Chem., 1975, 88,175).

ArM   +   Te $\longrightarrow$ ArTeM

    M = Li, MgX.

Aerobic oxidation of such solutions produces diarylditellurides, Ar Te , in moderate yield, although Seebach and
Beck have obtained diphenylditellurides quantitatively by
this method (W.S.Haller and K.J.Irgolic, J. organometal. Chem.,
1972,38,97; P.Schult and G.Klar, Z. Naturforsch. b, 1975,30,
43; N.Dereu, J.L.Piette, J.Van Cuppenolle and M.Renson,
J. heterocyclic Chem., 1975,12,423). No synthetic procedure
has been devised for the preparation of unsymmetrical
diarylditellurides, ArTeTeAr', although Dance, McWhinnie
and Jones (J. organometal. Chem., 1977,125,291) have proposed
that such species are present in mixtures of the two
symmetrical compounds.
Diarylditellurides are cleaved by a variety of reagents.
The reaction of diphenylditelluride with lithium metal

produces lithium phenyltelluride (K.J.Irgolic et al., J. organometal. Chem., 1975,88,175), whilst the corresponding sodium salt is best obtained by the reaction with sodium borohydride in methanol or ethanol/benzene (N.Dereu et al., J. heterocyclic Chem., 1975,12,423; S.R.Buzilova, I.D.Sadekov, T.V.Lipovich, T.M.Filippova and L.I.Vereshchapin, Zh. Obshch. Khim., 1977,47,199).

$$\text{ArTeLi} \quad \xleftarrow{\quad \text{Li} \quad} \quad \text{Ar}_2\text{Te}_2 \quad \xrightarrow{\quad \text{NaBH}_4 \quad} \quad \text{ArTeNa}$$

Elemental mercury inserts into the tellurium-tellurium bond of diarylditellurides. Thus, shaking di-p-ethoxyphenyl-ditelluride with mercury in benzene for 48h. affords bis(p-ethoxyphenyltelluro)mercury, m.p. 110-112° as a red powder (N.G.Dance et al., J. organometal. Chem., 1978,152, 175). Reaction with triorganostannanes (N.G.Dance, W.R. McWhinnie and C.H.W.Jones, J. organometal. Chem., 1977,125, 291) and $[(C_5H_5)Fe(CO)_2]_2$ (E.D.Schermer and W.H.Baddley, J. organometal. Chem., 1971,27,83) also results in metal-metal bond cleavage with the formation of tin- and iron-tellurides respectively.

$$\text{Ph}_2\text{Te}_2 \; + \; \text{Ph}_3\text{SnH} \; \xrightarrow{\quad 60-70^\circ \quad} \; \text{Ph}_3\text{SnTePh}$$
$$\text{m.p. } 91-93^\circ$$

$$[(C_5H_5)Fe(CO)_2]_2 \; + \; \text{Ph}_2\text{Te}_2 \; \longrightarrow \; (C_5H_5)Fe(CO)_2\text{TePh}$$
$$\text{green crystals m.p. } 66^\circ$$

Several other transition metal aryltelluride derivatives have been prepared by substitution reactions.

$$C_7H_7Mo(CO)Br + PhTeLi \; \xrightarrow{\quad CH_2Cl_2 \quad} \; C_7H_7(CO)Mo \underset{\underset{\text{Ph}}{\text{Te}}}{\overset{\overset{\text{Ph}}{\text{Te}}}{\diamond}} Mo(CO)C_7H_7$$

$$\text{m.p. } 118^\circ$$

(S.R.Buzilova et al., Zh. Obshch.   Khim., 1977,47,1999;
D.Mohr, H.Wienand and M.Ziegler, J. organometal. Chem.,
1977,134,281; Angew. Chem., internat. Edn., 1977,16,261).

$$(C_5H_5)_2MCl_2 \ + \ 2ArTeM' \longrightarrow (C_5H_5)_2M(TeAr)_2$$

M = Nb, Mo, W, Ti, Zr;   M' = Li, MgX.

$$(C_5H_5)(Bu_3P)Ni^+Cl^- \ + \ ArTeNa \longrightarrow (C_5H_5)(Bu_3P)NiTeAr$$

(M.Sato and T.Yoshida, J. organometal. Chem., 1975,87,217;
1974,67,395; 1973,51,231; 1975,94,403).
Aryltellurenyl halides, ArTeX, are only obtained from the
reaction of diarylditellurides with aryltellurium(IV)
trihalides when Ar = 2-biphenylyl and X = Br or when Ar =
C_6H_5 and X = I. The usual products of such reactions are
the diaryltellurium(IV) dichloride and elemental tellurium
(J.Meinwald, D.Dauplaise, F.Wudl and J.J.Hauser, J. Amer.
chem. Soc., 1977,99,255; P.Schulz and G.Klar, Z.Naturforsch.
b, 1975,30,40,43). Although aryltellurenyl halides may be
generated from the diarylditelluride by careful oxidation
with halogen and used further in situ, only those compounds
with a stabilising carbonyl group in the *ortho* position of
the aryl group or those with a very bulky group are isolable
by this method. However, if the diarylditelluride and
halogen are combined in a solvent in which the starting
materials are soluble and the aryltellurenyl halide insoluble
(eg. benzene, toluene, hexane, petroleum ether, carbon
tetrachloride, acetic acid), then the blue-violet to black
iodides and red-brown to black bromides may be isolated.
The iodides, once isolated, are rather insoluble, probably
polymeric, and stable for several weeks in dry air. The
corresponding bromides, however, are not very stable even
in the solid state, and become discoloured after one day.
Whether or not a particular aryltellurenyl halide can be
isolated depends on its thermal stability, which decreases
in the order: $3,4-(MeO)_2C_6H_3$, $4-MeC_6H_4 < C_6H_5$ and Cl < Br < I.
The decomposition route appears to be one in which a facile
migration of an aryl group in the associated aryltellurenyl
halide occurs. Such migration should be facilitated by small
halogen atoms (allows the tellurium atoms to be close to
each other) and by substituents such as MeO on the benzene

394

ring. Thus, 2-biphenylyltellurenyl bromide is monomeric in solution, and does not undergo aryl group migration to form the diaryltellurium dichloride and tellurium. Complexation by thiourea or selenourea ligands increases the stability (O.Vikans, Acta chem. Scand., 1975,29A,150,152,787), but the presence of an *ortho*-carbonyl substituent on the aryl group greatly enhances stability due to the presence of intramolecular carbonyl oxygen → tellurium coordination (M.Baiwir, G.Llabres, O.Dideberg, L.Dupont and J.L.Piette, Acta crystallogr., 1974,B30,139. Such compounds result from the ready cleavage of benzotellurophane by aqueous hydrogen halides at room temperature (J.M.Talbot, J.L.Piette and M.Renson, Bull. Soc. chim. France., 1976,294):

$$\text{(benzotellurophane with C=O)} \xrightarrow[X = Cl,Br,I]{Et_2O/H_2O/HX} \text{(aryl-COMe, TeX)}$$

or when 2-carboxyphenylaryltellurides are treated with dichloromethylbutylether at 100° in the presence of zinc chloride (J.L.Piette, P.Thibaut and M.Renson, Chem. Ser., 1975,8A,117).

$$\text{(aryl-CO_2H, TeAr)} \xrightarrow[ZnCl_2]{Cl_2CHOBu} \text{(aryl-COCl, TeAr)} \xrightarrow{100°} \text{(aryl-COAr, TeCl)}$$

Phenyltellurenyl bromide, decomp. 70-75°; phenyltellurenyl bromide-selenourea, decomp. 195°; o-Methylcarboxyphenyltellurenyl bromide, m.p. 115°.

Simple aryltellurols such as PhTeH do not as yet appear to have been isolated (methanetellurol has been prepared C.W.Sink and A.B.Harvey, J. chem. Phys., 1972,57,4434; K.Hamada and H.Morishita, Synth. React. inorg. met.-org. Chem., 1977,7,355)). However, 8-hydrothio-, 8-hydroseleno- and 8-hydrotelluro-1-hydrotelluronaphthalenes have been obtained by protonation of the corresponding dilithium salts in tetrahydrofuran at -25° (J.Meinwald, D.Dauplaise and J.Clardy, J. Amer. chem. Soc., 1977,99,7743; J.Mainwald, D.Dauplaise, F.Wudl and J.J.Hauser, ibid., 1977,99,255).

Diaryltellurides are produced by the dechlorination of
diaryltellurium dichlorides by anthracene in refluxing
methylethylketone in the presence of silver nitrate
(M.Albert and S.Shaik, J. chem. Soc. Perkin I, 1975,1223).
However, sodium sulphide monohydrate, copper powder and
Raney nickel are more common reagents for the reduction
of diaryltellurium dihalides, and both symmetrical and
unsymmetrical diaryltellurium compounds may be obtained
by such procedures (I.D.Sadekov, A.Y.Bushkov and V.I.
Minkin, Zh. obshch. Khim, 1973,43,815; 1977,47,631;
N.Petragnani, L.Torres and K.J.Wynne, J. organometal. Chem.,
1975,92,185; J.Bergman, Tetrahedron, 1972,28,3323).

Several other methods are available for the synthesis of
unsymmetrical diaryl- and arylalkyltellurides. The PhTe
anion (generated from diphenylditelluride and sodium in
liquid ammonia) reacts with arylhalides under irradiation
(A.B.Pierini and R.A.Rossi, J. organometal. Chem., 1979,
168,183). Phenylalkyltellurides may be prepared similarly
using PhTeLi in tetrahydrofuran, or by the alkylation of
aryltellurenyl halides using Grignard or organocadmium
reagents (N.Petragnani, L.Torres and K.J.Wynne, J. organometal. Chem., 1975,92,185; K.J.Irgolic, P.J.Busse, R.A.
Grigsby and M.R.Smith, J. organometal. Chem., 1975,88,175;
D.Seebach and A.K.Beck, Ber., 1975,108,314; J.L.Piette,
R.Lysy and M.Renson, Bull. Soc. chim. France, 1972,3159).

$$PhTeLi + RX \xrightarrow{THF} PhTeR$$

$$ArTeBr + RMgX \xrightarrow{THF} ArTeR$$

The phenylalkyltellurides, $PhTeC_nH_{2n+1}$, are yellowish oils,
which are stable when stored in the dark, but decompose in
light, probably to diphenylditelluride.

Aryltelluride anions also react with alkynes (S.R.Buzilova,
L.I.Vereshchagin, I.D.Sadekov and V.I.Minkin, Zh. obshch.
Khim., 1976,46,932; S.R.Buzilova, I.D.Sadekov, T.V.Lipovich,
J.M.Fillipova and L.I.Vereshchagin, ibid., 1977,47,1999),
chloro compounds (N.Derev, J.L.Piette, J.Van Coppenolle and
M.Renson, J. heterocyclic Chem., 1975,12,423), and carboxylic
anhydrides (J.L.Piette and M.Renson, Bull. chem. Soc. Belges.,
1970,79,367,383) to afford tellurides with a functionally-
substituted alkyl chain.

$$MeOC_6H_4TeNa \ + \ R-C{\equiv}C-CO-R' \ \longrightarrow \ MeOC_6H_4Te\overset{\displaystyle R}{\overset{|}{-}}C{=}CHCOR'$$

$$RC_6H_4TeNa \ + \ ClCH_2CH_2CO_2H \ \longrightarrow \ RC_6H_4TeCH_2CH_2CO_2H$$

$$PhTeNa \ + \ (RCO)_2O \ \longrightarrow \ RCOTePh \ + \ RCO_2Na.$$

(b) Aryltellurium(IV) Compounds

Tetraphenyltellurium, prepared by the reaction of lithium halide-free phenyllithium and tellurium(IV) chloride, reacts exothermically with *tert*-butylthiol at room temperature to afford diphenyltelluride, benzene and di-*tert*-butyldisulphide quantitatively. Thermal decomposition (140°, sealed tube, in vacuo) also yields diphenylditelluride (92%) along with biphenyl (89%) and benzene (10%) (D.H.R.Barton, S.A.Glover and S.V.Ley, Chem. Comm., 1977,266).

Mono-, di- and triphenyltellurium(IV) chlorides are conveniently prepared by the reaction of tellurium(IV) chloride with benzene in the presence of the appropriate quantity of aluminium trichloride (W.H.H.Guenther, J. Nepywoda and J.Y.C.Chu, J. organometal. Chem., 1974,74,79).

$$TeCl_4 \ + \ (n{+}1)AlCl_3 \ + \ nC_6H_6 \ \longrightarrow \ Ph_nTe_{4-n}$$

n = 1, 2 or 3.

Other arylating agents which have been used include a wide variety of arylmercury, arylsilicon and aryltin compounds (K.J.Irgolic, J. organmetal. Chem., 1978,158,285; I.D.Sadekov and A.A.Maksimenko, Zh. obshch. Khim., 1977,47,1918; R.C.Paul, K.K.Bhagin and R.K.Chadha, J. inorg. nucl. Chem., 1975,37, 2337).

$$TeCl_4 \ + \ 2PhHgCl \ \longrightarrow \ Ph_2TeCl_2$$

$$TeCl_4 \ + \ Me_3ArSi \ \longrightarrow \ ArTeCl_3$$

$$TeCl_4 \ + \ Ph_4Sn \ \longrightarrow \ PhTeCl_3$$

Diaryltellurium dihalides are also obtained by the oxidation of diaryltellurides using sulphuryl halides (N.Petragnani, L.Torres and K.J.Wynne, J. organometal. Chem., 1975,92,185; I.D.Sadekov, A.V.Bushkov, V.L.Pavlova, V.S.Yur'eva and V.I.Minkin, Zh. obshch. Khim., 1977,47,1305) or elemental halogens (K.J.Irgolic, P.J.Busse, R.A.Grigsby and M.R.Smith, J. organometal. Chem., 1975,88,175). The cleavage of diaryltellurides with bromine or iodine in chlorinated hydrocarbons yields aryltellurium tribromides and tri-iodides (C.Knobler and J.D.McCullough, Inorg. Chem., 1977, 16,612; P.Schulz and G.Klar, Z. Naturforsch. b, 1975,30,265). However, cleavage by excess methyl iodide produces a mixture of arylmethyltellurium diiodides and aryldimethyltelluronium iodides. The corresponding fluorides are generally prepared by substitution using silver fluoride in refluxing toluene (F.J.Berry, E.H.Kustan, M.Roshani and R.C.Smith, J. organometal. Chem., 1975,99,115). Diaryltellurium difluorides may also be obtained by treating diaryltelluroxides with 40% aqueous hydrofluoric acid (I.D.Sadekov, A.Y.Bushkov, V.L. Pavlova, V.S.Yur'eva and V.I.Minkin, Zh. obshch. Khim., 1977, 47,1305), or by passing sulphur(IV) fluoride through a refluxing benzene solution of a diaryltellurides or a diarylditellurides (I.D.Sadekov, Ya.A.Bushkov, L.N.Markovskii and V.I.Minkin, Zh. obshch. Khim., 1976,46,1660). Triphenyltelluronium chloride, m.p. 250°; triphenyltelluronium bromide, decomp. 261°; diphenyltellurium difluoride, m.p. 153°; diphenyltellurium dichloride, m.p. 161°; p-methoxy-phenyltellurium trifluoride, m.p. 176-178°.

The chloride ion in triphenyltelluronium chloride is readily replaced by other anions including CN^-, N_3^-, NCO^-, NCS^-, $NCSe^-$, and the nitrosylate anions $[RC(NO)_2]^-$ (R = Me, Ph) (R.F.Ziolo and K.Pritchett, J. organometal. Chem., 1976, 116,211; J.Kopf, G.Vetter and G.Klar, Z. anorg. allg. Chem., 1974,409,285). Analogous diaryltellurium dipseudohalides are obtained by substitution (T.N.Srivastava, R.C.Srivastava and M.Singh, J. organometal. Chem., 1978,160,449), or by oxidative-addition (F.H.Musa, W.R.McWhinnie and A.W.Downs, J. organometal. Chem., 1977,134,C43; F.H.Musa and W.R. McWhinnie, ibid., 1978,159,37). However, the reaction of diphenyltellurium dichloride with excess sodium thiocyanate in near-boiling aqueous solution yields $(Ph_2TeNCS)_2O$ (C.S.Mancinelli, D.D.Titus and R.F.Ziolo, J. organometal. Chem., 1977,140,113).

Aryltellurium trihalides readily accept further halide ion

to form complex ArTeX$_4^-$ anions (N.Petragnani, J.V.Comasseto
and Y.Kawano, J. inorg. nucl. Chem., 1976,38,608), and with
cold water to afford arene-telluric acid halides. Treatment
of p-ethoxyphenyltellurium trichloride with 10% aqueous sodium
carbonate solution at room temperature, however, yields the
arene-tellurinic acid, which on treatment with dilute acetic
acid is converted into the arene-tellurinic acid anhydride
(P.Thavornyutikarn and W.R.McWhinnie, J. organometal. Chem.,
1973,50,135).

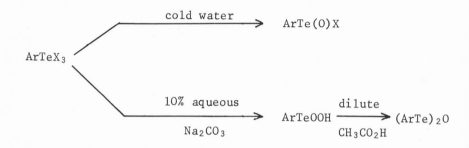

Phenyltellurinic chloride, m.p. 280°; p-ethoxyphenyltellurinic
anhydride, decomp. 234-238°.

Diaryltellurium bisalkoxides are obtained in good yield by
substitution of the corresponding dihalide using the sodium
alkoxide with the related alcohol as solvent. Diphenyltellurium bisethoxide undergoes transesterification with
para-cresol and diols at room temperature, but ethane and
propane dithiol cause reduction to diphenylditelluride
(M.Wieber and E.Kaunzinger, J. organometal. Chem., 1976,129,
339).

Several methods are available for the synthesis of
diaryltellurium dicarboxylates : the reaction of a diaryltelluroxide with the carboxylic acid or anhydride (I.D.
Sadekov, A.A.Maksimenko, A.I.Usachev and V.I.Minkin, Zh.
obshch. Khim., 1975,45,2562; S.Tamagaki, I.Hatanaka and
S.Kozuka, Bull. chem. Soc. Japan, 1977, 50,250); trans-ester-
ification using the easily accessible diphenyltellurium
diacetate (N.Dance and W.R.McWhinnie, J. organometal. Chem.,
1976,104,317); the reaction of diaryltelluroxide, a
carboxylic acid and silver oxide at room temperature, or
of diaryltellurium dichloride, a carboxylic acid and silver
oxide in boiling benzene, or by passing a solution of

diaryltellurium dichloride in ethanol/tetrahydrofuran over an
Amberlite IR45 resin in the carboxylate form (N.Petragnani,
J.V.Comasseto and N.H.Verella, J. organometal. Chem., 1976,
120,375), and by the oxidation of diaryltellurides using
lead tetraacetate (B.C.Pant, J. organometal. Chem., 1974,
65,51).

Aryltelluronium ylides have been obtained by the same two
routes used for the selenium analogues, viz. reaction of
diphenyltelluride with diazo-1,2,3,4-tetraphenylcyclopenta-
diene (B.H.Freeman and D.Lloyd, Chem. Comm., 1970,924), and
the reaction of diaryltellurium dihalides with 4,4-dimethyl-
cyclohexa-2,5-dione in benzene in the presence of
triethylamine (N.N.Magdesiera, R.A.Kyandzhetsian and O.A.
Rakitin, Zhur. org. Khim., 1975,11,2562; 1976,12,36; I.D.
Sadekov, A.I.Usachev, A.A.Maksimenko and V.I.Minkin, Zh.
obshch. Khim., 1975,45,2563).
Diphenyltelluronium 2,3,4,5-tetraphenylcyclopentadienide,
m.p. 175°.

Guide to the Index

This index is constructed in a similar manner to the volume indexes of the first edition of the Chemistry of Carbon Compounds. However, to make the index easier to use, more descriptive entries have been made for the commonly occurring individual, and groups of chemicals.

The indexes cover primarily the chemical compounds mentioned in the text, and also include reactions and techniques, where named, and some sources of chemical compounds such as plant and animal species, oils, etc.

Chemical compounds have been indexed alphabetically under the names used by authors, editing being restricted to ensuring uniformity of entries under the same heading. In view of the alternative nomenclature that can often be used, a limited amount of cross-referencing has been done where it is considered to be helpful, but attention is particularly drawn to Convention 2 below.

For this and the succeeding volumes, the indexing conventions listed below have been adopted.

1. *Alphabetisation*
(a) The following prefixes have not been counted for alphabetising:

n-	*o-*	*as-*	*meso-*	D	*C*
sec-	*m-*	*sym-*	*cis-*	DL	*O-*
tert-	*p-*	*gem-*	*trans-*	L	*N-*
	vic-				*S-*
		lin-			*Bz-*
					Py-

Some prefixes and numbering have been omitted in the index, where they do not usefully contribute to the reference.

(b) The following prefixes have been alphabetised:

Allo	Epi	Neo
Anti	Hetero	Nor
Cyclo	Homo	Pseudo
	Iso	

402

(c) A letter by letter alphabetical sequence is followed for entries, firstly for the main entry, followed by the descriptive entry. The only exception to this sequence is the placing of plural entries in front of the corresponding individual entries to prevent these being overlooked by a strict alphabetical sequence which could lead to a considerable separation of plural from individual entries. Thus "butanes" will come before *n*-butane, "butenes" before 1-butene, and 2-butene, etc.

2. *Cross references*

In view of the many alternative trivial and systematic names for chemical compounds, the indexes should be searched under any alternative names which may be indicated in the main body of the text. Only a limited amount of cross-referencing has been carried out, where it is considered that it would be helpful to the user.

3. *Esters*

In the case of lower alcohols esters are indexed only under the acid, e.g. propionic methyl ester, not methyl propionate. Ethyl is normally omitted e.g. acetic ester.

4. *Derivatives*

Simple derivatives are not normally indexed if they follow in the same short section of the text.

5. *Collective and plural entries*

In place of "– derivatives" or "– compounds" the plural entry has normally been used. Plural entries have occasionally been used where compiunds of the same name but differing numbering appear in the same section of the text.

6. *Main entries*

The main entry of the more common individual compounds is indicated by heavy type. Multiple entries, such as headings and sub-headings over several pages are shown by "–", e.g., 67–74, 137–139, etc.

INDEX

416

418

420

438